Fei He, Derong Li, Wei Quan, Shufeng Wang, Zhiyi Wei, Heping Zeng
Advances in Ultrafast Optics

Advances in Optical Physics

Editor-in-Chief
Jie Zhang

Volume 6

Fei He, Derong Li, Wei Quan, Shufeng Wang,
Zhiyi Wei, Heping Zeng

Advances in Ultrafast Optics

Edited by
Zhiyi Wei

DE GRUYTER

上海交通大學出版社
SHANGHAI JIAO TONG UNIVERSITY PRESS

Physics and Astronomy Classification Scheme 2010
32.80.Rm, 42.62.Eh, 42.65.Re, 74.25.nd, 78.47.J–, 78.47.+p

Editor
Prof. Zhiyi Wei
The Institute of Physics, Chinese Academy of Sciences
No. 8, 3rd South Street, Zhongguancun
100190, Haidian District
Beijing
China

ISBN 978-3-11-030436-7
e-ISBN (PDF) 978-3-11-030455-8
e-ISBN (EPUB) 978-3-11-038283-9
Set-ISBN 978-3-11-030456-5

Library of Congress Control Number: 2018934818

Bibliographic information published by the Deutsche Nationalbibliothek
The Deutsche Nationalbibliothek lists this publication in the Deutsche Nationalbibliografie; detailed bibliographic data are available on the Internet at http://dnb.dnb.de.

Cover image: Ellende/iStock/Thinkstock
Typesetting: le-tex publishing services GmbH, Leipzig
Printing and binding: CPI books GmbH, Leck

www.degruyter.com

The series: *Advances in Optical Physics*

Professor Jie Zhang, Editor-in-chief, works on laser-plasma physics and has made significant contributions to development of soft X-ray lasers, generation and propagation of hot electrons in laser-plasmas in connection with inertial confinement fusion (ICF), and reproduction of some extreme astrophysical processes with laser-plasmas. By clever design to enhance pumping efficiency, he and his collaborators first demonstrated saturation of soft-X-ray laser output at wavelengths close to the water window. He discovered through theory and experiments that highly directional, controllable, fast electron beams can be generated from intense laser plasmas. Understanding of how fast electrons are generated and propagated in laser plasmas and how the resulting electron beams emit from a target surface and carry away laser excitation energy is critical for understanding the fast-ignition process in ICF. Zhang is one of the pioneers of simulating astrophysical processes by laser-plasmas in labs. He and his collaborators used high-energy laser pulses to successfully create conditions resembling the vicinity of a black hole and model the looptop X-ray source and reconnection overflow in solar flares.

Because of his academic achievements and professional services, Professor Zhang received Honorary Doctors of Science from City University of Hong Kong (2009), Queen's University of Belfast (2010), University of Montreal (2011) and University of Rochester (2013). He was elected member of CAS in 2003, member of German Academy of Sciences Leopoldina in 2007, fellow of the Third World Academy of Sciences (TWAS) in 2008, foreign member of Royal Academy of Engineering (FREng) of the UK in 2011 and foreign Associate of US National Academy of Sciences (NAS) in 2012. He is the Vice President of the Chinese Academy of Science (CAS), and also a strong advocate and practitioner of higher education in China.

Preface

After a three years' effort by many top-tier scientists, the book series ***Advances in Optical Physics (English version)*** is completed.

Optical physics is one of the most active fields in modern physics. Ever since lasers were invented, optics has permeated into many research fields. Profound changes have taken place in optical physics, which have expanded tremendously from the traditional optics and spectroscopy to many new branches and interdisciplinary fields overlapping with various classical disciplines. They have further given rise to many new cutting-edge technologies:

- For example, nonlinear optics itself is an interdisciplinary field, which has been developing since the advent of lasers and it is significantly influenced by various technological advances, including laser technology, spectroscopic technology, material fabrication and structural analysis.
- With the rapid development of ultra-short intense lasers in the past 20 years, high field laser physics has rapidly developed into a new frontier in optical physics. It contains not only rich nonlinear physics under extreme conditions, but also has the potential for many advanced applications.
- Nanophotonics, which combines photonics and contemporary nanotechnology, studies the mechanisms of light interactions with matter at the nanoscale. It enjoys important applications such as in information transmission and processing, solar energy, and biomedical sciences.
- Condensed matter optics is another new interdisciplinary field, which is formed due to the intersection of condensed matter physics and optics. Here, lasers are used as probes to study the structures and dynamics of condensed matter. In addition, discoveries from condensed matter optics research can be applied to produce new light sources, detectors, and a variety of other useful devices.

In the last 20 years, with the increasing investment in research and development in China, the scientific achievements by Chinese scientists also became increasingly important. These are reflected by the greatly increased number of research papers published by Chinese scientists in prestigious scientific journals. However, there are relatively few books for a broad audience – such as graduate students and scholars – devoted to this progress at the frontiers of optical physics.

In order to change this situation, three years ago, Shanghai Jiao Tong University Press discussed with me and initiated the idea to invite top-tier scientists to write the series of ***Advances in Optical Physics***. Our initial plan was to write a series of introductory books on recent progress in optical physics for graduate students and scholars. It was later expanded into its current form. The first batch of the series includes eight volumes:

https://doi.org/10.1515/9783110304558-201

- *Advances in High Field Laser Physics*
- *Advances in Precision Laser Spectroscopy*
- *Advances in Nonlinear Optics*
- *Advances in Nanophotonics*
- *Advances in Quantum Optics*
- *Advances in Ultrafast Optics*
- *Advances in Condensed Matter Optics*
- *Advances in Molecular Biophotonics*

Each volume covers a number of topics in the respective field. As the editor-in-chief of the series, I sincerely hope that this series is a forum for Chinese scientists to introduce their research advances and achievements. Meanwhile, I wish these books are useful for students and scholars who are interested in optical physics in general, one of these particular fields, or a research area related to them. To ensure these books could reflect the rapid advances of optical physics research in China, we have invited many leading researchers from different fields of optical physics to join the editorial board. It is my great pleasure that many top tier researchers at forefronts of optical physics accepted my invitation and made their contributions in the last three years.

Almost at the same time, De Gruyter learned about our initiative and expressed their interest in introducing these books written by Chinese scientists to the rest of world. After discussion, De Gruyter and Shanghai Jiao Tong University Press reached an agreement to co-publishing the English version of the series. At this moment, on behalf of all authors of these books, I would like to express our appreciation to these two publishing houses for their professional services and support to sciences and scientists. In particular, I would like to thank Mr. Jianmin Han and his team for their great contribution to the publication of this book series.

At the end of this preface, I must admit that optical physics itself is a rapidly expanding forefront of science. Due to the nature of the subject area, this series can never cover all aspects of optical physics. However, what we can do – together with all authors of these books – is to try to pick up the most beautiful 'waves' from the vast science ocean to form this series. By publishing this series, it is my cherished hope to attract minds of a younger generation into the great hall of optical physics research.

Professor Jie Zhang
Editor-in-chief

Contents

Zhiyi Wei, Shaobo Fang

1 Ultrafast ultrahigh-intensity laser pulses

Generation of high intensity laser pulses has been regarded as one of the most important research topics since the invention of lasers. High intensity lasers are normally constructed using the master oscillator power amplifier (MOPA) configuration to boost the energy of short laser pulses. This configuration has been used in various locations such as the Shenguang facilities in China and the National Ignition Facility (NIF) at the Lawrence Livermore National Laboratory (LLNL) in the USA; the latter is currently the largest laser facility in the world. This type of giant laser facility is usually employed for high-cost and large-scale scientific projects involving complex technology and only a few countries can perform this type of laser research. Because of the low repetition rate and long pulse duration of the achieved laser pulses, these laser facilities are not suitable for large-scale applied research. Furthermore, the output laser intensity is limited. For example NIF, constructed in 2009, includes 192 ultraviolet (UV) beams with the total energy up to 1.8 megajoules (MJ, 10^6 J). However, the peak power is only approximately 500 terawatts (TW; 10^{12} W) [1] because the pulse duration is at the nanosecond (ns; 10^{-9} s) level; the separation time between two pulses is as long as a few hours. At the other end of the scale, several research institutes have developed tabletop laser systems within regular-scale labs through the amplification of ultrashort-pulse lasers at the femtosecond (fs; 10^{-15} s) level. These lasers have peak powers at the petawatt (PW; 10^{15} W) level [2–4], a repetition rate of 1 hertz (Hz) [5], and a focus intensity reaching 10^{22} W/cm^2 [6]. As a result of the 1991 discovery of the Kerr-lens mode-locking (KLM) phenomenon by the Sibbett group in the UK [7], along with the development of chirped pulse amplification (CPA) technology by the US-based Mourou group in 1985 [8], and research on ultrashort pulses, ultrahigh-intensity lasers have been advancing at an unprecedented pace since the late 1980s. New scientific records associated with significant breakthroughs have been made on a continuous basis. Ultrashort-pulse laser research has developed in two explorative directions. One path of development targets extremely short pulse duration promoted by mode-locking technology. Currently, laser pulse duration can reach the few-cycle level; in addition, attosecond (as; 10^{-18} s) laser pulses can be achieved using new physical mechanisms [9–11]. The second research trend is the ongoing extension of the laser peak power limit. An increasing number of femtosecond ultrahigh-intensity laser facilities with peak powers at the terawatt or even petawatt level have been developed [2–5], serving as powerful tools for in-depth research on ultrafast, ultrahigh-intensity laser pulses. Ultrashort pulses and ultrahigh-intensity

Zhiyi Wei, Shaobo Fang, Beijing National Laboratory for Condensed Matter Physics, Institute of Physics, Chinese Academy of Sciences, Beijing 100190, China

https://doi.org/10.1515/9783110304558-001

laser technology are not only extensively employed in the fields of micromanufacturing [12–14] and medicine [15, 16] but have also been successfully and prominently implemented in various research areas, such as the exploration of atomic and molecular motion patterns [17, 18], laboratory simulations in astrophysics [19], and precision spectroscopy [20]. These applications have facilitated the emergence of many new subfields and groundbreaking scientific achievements. One of the two most representative events was the award of the 1999 Nobel Prize in Chemistry to the American researcher, Prof. A. H. Zewail. This prize was awarded for work on chemical kinetics using femtosecond laser pulses. The second representative achievement was the joint sharing of one half of the 2005 Nobel Prize in Physics by Profs. J. L. Hall and T. W. Hänsch; this prize was awarded for their achievements regarding the femtosecond laser frequency comb technique. With state-of-the-art features and innovative applications, ultrashort pulses, ultrahigh-intensity lasers have become crucial tools in the field of optical physics. Therefore, this chapter will first briefly review the essential technology and associated progress in research.

1.1 Generation of ultrashort laser pulses

Since the first laser appeared on the historical stage, the generation of shorter and shorter laser pulses has been one of the most competitive scientific and technological frontiers. In 1961, the concept of Q-switching was proposed, a method that compresses all the optical radiation energy into a very short pulse [21–23]. In 1962, the first Q-switched laser with an output peak power of 600 kW and a pulse duration on the order of the 10^{-7} s [24] was invented. During the coming years, the electro-optic Q-switched, the acousto-optic Q-switched, and the saturable absorption Q-switched [25] techniques were invented in succession. The output power of lasers was rapidly increasing, and the pulse compression techniques made great progress. By the 1980s, thanks to the Q-switched technique, generation of ultrashort laser pulses was further compressed to the order of nanoseconds (10^{-9} s, ns), corresponding to giant pulses with gigawatts (10^9 W, GW) peak power. However, due to an intrinsic bottleneck of the compression mechanism, generating even shorter laser pulses using Q-switched technology became very challenging.

The mode-locking technique, which had been developed rapidly since 1964, became a more effective technique to generate ultrashort laser pulses. In 1965, Mocker and Collins obtained picosecond laser pulses in rubies by passive phase locking on multiple longitudinal modes in the laser cavity [26]. Afterwards, a variety of new mode-locking techniques further reduced the pulse durations down to subpicosecond level and even to the order of femtoseconds [27, 28]. In 1981, Fork et al. generated 90-fs ultrashort pulses for the first time by using the collided pulse mode-locking (CPM) technique in a dye laser [27]. This breakthrough represented the first generation of femtosecond lasers, which also made the dye laser a hot topic in research and the routine method in obtaining femtosecond pulses. In 1985, Valdmanis and Fork et al. further pushed the CPM techniques and obtained 27-fs ultrashort pulses in a ring dye laser [29]. Two years later, Fork et al. upscaled the output energy, and by combining the spectrum broadening in the fiber and the dispersion compensating with a grating-prism pair, generated a long-standing world record of a 6-fs pulse duration [30], which is the best performance of the dye laser so far in generating ultrashort laser pulses. However, the dye laser, even as the first generation ultrafast femtosecond laser, has not been widely used, because the performance is quite unsteady, the output power is very low, the dye has to be replaced frequently, and daily operation is inconvenient.

At the same time, people kept exploring new gain media for a new breakthrough in mode-locked lasers. Subsequently, Ti:sapphire, Cr:forsterite, Cr:YAG, Cr:LISAF, Yb:YAG and supplementary tunable solid laser gain materials were developed, forming the foundation for the practical development of ultrafast femtosecond lasers. In 1986, Moulton invented a titanium doped sapphire laser crystal [31], which successfully led to the rapid development of ultrafast femtosecond laser research and its general application. The Ti-doped sapphire lasers, with a wide tuning range, high laser gain, very short pulse duration, and robust performance, have become the most commonly used ultrafast laser. We therefore start with the Ti:sapphire lasers as a way to introduce the generation and development of ultrafast lasers.

1.1.1 Kerr-lens mode-locked Ti:sapphire lasers

Sapphire (Al_2O_3) crystal, doped with titanium ions, is the best broadband laser gain medium, with excellent physical and chemical properties. It has the advantages of high thermal conductivity, high wear resistance, and high damage threshold, as well as being suitable for high average power operation. The upper fluorescence lifetime is 3.2 μs, the stimulated emission cross section is 10^{-19} cm^2, and the fluorescence wavelength covers the range of 600–1100 nm, which can theoretically support a pulse width of 2.7 fs. This gives it, like the dye laser, the potential to produce very short laser pulses. In addition, the wide absorption range and high absorption efficiency of the 400–600 nm spectrum make it well suited for laser pumping. Table 1.1 lists the main characteristics of titanium doped sapphire laser crystals. Because of the superior properties of the solid medium with respect to liquid dyes, the crystal has been widely used in femtosecond laser mode-locking and amplification since its appearance in 1985.

In 1991, Professor W. Sibbett of the University of St Andrews in Scotland used the laser structure shown in Figure 1.1 to achieve stable passive mode-locking using the nonlinear effect of the gain medium without inserting any modulation elements in the cavity [7]. In the figure, M1 and M2 are a pair of confocal concave mirrors for focusing the oscillating laser to form a high gain in the gain medium, M3 is the total reflection mirror, OC is the output coupling mirror, P1 and P2 are a pair of high dispersion prisms to compensate for the dispersion caused by the gain medium, and BF is

Tab. 1.1: Characteristics of Ti-doped sapphire crystals.

Parameters	Values
Refractive index	$n = 1.76$ (800 nm)
Fluorescence lifetime	3.2 μs
Absorption spectrum	400–600 nm
Fluorescence wavelength	600–1100 nm
Peak emission wavelength	~790 nm
Stimulated emission cross section	4.1×10^{-19} cm^2
Quantum efficiency	$\eta_Q \approx 1$
Saturation energy density	0.9 J/cm^2
Damage threshold	10 J/cm^2
Thermal conductivity (room temperature)	46 W/mK

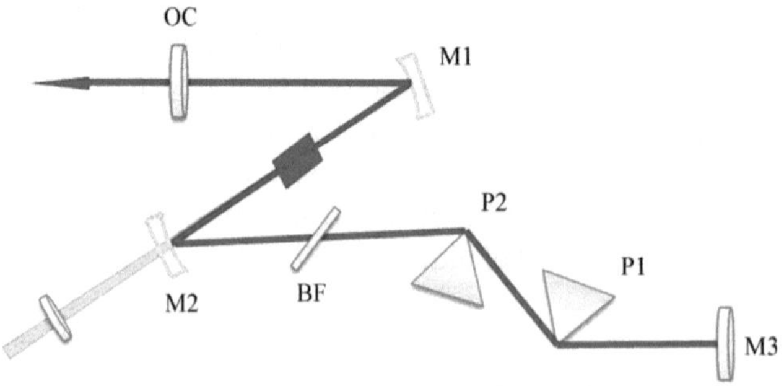

Fig. 1.1: Kerr-lens mode-locked titanium sapphire oscillator.

the birefringence filter to tune the laser wavelength. In the case of optimized dispersion compensation and cavity structures, other complex mode-locking devices and techniques are not required in the cavity, and they have obtained a stable self-mode-locked laser pulse output of 60 fs via an external perturbation. In fact, since 1965, self-mode-locking has been observed in He-Ne, copper vapor, and Nd:YAG lasers [23]. However, since these self-mode-locked pulse sequences are very unstable, they have not attracted much attention. After the discovery of the Ti:sapphire laser self-locking phenomenon and the excellent results, laser self-mode-locking attracted attention and became the most significant milestone in the history of ultrafast lasers. The fundamental principle is that when the laser peak power reaches GW (10^9 W) or more, the self-locking mode is a new mode-locking mode caused by the Kerr nonlinear effect of the gain medium. The refractive index of the medium n varies with the intensity of the incident laser. As shown in Figure 1.2, under the action of the strong focusing laser, the high-power pulsed laser beam will self-focus due to the nonlinear refractive index. If an aperture is present near the focal point, the high power pulse portion can be fully passed and the low power DC portion is depleted. The gain aperture formed by the finite pump region after the focusing of the solid laser pump achieves the aperture

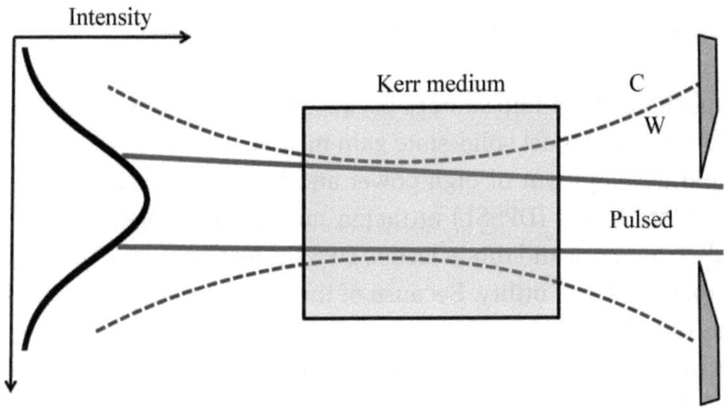

Fig. 1.2: Kerr-lens locking mechanism.

effect, while the Kerr nonlinear response time is less than 1 fs. The combination results in a novel 'fast' saturated absorption effect. Kerr-lens mode-locking (KLM) is a kind of passive mode-locking scheme [32], because the main principle is the Kerr nonlinear effect.

The Kerr-lens mode-locking mechanism opens a new chapter for solid femtosecond laser research. According to the uncertainty principle (for the spectrum width and the pulse time width), a wide enough spectrum is required to produce a shorter laser pulse. At the same time, to achieve the limit of the Fourier transform pulse width, it also needs to compensate for dispersion. By compensating the overall laser dispersion and increasing the output spectral bandwidth, it is possible to generate narrower femtosecond pulses. Normally an optimal Ti:Sapphire laser may include a thin gain medium, prisms, broadband coated mirrors, and chirped mirrors (chirped mirrors, CMs).

In 1993, Asaki et al. [13] obtained a 11 fs laser pulse by using thin Ti:sapphire crystals with high doping concentrations and using the fused silica prisms for dispersion compensation. For laser pulses shorter than 10 fs, crystals and prisms are not well matched for third- and fourth-order dispersion, due to the wide spectral bandwidth. Therefore, it is difficult to obtain a pulse width of 10 fs or less by performing the dispersion compensation with the prism pair. In 1994, Szipöcs and Krausz proposed the use of new chirp dielectric reflectors to compensate for the dispersion [34]. In 1995, Stingl et al. used wide-band chirped mirrors to compensate for dispersion without other dispersion compensation elements, resulting in a laser pulse of 8 fs [35]. One year later, Xu et al. [7] further optimized the ring cavity to obtain 7 fs pulses. In this paper, a combination of semiconductor saturable absorption mirrors (SESAM), dual chirped mirrors (DCM), and low-dispersion CaF_2 prisms are used to compensate for the combination of these techniques. Several groups in Switzerland, Germany, and the United States have implemented a pulse of about 5 fs [36–41], which represents the shortest pulse duration of Ti:sapphire oscillators.

1.1.2 Diode-pumped all-solid-state ultrafast lasers

The laser diode is an apt choice for pump sources because the emission wavelengths can cover the absorption peaks of most solid-state gain media. Since the 1980s, with the emergence and rapid development of high-power and high-brightness LDs, the diode-pumped all-solid-state lasers (DPSSL) attracted much attention. Combining the advantages of solid-state lasers and diode lasers, DPSSLs feature high efficiency, low cost, stability, compactness and utility. Because of these advantages, the DPSSL technology stood out as one of the notable landmarks in laser development. In 1992, Keller et al. demonstrated a stable mode-locked Nd:YLF laser with a novel SESAM [42]. The parameters of SESAM, such as wavelength, saturation fluence $F_{\mathrm{sat,A}}$, modulation depth ΔR, nonsaturable loss ΔR_{ns} and relaxation time τ_{A}, and so on, can be flexibly designed to meet various needs of different mode-locked solid-state lasers. However, the SESAM was not critical to the cavity design, resulting in a simple oscillator cavity.

The typical cavity design of SESAM mode-locked solid-state lasers is shown in Figure 1.3. Combining SESAM and dispersive compensating devices (such as GTI mirror or prisms), it was easy to obtain femtosecond pulses. First, the major problem in SESAM mode-locked solid-state lasers was the Q-switched mode-locking instead of stable mode locking, leading to the instability of the mode-locked operation. In 1999, Hönninger et al. studied these problems theoretically and experimentally [43]. The results showed that the following measures could solve this issue: (1) Increasing the pulse energy by increasing the pump power to turn Q-switch mode-locking operation into continuous mode locking. (2) Using the output coupler mirror with lower transmission to increase the intra-cavity pulse energy. (3) Increasing the cavity length. This method leads to the decreasing of the repetition rate and increasing of the pulse energy. (4) Decreasing the spot size on the SESAM to reduce the saturable energy of the SESAM and gain medium. (5) Using the SESAM with lower modulation depth. Through these methods, solid-state lasers, even with a long upper energy lifetime gain medium, could reach stable continuous-wave mode-locking operation. Curved mirrors decrease the spot sizes of SESAM and gain media in a typical SESAM mode-locked cav-

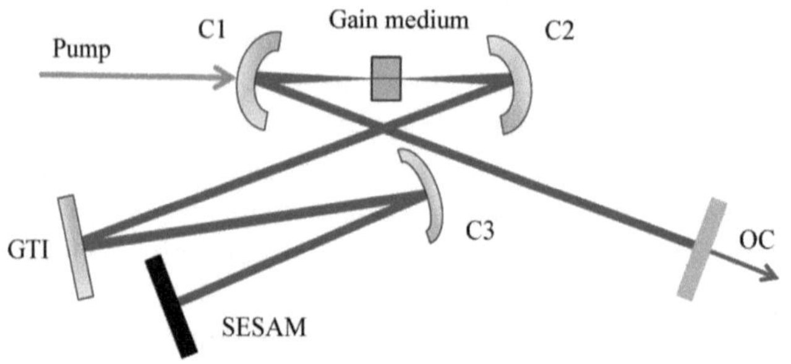

Fig. 1.3: Cavity structure of a passively mode-locked all-solid-state laser with SESAM.

ity, and the parameters of the mirrors should be chosen based on the experimental conditions. Owing to the appearance of SESAM, ultrafast all-solid-state lasers entered into a rapid development stage. Until now, with the exception of Kerr-lens mode locking, SESAM was a favorite candidate mode-locking device for passively mode-locked all-solid-state lasers. In 1998, a self-starting SESAM mode-locked Ti:sapphire laser was demonstrated in 6.5 fs pulses [37]. In 2000, Spühler et al. achieved a picosecond Nd:YAG laser by side pumping with 27 W average output power [44]. Using the thin disk concept, U. Keller's group demonstrated an 272 W Yb:YAG thin disk laser in 1.07 ps pulses, with 272 W of output power and the corresponding pulse energy of 80 μJ (and peak power of 66 MW) [45]. This means that ultrafast laser pulses with a pulse energy of a few hundred microjoules could be obtained directly from an oscillator without cavity dumping or amplifier technologies. In 1.5 μm communication wavelength, W. Sibbett's group demonstrated SESAM mode-locked Cr:YAG laser emitting 230 fs pulses [46]. At 2.4 μm wavelength, in 2012 Sorokin et al. achieved 130 fs pulses from Cr:ZnSe and Cr:ZnS solid-state lasers [47]. The passive mode-locking of a 2.1 μm Tm, Ho:YAG picosecond laser with GaInAs- and GaSb-based SESAM was investigated by Yang et al. in 2013 [48].

Since the beginning of the 21st century, a few novel saturable absorbers were found, including single wall carbon nanotubes (SWCNT), graphene and topological insulators (TI). Furthermore, many new solid-state host media that can be pumped directly by laser diodes were found, such as garnet (YAG, YGG, GGG), sesquioxide (Y_2O_3, Lu_2O_3, Sc_2O_3), sapphire (Al_2O_3), vanadate (YVO_4, $GdVO_4$, $LuVO_4$), double tungstate (KGW, KYW, KLW), double molybdate (NYM, NLM), borate (YCOB, GdCOB), metasilicate (LSO, GYSO, GSO, YSO), fluoride (CaF_2, $YLiF_4$), and so on. Because of the different properties, these host materials perform various laser capabilities by doping different ions. The usual doping ions included trivalent rare earth ions (Nd^{3+}, Er^{3+}, Ho^{3+}, Tm^{3+}, Yb^{3+}), divalent rare earth ions (Sm^{2+}, Dy^{2+}, Tm^{2+}), transition metal ions (Cr^{4+}, Cr^{3+}, Cr^{2+}, Ti^{3+}, V^{2+}), metal ions (Fe^{2+}) and actinoid U^{3+} and so on. With the proper combination of host materials and doping ions, lasers were obtained based on hundreds of solid-state gain media, whose wavelengths covered the visible to mid-infrared range. Detailed information is not introduced here, due to limited space.

Nowadays, the main research on all-solid-state technology is mainly focused on how to obtain ultrafast laser sources with shorter pulses, higher energy, and broader spectra. Benefiting from the broad emitting spectrum of Yb^{3+} doped gain media, Yb^{3+} doped mode-locked lasers in sub-100 fs pulses were widely studied by properly compensating dispersion and using SESAM mode-locking or Kerr-lens mode-locking methods [49–73]. Figure 1.4 shows the principle of sub-100 fs lasers based on Yb^{3+}doped gain media by SESAM and Kerr-lens mode-locking [67].

In Figure 1.4, a 7 W fiber coupled laser diode at 970 nm was used as the pump source, and the core diameter of the fiber was 50 μm. The pump light was focused by a 1:1 image system into the crystal. A 5% doped Yb:YGG crystal was used as the gain medium with a length of 4.5 mm. The cavity was an X-type structure. M1 and M2

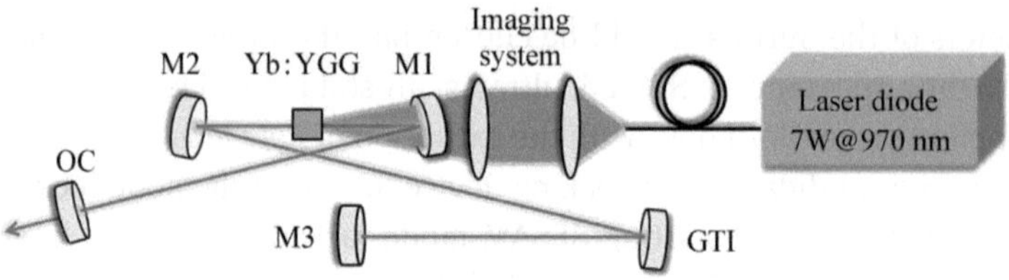

Fig. 1.4: 88 fs Kerr-lens mode-locked Yb:YGG laser.

were curved mirrors with the radius of curvature of 75 mm, and the transmission of the output coupler was 0.4%. A total GDD of $-1000\,fs^2$ was used to compensate the dispersion of crystal and mirrors by a piece of GTI mirror. By optimizing the cavity, the stable mode-locking operation was obtained with an average power of 104 mW, a pulse width of 88 fs and a central wavelength of 1042 nm.

However, there are always thermal effects in conventional bulk lasers. The thermal effects not only limited the output power, but also decreased the beam quality of the lasers. Even worse, the thermal stress in the crystal could damage the crystal. Thus the thermal effect prompted resistance by solid-state lasers toward high power levels. To solve this problem, some measures were taken, such as using a multicrystal slab and thin disk structures. A typical bicrystal all-solid-state laser is shown in Figure 1.5. To obtain higher output power, the pump power should be increased. To solve the heat removal capabilities of all-solid-state lasers, Calendron et al. presented a bicrystal structure pumped by two sides [70]. An average continuous output power of 24 W was obtained with fundamental mode operation. Meanwhile, the mode-locking operations were obtained at a normal dispersion and a negative dispersion range with output powers of 17 W and 14.6 W respectively, and the pulses were compressed to 470 fs

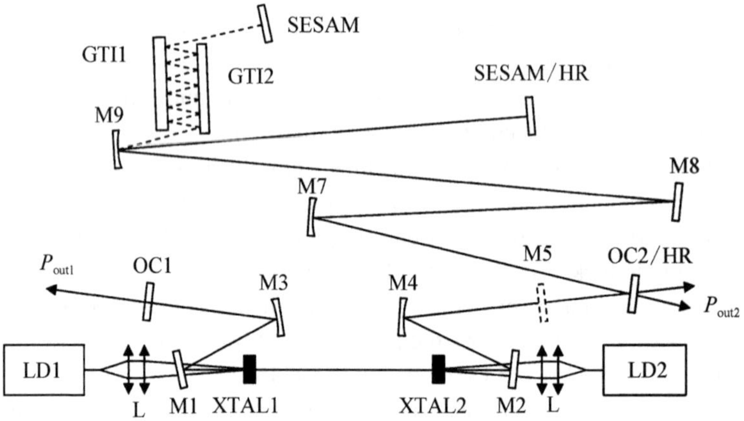

Fig. 1.5: High power femtosecond Yb:KYW laser based on bicrystal resonator.

and 450 fs. These results were the highest output power that have been obtained from bulk Yb^{3+}-doped gain media.

Because of its complex cavity structure and difficult adjustment, current bicrystal technology isn't widely used. Another type of solid-state high power lasers is based on thin disk technology. Just as the name implies, the gain medium is always a very thin disk structure, typically with the thickness of 100–400 μm and the diameter of tens of millimeters. Thin disk construction is characterized by a large ratio between surface area and thickness. One of the faces of the gain medium is coated with antireflection film and the other side is coated by a high reflection film for both the pump laser and laser. The surface coated with high reflection film is directly mounted on a water-cooled heat sink to efficiently cool the crystal. For a quasi-three-level system, it is difficult to obtain high power and high efficiency, since the population inversion can decrease with the increased temperature, owing to the sublevel energy being close to ground energy. In addition, the diameter of the pump beam on the gain medium is much larger than the thickness of the gain medium, and the heat grade can be considered to be one-dimensional and perpendicular to the surface of crystal, which efficiently reduces the thermal effects on the gain medium. Therefore the appearance of thin disk lasers solved the problems in conventional solid-state lasers, such as low output power, beam distortion, and low efficiency. However, limited by the doping level and thickness of gain media, the single absorption is very low. To solve this problem, Giesen rendered the multipump structure in 1994 [71]. The absorption efficiency was increased by multiple passing of the gain medium, usually 16 or 24 passes. Figure 1.6 shows the typical 16-pass pump structure.

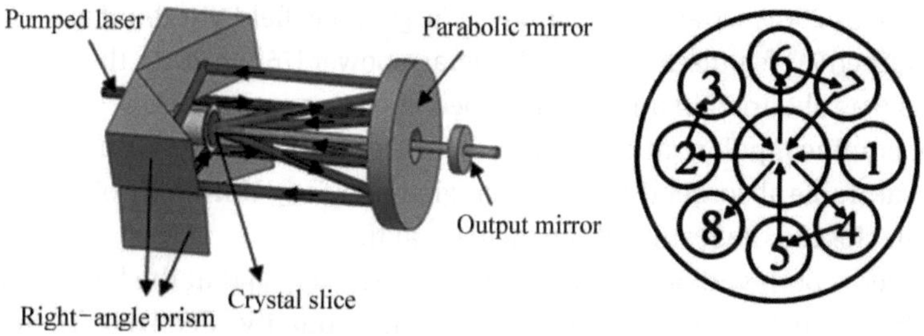

Fig. 1.6: Typical pump design for 16 absorption passes.

Based on this pump structure, the typical thin disk oscillator is shown in Figure 1.7. The self-phase modulation (SPM) was weak even in high power operation, because of the small thickness of gain media. So an additional plate with high refractive index was always inserted in the cavity, such as a YAG plate, to increase the SPM effect.

In 2013, Diebold et al. demonstrated a stable mode-locking operation of 62 fs pulses and 5.2 W average power based on the cavity structure shown in Figure 1.7 [72], which means that the thin disk lasers entered into a sub-100 fs era.

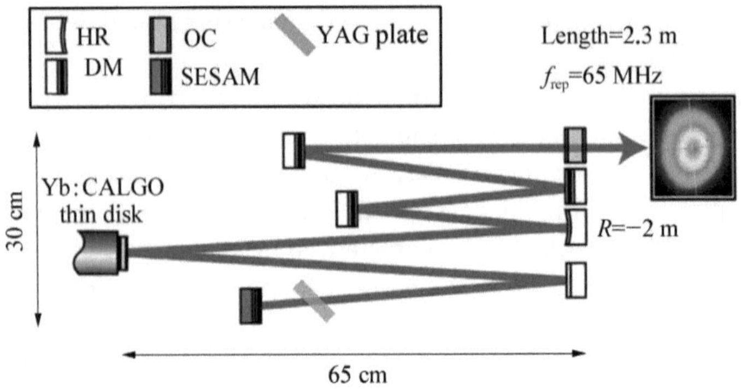

Fig. 1.7: Yb:CALGO thin disk laser in 62 fs pulses.

1.1.3 Ultrafast fiber lasers

As a typical representative of the third generation of mode-locked lasers, fiber lasers have many merits, such as low threshold, high efficiency, compact structure, light weight, stablility and reliability, no need for water cooling, and the advantages of compatibility with modern communication. At present, ultrafast fiber laser technology is one of the important research topics. As early as 1989, mode-locking picosecond lasers with optical fiber laser as the gain medium had been reported [73]. Four years later, the optical fiber capable of generating 77 fs pulses was demonstrated [74]. In recent years, ultrafast fiber laser technology has shown remarkable development. The nonlinear polarization rotation technique is preferred in obtaining ultrashort pulses for the fiber oscillators. In 2009, Ortac et al. employed large mode field ytterbium doped fiber [75], and obtained 711 fs pulses with 9 mW average power [76]. In 2010, the shortest pulse of 37.4 fs was obtained in an erbium-doped fiber ring cavity design through the adaptation of nonlinear polarization rotation technology [77]. In 2014, with the implementation of saturable absorber (SESAM), Shanghai Jiao Tong University obtained a 41.9 fs pulse [78], close to the results of the theoretical limit.

The optical fiber is born with a micron-sized core diameter, so that the light field can be bounded tightly in it, bearing the stronger nonlinear quality effectively. Therefore, optical fiber lasers showed richer mode-locking mechanisms. The common three mode-locking methods are: nonlinear loop mirror, mode-locking nonlinear polarization rotation mode-locking, and semiconductor saturable absorber. According to the net dispersion in cavity, mode-locking mechanisms can be distributed as shown in Figure 1.8, including soliton pulse locking all-normal dispersion (ANDi) mode-locking, dispersion management mode-locking, and self-similar pulse mode-locking.

Figure 1.9 shows the typical all-normal-dispersion ultrashort pulse mode-locking ytterbium doped fiber laser setup, where the cavity arrangement includes fiber path and free spatial path. The fiber path has about 25 cm of single-mode, highly-doped Yb gain fiber (CorActive company Yb-125, with an absorption coefficient of more than 350

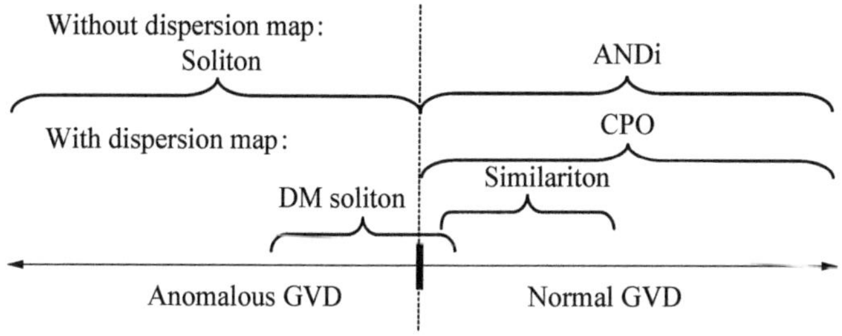

Fig. 1.8: Fiber mode-locked laser types (based on net dispersion and dispersion maps in the cavity).

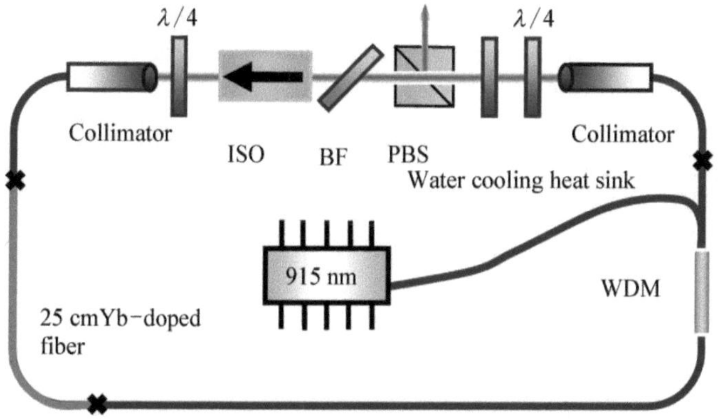

Fig. 1.9: Schematic diagram of all-normal-dispersion mode-locked femtosecond fiber laser experiment.

db/m at 915 nm) wavelength division multiplexer (WDM) with a SMF-28e tail fiber, and another fiber with a single mode fiber (HI1060). The total cavity length was 3 m or so.

The free spatial path includes the wave plate, the polarization dispersion prism (PBS), and a birefringent filter (BF) as well as additional intensity modulation components. Rotating the birefringent filter periodically causes a mode-locked spectrum.

Figure 1.10 shows the stable repetition rate frequency and spectral tuning in the mode-locked state. The autocorrelation curve is shown in Figure 1.11 (a), and the pulse width is about 5.1 ps in accordance with the Gaussian pulse shape. A pair of transmission gratings with group density of 600 lines/mm is used to compensate dispersion outside the cavity. By optimizing the grating space, the pulse was compressed to about 210 fs, and the corresponding spectrum curve is shown in Figure 1.11 (b). The spectrum supports 200 fs transform-limited pulses. Therefore, the compressed pulse is close to the theoretical limit of pulse width.

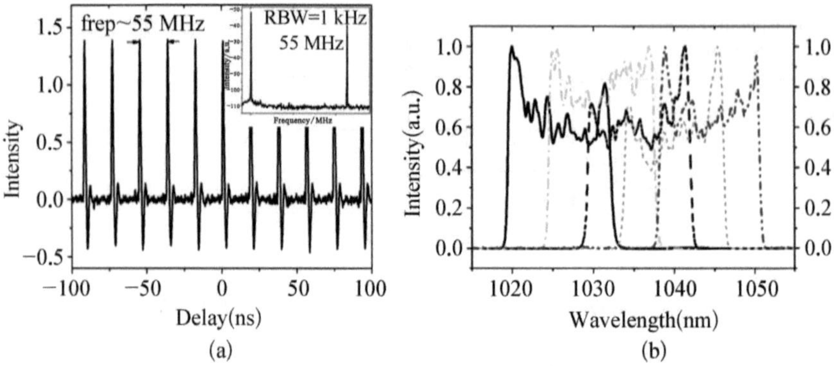

Fig. 1.10: The pulse repetition frequency of (a) the spectrum tuning curve (b) all-normal-dispersion mode-locked femtosecond fiber laser output.

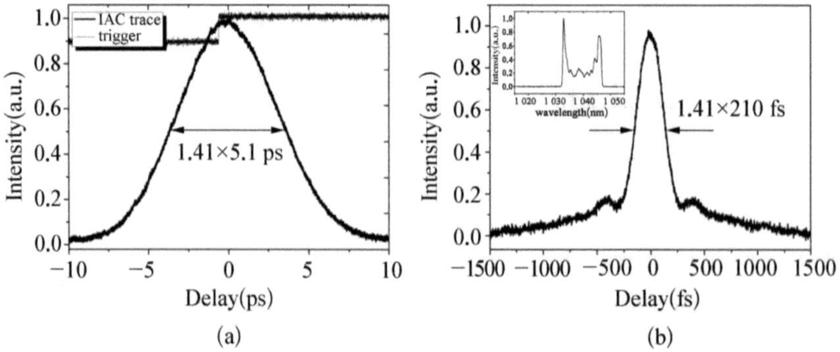

Fig. 1.11: (a) Intensity autocorrelation curve of femtosecond fiber laser output pulse width and (b) pulse width after compression.

1.2 Amplification of ultrashort pulses

1.2.1 Chirped pulse amplification technique

The pursuit of high-intensity lasers has always been one of the most important research directions of laser science. Since the invention of the laser in 1960, Q-technology can only achieve a nanosecond pulse width, so the corresponding increase in laser intensity has been very slow. The KLM technique not only directly compresses the pulse width to the femtosecond regime, but also has a higher average power relative to the dye laser. Consequently, the corresponding peak value has been dramatically improved. However, further improvement of the peak power by the direct amplification of energy would result in damage to the optical element and become a bottleneck in the development of high-intensity lasers [79]. In 1985, the CPA technology was invented, which cleverly solved the problem of limited laser intensity enhancement [8]. Since then, the laser peak power and intensity have increased by nearly ten orders of magnitude.

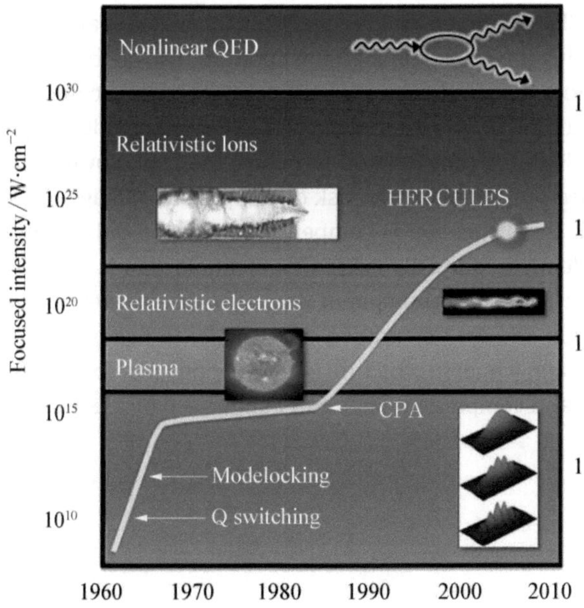

Fig. 1.12: Development of laser focusing power density.

Meanwhile, large-energy lasers can be built in a table size in the general university level of the laboratory. As shown in Figure 1.12, CPA technology plays an important role in the development of ultrastrong lasers, and the corresponding focus laser intensity. At present, some countries have even used CPA technology to build a laser system as the basic means of laser fusion projects. Some laboratories have also built CPA laser systems with petawatt peak power (10^{15} W, PW), such as the 1.5 PW, 450 fs neodymium glass system at Lawrence Livermore National Laboratory [80], the 1.16 PW laser system at the Institute of Physics, Chinese Academy of Sciences, which developed a 1.16 PW laser device [2], and the 1.5 PW Ti:sapphire CPA system at the Korea Advanced Photonics Research Institute [3].

In a typical CPA laser system, as the laser pulse peak power increases, the refractive index of the medium n is dependent on the intensity of the incident laser, resulting in a nonlinear refractive index that will spread the wave front. This effect can be expressed as the B integral:

$$B = \frac{2\pi}{\lambda} \int_0^L n_2 I dl \, . \tag{1.1}$$

For a Gaussian beam, the self-focusing effect occurs in the medium when the B-integral reaches the order of π. As a result of the higher power, the propagation medium is destroyed.

Before the advent of CPA technology, expanding the spot and increasing the medium diameter reduced the high peak power density, causing damage. When amplifying ultrashort pulses, the CPA technique avoids generating too high a peak power in the medium without increasing the beam and medium aperture. The method comprises the following steps: before the amplification, the femtosecond or picosecond pulse is introduced into a certain dispersion, and the pulse width is extended to picoseconds or even nanoseconds in the time domain to reduce the peak power, and then the amplification is carried out, reducing the risk of component damage. After the higher energy is obtained, the dispersion is compensated and the pulse width is compressed to the order of femtoseconds. By using CPA technology, it is possible to avoid nonlinear effects such as system component damage caused by excessive peak power during amplification and gain saturation effect. Refer to Figure 1.13 for the CPA's basic schematic diagram, which in structure can be divided into four parts: oscillator, expander, amplifier and compressor. Figure 1.14 shows a typical structure of the overall system.

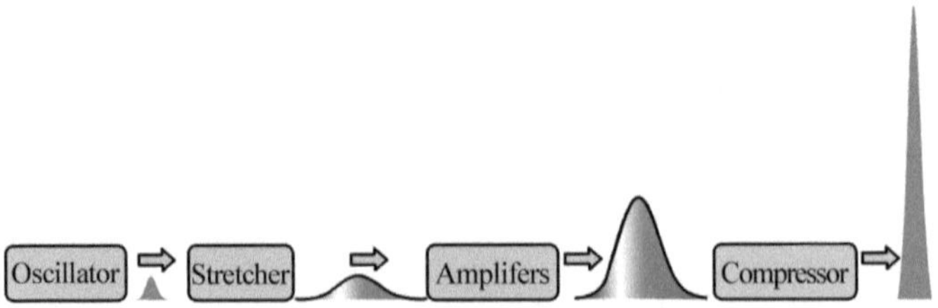

Fig. 1.13: Schematic of chirped pulse amplification technology.

In each CPA unit system, the femtosecond laser oscillator is used to provide seed laser pulses. The output power is generally in the range of a few milliwatts to hundreds of milliwatts, with a repetition rate of 80–100 MHz. The single pulse of energy is generally only a few nJ to several nJ. The stretcher is used to stretch the seed pulse before the amplifier. In the early CPA experiments [81], the positive dispersion of optical fibers was used to broaden the pulse, the negative dispersion was compensated, and the pulse width was compressed using a Treacy-type grating [82]. This broadening compression scheme easily compensated for the second-order dispersion, but the third-order dispersion was mismatched. At present, pulse stretching techniques widely used are grating and bulk material expansions, such as the reflective Martinez expansion [83] and the aberration-free Öffener expansion [84], Stone Stretcher [85], and so on. With the development of chirped mirror coating technology, chirped mirrors have been widely used for pulse stretching in recent years [86]. The amplifiers are the core of the CPA system to boost the energy of the chirped laser pulse and can be divided into the preamplifier and the main amplifier. Pre-amplification can be achieved by regenerative amplification [87], multiplying amplification [88] and optical parametric amplification [89]. The main function is to amplify the single pulse energy from about nJ. The magnitudes are scaled to mJ, and the repetition rate is reduced from 100 MHz to a few Hz to several

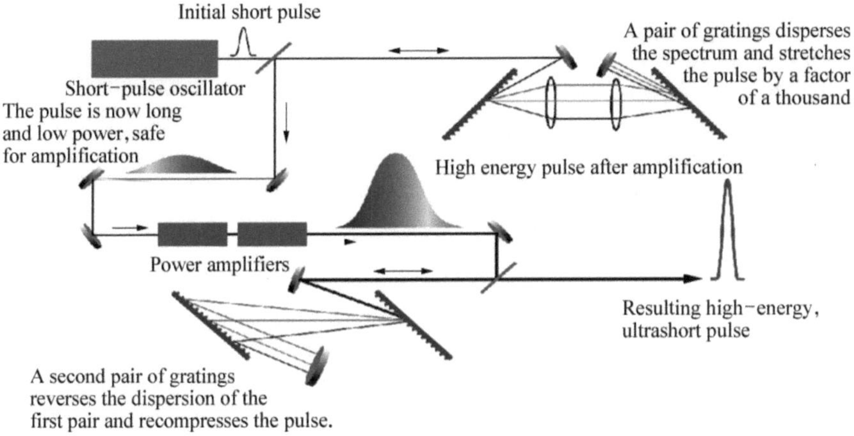

Fig. 1.14: Principle diagram of chirped pulse amplification.

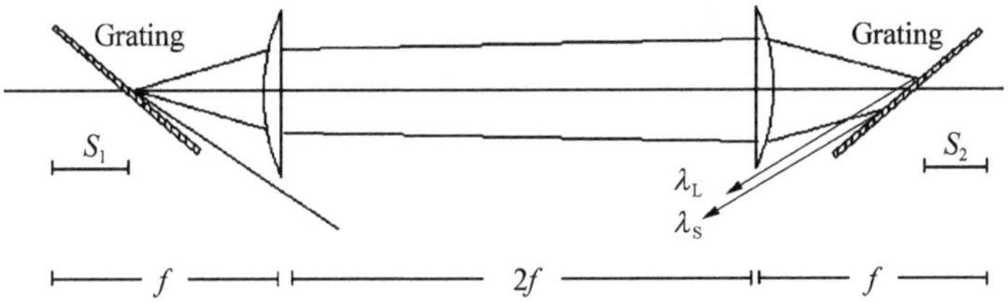

Fig. 1.15: Schematic of transmission Martinez pulse stretcher.

kHz. The type of the main amplifier is mainly multipass amplification and optical parametric amplification. The final compressor can be a Treacy-structured grating pair, a prism-to-compressor, or a chirped mirror. Next, we present a detailed description of each unit.

A typical Martinez stretcher is shown in Figure 1.15 and has been widely used in systems that output 100 fs laser pulses. For the system that generates sub-50 fs pulses after the compression, a reflective Martinez structure stretcher (shown in Figure 1.16) can be used to avoid detrimental dispersion caused by lenses. As a result of the effective reduction of the material dispersion, it is possible to successfully and reliably generate an amplified compression pulse as small as 30 fs [90].

Using the ray-tracing method, which is also labeled in Figure 1.16, we can calculate the total phase shift of the light at different frequencies in the system as:

$$\varphi(\omega) = \frac{\omega}{c}(C + A - D) + \frac{2\pi G}{d}\tan(\gamma - \theta_0 - \theta_4) + \frac{2\pi}{d}(G_0 - G)\tan(\gamma - \theta_0)\,. \tag{1.2}$$

The last term in the equation is the phase correction factor, which is increased in consideration of the aberration of the grating. Dispersion can be easily calculated from Eq. (1.2).

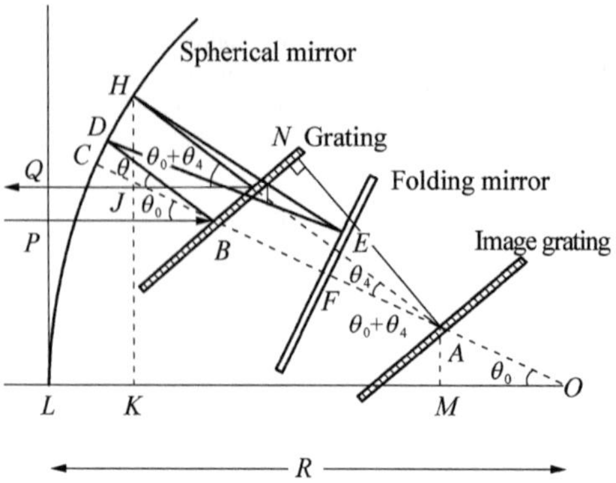

Fig. 1.16: Schematic of equivalent reflective Martinez structure spreader and ray-tracing model [90].

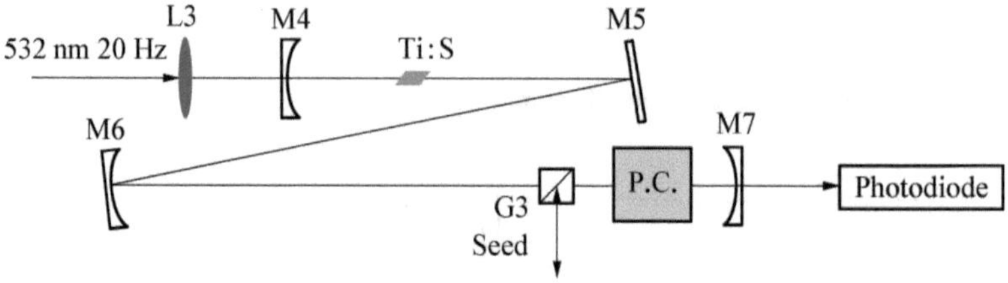

Fig. 1.17: Schematic diagram of the regenerative amplifier.

Inside a regenerative amplifier, the stretched seed pulse circulates in the amplifier cavity several times to absorb sufficient energy before the amplified pulse is ejected out of the cavity. The regenerative amplifier itself is also a Q-switched laser, and the resonant cavity may include only two mirrors, or multiple mirrors; see Figure 1.17 for the typical structure. The principle of the cavity is to maintain high energy efficiency, but also to prevent damage to cavity components within the laser. The optical beam has a smaller spot size on the gain medium and a larger spot size on the switch components (e.g. Pockels cells). This is because the gain field has a high value (typically 5 to 10 GW/cm^2), whereas a Pockels cell (usually made of KDP crystals) has a low value of 500 MW/cm^2. Since the regenerative amplifier is composed of a stable resonant cavity, it has the advantages of high amplification efficiency, stable operation, and good output beam quality, and can obtain the fundamental mode output of TEM_{00}.

The multipass amplifier uses a few faceted mirrors so that the seed pulse passes several times through the amplifier gain medium [91]. Figure 1.18 shows a typical multipass amplifier, which differs from a regenerate amplifier in that there is no resonant cavity and therefore the structure is simple. The intracavity switching elements are not required in the multipass amplifier. The seed pulse passes through the menu elements only once (in Pockels and Glan's prisms), thus greatly reducing the dispersion of the material in the CPA system. When a multipass amplifier is used as a primary amplifier, a focusing lens is typically used to fold the seed pulse and focus it on the gain medium to fully absorb energy. The main drawback of multipass amplifiers is

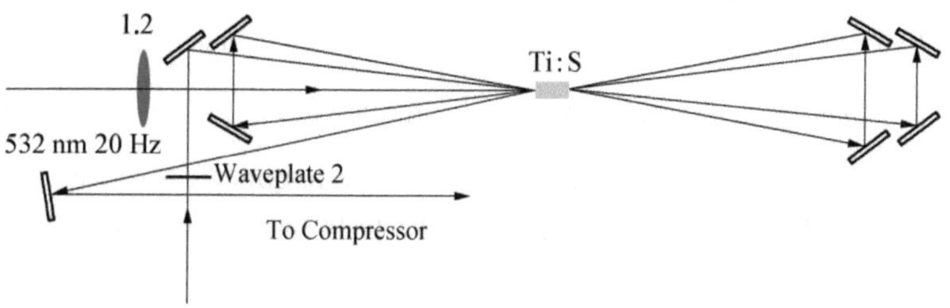

Fig. 1.18: Schematic diagram of the multipass amplifier.

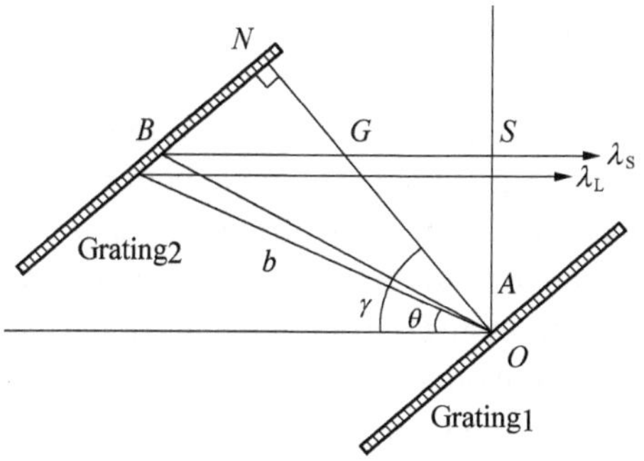

Fig. 1.19: Negative dispersion grating for ray tracing.

that they do not guarantee good colinear coupling of the seed light to the pump light, resulting in distortion of the beam quality. In addition to a certain multipass amplifier, higher energy amplification often needs to increase the pump's energy, which tends to enhance the thermal lens effect and destroy the amplifier stability. The compressor is achieved by using parallel pairs of gratings, providing dispersion that is close to the absolute value, as opposed to the spreader symbol. Figure 1.19 shows the typical structure. The ray tracing method can be used to calculate the phase shift [82] as follows:

First of all, simple light can be drawn down the ABS path geometry at length P:

$$P = b \times (1 + \cos\theta) = \frac{G}{\cos(\gamma - \theta)} \times (1 + \cos\theta)\,. \tag{1.3}$$

In addition to the actual phase, a phase correction factor must also be considered, since the diffracted light of the first grating is not simply reflected, but diffracted, when collimated by the second grating. There is also a phase difference due to the difference in diffraction positions, i.e. the beam does not follow Fresnel's law because of the diffraction enhancement produced on the various nicks, in addition to the geometric path length difference between the different wavelength components. Therefore, the phase of the light passing through the grating pair should be the phase shift caused by the geometric path, minus the phase shift caused by the diffraction. If the vertical point N is used as a reference point, the phase correction factor for any one wavelength component can be written as the number of indentations between BNs:

$$2\pi \times \frac{G\tan(\gamma - \theta)}{d}\,. \tag{1.4}$$

Then the total phase is:

$$\varphi(\omega) = \frac{\omega}{c} b \times (1 + \cos\theta) - \frac{2\pi G}{d} \tan(\gamma - \theta) . \tag{1.5}$$

Next, the group delay is the derivative of the phase versus frequency. Note that the derivative of the path in the first term is exactly the same as the derivative of the second term, i.e.

$$\frac{\omega}{c}\frac{d}{d\omega} [b \times (1 + \cos\theta)] = \frac{d}{d\omega}\left[\frac{2\pi G}{d} \tan(\gamma - \theta)\right] . \tag{1.6}$$

In this way, a simple group delay formula can be obtained:

$$\tau = \frac{d\varphi}{d\omega} = \frac{b \times (1 + \cos\theta)}{c} . \tag{1.7}$$

This formula implies that the group delay obtained by the pulse in the raster pair is equal to the time delay caused by the geometric path. With this formula, both second-order and third-order dispersion can be obtained in turn. Since this grating pair provides negative group delay dispersion, it is often used to compensate for the positive chirp from the material in the pulse, thereby compressing the pulse. For historical reasons, such grating pairs are known as pulse compressors.

The invention of CPA technology has greatly improved the peak power of ultrashort pulse lasers, which makes it possible to realize the ultrastrong laser output of TW and PW at the laboratory bench. The current peak power density after focusing is expected to reach the order of 10^{22} W/cm^2 [6]. Such high power densities can produce unprecedented conditions such as strong electric fields, strong magnetic fields, and high pressure fields under laboratory conditions, and provide the basis for the research of strong field physics and high-energy density physics. X-ray lasers, laboratory astrophysics, laser particle accelerators, inertial confinement fusion, and other research fields are extremely important applications. As shown in Figure 1.12, with the in-depth development of application research, the performance of laser devices and parameters will continue to put forward new demands, as ultrashort laser technology becomes the driving force for sustainable development.

1.2.2 All-solid-state ultrafast and ultraintense lasers

The average power and pulse energy of all-solid-state laser oscillators are limited and not high enough for many applications. In order to get ultraintense lasers, amplification technologies need to be used. Regenerative amplification, chirped pulse amplification (CPA) and power amplification are common technologies. Figure 1.20 shows

Fig. 1.20: A general scheme of ultrashort and ultraintense systems.

a general scheme of ultrashort and ultraintense systems. Ultrashort pulses from an all-solid-state oscillator first pass through a pulse picker to reduce the repetition rate, and then are injected into a regenerative amplifier. In the regenerative amplifier, pulse energy could be enhanced by an order of 10^6 times, from a few nJ to a few mJ. Finally, the pulses are further amplified with power amplifiers to get an ultraintense laser. For femtosecond pulses, CPA has to be used to improve the amplification efficiency and avoid the damage induced by high intensity. The pulses are stretched before amplifications and compressed afterward to obtain ultrashort and ultraintense lasers.

CPA was first implemented in Nd-doped crystal lasers. With the development of diode-pumped laser technology, the all-solid-state laser with CPA technology has become an important aspect in the study of ultrashort and ultraintense lasers. In 2003, Kawanaka et al. achieved an amplified laser with a repetition rate of 10 Hz and pulse energy of 30 mJ from a diode-pumped thin-disk Yb:YLF regenerative amplifier [92]. In 2011, more than 100 mJ of pulse energy was realized in Japan by using cryogenic cooling technology in a similar regenerative amplifier [93]. Based on the characteristics of Nd-doped laser media, Ai et al. obtained an amplified laser with a pulse energy of 20.8 mJ, pulse duration of 12.9 ps, and a repetition rate of 10 Hz by using a bulk Nd:YLF crystal as the laser medium, a regenerative amplifier, and a five-pass main amplifier [94]. In 2013, Noom et al. demonstrated a high-energy laser system with a

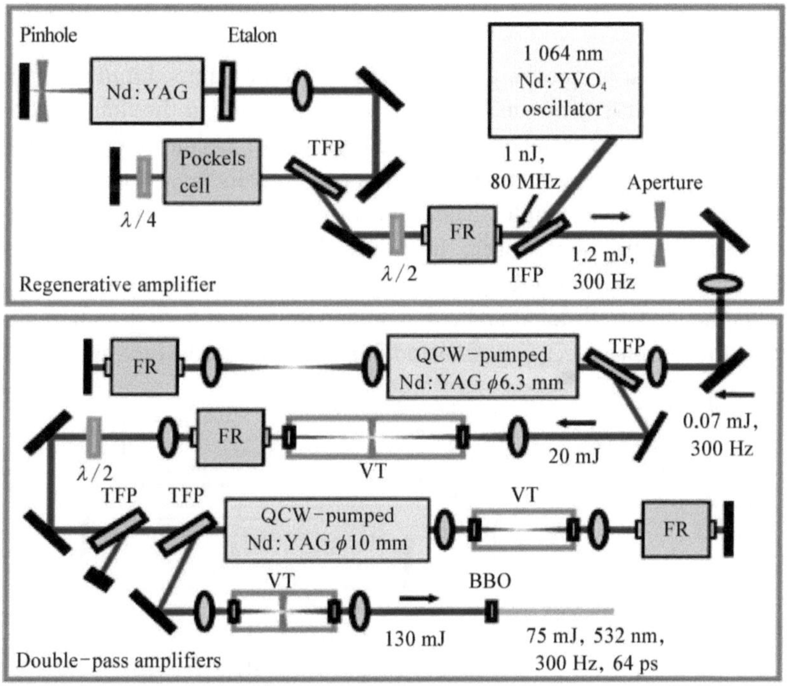

Fig. 1.21: 130 mJ, 64 ps, 300 Hz high-energy picosecond laser system [97].

pulse energy of 130 mJ, pulse duration of 64 ps, and a repetition rate of 300 Hz by using a Nd:YAG crystal and a similar scheme as the one mentioned above. The system is shown in Figure 1.21 [95]. A Nd:YVO_4 ultrashort oscillator with 1 nJ and 80 MHz was used to provide a seed pulse. Then the picked seed pulse was injected into a regenerative amplifier and amplified to be 1.2 mJ with a repetition rate of 300 Hz. Finally, the output of the picosecond laser with up to 130 mJ was achieved after two stages of power amplifiers. With a BBO crystal as the frequency-doubling medium, a green picosecond laser at 532 nm could be realized with pulse energy of 75 mJ and pulse duration of 64 ps.

In recent years, by combining CPA technology and laser media with broad emission spectra, important progress has been made on all-solid-state femtosecond ultraintense lasers. In 2013, Alizée Mareczko et al. achieved an amplified laser with a pulse energy of 100 mJ, a repetition rate of 10 Hz, and pulse duration of 640 fs based on a Yb:CaF_2 crystal [96]. In 2014, Kessler et al. obtained ultrashort and ultraintense lasers with pulse energy as high as 16.6 J and pulse duration of 150 fs at a repetition rate of 1 Hz pumped by a high-energy laser with POLARIS (Germany) [97]. With the gradually stronger demand of high-power ultrashort lasers in the field of laser processing, many companies such as Coherent, Spectra-Physics, TRUMPF, Rofin, and Guoke Laser have also launched high-power ultrashort laser products and industrial processing systems. It is expected that all-solid-state picosecond and femtosecond lasers, which are important tools for industrial processing and other applications in the future, will be developed rapidly.

To achieve higher output power, a huge power has to be pumped into the gain medium, and lots of heat is generated in the gain medium due to the limit of conversion efficiency. Thus the thermal effect of laser crystals will become more serious, which limits the further enhancement of the output power. Thus, how to design laser systems and to dissipate the accumulated heat effectively becomes an important problem for all-solid-state ultraintense lasers. In the last few years, several approaches such as bulk, thin-disk and innoslab materials are commonly used in laser systems. Bulk material is the most traditional and common one in current laser design. In order to overcome the thermal effect, innoslab and thin-disk gain media are studied and

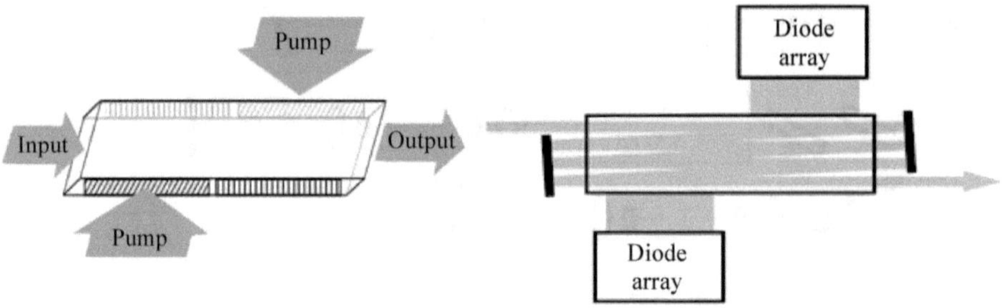

Fig. 1.22: The 'Z' type side-pumped structure with innoslab material.

often used. For innoslab material, the 'Z' type side-pumped structure (shown in Figure 1.22) guides the temperature gradient in the direction of thickness, which is parallel to the direction of light, and could effectively avoid thermal lens and thermal distortion along the thickness. But thermal distortion still exists in other directions, and the unused gain part of the crystal will reduce the conversion efficiency and increase the possibility of parasitic oscillation. In 2014, Young et al. achieved an amplified laser with an output power of 225 W, pulse energy of 450 μJ, and pulse duration of 11 ps based on a Nd:YAG innoslab structure [98]. For thin-disk material, it has a very thin thickness (generally less than 100 μm) with the temperature gradient perpendicular to the direction of the disk. So the thermal lens can be avoided and other thermal effects could be avoided. But laser systems with thin-disk material generally have to use complex multipass or regenerative structures to achieve high power because of their low single-pass gain.

The most significant characteristic of the thin-disk laser is the large ratio between surface area and thickness. In order to cool effectively, the back surface of the disk gain medium is directly fixed on a water-cooling heat sink as shown in Figure 1.23. The disk structure of the gain medium allows it to support high pump power density without obviously increasing temperature, which provides the possibility of achieving high power for the quasi-three-level laser media. In addition, the temperature gradient of the disk medium is almost one-dimensional and perpendicular to the crystal surface because the beam diameter on it is far bigger than its thickness. Thus, the thermal lens is reduced significantly. But a complex multipass pump structure (as shown in Figure 1.6) has to be used to obtain high output power and good beam quality due to the low single-pass gain of the disk medium. In 2012, Saraceno et al. achieved an average power of 275 W and pulse duration of 583 fs thin-disk laser in a vacuum chamber [99], which is the highest power for oscillators up to now. In 2014, Saraceno et al. got the highest pulse energy of 80 μJ [100] from a thin-disk oscillator after the work by

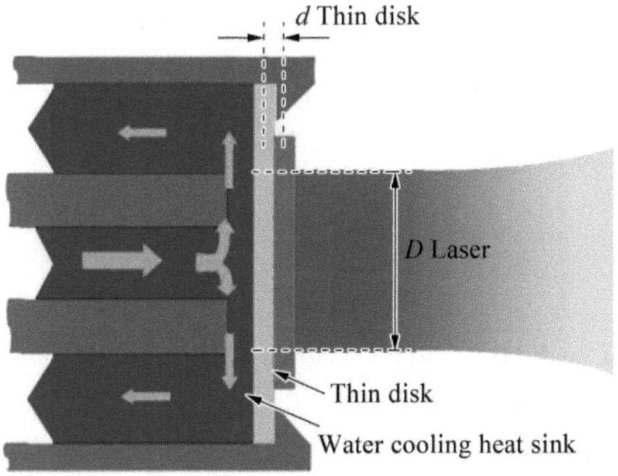

Fig. 1.23: The cooling structure of thin-disk media.

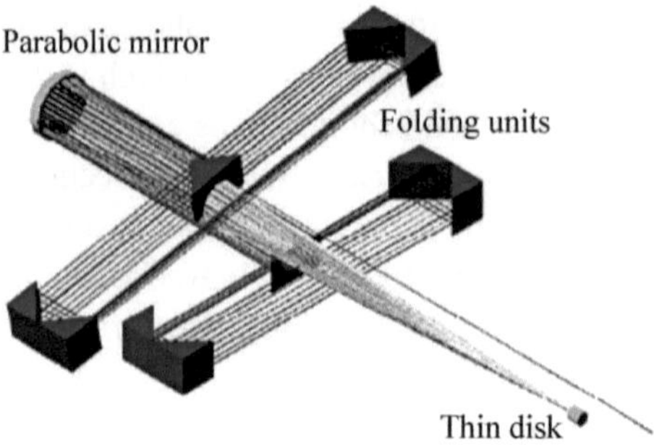

Fig. 1.24: The scheme of multipass thin-disk amplification [103].

Brons et al., which reported a KLM thin-disk oscillator with an output power of 230 W and a pulse duration of 330 fs.

Due to the high-power density of the cavity, further improvement of the output power from thin-disk oscillators will be limited. Thus, thin-disk amplifiers are studied and rapidly developed for high power. In 1997, Hönninger et al. first realized a diode-pumped thin-disk Yb:YAG regenerative amplifier and achieved a pulse energy of 180 μJ at a repetition rate of 750 Hz [101]. In 2001, Müller et al. obtained 1 mJ of pulse energy at a 10 kHz repetition rate by using thin-disk amplification [102]. Two years later, they designed a thin-disk multipass amplification for the first time, as shown in Figure 1.24, and achieved an amplified result of 11.6 mJ energy at 1 kHz [103]. In 2008, TRUMPF Laser successfully realized a picosecond thin-disk regenerative amplifier with an output pulse energy up to the mJ level, pumped by a kHz laser [104]. In 2013, Negel et al. reported a result of 1.1 kW and 1.38 mJ from a multipass Yb:YAG thin-disk amplifier, which is the highest output power to date [105].

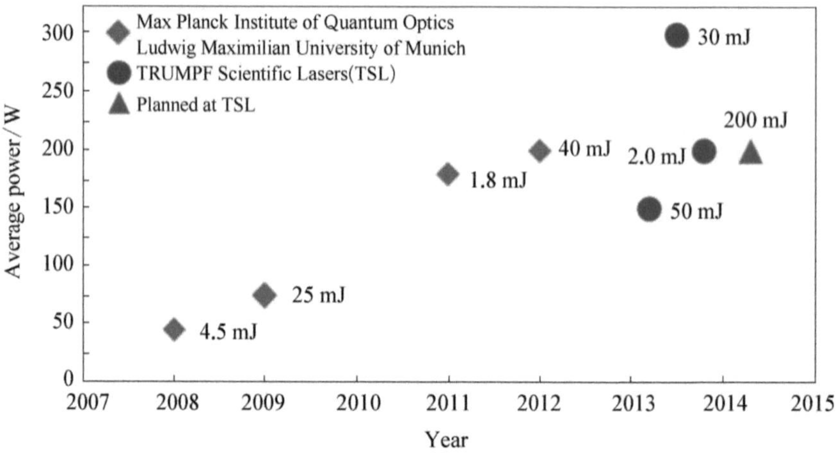

Fig. 1.25: Progress on picosecond thin-disk amplifiers.

Since the concept of the thin-disk was proposed in 1994, its excellent characteristics of high-power, high-pulse energy, and good beam quality are more outstanding than other type lasers. The effective cooling capacity makes the thermal effects and distortion very small even in conditions of high-power pumps. The output power of thin-disk amplifiers could be as high as hundreds of watts or even kilowatts, and the pulse energy can also be improved to be dozens of mJ. Figure 1.25 shows the main progress on picosecond thin-disk amplifiers in recent years.

1.2.3 Ultrashort and high-intensity fiber lasers

Due to the thermo-optic effect restricting power scaling, many novel methods have been proposed based on the increased dimensions of gain media, such as the z-type pumped slab, thin disk, and large core fiber, etc. For slab gain media, z-type side-pumped modules make the temperature gradient occur toward the thickness and the propagation of the laser beam parallel to the temperature gradient, which could effectively avoid the effects of thermal lens and thermal-optical distortion. However, thermal-optical distortions along other directions still exist. In the unabsorbed regime, it could reduce the laser efficiency and increase the possibility of parasitic oscillation.

For thin disks, their very thin thickness (< 100 um) is smaller than the focal point of the pump laser in gain media. Because the direction of the temperature gradient is perpendicular to the disk, the effects of thermal lens and thermal distortion, as well as thermal induced birefringence, could be effectively reduced.

The main problem for thin disk amplifiers is that they are weak gain for a single pass, which necessitates multipass and/or regenerative amplification for high-power laser output. In contrast to thin disk amplifiers, due to their long gain length, optical fibers could provide very large single-pass gain and their large surface-to-volume ratio improves the heat dissipation along the direction of the laser beam propagation. In addition, fibers also have other advantages: (1) The waveguide structure of fiber ensures good beam quality and effectively suppresses the thermal distortion; (2) the waveguide structure guides the pump and seed laser propagation along the fiber, and improves the single-pass gain; (3) up to now, the high-power fiber lasers have already employed Yb doped fibers as gain media, which are directly pumped by the available high-power diode laser; in addition, due to the low quantum defect (10%), a high optical-optical efficiency of 80% can be obtained.

However, fiber also has its own defects, which include high nonlinearity. Thus, the main restriction for the generation of high-power lasers is the further improvement of the peak power in the femtosecond amplification domain.

Even though there are disadvantages associated with the three methods, slab, thin disk, and fiber lasers still lead the way in generating high-power lasers. In fact, these three types of lasers have already realized thousands of watts of output. So for current

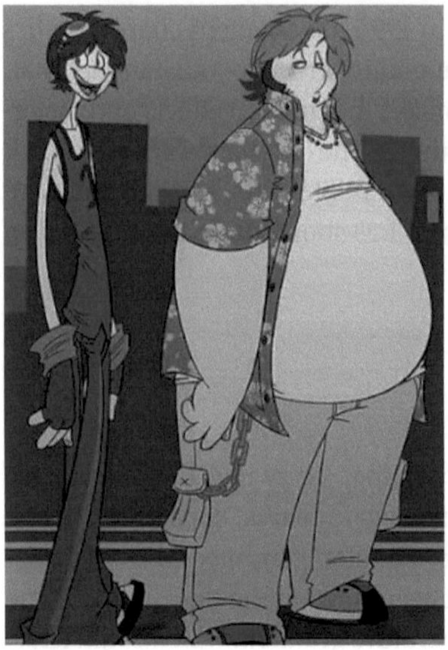

Fig. 1.26: Thin disk or fiber, which one is more suitable for femtosecond lasers?

applications, what kind of high-power laser is chosen as an outstanding and usable source, as shown in Figure 1.26, remains unclear.

From the above discussion, the critical problem in fiber lasers is nonlinear effects, which are proportional to the inherent length of action and intensity, and inversely proportional to the mode field diameter (MFD). Thus, it is beneficial to lower the nonlinear effect by increasing the MFD and reducing the inherent action length. In the last ten years, with the advent of large mode areas and large airhole pitch photonic crystal fibers (LMA-PCF), the nonlinearity of fiber has been efficiently suppressed, and high-power femtosecond fiber lasers have been rapidly developed using LMA-PCF.

As a leading authority on ultrafast fiber optics, Professor Andreas Tünnermann at Friedrich-Schiller-Universität has made many contributions in high-power CPA fiber lasers at high repetition rates and high peak power fiber laser with high energy.

As early as 2002, Andreas Tünnermann's group carried out studies on the high-power fiber laser using LMA-PCF. One stage of PCF amplifier was developed with a Nd:glass oscillator injecting as a seed laser, and a 10 W output power with a pulse duration of 80 fs at 75 MHz, obtained after compression [106]. They subsequently proposed many experimental schemes and new records have been achieved in average power, single pulse energy, and peak power [107–111]. To date, the highest record is 830 W with a pulse duration of 640 fs at 78 MHz, reported using the LMA-PCF as gain medium and using CPA technology, in 2010 [110].

As shown in Figure 1.27, an all-solid-state Yb:KGW femtosecond fiber laser was employed as the seed source, and gratings were used in the stretcher-compressor in this system. In 2011, 1 mJ with a pulse duration at 30 kHz was demonstrated by Haedrich, in which a Ti:sapphire laser oscillator was used as the seed source and nonlinear stretch-

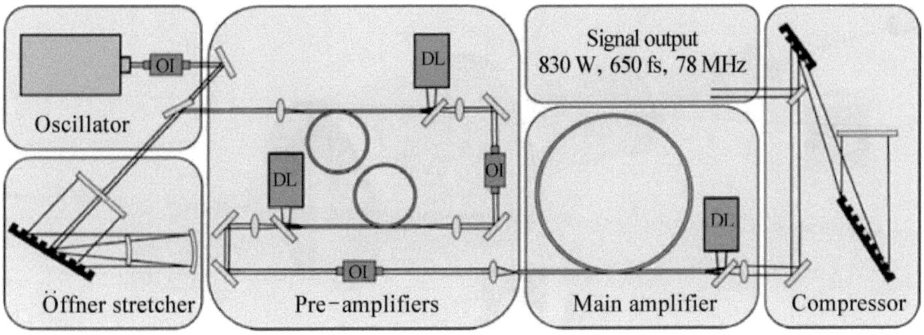

Fig. 1.27: Fiber chirped pulse amplification system with 830 W output power.

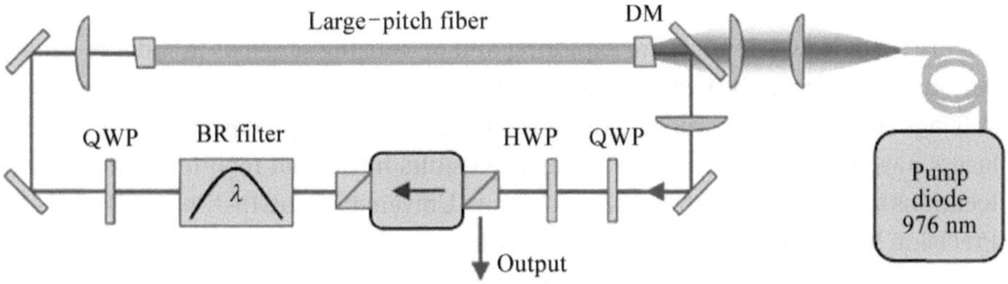

Fig. 1.28: High-power mode-locked femtosecond PCF fiber oscillator.

ing in nonlinear fiber, as well as CPA technology, was adopted [112]. In addition, by utilizing a combination of the technology of fiber amplification and optical parametric chirped pulse amplification (OPCPA), more than 100 μJ with a pulse duration of sub-5-fs, which corresponds to 12 GW of peak power, and a fiber amplifier at 96 kHz, has also been demonstrated.

Except for the progress in amplification above, a high-power mode-locked femtosecond fiber laser oscillator has also been realized based on LMA-PCF as the gain medium. As shown in Figure 1.28, a high-power mode-locked all-normal dispersion ytterbium fiber oscillator, which delivered 66 W at 78 MHz with a pulse duration of 91 fs, was reported by M. Baumgartl in 2012, in which a LMA-PCF was adopted for the gain and nonlinear polarization evolution [113]. This experimental scheme is very straightforward in order to generate high-power femtosecond output and it is also very promising as a seed source for fiber laser amplifiers with hundreds of watts with its compact design and reliable performance.

These experimental results presented above approach the limitation of the power and energy of fiber lasers. Based on the available fiber amplification and coherent synthesis technology, G. Mourou proposed a scheme that delivers 10 J at 10 kHz, in order to meet the needs of the application of the next generation particle accelerator in 2013 [114]. For such a large program and scheme, there are major challenges of design and optimization of the high-power fiber amplifier and improvement in performance

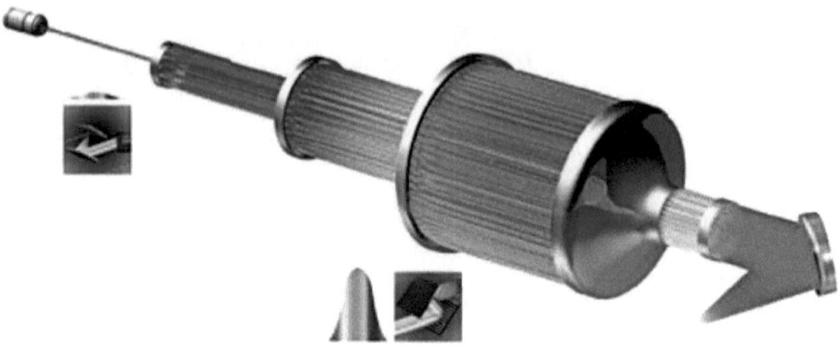

Fig. 1.29: Experiment scheme of the next generation coherent combination.

(Figure 1.29). Now scientists are exploring and optimizing schemes in order to satisfy various research needs. Even though our domestic research on ultrashort fiber lasers still lags behind other developed foreign countries, many major developments on fiber lasers have been achieved and satisfactory results have been reported to date, from locations such as Tianjin University, Tsinghua University, Xi'an Institute of Optics and Precision Mechanism of CAS, East China Normal University, and so on [115–118].

The above-mentioned research and breakthrough progress on femtosecond fiber laser complements those on an international scale. As early as 1961, Snitzer proposed the idea of putting fiber into the research of the fiber laser and fiber amplifier and first demonstrated a glass fiber laser in 1964 [119, 120].

However, due to the technical restrictions in preparation, further development of fiber lasers has been very slow. It was not until the 1970s and 1980s, with the successful development of low loss silica fiber and doped fiber, as well as the maturing technology of fiber components, that scientists began to focus attention on fiber laser research.

In the early 1990s, due to the rapid development of optical fiber communication, Er-doped fiber lasers attracted intense interest. Many groups were devoted to studying Er-doped fiber lasers and many outstanding results were obtained in dispersion management [121–123], high pulse energy [124], the mechanism of mode-locking [125, 126], and the compression of short pulses [77, 127]. Although the Er-doped fiber laser is compact and stable in performance, there exist disadvantages when it functions in high-power and high-energy scenarios. In the late 1990s, the first mode-locked Yb-doped fiber laser was demonstrated [128], which indicated the arrival of the high-power and high-energy fiber laser. In Yb-doped fiber lasers, there are many mode-locked mechanisms that have been demonstrated [129–132]. Based on the management of dispersion, mode-locked Yb-doped femtosecond fiber lasers at high repetition rates have been widely studied [133, 134]. The advent of double-cladding fiber and PCF further enriched the high-power femtosecond fiber lasers [110–113, 135–137].

1.2.4 Optical parametric chirped pulse amplification technology

Although ultraintense laser pulses with peak powers of more than 1 PW have been produced using the current CPA technology, they have the following disadvantages: first, the gain-narrowing effect during the amplification process has a harmful effect on the available shortest pulse duration after recompression, and this effect becomes greater with the improvement of amplified energy. Therefore, it is very difficult to obtain recompression laser pulses with extremely short durations. Secondly, the accumulation and magnification of ASE in the amplification process reduces the contrast ratio of the amplified laser pulses. Last but not least, since the gain of single-pass structure is low, a multistage multipass is necessary in the amplification process, resulting in more complex experimental devices and higher costs.

To overcome the disadvantages of traditional CPA technology, Dubietis et al. introduced the principle of optical parametric chirped pulse amplification (OPCPA), which employed the parameter amplifier based on a nonlinear process instead of the amplifier based upon the gain medium of CPA in 1992 [146]. Using the OPCPA technology with BBO crystals, they obtained amplified pulses with 70 fs pulse duration and pulse energy of 65 μJ, and peak power of 0.9 GW.

Different from traditional CPA technology, OPCPA is a nonlinear process based on OPA technology using nonlinear crystals (KDP, BBO, KTP etc.) as the gain medium and high-energy lasers satisfying the nonlinear phase matching condition as the pump source to amplify the signal pulses. The OPA technology has been widely used in producing laser pulses with various tunable wavelengths, due to the advantages of nonlinear crystals, such as wide transmission and tunable spectral range, high single-pass gain, no thermal deposition, and so on. Ross at the UK Rutherford Appleton Laboratory combined the OPA and CPA in 1997 and proposed the concept of OPCPA [147]. OPCPA uses nonlinear crystals (KDP, BBO etc.) as the gain medium, replacing traditional gain media like neodymium glass and Ti:sapphire, resulting in the combination of high pulse energy in the CPA system and a wide spectral range in the OPA system. Moreover it can circumvent the gain-narrowing effect in the process of chirped pulse amplification. In addition, compared with the general CPA process, OPCPA employs the parametric process inside the crystal without the ASE effect so that the output has a higher signal-to-noise ratio. Since there is no thermal deposition process during the OPCPA process, the laser beam obtained from the OPCPA system has a very low B integral and smaller thermal distortion. Figure 1.30 shows one OPCPA scheme demonstrated by Ross et al. in 2000 [148]. They use a LBO crystal as the first stage OPA gain medium and a KDP crystal as the second stage OPA gain medium. The seed source is a commercial Tsunami oscillator, and it is stretched to 300 ps and then introduced into the amplifier for amplification by a single-mode fiber. The wavelength of the pump pulse is 527 nm, obtained by frequency doubling the central part of a beam emitted by the Vulcan laser device. The beam size of the pump pulse is 20 mm, the pulse energy is ~2.5 J, and the pulse duration is 600 ps. A small part separated from

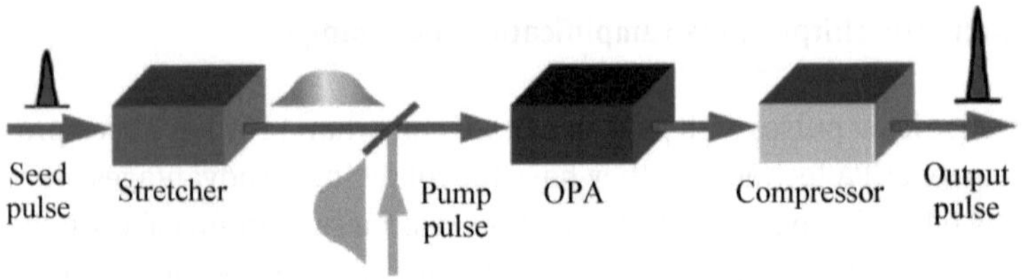

Fig. 1.30: Schematic diagram of OPCPA system.

the master pump pulse by a 9% beam splitter is shrunk to 1 mm. At the first stage of the pump pulse, the beam size of the seed pulse emitting from the fiber is 0.5 mm, and the two beams use noncollinear matching with the matching angle being 0.5°. After being amplified by the first stage, the beam size of the seed pulse is expanded to 10 mm and then the seed pulse enters into the second amplifier pumped by the residual 527 nm pulse whose beam size is 14 mm. After two-stage amplification, the seed pulse is amplified to 500 nJ. The conversion efficiency of the whole system is up to 20% and the gain is 10^{10}. Subsequent experiments achieved the contrast ratio of 1.5×10^8 for the amplified pulses, which is far better than that of the general CPA system. Its main advantages are described in the following aspects.

First, as OPCA is different from the amplifier based on the gain medium, OPCPA has no ASE that is inevitable in the amplifier on the basis of the gain medium. The nonradiative transition process during the stimulated emission in the conventional CPA system commonly results in a large amount of heat accumulated in the gain medium. Higher pulse energy can be obtained at the expense of repetition frequency. As a nonlinear process, the parametric process adopts a virtual energy level amplification without heat accumulation in the gain medium so that it removes the complexity and costs caused by the cooling system. Meanwhile, it can effectively improve the contrast ratio since there is no spontaneous emission noise produced during the radiative transition process.

Second, it can provide a large gain bandwidth and avoid the spectral narrowing effect in the CPA process so that it is able to obtain a pulse after the recompression amplification, achieving approximately the same pulse duration as the seed pulses. The gain crystal has an emission peak, which causes the gain narrowing effect because the gain of the wavelength at emission peak is higher than that of the wavelength deviating emission peak. The narrowed spectrum is insufficient to support short enough pulses, which then leads to the broadening of pulse duration and a decrease in peak power and power density. There is no 'emission peak' in a parametric process, and it is the central wavelength acquiring the gain that exactly satisfies the phase matching condition. In the meantime, as the seed pulse (or signal pulse) has a certain bandwidth, a definite phase mismatching range can obtain gain. Choosing a suitable nonlinear medium, phase matching angle, and noncollinear angle, can keep the whole wavelength band of seed pulse in the range of gain and acquire amplification with

Fig. 1.31: Schematic diagram of phase matching in noncollinear parametric process; the figure on the right illustrates the direction of wave vectors in the crystal.

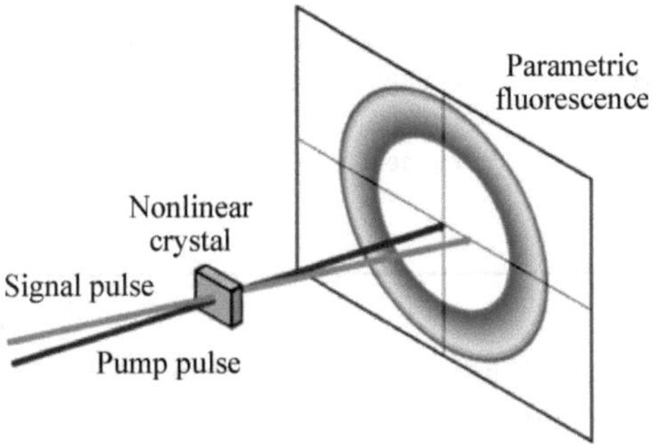

Fig. 1.32: Schematic diagram of parametric amplification setup.

a wide bandwidth, high gain, and no sacrifice to the spectral bandwidth. Figure 1.31 shows the relationship between the wave vectors satisfying the phase matching in the parametric process under noncollinear conditions. The figure on the right illustrates the relationship between the three waves inside the parametric crystal.

Third, the parametric process with single-pass structure can reach high gain and its structure is relatively simple. Figure 1.32 shows the schematic diagram of the parametric process. When the pump pulse and signal pulse are temporally synchronized and spatially overlapped, a single pass through the parametric crystal is able to achieve a gain with a 106 order of magnitude and amplify the laser with pJ energy to μJ energy directly. In contrast, traditional stimulated emission amplification in CPA needs to pass through the gain medium a dozen times or even several dozens of times (in multipass structures) or inject into the regenerative amplification cavity to acquire a gain with the same order of magnitude.

Fourth, the stimulated emission amplification only amplifies the specific wave band, while parametric amplification has the capacity to tune and transform wavelength. The feature of wavelength tuning in parametric processes is useful for investigating spectral characteristics of varied atoms and molecules, which have different absorption wavelengths. Thus the physical process occurring inside atoms and molecules can be studied by tuning the wavelength of signal pulses into a special range. In addition, the idler pulse with a longer wavelength whose bandwidth is wider than a signal pulse is also of value. Because idler pulses acquire the gain from quantum noise, their contrast ratio outclasses the signal pulse and equals the product of the contrast ratios of pump and signal pulses theoretically.

Tab. 1.2: Comparison on characteristics between OPA and OPCPA.

	CPA	OPCPA
Gain Bandwidth	~30 THz	> 100 THz
Single-pass Gain	< 10	~ 10^6
B Integral	High, > 1	Low, < 1
Thermal Load	Serious With thermal lens effect	Negligible No thermal lens effect
Laser Energy Storage Time	~microseconds (Relaxation Time)	Simultaneously (Strictly generated at the same time)
Background Noise	Spontaneous Emission Amplification	Parametric Fluorescence Amplification
Gain Wavelength	Stationary	Changeable
Stability	Good	–
Conversion Efficiency	High	–

Fifth, self-stability of the carrier-envelope phase (CEP) can be achieved in OPCPA. No matter how the seed source drifts, the phase of the idler pulse generated in a parametric process is definitely stable if the pump and signal pulses come from the same seed source.

Table 1.2 lists a comparison of the main characteristics between conventional CPA and OPCPA. Although OPCPA is capable of improving the contrast ratio of laser pulses, its parametric crystal also generates the parametric fluorescence stimulated by the pump pulse. The amplified parametric signal pulse pumped by the nanosecond pulse also has parametric fluorescence over a long timescale and the fluorescence has an effect on the contrast ratio of pulses [149]. To solve these problems, optical parametric amplification technology based on the ultrashort pump pulse has been recently developed [150, 151] and is shown in Figure 1.33. The duration of the pump pulse is similar to the signal in the range from several picoseconds to the femtosecond magnitude so that parametric fluorescence can be extremely inhibited in the timescale of hundreds of picoseconds or even tens of picoseconds and the contrast ratio is efficiently improved. Specifically, the OPA process idler pulse has a higher contrast ratio than that of an amplified signal pulse when the pump source is unconsumed and small signal approximation exists. This will be further discussed in the subsequent work.

Fig. 1.33: Schematic diagram of OPCPA system pumped by ultrashort pulse.

1.3 Carrier envelope phase of ultrashort pulses and related control

1.3.1 Ultrashort pulse compression

Besides the ultrafast temporal characteristics, the femtosecond laser also has the properties of ultrahigh peak intensity and an ultrabroadband spectrum, which extend its wide applications in strong field atomic and molecular physics [153], attosecond physics [154], ultrawide time-resolved spectroscopy [155], and optical frequency combs metrology [156]. Generally speaking, owing to the restrictions of spectral bandwidth and dispersion management, it is difficult to get the shortest pulse duration directly from the oscillator. To date, the most effective way to generate the shortest pulse duration has been the spectrum broadening of the amplified laser pulse, combined with external compression technology. Several pulse duration world records have been set with the technology from the beginnings of dye lasers. The main technologies of femtosecond lasers and the progress in shortest pulse duration are shown in Figure 1.34. In 1986, the shortest 6 fs pulse duration was achieved by American scientist Fork, who broadened the spectrum of the amplified dye laser with optical fiber, and compressed the pulse duration by grating and using a prism pair [30]. With the emergence and development of the Ti:sapphire laser, Baltuska and Wei, taking the homemade cavity-dumping femtosecond Ti:sapphire oscillator as the light source, on the basis of optical fiber spectrum widening, and by using two compression methods of a grating-prism pair and a chirped mirror-prism pair, broke the world record in two ways [157] in 1997. The pulse duration is less than 5 fs and the repetition frequency was

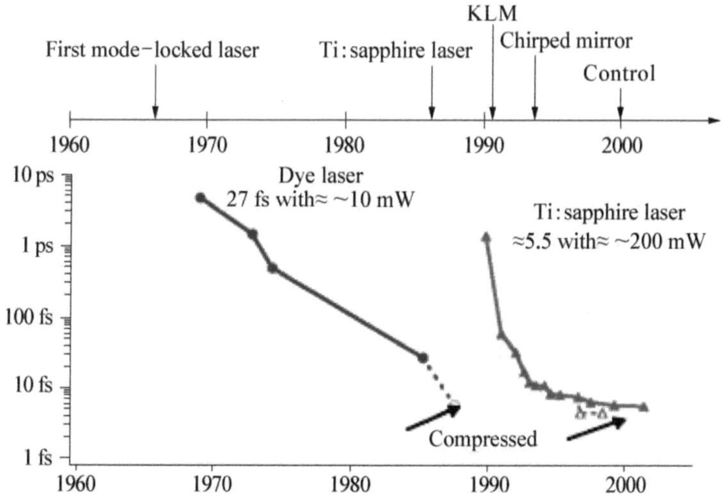

Fig. 1.34: The breakthrough technologies (above) and development of the shortest reported pulse duration (below).

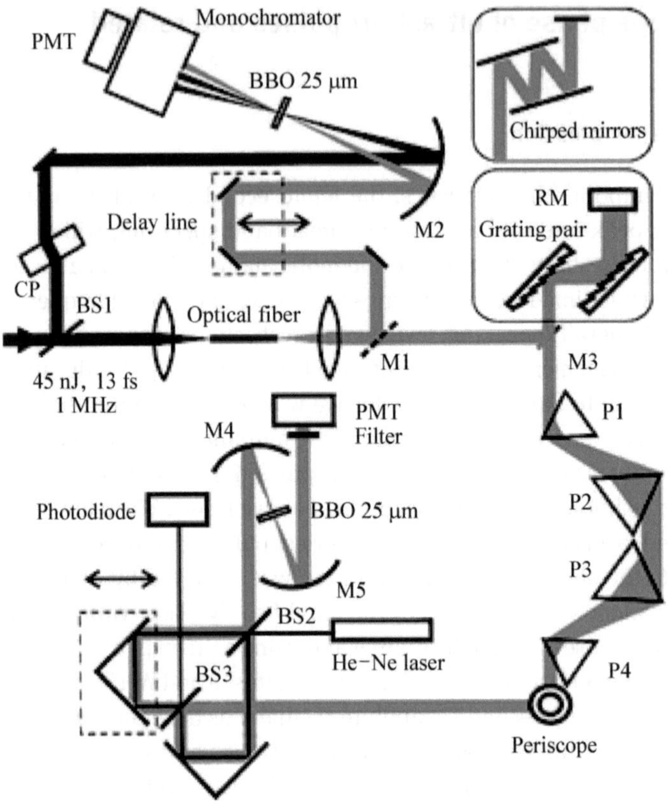

Fig. 1.35: Femtosecond cavity dumping compression system based on chirped mirrors and a grating-prism pair.

1 MHz, much higher than the conventional results of 1 kHz, the compression system shown in Figure 1.35. Afterward, Vienna University of Technology's Krausz [158] and University of Tokyo's Shirakawa [159] used the spectrum widening technologies of hollow fiber and nonlinear parametric amplifiers, and achieved the compressed 5 fs pulse durations one after another. In 1999, Morgner [160] used double-chirped mirrors and low dispersion prisms and Sutter [161] used chirped mirrors to control dispersion. They obtained 5 fs pulse duration directly from the oscillator. The four plans mentioned marked the results of the shortest pulse durations to a certain degree. The 5 fs interferometric autocorrelator curve of each plan is shown in Figure 1.36 [162].

During pulse compression, the most important factor in achieving the shortest pulse duration is the dispersion compensation of the stretched-out spectrum. This requires us to know the chirp of the spectrum first, and then we can design the corresponding compressor to compensate for dispersion, and therefore achieve an ultrashort pulse duration. Since self-phase modulation generates positive dispersion by using nonlinear media, the compressor should provide enough negative dispersion in

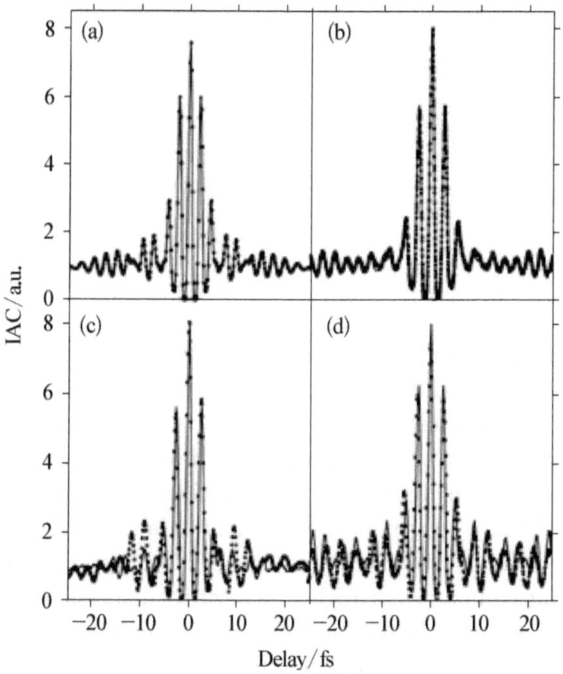

Fig. 1.36: 5 fs interferometric autocorrelator curve generated from four compression plans. (a) optical parametric amplifier, (b) cavity dumping, (c) hollow fiber, and (d) oscillator.

order to compensate dispersion. Several compressors can provide negative dispersion, such as a prism pair compressor, grating compressor, and chirped mirror compressor. The three compressors' features are illustrated below.

Generally speaking, the prism pair compressor has features of robust and high reflectivity, especially suitable for compressing the nanojoules pulse from the oscillator and millijoules pulse from the amplifier. A prism pair can be used to compensate the intracavity dispersion of the femtosecond oscillator; however, for a pulse duration less than 10 fs, owing to the broad spectrum bandwidth, the third-order dispersion and fourth-order dispersion generated from the crystal and prism cannot match each other. Therefore, it's difficult to obtain sub-10 fs pulses from a prism pair. In 1994, in order to reduce the effect of high order dispersion, Szipöcs and Krausz proposed a solution of adopting a newly designed chirped mirror to compensate for dispersion [34]. Moreover, the negative dispersion provided by a single prism pair is very small. It's impossible to indefinitely increase the distance between the prisms. To provide the needed dispersion, and close the distance between prisms, a double prism pair structure has been put to use. By using large refractive index media as the prism material and making the light pass through the prism pairs twice (Figure 1.37), the compressor

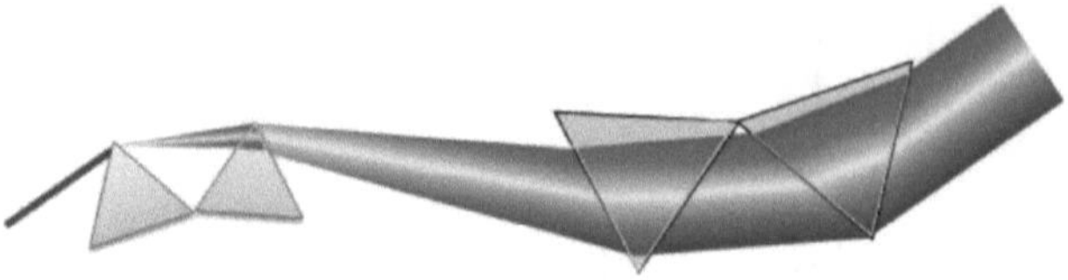

Fig. 1.37: A diagram of double prism pairs.

can provide enough dispersion, and is suitable for a sub-5 mJ pulse compression in the chirped pulse amplification system.

The grating pair compressor can provide enough second-order dispersion and third-order dispersion. Besides, it's technically easier to make large grating. To date, the most commonly used compressor in most terawatt amplifiers is the Treacy grating compressor, as shown in Figure 1.19. When the light passes through the parallel diffractive grating, different wavelengths of light have different diffraction angles and different optical paths. The optical path is longer and the time delay is larger for longer wavelengths. As a result, a grating pair can provide negative dispersion. By adjusting the separation between the gratings and incident angle, the negative dispersion can be changed. A grating pair compressor can be used to compensate for second-order and third-order dispersion originating from the oscillator and amplifier, and to compress the pulse duration to a femtosecond level.

On the other hand, to obtain a few-cycle laser pulse, as with the Ti:sapphire oscillator centered at 800 nm, the spectrum width should exceed 300 nm, which requires one to control dispersion precisely. The technical superiority of the chirped mirror compressor over the prism compressor and grating compressor is established. Figure 1.38 represents the principle using chirped mirror to compensate for dispersion. By overlapping reflective foils centered at different wavelengths, the penetration depth is different for each wavelength. The long wavelength's time delay is larger than that of the short wavelength, so it can provide negative dispersion. The number and depth of the coating layer can be calculated according to the system's dispersion. In this way, we can compensate for the dispersion of ultrabroad spectra. Chirped mirrors have a

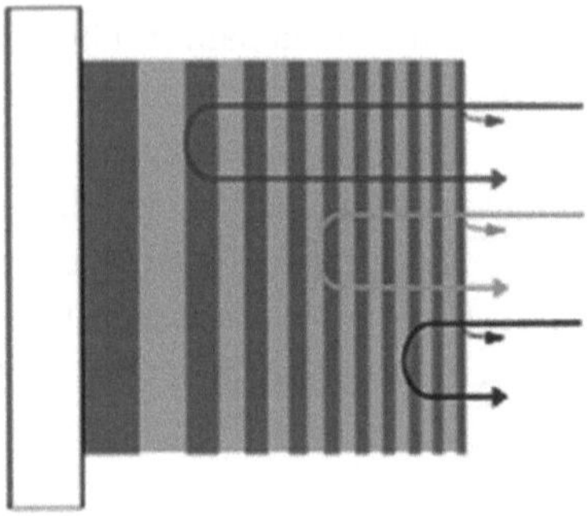

Fig. 1.38: A principle diagram of a chirped mirror.

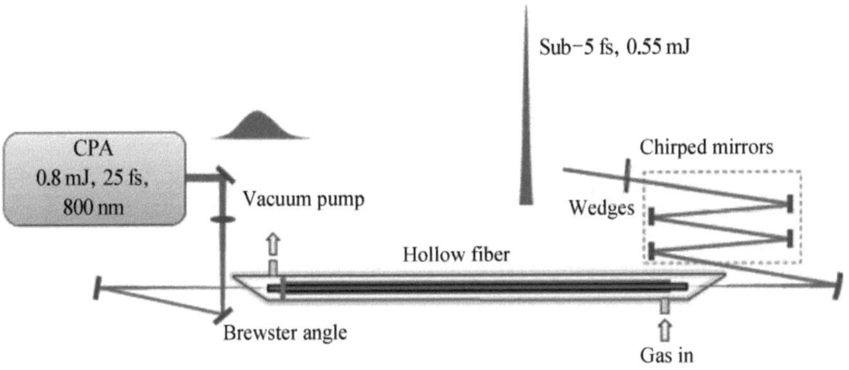

Fig. 1.39: A diagram of subcycle pulse generated from hollow fiber and chirped mirrors.

broad dispersion compensation bandwidth; pulses can be reflected several times in the same mirror to reduce the number of elements and the system volume.

Among the four compression schemes mentioned, since the hollow fiber scheme can obtain the highest single pulse energy, it is the most widely used pulse compression technology. Experiments confirmed that spectra that support less than 4 fs can be achieved from hollow fiber when the injected pulse duration is 20 fs. Zhang from the Institute of Physics at the Chinese Academy of Sciences [163] achieved 4 fs, kHz ultrashort pulses using similar methods; the experiment diagram is shown in Figure 1.39.

Notably, the three dispersion compensation methods have their own advantages and disadvantages. For specific systems, the combination of different compensation methods are used to generate the shortest pulse duration.

1.3.2 The measurement and controlling of the carrier envelope phase

For laser pulses shortened to a 10 fs magnitude, the relative positional relationship between the pulse envelope and the optical carrier becomes important. The time domain of pulses from a mode-locked femtosecond laser are depicted in Figure 1.40. The dash line and the solid line represent the pulse envelope and the optical carrier respectively. The interval φ_{CE} marked by two arrows is the phase between the envelope and the carrier, known as the carrier envelope phase (CEP). While the pulse propagates in the medium (solid or fiber), due to the different propagation velocities between the envelope and the carrier directly caused by the difference between group velocity and phase velocity, each pulse has a different CEP. As shown in Figure 1.40, from right to left, the CEP of two pulses are 0 and φ_{CE}. The differential value between the CEP of two pulses is defined as the carrier envelope offset, or CEO, a variation of the CEP. From the perspective of the frequency domain, we can get a more intuitive statement about the CEO. By Fourier transforming the mode-locked laser pulses, we can get the

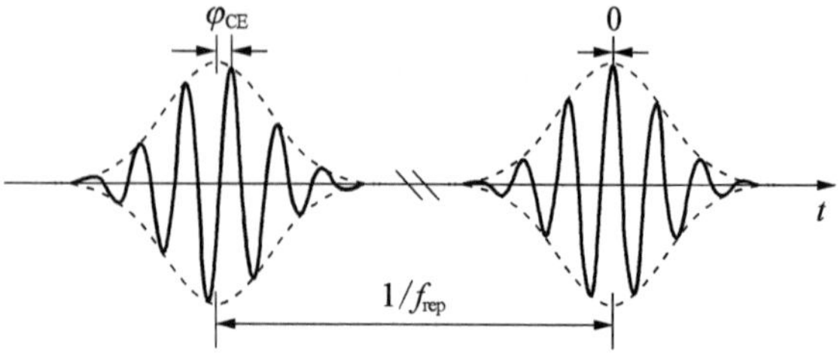

Fig. 1.40: Phase relationship between the carrier and the optical phase of the mode-locked laser in the time domain.

frequency domain diagram of the pulses, as shown in Figure 1.41. The laser pulses also cover a certain optical frequency range in the frequency domain, whose fine structure is made up of frequency teeth. According to the principle of the Fourier transform, the spacing between the teeth of the frequency combs represents the repetition rate f_{rep}. Assuming that the n-th frequency is f_n, with intervals of f_{rep} from f_n to a frequency of zero, we find that the carrier envelope offset frequency (f_{ceo}) is less than the f_{rep}. Therefore, the f_n can be expressed as the following formula with f_{ceo} and f_{rep}:

$$f_n = f_{ceo} + n \times f_{rep} \, . \tag{1.8}$$

As shown in Figure 1.41, this formula (1.8) is the most classic definition of the optical frequency comb, which can express any optical frequency with the repetition rate f_{rep} and the initial offset of the frequency comb of the mode-locked laser pulses. For the researcher in ultrafast optics, how to measure the CEO of a few-cycle femtosecond oscillator is an important problem. As early as 1996, Xu et al. presented a preliminary proposal in order to obtain the varying information of CEO in a frequency domain based on the cross-correlation characteristic of adjacent pulses. In 1999, Telle et al. demonstrated some measurement technologies that turned the offset of the phase into the measurement of the frequency. After the processing of frequency conversions, such as difference frequency and sum frequency, two similar frequencies are generated. Through the beat frequency of those two frequencies, the CEO signal can be measured. The f-2f self-reference technique is one of the simplest methods. As shown in

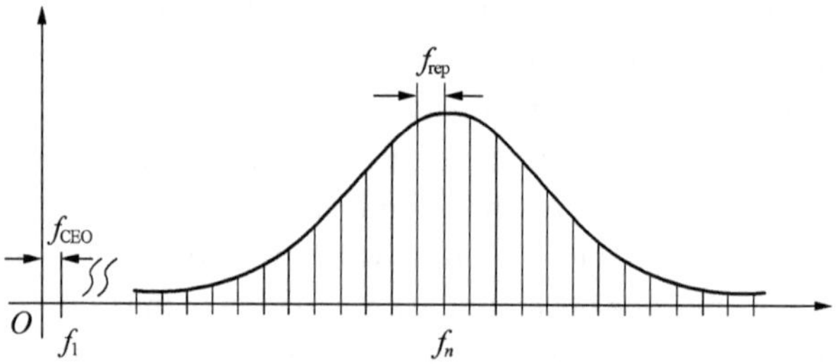

Fig. 1.41: Carrier envelope offset (CEO) of the mode-locked pulse in a frequency domain.

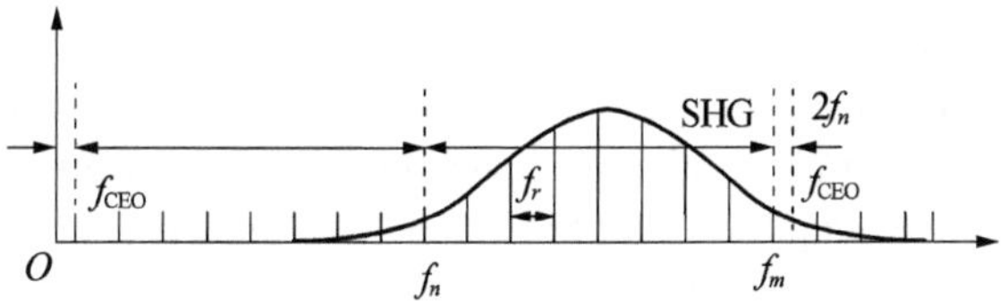

Fig. 1.42: Principle of measuring the CEO signal based on the self-reference method.

Figure 1.42, in an octave-spanning spectrum, the low frequency is first doubled, so the CEO signal can be obtained through the beat frequency. Assuming a repetition rate of f_r, according to the above formula (1.8), the frequency of the long-wavelength part can be expressed as $f_n = nf_r + f_{CEO}$ and the short-wavelength can be shown as:

$$f_m = 2nf_r + f_{CEO} \,. \tag{1.9}$$

So the frequency of the doubled frequency laser is $2(nf_r + f_{CEO})$. The f_{CEO} can be obtained through the beat frequency technique, as shown in the following formula:

$$2f_n - f_m = 2(nf_r + f_{CEO}) - (2nf_r + f_{CEO}) = f_{CEO} \,. \tag{1.10}$$

Additionally, the method for measuring the CEO signal with the beat frequency technique of doubling and tripling the frequency is known as the 2f-3f technique.

In 2000, Jones et al. implemented the CEP locking femtosecond oscillator based on the self-reference method, which is the most classic and revolutionary among ultrafast lasers and optical frequency measurements. In 2004, Fuji et al. proposed a difference frequency method for f_{CEO} measurements, which is shown in Figure 1.43. Based on the supercontinuum spectrum spanning 480~1100 nm, obtained by injecting the pulsed laser emitting from an oscillator into a segment of PCF, the CEO canceled each other out. This is the CEO self-stability difference frequency laser. Because the broadened fundamental spectrum still has this f_{CEO}, the f_{CEO} is obtained after the beat frequency, and the method is also called 0-f [157]. In 2005, by extending the bandwidth of the oscillator output from the mode-locked laser, the f_{CEO} was obtained by using the PPLN crystal without a PCF for the difference frequency measurement. This method makes f_{CEO} locking stable in the long term and causes a great increase in the output

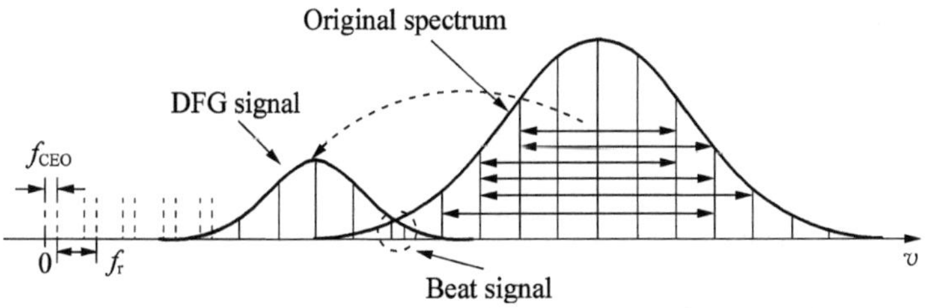

Fig. 1.43: Measurement principle of the difference frequency for f_{CEO}.

power. The overall performance parameters were significantly better than the results based on self-reference, and this scheme is referred to as the 'monolithic CEP stability scheme' [158]. In 2009, Wei Zhang, Zhi-yi Wei, and other researchers at the Institute of Physics, Chinese Academy of Sciences, reported that the plastic oscillator produced spectra and further enhanced the participation frequency spectrum difference of the frequency process component strength improvement scheme. By reasonable design of PPLN crystal parameters, this method greatly enhanced the CEO locking stability with nine hours continuous operation. Compared to previous f_{CEO} locking solutions and optical frequency combs, this improved scheme in principle supports better CEO locking results and output power. This scheme has good application prospects in attosecond generation, optical frequency comb technique, and optical frequency measurement.

In recent years, the forward feedback control technique for CEO locking proposed by Koke et al. has provided some interesting research directions [160–162], as depicted in Figure 1.44. The output from a femtosecond oscillator was diffracted by an acousto-optical frequency shifter (AOFS), then the CEO of the zero-order diffraction laser and first-order diffraction laser were measured by two sets of f-2f interferometers. The zero-order diffraction laser was injected into an in-loop and the f_{CEO} detected by an APD was used to mix frequencies with a synthesizer. After frequency mixing processing, the control signal was obtained and fed back to the AOFS. The first diffraction light

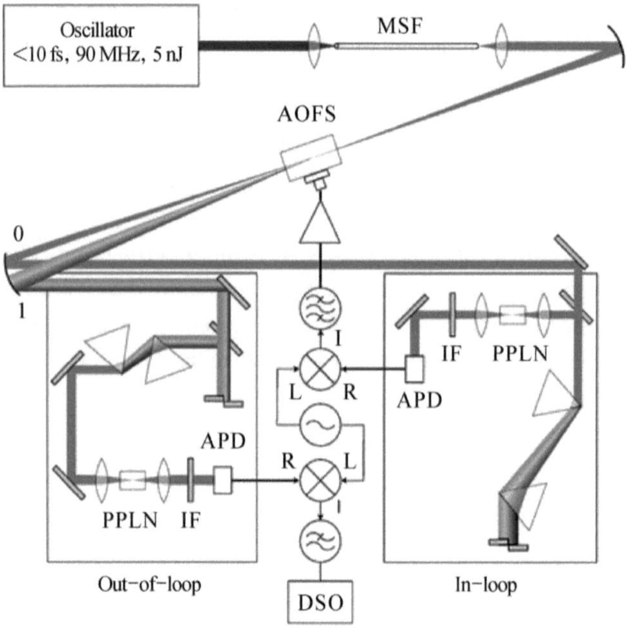

Fig. 1.44: Principle of the front feedback circuit for CEO signal.

was injected into an out-loop to measure and observe the variation of the f_{CEO}. The advantage of the scheme is that the CEP can be phase-locked and it is only necessary to maintain a mode-locked oscillator state without adjusting the pump power of the oscillator via an acoustic optical modulator (AOM).

With the rapid development of CPA techniques and pulse width compression schemes, more and more laboratories can produce few-cycle femtosecond laser pulses, which play a very important role in the field of strong field physics. Especially in high harmonic and attosecond generation experiments, the few-cycle femtosecond laser has become an indispensable tool. When high-energy few-cycle femtosecond lasers are applied in the attosecond [163], above-threshold ionization [164], and other types of experiments, the stability of the CEP must be considered in many sensitive experimental studies, since the CEP of each pulse has a great effect on the electric field intensity. Therefore, the measurement and control of the CEP has become one of the major frontiers in the field of ultrafast lasers in recent years.

The CEO locked oscillator provided a prerequisite for the CEO locking of the CPA amplification. Due to the high repetition frequency of the oscillator, the locking in this section was called the fast loop. Due to the directional instability, temperature fluctuation, vibration, and pump energy jitter, even for the CEP locked oscillator, the CEP of the amplification is slowly drifting and further CEP locking of the amplifier is necessary. Due to the low repetition frequency of amplifier, the CEP locking of amplifier is also known as the slow loop. In 2002, Kakehata et al. put forward a spectral interferometry schematic to measure the CEP of low repetition frequency amplifiers. In 2003, Baltuska accomplished the CEP locking of an amplifier by an f-2f spectral interference method. However, at that moment, the CEP locking scheme, which only loaded the feedback signal to an AOM to control the CEO of the oscillator, had not separately re-

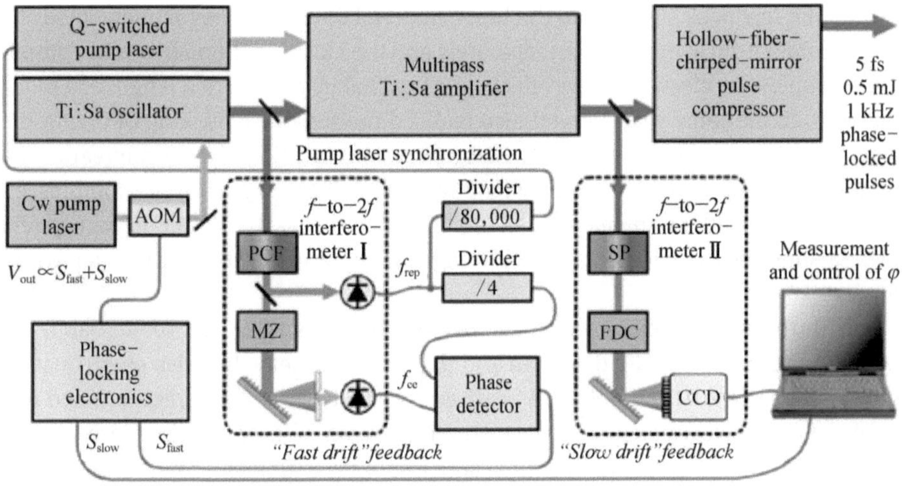

Fig. 1.45: Schematic of realizing the CEO locking of amplifier used by Baltuska.

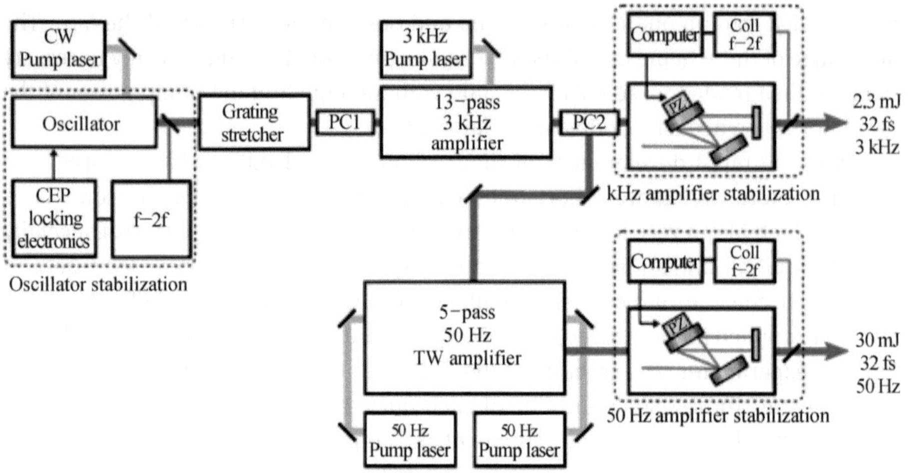

Fig. 1.46: Schematic of the experiment setup that locked the CEO signal of the oscillator and the amplifier separately.

alized the locking of the oscillator and amplifier, as shown in Figure 1.45. With the development of a difference frequency method to measure and lock CEP technology for the oscillator, the CEP locking scheme for the few-cycle scale of femtosecond CPA amplification system has also become more mature. However, it is not preferable to load the CEO of the oscillator and the amplifier on the oscillator together in obtaining the locking time and results. After 2009, a new scheme was proposed to separately lock the oscillator and the amplifier. Zenghu Chang's team loaded the feedback signal on the grating stretcher, and the CEP signal of the oscillator and amplifier were locked separately by fine-tuning the grating spacing [168, 169]. In 2011, Anderson attached the feedback signal onto the block material stretcher. They implemented the CEP of 5 mJ output energy locking separately and the 1 kHz repetition frequency amplifier system by the electrostrictive effect of PZT to change the inserted length of a block material. In the same year, Hergott also realized the separately locking CEO with the electro-optic modulation effect of a $LiNbO_3$ crystal. Additionally, we can realize the phase locking of the CPA system by loading the feedback signal on the compressor. For example, in 2011 Gademann et al. implemented the 50 Hz TW CEP locked system by adjusting the distance between the compressor gratings with a PZT, as shown in Figure 1.46.

With the further advance of the above technologies [172, 172], the research applied in implementing CEP detection and the locking scheme has also seen constant development. In 2011, Adolph et al. realized the single shot, high precision, and real time absolute measurements. In this method, one path of laser was used for feedback control, and the other one for real-time single shot measurement. They implemented the CEP measurement and locking of the amplifier system with high precision (Figure 1.47). But the cost of those two sets of microchannel plate detectors (MCP) is not affordable for typical laboratories, and hence the promotion of the method was limited.

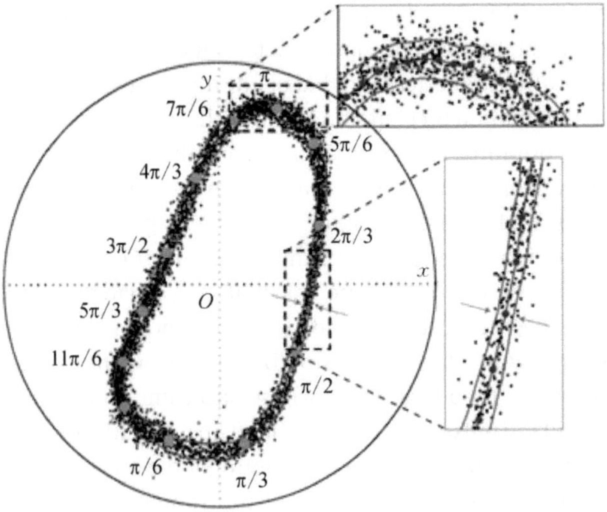

Fig. 1.47: The maps of CEP electronic measurements based on the above threshold ionization of a MCP.

In terms of the scheme that broadens the spectrum to obtain the few-cycle femtosecond pulse based on the self-phase modulation, we also need to consider the phase change introduced by the hollow fiber and the gas nozzle for the broadened spectrum. We can realize the CEP locking based on the nonlinear effects of the broad spectrum processing in the inflatable hollow optical fiber, gas nozzle, and the plasma channel, thus obtaining higher precision CEP locking. Moreover, it plays a very important role in high harmonic and attosecond experiments.

1.3.3 Coherent synthesis in the laser optical field

Coherent synthesis is widely applied in various optical domains. For example, in optical communication, coherent synthesis can be used to boost the transmission bandwidth of information; the coherent synthesis of a laser array can get a much higher power output; and coherent synthesis can be used to control the waveform of an optical field. With regard to pathways, it can generally be divided up into space domain and frequency domain coherent synthesis. Here, we discuss independent amplitude and the phase control of different spectral components of wide spectrum ultrashort pulses, in order to get single optical cycle, even sub-cycle pulses, with an even shorter pulse duration and manageable waveform. Both the control of the amplitude envelope of the optical field amplitude and the control of the chirp are more mature techniques. If we add the measurement and control of the absolute pulse phase, we can realize a complete control of the optical field.

The significance of coherent synthesis is that synthesis of different phase-locking pulses can not only improve the amplitude of the optical field, but also generate synthesized pulses whose pulse duration is shorter than that of any input pulse. It breaks through the laser gain medium limitation on emission bandwidth. In principle it can realize a subfemtosecond pulse duration. For example, multiple phase-locking femtosecond pulses of different wavelengths can be generated by various pathways such as optical parametric amplifiers, and coherent synthesis, which can generate cycle, even subcycle, pulses whose durations are shorter than any input pulse. Coherent synthesis also breaks through the chirp mirror coating process limitation on single broadband pulses. For example, the spectrum can be broadened by the hollow fiber and compressed by the chirp mirror to generate a cycle pulse. Due to the coating process limitation, dispersion compensation for a laser pulse whose spectrum is more than one octave is really difficult. But if the pulse spectrum is divided up into multiple spectral components, and the compensate dispersion for and control of the phase of relatively shorter spectral components is independent, subcycle ultrashort and ultrahigh-intense pulses can be generated.

The fundamental scheme of coherent synthesis is as follows: first, a supercontinuous laser source with a very broad spectrum should be available, which can be divided up into several channels using dispersion and split elements, according to its spectral components; then the amplitude and phase control of the spectral components of every channel should be conducted independently; finally the optical field of every spectral component should be coherently superimposed, in order to synthesize the optical field with an expected waveform. According to the principle of the Fourier transform, any extraordinary, complex waveform of an optical field can be obtained by the superimposition of a series of sine or cosine waves. Similarly, by the accurate adjustment of the spectral components of the ultrabroadband continuous spectrum of the optical field, it is possible to synthesize optical pulses with arbitrary waveforms, as shown in Figure 1.48, which absolutely includes cycle and even sub-cycle ultrashort pulses.

Figure 1.49 shows the frequency domain and time domain distribution of an ultra broadband spectrum optical field. The frequency domain modulation resolution $\delta\omega$ determines the time window; the modulation bandwidth Ω determines the time resolution δt that the optical field modulation in the time domain can achieve. When the modulation bandwidth is shorter than the carrier wave frequency, i.e. $\Omega < \omega_0$, resolution is about $\delta t \sim 2\pi/\Omega$, which shows modulation has the capacity to obtain cycle pulse bandwidth in the scale of pulse time-domain envelope. When modulation bandwidth is more than carrier frequency, i.e. $\Omega \geq \omega_0$, time-domain modulation res-

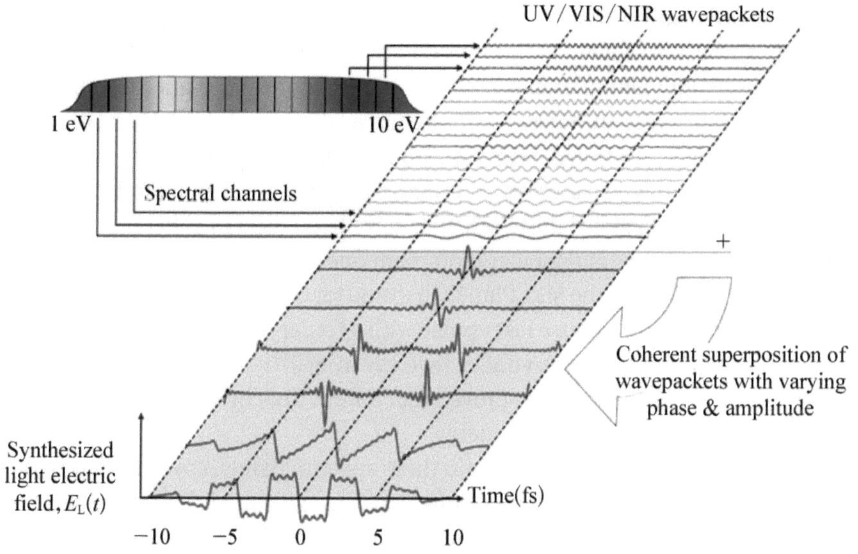

Fig. 1.48: Schematic diagram of coherent synthesis of arbitrary waveform.

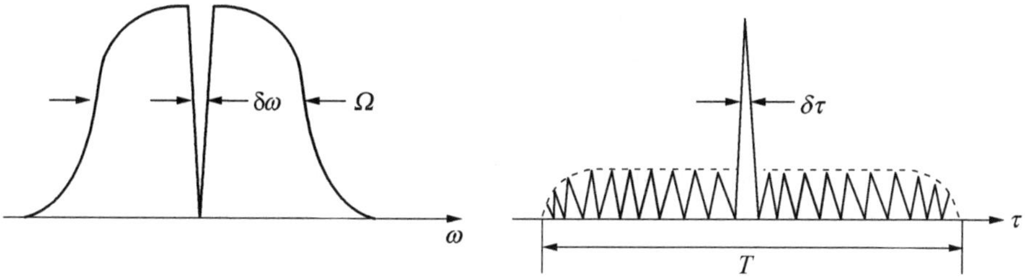

Fig. 1.49: The Fourier transform relation between modulation resolution δt and frequency domain bandwidth of a cycle optical pulse.

olution δt is less than $2\pi/\Omega$ (T_0), allowing us to control the optical field waveform in an optical cycle.

Therefore, in order to manipulate the optical field waveform, a coherent synthesis system includes at least the following three elements:

(1) It should have a coherent laser source that can generate at least one octave spectrum, i.e. $\Omega \geq \omega_0$.
(2) It should have the split and synthesis system to realize the phase modulation of single amplitude.
(3) It should have the waveform measurement arrangement to accurately measure the synthesized optical field.

The basis of a coherent synthesis of optical fields is the coherent laser source that reaches and even exceeds one octave spectrum in order to measure the pulse carrier envelope phase. Based on the above principles, the supercontinuum sources used for coherent synthesis are divided into three types: (1) use gas-filled hollow fiber to broaden the spectrum to more than one octave; (2) use optical parametric amplification to generate femtosecond pulses of different wavelengths that have a stable carrier envelope phase; and (3) generating multiple Stokes and anti-Stokes sidebands through stimulated Raman scattering, in order to obtain a spectrum bandwidth covering more than one octave. All these schemes have been used in coherent synthesis of laser fields. The first type is the most prominent one.

In 2011, Wirth et al. from the Max Planck Institute for Quantum Optics used a static inflatable hollow fiber to broaden a laser pulse with a center wavelength of 780 nm and pulse duration of 25 fs to more than one octave covering 270 to 1100 nm. The result is shown in Figure 1.50. The experiment proved that hollow fiber broadens the spectrum, to support multiple μJ, even mJ pulse energy.

The broadened spectrum was split into three spectral bands, covering 700~1100, 500~700, and 350~500 nm. The delay line unit consists of a pair of right angle mirrors, a piezoelectric ceramic platform, and a manual platform. The manual platform provided a rough delay adjustment, whose accuracy is 10 μm. The piezoelectric ceramic provided a 25 nm accurate adjustment resolution and locked the delay line according to the spectral interference fringes of the adjacent channels. After chirp mirror compression of every channel, the duration was 6.8 fs, 5 fs, and 4.5 fs respectively. After coherent synthesis, they obtained an ultrashort pulse with a center wavelength of 710 nm, a pulse duration of 2.1 fs, which is only a 0.88 optical cycle. Figure 1.51 shows the schematic diagram, experiment photograph, and spectrum of every channel in this experiment. In 2013, they further used a four-channel split and synthesized an

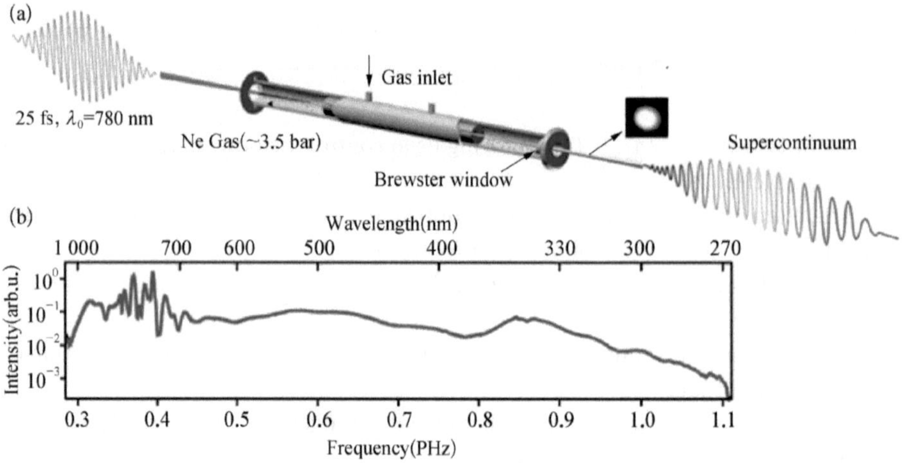

Fig. 1.50: Schematic diagram of hollow fiber broadening experimental setup that Wirth et al. from the MPQ used, with experimental results.

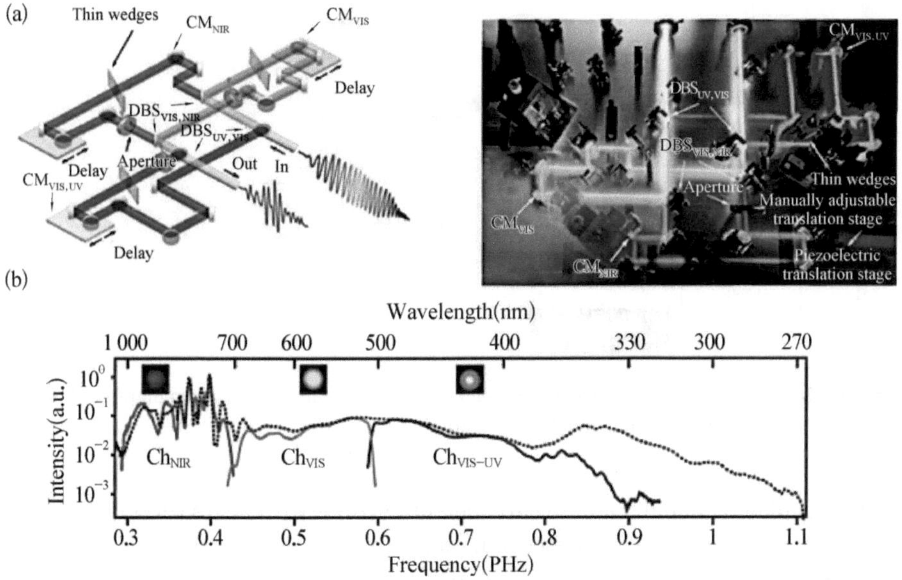

Fig. 1.51: (a) Schematic diagram and experimental photogram of MPQ's three-channel optical coherent synthesis; (b) the transmitted spectrum of every split element.

ultrashort ultrafast pulse with a center wavelength of 636 nm, pulse duration of 1.7 fs, which is a 0.8 optical cycle.

In 2013, Shaobo Fang et al. from the Center for Free-Electron Laser Science in Germany used optical parametric chirped pulse amplification combined with the optical parametric amplification technique, and obtained a super broadband coherent spectrum covering 870 nm to 2.15 μm. Finally, the pulse of three channels was compressed by chirped mirrors, and was coherently synthesized. They obtained near-infrared output pulse with a pulse duration of 1.9 fs and a pulse energy of 200 μJ. Figure 1.52 shows the schematic diagram of the light path of their coherent synthesis.

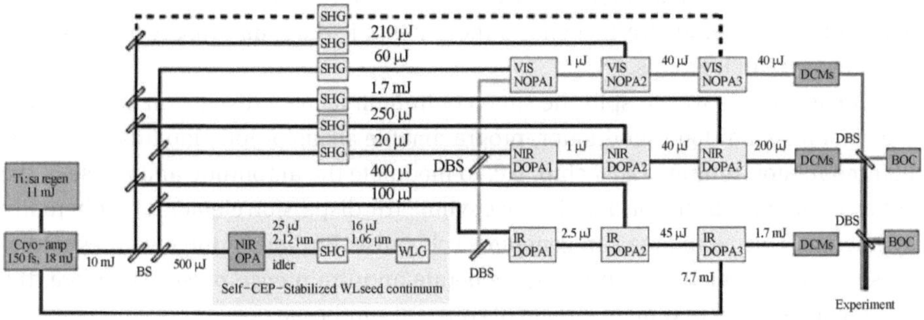

Fig. 1.52: Schematic diagram of multichannel coherent synthesis of CFEL.

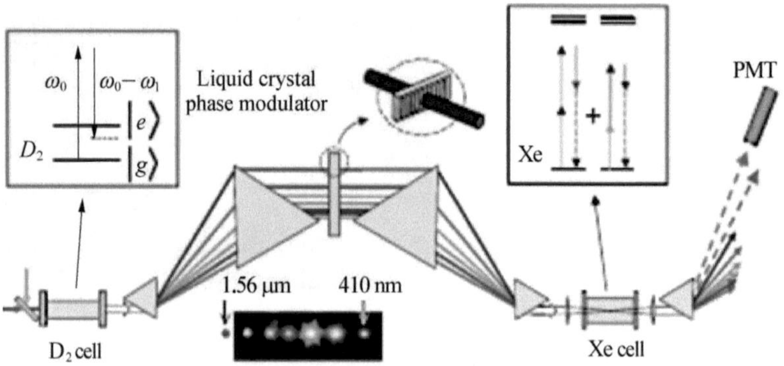

Fig. 1.53: Schematic diagram of generating a super broad spectrum by Harris from Stanford University using the technique of stimulated Raman scattering.

Using stimulated Raman scattering to coherently synthesize is also a very effective way to generate ultrashort, even attosecond, pulses. In 2005, S. E. Harris' research group at Stanford University used the technique of pulses of different frequencies to stimulate deuterium molecules to vibrationally transition, to generate Raman sideband, and multiple Stokes and anti-Stokes sidebands. A liquid crystal spatial light modulator can be used to modulate the phase of these Raman sidebands, and can coherently synthesize a cycle pulse. Figure 1.53 shows the schematic diagram of the arrangement they used. By adjusting the Raman sideband phase of seven wavelengths, i.e. 1.56, 1.064, 807, 650, 544, 468, and 410 µm, they obtained an output pulse with a pulse duration of 1.6 fs and pulse interval of 11 fs, which was the shortest pulse in the optical wave band.

It is worth mentioning that the U. Keller research group used cascade hollow fiber to broaden the spectrum, then used a spatial light modulator to adjust every spectrum component, and realized a 3.8 fs near cycle pulse in 2003. Then the record was broken by 3.8 fs, 3.4 fs, 2.8 fs, etc. In 2007, the Mikio Yamashita Group from Hokkaido University adopted a more complex scheme and further obtained a 2.6 fs pulse, including only 1.3 cycles, which is much closer to the pulse duration limitation in the visible-near infrared scale. Figure 1.54 shows the schematic diagram of light path used by these two research groups.

The technique was to split the supercontinuum using a dispersion element, liquid crystal spatial light modulator, programmable acousto-optic modulator, molecular modulator, rotating CaF_2 chip, etc. to modulate the amplitude and phase of the supercontinuum spectrum, and then use symmetric dispersion elements to coherently synthesize. This technique can generate cycle pulse. But due to the low threshold of these dispersion elements, it cannot generate an ultraintense pulse. Therefore, the MPQ scheme is the new main stream of coherent synthesis.

Some problems that need attention in designing the light path of coherent synthesis are as follows: (1) The shorter the pulse wavelength is, the more dispersion the

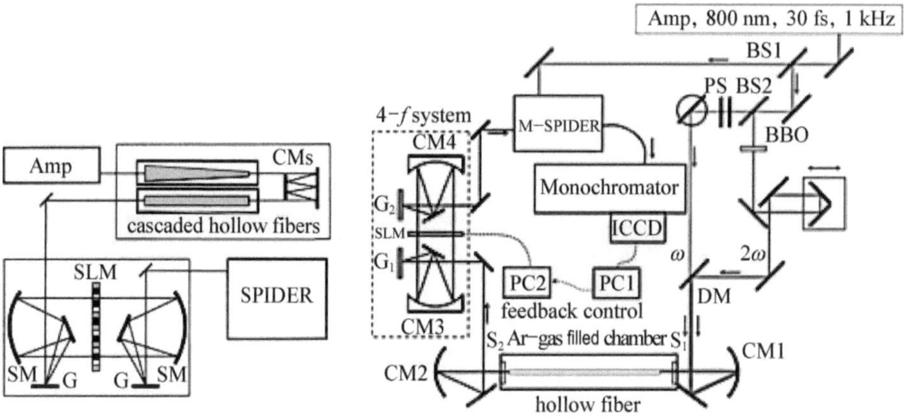

Fig. 1.54: (a) U. Keller research group and (b) the schematic diagram of light path of the Mikio Yamashita Group from Hokkaido University, using hollow fiber to broaden the spectrum and combining with a liquid crystal spatial light modulator to coherently synthesize.

pulse introduces. Therefore, the beam splitter of shorter wavelength is harder to produce. We should first reflect the shorter wavelength. (2) For the synthesis of a center wavelength of 710 nm and a subcycle optical field, the locking resolution of relative delay should be less than 0.2 fs, corresponding to a 69 nm optical path difference in space. Therefore, the light path should combine passive stabilization with active stabilization, i.e. integrate design at the bottom and keep the temperature constant throughout the water. Then feedback locks the relative delay of the light path by active stabilization measures such as spectral interferometric, etc. (3) When we split the spectral component that covers an octave, we should pay attention to the proportion of different spectral components. The intensity difference between different components should not be too much, ideally within 30 dB, in order to make sure that the weakest spectral component contributes to the optical field in coherent synthesis.

Pulse measurement and diagnosis in the process of coherent synthesis can use the transient grating type frequency resolved optical grating measurement system (TG-FROG). It includes two steps: (1) measure the pulse characteristic of every channel before and after dispersion compensation to confirm the design parameters of the chirp mirror and pulse duration compression effect; and (2) measure the coherently synthesized optical field. Because of the dispersion introduced by the nonlinear material, in the aspect of time characteristic of coherently synthesized optical field measurement, it is impossible to derive the real waveform of a coherently synthesized optical field. Figure 1.55 (a) shows the schematic diagram of light path of TG-FROG measurement, while (b) and (c) are the spectral interference pattern when the relative delay of every channel is nonzero and zero. It can be seen that accurately locking the delay to zero is extremely important for coherent synthesis. In this technology, Huang et al. at Massachusetts Institute of Technology (MIT) used balanced optical cross-correla-

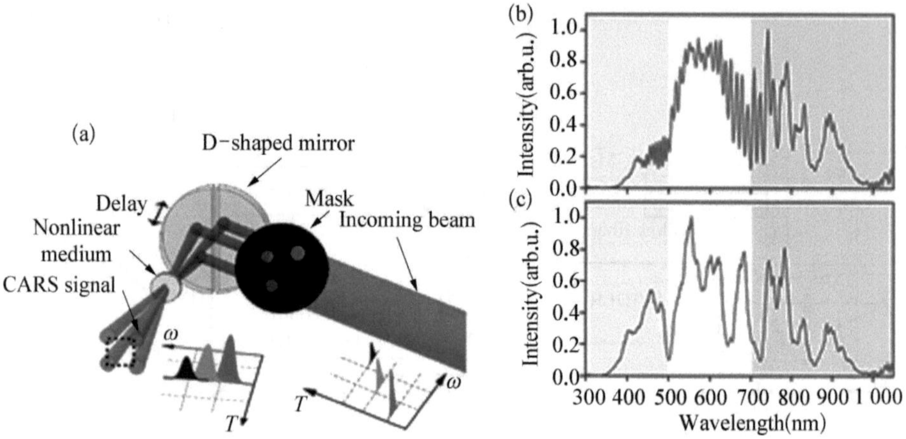

Fig. 1.55: (a) The schematic diagram of the light path of TG-FROG arrangement; (b) the spectral interference pattern when delay time is not totally coincident; (c) the totally coherently spectral pattern when the relative delay of different channels is zero.

tion (BOC) to achieve feedback lock, and the mean square root of the timing jitter of feedback lock is less than 30 as.

Since the coherently synthesized pulse has a duration ranging from multiple femtoseconds to attoseconds, the attosecond streak camera can be used to measure the optical field characteristic for generating a single attosecond pulse. Its advantage is that the pulse travels only a short distance of inert gas, so the generated dispersion is negligible. Figure 1.56 shows a typical measurement result (a), information about inverse photoelectric field (b), and the coherently synthesized pulse duration (c).

To generate sub-femtosecond and even attosecond ultrashort ultraintense pulses, coherent synthesis is an important frontier marker and direction for ultrafast laser science.

The obtained results not only indicate the control of laser optical field, advancement from amplitude modulation to photoelectric waveform, but also promise many applications such as sub-femtosecond time resolved detection, manipulation of atom and molecular excited states, and attosecond electronic dynamic evolution in nanoscale in ultraviolet to near-infrared regions. Coherently synthesized subcycle ultrashort ultraintense lasers can also serve as the drive optic field, and have more intense peak power and contrast. Furthermore, they can generate much broader and more intense supercontinua in cut-off regions. In addition, compared to ordinary ultrashort pulse waveform optical fields, the coherently synthesized special waveform ultrashort pulses can greatly enhance electron recollision energy and improve high harmonic conversion efficiency, thus enhancing cut-off energy regions [183].

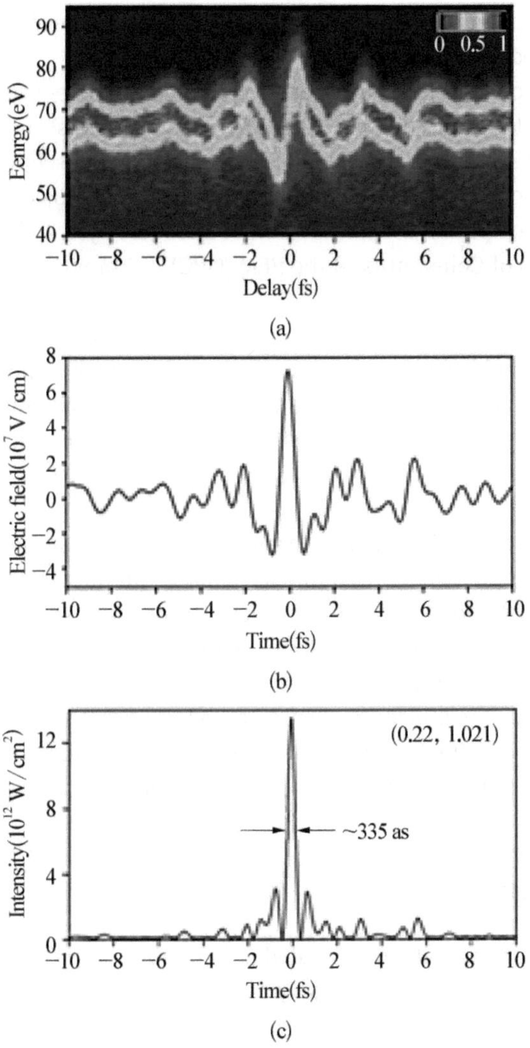

Fig. 1.56: (a) The spectral diagram scanned by the attosecond streak camera, (b) information about inverse photoelectric field, (c) the FWHM (full width half maximum) of pulse duration of a coherently synthesized photoelectric field.

1.4 The main categories of ultrafast lasers

Nowadays, ultrashort pulse lasers, especially the femtosecond lasers, are widely used in various scientific researches. The commonly used ultrashort pulse lasers can be summarized under three main categories according to their different applications:

(1) High repetition rate ultrafast lasers. Represented by femtosecond pulse laser oscillators, their repetition rate is mainly in the range of 10 MHz to 1 GHz. This kind of laser is widely applied in biomedicine, multiphoton microscopy, THz imaging, pump-probe experiments, and other scientific research. Its typical advantages are simple operation and stable performance. There are several representative prod-

ucts such as the Mai Tai and Tsunami produced by Spectra-Physics, and the Mira and Chameleon Series produced by Coherent.

(2) High average power ultrashort pulse lasers. These lasers produce millijoule-level pulse energy with repetition rates from 1 kHz to 1 MHz. This kind of laser is commonly utilized to study ultrafast phenomena, femtosecond laser spectroscopy, material micromachining, biochemistry, and other interdisciplinary subjects. Representative products are the Spitfire Ace and Unison of Spectra-Physics, the Legend, Libra, and RegA series of Coherent, and the FEMTOPOWER products of Femtolasers.

(3) High peak power ultrashort pulse lasers. Applications using these lasers are mainly concentrated in strong field physics research. Their typical peak power is greater than 20 TW and the repetition rate is less than 10 Hz. French company Thales and Amplitude Technologies occupy the majority of this kind of product in the global market.

These classes of femtosecond lasers are utilized mainly in some national laboratories due to their large cost and specific applications. Therefore its demand quantity is relatively small. The first- and second-type femtosecond lasers have a broad range of applications, and they can be fitted to the secondary development of microscopes, laser scalpels, ranging trackers, industrial processing, and other equipment; therefore the demand is enormous. In recent years, the optical frequency comb has become a new class of products in the large family of ultrashort pulse lasers due to its revolutionary applications in the measurement of physical quantities and in precision metrology.

1.4.1 High repetition rate ultrafast lasers

The pulse repetition frequency f_{rep} of an oscillator is related to the length of the laser cavity. Approximately, they have the following relationship:

$$f_{rep} = c/2L \,. \tag{1.11}$$

In this formula, c is the light speed and L is the cavity length. Ultrafast laser oscillators with different repetition rate have different characteristics, and can be applied to different areas. In order to produce a hundred nanojoule-level laser output, we can reduce its repetition rate by adopting a long cavity to increase the laser pulse energy. A low repetition rate laser usually contains a long cavity to several meters. In order to reduce its volume, a well-designed telescope scheme is commonly used to fold the optical path. In contrast, it is possible to obtain a high repetition rate pulse laser using a short cavity. For example, 1 GHz repetition rate corresponds to 30 cm cavity length. It has a compact structure, good stability, and a wide range of applications.

In 1998, Ramaswamy et al. [208] and Stingl et al. [209] proposed the concept of high repetition frequency oscillators. Instead of using traditional prism pairs, they

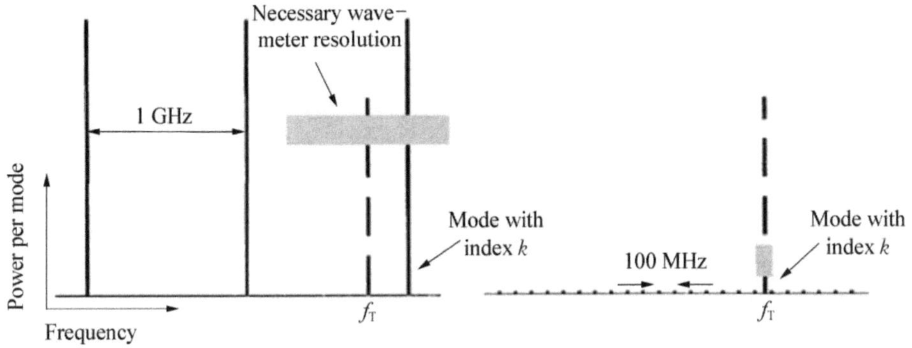

Fig. 1.57: Frequency measurement advantage using a 1 GHz femtosecond laser (left) compared with a 100 MHz laser (right). Height of the solid line represents the power of each comb, and the gray area represents the resolution required in measurement.

introduced a more compact configuration to compensate dispersion of the resonant cavity. In the beginning of the 21st century, the development of optical frequency combs [210–213] greatly promoted the development of high repetition frequency ultrashort pulse lasers. Since the repetition frequency of each comb is directly related to its laser power, the higher repetition frequency lasers have more advantages than the lower ones do in the application of optical frequency combs. For a 1 GHz femtosecond laser optical frequency comb, its single comb carries 100 times the power of that of a 100 MHz laser (Figure 1.57). The advantage of its power is quite effective in improving its beat signal-noise ratio in frequency measurement [214].

Obtaining high repetition rate ultrashort pulse lasers is more difficult. Generation of ultrashort pulse lasers depends on continuous mode locking technology. This method requires a saturable absorber (e.g. SESAM), or uses the Kerr-lens effect of crystal to start mode locking and the pulse energy density within the resonant cavity exceed a certain threshold. Assuming that the average power is constant, this leads to lower pulse energy. Researchers have created a number of techniques for obtaining high repetition rate ultrashort pulse lasers, such as the Kerr-lens mode-locked Ti:sapphire oscillator, the passively mode-locking technique using a semiconductor saturable absorber mirror (SESAM), and the harmonic mode-locking technique in fiber lasers [215].

In 1999, A. Bartels et al. demonstrated a Kerr-lens mode-locked femtosecond Ti:sapphire laser emitting 23 fs pulses at a 2 GHz repetition rate. Compared with that of conventional 100 MHz systems, this oscillator showed several advantages in optical time-resolved spectroscopy [216]. In 2008, they obtained a passively mode-locked Ti:sapphire laser operating at a repetition rate of 10 GHz (Figure 1.58) [217].

Passively mode-locked femtosecond fiber lasers have also been demonstrated with a repetition rate in excess of 2 GHz and 19.5 GHz [218, 219]. They reflect an important application prospect due to their compact configuration and remarkable en-

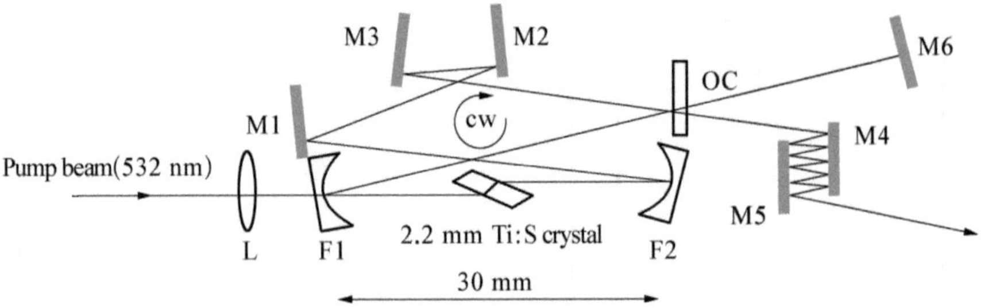

Fig. 1.58: Cavity design of the 2-GHz mirror-dispersion-controlled Ti:S ring oscillator [217].

vironmental stability. For high repetition rate solid-state lasers, the U. Keller research group has made outstanding contributions. In 1999, a passively SESAM mode-locked $Nd:YVO_4$ laser was realized with a repetition rate of 29 GHz, and a more compact resonant cavity was adopted. In 2002, the repetition frequency of a $Nd:YVO_4$ laser was pushed up to nearly 160 GHz (Figure 1.59), reaching the fundamental limit of the repetition rate that is determined by the pulse duration of several picoseconds and thus by the spectral bandwidth of the laser crystal [220, 221].

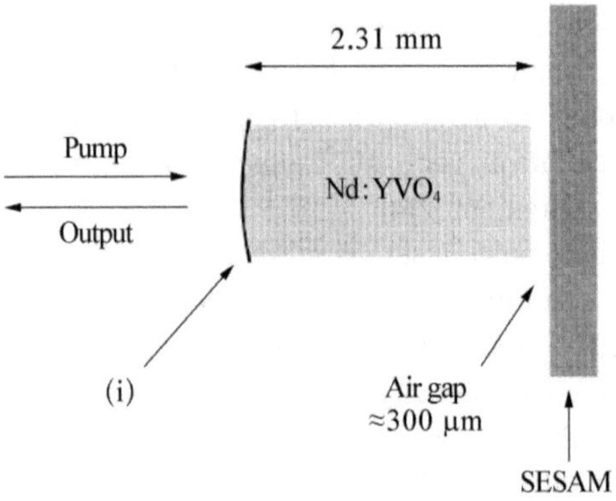

Fig. 1.59: Schematic miniature laser setup, as used for repetition rates > 20 GHz [221].

Recently, all-solid-state femtosecond lasers are widely regarded as an excellent candidate for a robust optical frequency comb source. In comparison with the Ti:sapphire laser, all-solid-state lasers have many special advantages, such as high power, high repetition rate, low noise, and low-loss cavity. They can be directly pumped by diode, with up to 50% efficiency. Current ytterbium-doped and erbium-doped fiber lasers can also be directly diode-pumped, but they usually featture pulse, lower output power and higher loss cavity. Compared with all-solid-state lasers, these factors lead to a greater inherent noise in the fiber optical frequency comb [224, 225]. Therefore, the study of solid-state optical frequency combs is of great importance in both research and application areas. Based on the requirements for solid-state optical frequency combs, the high repetition rate solid-state femtosecond laser has been developed

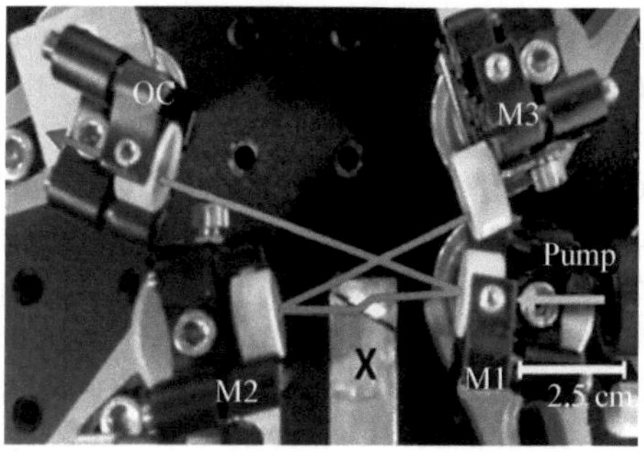

Fig. 1.60: Photograph of the demonstrated 1.2-GHz repetition rate Yb:KYW laser [227].

rapidly. Using good laser performance crystal such as Yb:KYW, Yb:KGW, several GHz solid-state femtosecond lasers were reported successively. In 2009, Wasylczyk et al. obtained a passively mode-locked, diode-pumped Yb:KYW oscillator (Figure 1.60) in a four mirror ring cavity configuration with 1 GHz repetition rate [226]. In 2012, using a carbon nanotube saturable absorber mirror as a passive mode-locking component, Yang et al. implemented a 1.2 GHz repetition rate diode-pumped Yb:KYW laser [227].

In the same year, Endo M. et al. developed a 4.6 GHz repetition rate, diode-pumped Kerr-lens mode-locked Yb:KYW oscillator [228].

Nowadays, a variety of new solid-state optical frequency combs are continuously emerging. The investigation and development of solid-state optical combs accelerated the development of high repetition rate lasers. Meanwhile, scientists are still devoting themselves to investigations of other applications of high repetition rate lasers.

1.4.2 High average power ultrafast lasers

Generally high average power ultrafast lasers have high repetition rates (typically above 1 kHz). The main gain materials used are Ti:sapphires. In 1993, Rudd et al. obtained 55 fs pulses at 1 kHz repetition rate with 350 μJ pulse energy, by seeding 40 fs pulses (produced by a Kerr mode-locking [KLM] Ti:sapphire oscillator) into a Ti:sapphire regenerative amplifier [229]; Lenzner et al. realized the output of laser pulses of 1 kHz, 18 fs, 100 μJ utilizing a four-pass Ti:sapphire CPA system in 1995 [230]; at the same time Backus et al. achieved 1 kHz, 21 fs, 1 mJ pulse output by using a multipass CPA system with a three-mirror ring amplifier [231]. S. Santania et al. realized laser pulse output of 1 kHz, 20 fs, 1.5 mJ by using an eight-pass Ti:sapphire CPA system in 1997 [232]. Hentschel et al. realized laser pulse output of 1 kHz, 20 fs, 20 mJ by using a ten-pass Ti:sapphire CPA system in 2000, with a peak power of 0.1 TW [233]. In the same year, Zeek et al. realized a 1 kHz, 15 fs, 1 mJ pulse output by using a multipass CPA system with a three-mirror ring amplifier [234]. In 2002 Cheng et al. realized laser pulse output of 1 kHz, 17.5 fs, 2 mJ by using a ten-pass Ti:sapphire CPA system [235].

To obtain more powerful pulses, a several-stage amplifier was introduced to enhance pulse energy. In 1997 Backus et al. realized a laser pulse output of 1 kHz, 20 fs,

4 mJ, with 0.2 TW peak power and 4 W average power of [236], by adding one stage of amplifier to the base of a multipass CPA system with a three-mirror ring amplifier [231], producing four times more average power than using one stage of amplifier. In 1998, Nabekawa et al. combined a prisms-compensated regenerative amplifier with a two-stage Ti:sapphire multi-pass amplifier to generate 1 kHz, 21 fs, 14 mJ, corresponding to 0.66 TW peak power and 14 W average power [237]. In 2000, Bagnoud et al. utilized one stage of regenerative amplifier and three stages of multi-pass amplifier CPA system to produce 1 kHz, 20 fs, 20 mJ, pulses with 1 TW peak power and 20 W average power; this is the highest peak power and average power output for a 1 kHz CPA system [238]. Because of the gain narrowing effect, pulse duration is limited to about 20 fs. Using a programmable acousto-optical diffraction filter to compensate for the gain narrowing effect and spectrum distortion, in 2003 Seres et al. obtained 1 kHz, 3.4 mJ, 9.4 fs pulses with 0.3 TW peak power, which was the first time that anyone realized a 1 kHz Ti:sapphire CPA system of TW-level peak power and sub-10 fs pulse duration [239].

In fact, there are various kinds of CPA systems and different gain materials besides Nd:glass and Ti:sapphire. By utilizing Alexalldrite, Cr:LiSAF, KrF, etc. as gain material in CPA systems, we can also achieve high-energy and high peak power pulses.

1.4.3 High peak power ultrafast laser pulses

High peak power ultrafast laser pulse systems have the following typical characteristics: a 10 Hz repetition rate, micro-Joule level pulse energy, and sub-100 fs pulse duration (for now usually less than 30 fs). This kind of device can generate pulses with more than 100 TW peak power. The pulse can be focused to achieve an extreme high intensity due to its better beam quality than Nd:glass amplifier systems. This kind of device needs less space and is called a CPA system because it is installed on laboratory optical breadboards. As they are cheaper and more broadly applied, these devices are the main development direction of research on CPAs.

In 1991, A. Sullivan et al. produced 5 Hz, 95 fs pulses with 3 TW peak power by utilizing four-stage Ti:sapphire amplification [241]. In 1994, Barty et al. obtained 10 Hz, 30 fs pulses with 4 TW peak power by controlling the amplitude and phase of the amplified pulse [242]. In 1996, A. Sullivan et al. generated 10 Hz, 120 fs pulses with 9.8 TW peak power by using more powerful frequency doubled Nd:glass as the pump [214]. In the same year, Barty et al. achieved 50-Hz, 18-fs pulses with 4.4 TW peak power by reshaping the amplified pulse and compensating dispersion at a higher order [244]. Using a stretcher and compressor system with no aberration, Chambaret et al. realized concise compensation of dispersion by controlling the phase of spectra, which resulted in 10 Hz, 32 fs pulses with 25 TW peak power. In addition, the output laser has good beam quality, leading to a focused intensity of $5 \times 10^{19}\,\mathrm{W/cm^2}$ [245]. In 1998, K. Yamakawa and Barty et al. at the University of California came up with the concept of depressing gain narrowing and gain saturation utilizing etalon and compensated pulse dispersion by introducing the third-order dispersion through multilayer media. Amplified by a two-stage amplifier, the laser pulse is compressed to 16 fs at 10 Hz repetition rate with 10 TW peak power. It was the first time anyone realized a high-power conversion efficiency over 90% [246]. In the same year, K. Yamakawa adopted three-stage Ti:sapphire amplifier to produce 10 Hz, 19 fs pulses with 100 TW peak power, reaching the efficiency limit of quantum conversion [247]. In 2000, Bonlie et al. obtained 75 fs pulses with 200 TW peak power and 10^{21} W/cm^2 focused intensity utilizing large aperture Ti:sapphire as gain material [248]. In 2003, a research group at the Japanese Institute of Atomic Energy adopted a four-stage Ti:sapphire amplifier to produce 33 fs pulses with 850 TW peak power [249].

Recently, Chinese researchers have made significant achievements in compact ultrafast lasers. Since 2004, the China Academy of Engineering Physics, the Institute of Physics of CAS, and the Shanghai Institute of Optics and Fine Mechanics of CAS have realized femtosecond pulses with peak power of 286, 355, and 890 TW [250–252]. To obtain a pulse output of a higher energy level, researchers at the Institute of Physics of CAS have invented two types of technology ranging from seed sources to diagnostic systems [253–255] and improved stability of preparts. Breakthroughs were made in 2011 on the experiment set of 'XL-III'. They realized a laser pulse output of 27.9 fs pulse duration, 1.16 PW peak power with high contrast ratio. This work broke the record of pulse peak power based on Ti:sapphire amplifiers. They adopted double CPA and a noncollinear OPA to realize a high contrast ratio of 10^{10} in 'XL-III'. Figure 1.61 is a photo of part of the experiment set 'XL-III'. Figure 1.62 is the experiment layout.

Fig. 1.61: Photo of part of experiment set 'XL-III'.

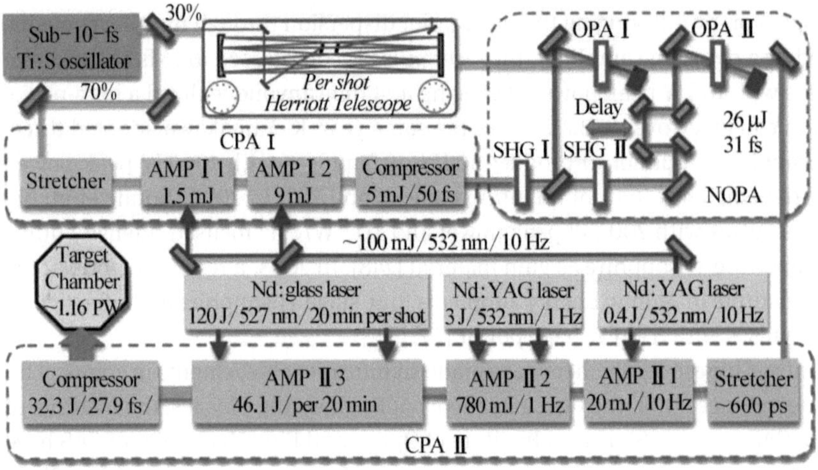

Fig. 1.62: Layout of experiment set 'XL-III' based on DCPA and NOPA.

High peak power femtosecond laser devices are widely applied in intense field physics, strong energy density physics, etc. Customers of such devices are usually key national laboratories because of the huge expense and unique applications. Although such de-

Fig. 1.63: BELLA PW laser device installed at the Berkeley National Laboratory in Lawrence, California, USA.

vices in China have reached the same level in peak power as the international advanced level, the market in this domain is dominated by foreign products, e.g. Thales and Amplitude Technologies from France. Figure 1.63 is the experiment set of BELLA, built by the Lawrence Berkeley National Laboratory in California, USA. This device outputs 1 Hz, 40 fs pulses with 1 PW peak power. Utilizing this device, the United States Department of Energy plans to promote research of miniaturized particle accelerators through the process of laser wakefield acceleration to obtain electrons with high energy, which can be implemented in collider, imaging and medical applications.

1.5 Noise suppression of ultrahigh-intensity lasers

Due to the emergence of CPA technology, ultrahigh-intensity lasers have made remarkable progress in the past 20 years. As mentioned above, the highest peak power has been obtained using the latest CPA technology, and the intensity after focusing is 10^{21}~10^{22} W/cm^2. The EU is building ELI with peak power of more than 200 PW [236]. At such a high density, the pulse contrast of the laser pulse becomes a very important parameter for strong field physics. The contrast is defined as the ratio of the prepulse of the laser front and the intensity of the background noise with respect to the main pulse [237, 238].

In the conventional CPA laser system, background noise is inevitably present in the amplified laser pulse due to factors such as the ASE and extinction ratio of the EL element. If no other technique is adopted to improve the contrast, the final pulse contrast is generally only 10^6 or so. At present, the peak intensity of an ultrahigh-intensity laser is higher than 10^{17} W/cm^2, and the intensity of the pre-pulse and ASE exceeds 10^{11} W/cm^2. Such an intensity is sufficient to ionize the material to form plasma. Thus, the prepulse or ASE interacts with the material before the main pulse interacts with the substance. The existence of pre-plasma changes the initial state of the interaction between the main pulse and the material [239, 240]. In experiments where the laser interacts with a solid target, poor contrast can alter the energy spectrum structure of the ions [241] and the exit spatial distribution [242]. In the laser tail wakefield electron acceleration experiment, the existence of the prepulse makes the beam directionality worse [243]. In the X-ray generation experiment where the laser interacts with the cluster, poor contrast can lead to a sharp reduction in X-ray yield [244].

The main laser noise sources in the time domain are shown in Figure 1.64, which mainly include a nanosecond-scale prepulse, amplified spontaneous emission, parametric fluorescence at the hundred-picosecond scale, and high-order dispersion at the picosecond scale. In order to obtain a high-contrast amplified pulse, several contrast enhancement techniques have been proposed in recent years, such as high-energy and high-contrast seed implantation [245], saturable absorber [246], cross polarization technique [247, 248], optical parametric chirped pulse amplification techniques [138], ultrashort pulse pumped optical parametric amplification [235, 249],

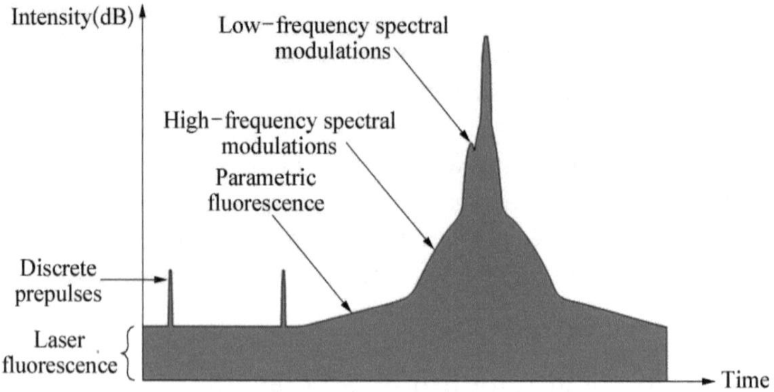

Fig. 1.64: Laser pulse time domain contrast.

plasma mirror technology [250, 251], and so on. This section will focus on the principles and application of these contrast enhancement techniques.

1.5.1 Spatial filtering and temporal filtering

Spatial filtering is the first technique used to reduce the spontaneous emission of laser (ASE) in an amplified laser. Figure 1.65 shows the basic principles.

Fig. 1.65: Schematic diagram of spatial filtering.

The different frequencies of light through the focus lens are gathered to a different location, placing a tiny pinhole at the lens focus can block the main ASE frequency. As a result the ASE spatial filtering is greatly reduced, which can effectively improve the pulse contrast.

Unfortunately the spatial filtering also attenuates the main pulse significantly. In contrast temporal filtering introduces less loss to the main pulse. The basic principle of temporal filtering is shown in Figure 1.66.

Temporal filtering uses two Pockels boxes (equivalent to electro-optical crystal) and two polarizers. By varying the instantaneous half-wave voltage within 4–7 ns for the Pockels box with 4–7 ns, the main pulse's polarization direction can be changed. The polarization of the small pulses that sit about 12 ns from the main pulse remains unchanged. Consequently the small pulses are isolated by the polarizers. Meanwhile

Fig. 1.66: Schematic diagram of time filtering.

because the application of half-wave voltage time is very short, ASE is also partially isolated. After temporal filtering, the main pulse is relatively clean.

1.5.2 Suppression of nanosecond prepulses

The seeds to the CPA system are generally provided by mode-locked laser oscillators with a 100 MHz repetition rate. The interval between adjacent pulses is about 10 ns. Due to thermal effects during the amplification process and the repetition rate of the pump laser itself, it is often necessary to first reduce the repetition rate of the amplified laser using a combination of a Pockels cell and a polarizing element. However, the extinction ratio of the polarizing element is limited (typically 100:1), which results in the presence of a prepulse of nanosecond time scale before and after the main pulse. In the case of the Ti:sapphire regenerative amplifier CPA system, the seed pulse is amplified once each time in the regeneration chamber and then is leaked out once through the polarizing element, so that there is a nanosecond prepulse on the leading edge of the amplified pulse.

The nanosecond prepulse allows the amplified laser pulse to be suppressed by the combination of the Pockels cell and the polarizing element. As shown in Figure 1.67, when the amplified pulse passes through the Pockels box, the high-voltage time window will fall on the amplified pulse by controlling the delay and width of the high-voltage signal. Nanosecond prepulse contrast increase depends mainly on the extinction ratio of the polarizing element and Pockels box's electro-optical switch rise time. For polarizers with an extinction ratio of 100:1, it is theoretically possible to increase the contrast of nanosecond prepulses by two orders of magnitude.

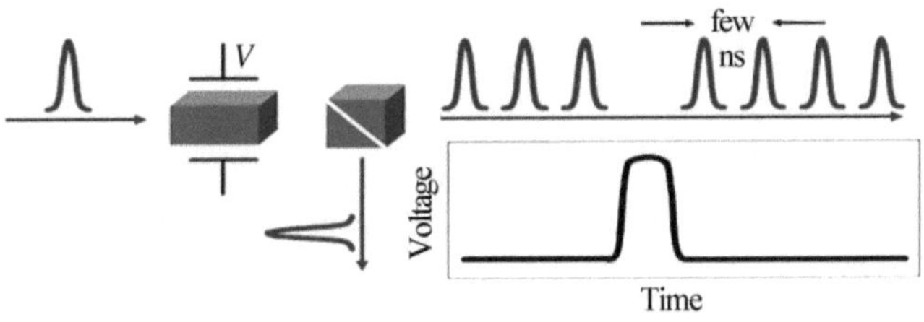

Fig. 1.67: Schematics of suppressing nanosecond prepulses.

1.5.3 Improve pulse contrast through saturable absorbers

Saturated saturable absorber (e.g. a Schott glass) has a power-dependent transmittance. Higher peak power of the main pulse corresponds to higher transmittance. Because of the low transmittance of the prepulse and the peak power of the ASE, the prepulse and ASE can be filtered out after the laser pulse passes through the material to achieve contrast enhancement.

Since the peak power of the seed pulse is only on the order of MW, the saturable absorber absorbs the seed pulse very strongly. The use of a saturable absorber to enhance the magnification of laser contrast usually follows three options: (1) do not expand the seed pulse directly amplified, and then pass through the saturated absorber to obtain clean seed pulse prior to the conventional CPA [252, 253]; (2) add a saturable absorber to the two-stage CPA [254]; and (3) add a saturable absorber directly behind the amplified lasers output [255].

Figure 1.68 is an experiment conducted by Itatani et al. at the University of Michigan using the first scheme [252]. The femtosecond seed pulse with 3 nJ pulse energy is directly amplified and then filtered using a saturable absorber to obtain a clean 1 μJ pulse, and sent into the CPA. With 10^4 gain, 4 mJ amplified laser pulses are achieved with 10^7 pulse contrast two orders of magnitude higher than the conventional CPA.

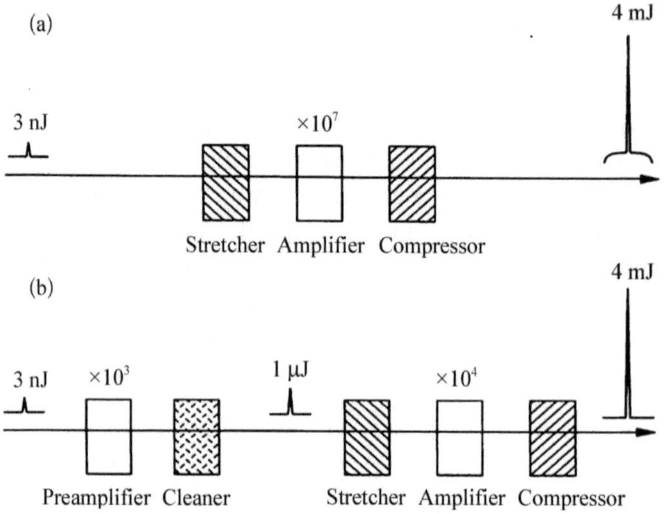

Fig. 1.68: Schematics of (a) the original CPA laser concept and (b) the modified CPA laser concept with injection of clean μJ pulses for the ASE suppression.

1.5.4 Contrast and energy improvement of injected seeds

The process of femtosecond laser amplification is accompanied by the presence of ASE, especially in pre-amplification systems with high gain. When nJ seed pulses are directly amplified to the mJ level, the resulting 10^6 gain leads to large ASE accumulation and amplification [256]. The use of high-energy high-contrast seed pulse in the follow-up amplifier stage can reduce the required gain and the associated ASE and greatly improve the pulse contrast.

Figure 1.69 shows a typical technical roadmap for increasing contrast in this manner. First, the unexpanded seed pulses are directly amplified from nJ level to μJ level. The amplified seed pulse is then filtered using a contrast enhancement technique to generate high-energy, high-contrast seed pulses for the CPA amplifier.

In 1998, Itatani et al. used an experimental optical path as shown in Figure 1.69 to improve the pulse contrast [252]. The main idea was to increase the contrast and intensity of the seed laser in order to obtain a high-contrast laser output. This method includes three steps: (1) the unexpanded seeds are directly amplified from nJ to μJ level; (2) the seed is filtered by a saturable absorber to generate clean seed; and (3) the clean seed is injected into the traditional CPA amplifier. Because the seed pulse energy is increased by three orders of magnitude, the amplifier gain is provided to the seed. Magnification is also reduced by several orders of magnitude, leading to less ASE. They used this method to increase the laser contrast from 10^5 to 10^7.

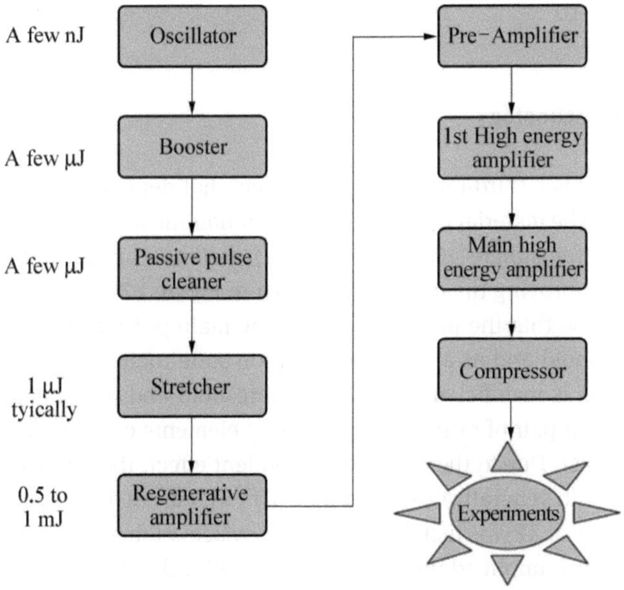

Fig. 1.69: The schematic diagram of the high-energy high-contrast seed injection technology for contrast enhancement.

1.5.5 Amplifier technology in ring cavity

ASE is mainly from the pre-amplification process, especially when the regenerative amplification technique is used. Therefore, it is important to suppress ASE to improve the pulse contrast. In 2002, Yanovsky et al. reported a ring cavity amplification scheme [257], which is shown in Figure 1.70.

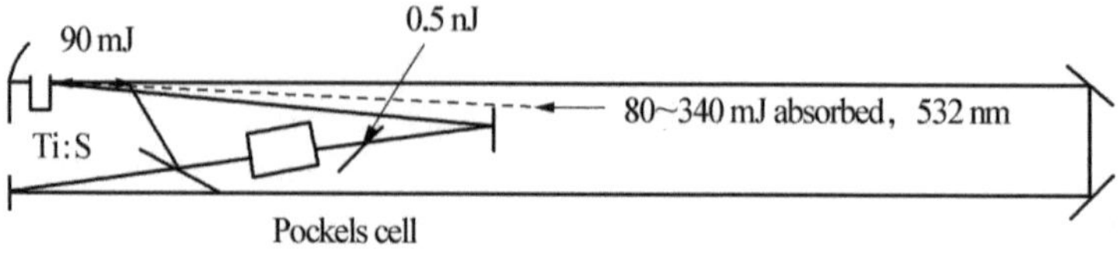

Fig. 1.70: Schematic of the amplifier in a ring cavity.

In this scheme, the total laser cavity length is 7.5 m, the concave mirror has a radius of curvature of 4 m, and the rest are plane mirrors. The Ti:sapphire crystal is placed near the concave mirror, and the seed laser and the amplified laser are respectively supplied by Pockels. 90-mJ amplified pulses are obtained with 340-mJ of 532-nm. The characteristic of this scheme is that the amplifier cavity length is longer than the traditional linear cavity. Because ASE is inversely proportional to the cavity length, the ASE in the ring cavity is relatively small. According to the results in the paper, the contrast ratio of the amplified pulse is 10^8.

1.5.6 Cross-polarized wave technology

A cross-polarized wave (XPW) is a third-order nonlinear effect that depends heavily on the laser peak power [54]. The material used has high-order nonlinear characteristics. These materials include atmosphere [258], inert gas [259], nonlinear crystal [260], and so on. The principle of improving the contrast is shown in Figure 1.71. The laser is focused on a BaF_2 crystal, so that the peak intensity of the main pulse can reach cross-polarization filter threshold and as a result polarization state rotates 90°. The intensity of prepulse and ASE is insufficient to reach the threshold and their polarization does not rotate. Thus, a pair of orthogonal polarizing elements can separate and enhance the pulse contrast. Due to the intensity-dependent effect, the intensity required to produce XPW should generally be greater than 10^{12} W/cm^2 in order to ensure a high conversion efficiency. XPW is usually used in the system of dual CPA with the first CPA producing mJ level amplified pulses, and then XPW technology is used enhance the pulse contrast with tens to hundreds of μJ energy. Finally these pulses are amplified by the second CPA to obtain a higher-energy, high-contrast pulse.

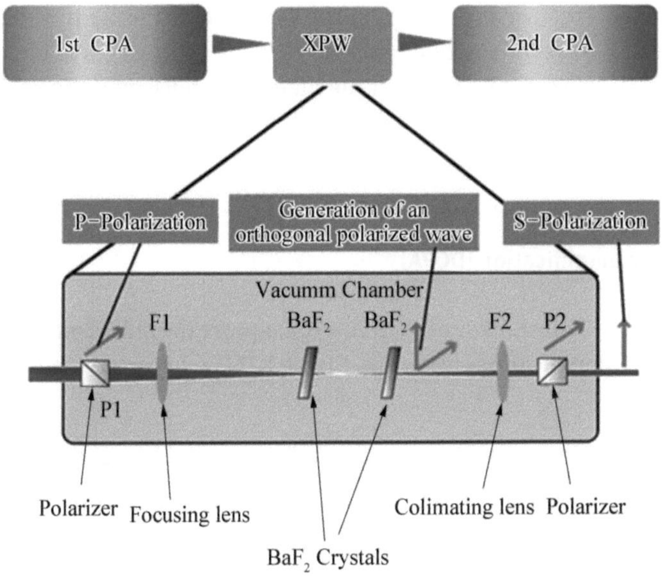

Fig. 1.71: Schematic diagram of cross-polarized wave technology for contrast enhancement.

The most successful experimental results were obtained by Jullien et al. In 2005, they used a BaF_2 crystal to obtain amplified pulses with the contrast of up to 10^{10} [260]. In their experiments, the laser system is a 1 kHz titanium doped sapphire CPA device, which generates 2 mJ, 42 fs pulses. They used a lens with 3 m focal length to focus the pulses onto 2-mm thick BaF_2 crystals, from which a 120 μJ output can be achieved with a conversion efficiency of 10%. The contrast information measured by the third-

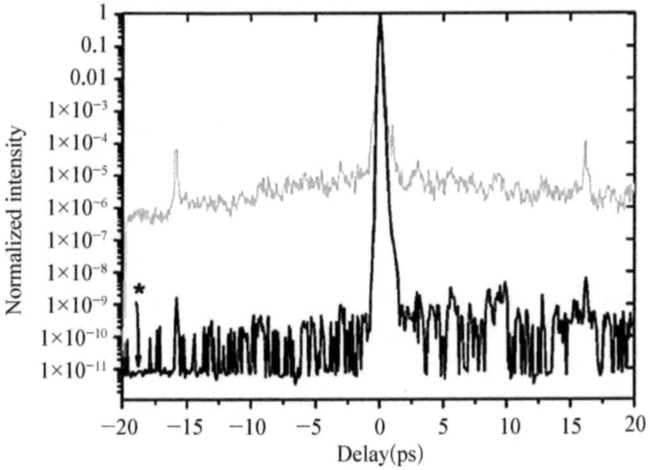

Fig. 1.72: The laser contrast curve obtained by BaF_2 crystal filter.

order correlation instrument is shown in Figure 1.72. The dark curve in the figure is the contrast curve without BaF_2 crystal. The light gray curve is the contrast curve after filtering with BaF_2 crystal. The contrast between the main laser pulse and the ASE is 10^{10}. In a recent report, they employed two BaF_2 crystals to further improve the contrast to 10^{11}.

1.5.7 Dual-chirped pulse amplification (DCPA)

Double-CPA was proposed by Kalashnikov et al. in 2005 to improve the pulse contrast, which consists of two CPA amplifications [258, 261]. Figure 1.73 (a) shows the scheme.

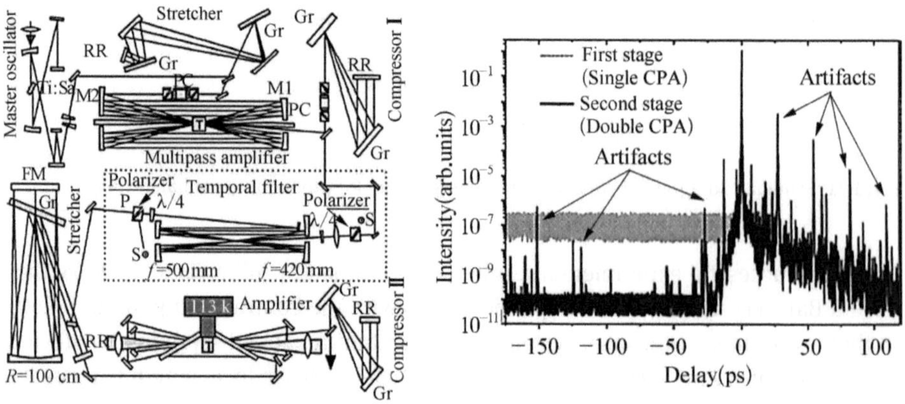

Fig. 1.73: (a) Dual-chirped pulse amplification (DCPA) schematic and (b) the resulting contrast curve.

The 200-fs pulse is stretched to 300 ps and then injected into the multipass amplifier. The multipass amplifier is pumped by a 532 nm laser with 60 mJ of energy. After the amplification, the seed energy reaches 2 mJ. After compression, the pulse duration is 35 fs. The compressed seed is filtered into a spatial filtering system to remove ASE. The filter system consists of a pair of orthogonal polarizers, two quarter wave plates, and one set of multipass amplifiers similar to the composition of the confocal cavity, as shown in the dashed box. The first quarter-wave plate rotates the pulse into an elliptically polarized light with a slight elliptical polarization, and then the pulse is injected into the confocal cavity to perform spatial filtering using the air, which takes place at the beam waist. The filtered seed laser was then stretched to 200 ps by a stretcher and then injected into a four-pass amplifier pumped by a 350 mJ, 532 nm laser, resulting in a saturation laser output of 100 mJ. The results show that the pulse contrast reached 10^{10}. Figure 1.73 (b) shows the measured results.

1.5.8 Ultrashort pulse pumped optical parametric amplifier technique

OPCPA can enhance the pulse contrast, but it can produce parametric fluorescence if pump pulse is nanosecond long, degrading the pulse contrast [262].

To solve this issue, an OPA technique based on ultrashort pulse pumping has been developed in recent years [263, 264]. The schematic diagram is shown in Figure 1.74. It differs from OPCPA in that the signal light is a femtosecond laser, and the pump pulse is also on the order of a few picoseconds, which greatly suppresses the propagation of parametric fluorescence that has the time scale of one hundred picoseconds or even tens of picoseconds. Therefore, pulse contrast can be effectively enhanced.

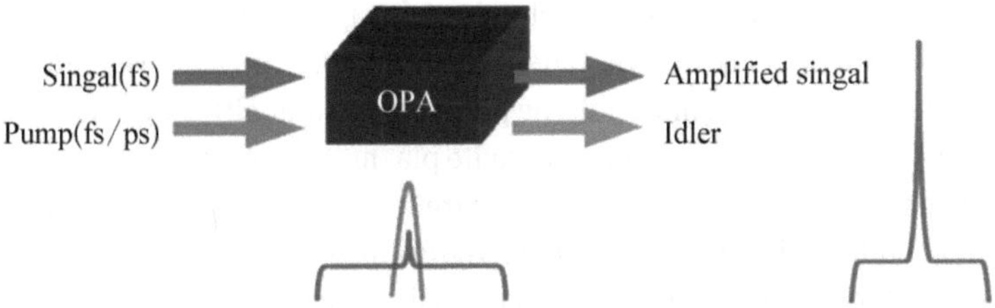

Fig. 1.74: Schematic diagram of ultrashort pulse pumped OPA.

In particular, if the pump pulse is not depleted, the idler pulse has a higher contrast than in the amplified signal pulse [265, 266]. Figure 1.75 (a) shows the experimental schematic diagram of Shah et al. In the experiment, they split the beam into two beams with 500-fs pulse duration. One beam serves as the signal and the other as the pump. Since the pump pulse is the second harmonic of the incident laser, its intensity is proportional to the square of the incident laser so that the intensity of the idle light gener-

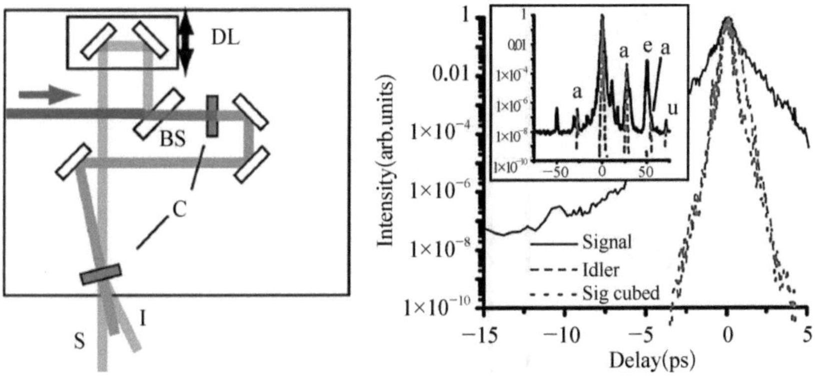

Fig. 1.75: (a) Experimental setup of optical parametric amplification and (b) ultrahigh contrast results.

ated by the OPA process is proportional to the third power of the incident laser. As can be seen from the contrast results in Figure 1.75 (b), the actually measured contrast of the idle pulse is indeed the third power of the signal pulse contrast. With this method, a very high contrast light can be obtained.

1.5.9 Plasma mirror technology

The contrast enhancement techniques described above are designed for the amplification part of the system. In some large-scale laser devices, the contrast enhancement technique can also be applied to the amplified laser output from the terminal, for example using self-induced plasma switching technology [250]. The working principle of the plasma mirror is shown in Figure 1.76. When a certain pulse intensity is incident on the surface of transparent optical material, the prepulse is transmitted out, and the intensity of the pulse front part is high enough to excite plasma. When the plasma density exceeds the critical electron density, the transparent medium instantly changes from a transparent solid to a plasma with the reflectivity close to 1, which reflects the main pulse. The magnitude of contrast enhancement using plasma mirrors is proportional to the ratio of the main pulse to the background noise reflectance. Plasma mirror technology is used to enhance the contrast with subpicosecond response speed, and can be used in any amplifier. Figure 1.77 shows the results obtained by Rödel et al. [10] in the German JETI laser device (10 TW). The contrast ratio of the amplified laser can be increased from 10^{-8} to 10^{-11} by selecting the appropriate dielectric material [267].

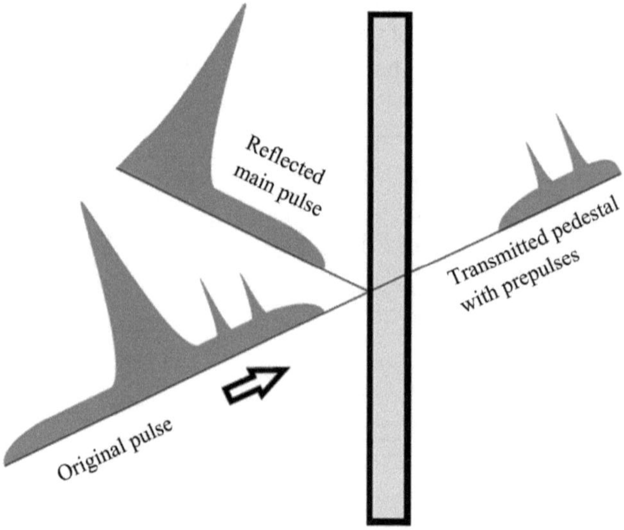

Fig. 1.76: Schematic diagram of the contrast enhancement principle of plasma mirror.

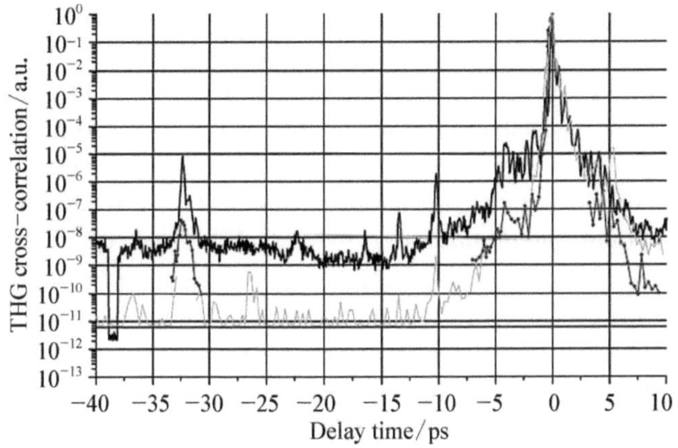

Fig. 1.77: Use of a plasma mirror to enhance the contrast results.

1.6 Prospects for ultrahigh-intensity laser development

Ultrahigh-intensity lasers have found many applications in defense, high-tech industries, cutting-edge science, medicine etc. These lasers have been developed in many research institutions, such as the Lawrence Livermore National Laboratory (LLNL), the United Kingdom's Rutherford Appleton Laboratory, Japan's Osaka University and Atomic Energy Research Institute, and the German Max Planck Institute of Quantum Optics. There are several companies that sell high-intensity lasers, such as Austria's HQLaser company, PHOTONICS in the United States, Switzerland's Time-Bandwidth, Germany's LUMERA, and Lithuania's EKSPLA. The market reaches more than 10 billion dollars and will be increased to 100 billion dollars within ten years. Ultrahigh-intensity ultrafast lasers directly pumped by diodes have become one of the new directions in laser science and technology.

In the next five to ten years, demand for ultrahigh-intensity lasers will appear in a number of areas. In basic research, femtosecond lasers will provide powerful support for THz spectroscopy, attosecond science, ultrafast spectroscopy, molecular regulation, ultrafast X-rays, etc. In the field of micromachining, they have great potential in material processing and device processing, and femtosecond lasers are an enabling tool in the field of medical and biological research. Ultrahigh-intensity lasers are of particular importance in the inertial confinement of nuclear fusion. The international community, including the United States, Japan, the European Union, Russia, and so on, are developing laser inertial confinement fusion technology. China also initiated a laser fusion ignition project as one of 16 major scientific and technological projects.

For more than ten years, all-solid-state ultrafast laser technology and its applications have been rapidly developed. Ytterbium doped (Yb) laser materials have extensive attention. Excellent laser materials, such as Yb:YAG, Yb:S FAP, Yb:CaF_2, Yb glass fiber and transparent ceramics, have greatly promoted the development and application of all-solid-state laser technology. For example, POLARIS, an all-solid-state (diode-pumped) PW laser constructed by Jena University in Germany, uses a $Yb:CaF_2$ crystal in 2008 to achieve a full LD pumped pulse output of 192 fs with 1 TW peak power. The field of laser fusion reached a milestone in 2014 as the laser generated 16.6-J, 150-fs pulses with 110-TW peak power. $Nd:Ca(Sr)F_2$ crystal materials have been given new life by the local coordination structure at the beginning of 2014, generating 103-fs pulses. This is the first time that ultrafast lasers based on Nd-doped crystal can generate femtosecond pulses.

After more than half a century of development, pursuit of high average power,and high peak power (ultrahigh-intensity ultrafast) laser technology is at the forefront of laser research, which manifests as the following two trends:

(1) High peak power, ultrafast laser amplification and application.
To develop high peak power, tabletop TW/PW/EW ultrafast laser technology provides new experimental means and extreme physical conditions. A practical and reliable ultrafast laser with 1~10 Hz repetition rate, 10~1000 TW peak power, 30 fs pulse duration is crucial for research in laser fusion clean energy, strong field physics, high-energy density physics, new electron accelerators, novel ion source, industrial processing, laser surgery, and many other applications.
(2) Generation, amplification, and application of high average power all-solid-state Yb femtosecond lasers.
Combined with CPA techniques and a large sized ceramic laser gain medium, 1 μm ultrafast lasers with 1~100 kHz repetition rate, 100-W average power, and < 300-fs pulse duration are of importance for applications, industrial processing, laser surgery, and other applied research. This involves the following types of key technologies:
    (a) Femtosecond lasers with all-solid-state laser pumping;
    (b) Precise dispersion compensation for broadening and compression in CPA;
    (c) Gain-narrowing control in the amplification;
    (d) High-performance gain crystal;
    (e) Semiconductor laser pumping coupling technology;
    (f) Innovative technology of cooling gain crystals;
    (g) All-solid-state high-power amplification technology; and
    (h) Beam quality control techniques.

Bibliography

[1] M. J. Edwards, J. D. Lindl, L. J. Atherton et al. Progress towards ignition on the National Ignition Facility. Phys. of Plasmas, **20** (2013): 070501.

[2] Z. H. Wang, C. Liu, Z. W. Shen. et al. High-contrast 1.16 PW Ti:sapphire laser system combined with a doubled chirped-pulse amplification scheme and a femtosecond optical-parametric amplifier. Opt. Lett., **36** (2011): 3194.

[3] T. J. Yu, S. K. Lee, J. H. Sung, et al. Generation of high-contrast, 30 fs, 1.5 PW laser pulses from chirped-pulse amplification Ti:sapphire laser. Opt. Express, **20** (2012): 10807

[4] Y. X. Chu, X. Y. Liang, L. H. Yu, et al. High-contrast 2.0 petawatt Ti:sapphire laser system. Opt. Express, **21** (2013): 29231.

[5] S. Laux, F. Lureau, C. Radier, et al. Suppression of parasitic lasing in high energy, high repetition rate Ti:sapphire laser amplifiers. Opt. Lett., **37** (2012): 1913.

[6] V. Yanovsky, V. Chvykov, G. Kalinchenko, et al. Ultra-high intensity-300-TW laser at 0.1 Hz repetition rate. Opt. Express, **16** (2008): 2109.

[7] D. Spence, P. N. Kean and W. Sibbett. 60-fsec pulse generation from a self-mode-locked Ti:sapphire laser. Opt. Lett., **16** (1991): 42.

[8] D. Strickland and G. Mourou. Compression of amplified chirped optical pulses. Opt. Commun., **56** (1985): 219.

[9] M. Hentschel, R. Kienberger, C. Spielmann, et al. Attosecond metrology. Nature, **414** (2001): 509.

[10] G. Sansone, E. Benedetti, F. Calegari, et al. Isolated single-cycle attosecond pulses. Science, **314** (2006): 443.

[11] K. Zhao, Q. Zhang, Y. Wu, et al. Tailoring a 67 attosecond pulse through advantageous phase-mismatch. Opt. Lett., **37** (2012): 3891.

[12] C. B. Schaffer, A. Brodeur, E. Mazur, et al. Micromachining bulk glass by use of femtosecond laser pulses with nanojoule energy. Opt. Lett., **26** (2001): 93.

[13] R. R. Gattass and E. Mazur. Femtosecond laser micromachining in transparent materials. Nat. Photonics, **2** (2008): 219.

[14] T. H. Her, R. J. Finlay, C. Wu, et al. Microstructuring of silicon with femtosecond laser pulses. Appl. Phys. Lett., **73** (1998): 1673.

[15] S. V. Patel, L. J. Maguire, J. W. McLaren, et al. Femtosecond laser versus mechanical microkeratome for LASIK. Ophthalmology, **114** (2007): 1482.

[16] K. Plamann, V. Nuzzo, O. Albert, et al. Femtosecond laser corneal surgery with in-situ determination of the laser attenuation and ablation threshold by second harmonic generation. Ophthalmic Technologies XVII, **66** (2007): 4261.

[17] A. H. Zewail, 4D ultrafast electron diffraction, crystallography, and microscopy. Annual. Rev. of Phys. Chem., **57** (2006): 65.

[18] M. Schultze, E. M. Bothschafter, A. Sommer, et al. Controlling dielectrics with the electric field of light. Nature, **493** (2013): 75.

[19] Q. L. Dong, S. J. Wang, C. Huang, et al. Plasmoid ejection and secondary current sheet generation from magnetic reconnection in laser-plasma interaction. Phys. Rev. Lett., **108** (2012): 215001.

[20] V. Gerginov, C. E. Tanner, S. A. Diddams, et al. High-resolution spectroscopy with a femtosecond laser frequency comb. Opt. Lett., **30** (2005): 1734.

[21] F. J. McClung and R. W. Hellwarth. Giant optical pulsations from ruby. J. Appl. Phys., **33** (1962): 828.

[22] J. L. Helfrich. Faraday Effect as a Q-Switch for Ruby Laser. J. Appl. Phys. **34** (1963): 1000.

[23] L. Hargrove, R. Fork and M. Pollack. Locking of He-Ne laser modes induced by synchronous intracavity modulation. Appl. Phys. Lett., **15** (1964): 4.

[24] W. E. Lamb. Theory of an optical maser. Phys Rev., 1964,**134**(6A): Al429.

[25] P. W. Smith. Mode-locking of lasers. Proc. of the IEEE, **58**(9) (1970): 1342.

[26] H. W. Mocker and R. Collins. Mode competition and self-locking effects in a Q-Switched ruby Laser. Appl. Phys. Lett., **7**(10) (1965): 270.

[27] R. Fork, B. Greene and C. Shank. Generation of optical pulses shorter than 0.1 psec by colliding pulse mode locking. Appl. Phys. Lett., **38** (1981): 671.

[28] A. M. Johnson and W. M. Simpson. Tunable femtosecond dye laser synchronously pumped by the compressed second harmonic of Nd:YAG. J. Opt. Soc. Of Am. B, **2**(4) (1985): 619.

[29] J. A. Valdmanis, R. L. Fork and J. P. Gordon. Generation of optical pulses as short as 27 femtoseconds directly from a laser balancing self-phase modulation, group-velocity dispersion, saturable absorption, and saturable gain. Opt. Lett., **10** (1985): 131.

[30] R. L. Fork, C. Cruz, P. Becker, et al. Compression of optical pulses to 6 fs by using cubic phase compensation. Opt. Lett., **12**(7) (1987): 483.

[31] P. F. Moulton. Spectroscopic and laser characteristics of Ti:Al_2O_3. J. Opt. Soc. of Am. B, **3**(1) (1986): 125.

[32] T. Brabec, C. Spielmann, P. Curley, et al. Kerr lens mode locking. Opt. Lett., **17**(18) (1992): 1292.

[33] Asaki MT, C. P. Huang, D. Garvey, et al. Generation of 11-fs pulses from a self-mode-lockcd Ti:sapphire laser. Opt. Lett., **18**(12) (1993): 977.

[34] R. Szipöcs, K. Ferencz, C. Spielmann, et al. Chirped multilayer coatings for broadband dispersion control in femtosecond lasers. Opt. Lett., **19**(3) (1994): 201.

[35] A. Stingl, M. Lenzner, C. Spielmann, et al. Sub-10-fs mirror-dispersion-controlled Ti:sapphire laser. Opt. Lett., **20**(6) (1995): 602.

[36] L. Xu, C. Spielmann, F. Krausz, et al. Ultrabroadband ring oscillator for sub-10-fs pulse Generation. Opt. Lett., **21** (1996): 1259.

[37] I. D. Jung, F. X. Kartner, N. Matsuchek, et al. Self-starting 6.5-fs pulses from a Ti:sapphire laser. Opt. Lett., **22** (1997): 1009.

[38] U. Morgner, F. X. Kartner, S. H. Cho, et al. Sub-two-cycle pulses from a kerr-lens mode-locked Ti:sapphire laser. Opt. Lett., **24** (1999): 411.

[39] D. H. Sutter, G. Steinmeyer, L. Gallmann, et al. Semiconductor saturable absorber mirror assisted Kerr-lens mode-locked Ti:sapphire laser producing pulses in the two-cycle regime. Opt. Lett., **24** (1999): 631.

[40] R. Ell, J. G. Fujimoto, V. Scheuer, et al. Generation of 5-fs pulses and octave spanning spectra directly from a Ti:sapphire laser. Opt. Lett., **26** (2001): 373.

[41] Y. Y. Zhao, P. Wang and Z. Y. Wei. Generation of 7-fs laser pulse directly from a compact Ti:sapphire laser with chirped mirrors. Science in China Series G, **50** (2007): 261.

[42] U. Keller and T. H. Chiu. Resonant Passive Mode-Locked Nd-YLF Laser. IEEE J Quantum Electron., **28** (1992): 1710.

[43] C. Hönninger, R. Paschotta, U. Keller, et al. Q-switching stability limits of continuous-wave passive mode locking. J. Opt. Soc. Am. B, **16** (1999): 46.

[44] G. J. Spühler, T. Sudmeyer, R. Paschotta, et al. Passively mode-locked high-power Nd:YAG lasers with multiple laser heads. Appl. Phys. B, **71** (2000): 19.

[45] C. J. Saraceno, F. Emaury, C. Schriber, et al. Ultrafast thin-disk laser with 80 μJ pulse energy and 242 W of average power. Opt. Lett., **39** (2014): 9.

[46] N. K. Metzger, C. G. Leburn, A. A. Lagatsky, et al. Femtosecond pulse generation around 1,500 nm using a GaInNAsSb SESAM. Opt. Express, **16** (2008): 18739.

[47] E. Sorokin, N. Tolsrik, K. I. Schaffers, et al. Femtosecond SESAM-model locked Cr:ZnS laser. Opt. Express, **20** (2012): 28947.

[48] K. Yang, D. Heinecke, J. Paajaste, et al. Mode-locking of 2 μm Tm, Ho:YAG laser with GaInAs and GaSb-based SESAMs. Opt. Express, **21** (2013): 4311.

[49] F. Druon, F. Balembois, P. Georges, et al. Generation of 90-fs pulses from a mode-locked diode-pumped $Yb^{3+}:Ca_4GdO(BO_3)_3$ laser. Opt. Lett., **25** (2000): 423.

[50] J. Boudeile, F. Druon, M. Hanna, et al. Continuous-wave and femtosecond laser operation of $Yb:CaGdAlO_4$ under high-power diode pumping. Opt. Lett., **32** (2007): 1962.

[51] F. Friebel, F. Druon, J. Boudeile, et al. Diode-pumped 99 fs $Yb:CaF_2$ oscillator. Opt. Lett., **34** (2009): 1474–1476.

[52] A. Agnesi, A. Greborio, F. Pirzio, et al. 40-fs $Yb^{3+}:CaGdAlO_4$ laser pumped by a single mode 350-mW laser diode. Opt. Express, **20** (2012): 10077.

[53] Y. Zaouter, J. Didierjean, F. Balembois, et al. 470-fs diode-pumped $Yb^{3+}:CaGdAlO_4$ laser. Opt. Lett., **31** (2006): 119–121.

[54] U. Griebner, S. Rivier, V. Petrov, et al. Passively mode-locked Yb:KLu $(WO_4)_2$ Oscillators. Opt. Express, **13** (2005): 3465.

[55] S. Rivier, A. Schmidt, C. Krankel, et al. Ultrashort pulse Yb:LaSc3 $(BO_3)_4$ mode-locked oscillator. Opt. Express, **15** (2007): 15539.

[56] A. Schmidt, S. Rivier, V. Petrov, et al. Continuous-wave tunable and femtosecond Mode locked laser operation of Yb:NaY $(MoO_4)_2$. J. Opt. Soc. Of Am. B, **25** (2008): 1341–1349.

[57] A. Garcia-Cortes, J. M. Cano-Torres, M. D. Serrano, et al. Spectroscopy and lasing of Yb-doped $NaY(WO_4)_2$: Tunable and femtosecond mode-locked laser Operation. IEEE J. Quantum Electron, **43** (2007): 758–764.

[58] M. Tokurakawa, A. Shirakawa, K. Ueda, et al. Ultrashort pulse generation from diode pumped mode-locked Yb^{3+}:sesquioxide single crystal lasers. Opt. Express, **19** (2011): 2904–2909.

[59] A. Yoshida, A. Schmidt, V. Petrov, et al. Diode-pumped mode-locked Yb:YCOB laser generating 35 fs pulses. Opt. Lett., **36** (2011): 4425.

[60] C. W. Xu, D. Y. Tang, J. Zhang, et al. Sub-100 fs pulse generation in a diode pumped $Yb:Sc_2$ SiO_5 laser. Opt. Commun., **294** (2013): 237.

[61] H. Liu, J. Nees and G. Mourou. Diode-pumped kerr-lens mode-locked $Yb:KY(WO_4)_2$ 1aser [J]. Opt. Lett., **26** (2001): 1723–1725.

[62] A. A. Lagarsky, A. R. Sarmani, A. C. T. Brown, et al. Yb^{3+}-doped YVO_4 crystal for efficient kerr-lens mode locking in solid-state lasers. Opt. Lett., **30** (2005): 3234–3236.

[63] M. Tokurakawa, A. Shirakawa, K. Ueda, et al. Diode-pumped sub-100 fs kerr-lens mode-locked $Yb^{3+}:ScO_3$ ceramic laser. Opt. Lett., **32** (2007): 3382–3384.

[64] M. Tokurakawa, A. Shirakawa, K. Ueda, et al. Diode-pumped 65 fs kerr-lens mode-locked $Yb^{3+}:Lu_2O_3$ and non-doped Y_2O_3 combined ceramic laser. Opt. Lett., **33** (2008): 1380–1382.

[65] S. Uemura and K. Torizuka. Sub-40-fs pulses from a diode-pumped kerr-lens mode-locked Yb-doped yttrium aluminum garnet laser. Jpn. J. Appl. Phys., **50** (2011): 010201.

[66] G. Machinet, P. Sevillano, F. Guichard, et al. High-brightness fiber laser-pumped 68 fs-2.3 W kerr-lens mode-locked $Yb:CaF_2$ oscillator. Opt. Len., **38** (2013): 4008–4010.

[67] J. W. Zhang, H. N. Han, W. L. Tian, et al. Diode-pumped 88-fs kerr-lens mode-locked $Yb:Y_3Ga_5O_{12}$ crystal laser. Opt. Express, **21** (2013): 29867–29873.

[68] H. T. Zhao and A. Major. Powerful 67 fs kerr-lens mode-locked prismless Yb:KGW oscillator. Opt. Express, **21** (2013): 31846–31851.

[69] Z. Gao, J. Zhu, W. Tian, et al. Diode-pumped 73-fs kerr-lens mode-locked $Yb:YCa_4O(BO_3)_3$ laser. OSA/CLEO, 2014: SM_4F. 1

[70] A. L. Calendron, K. S. Wentsch and M. J. Lederer. High power cw and mode-locked oscillators based on Yb:KYW multi-crystal resonators. Opt. Express, **16** (2008): 188318843.

[71] A. Giesen, H. Hugel, A. Voss, et al. Scalable Concept for diode-pumped high-power solid-state lasers. Appl. Phys. B, **58** (1994): 365–372.
[72] A. Diebold, F. Emaury, C. Schriber, et al. SESAM mode-locked Yb:$CaGdAlO_4$ thin disk laser with 62 fs pulse generation. Opt. Lett., **38** (2013): 3842–3845.
[73] J. Kafka, D. W. Hall and T. Baer. Mode-locked erbium-doped fiber laser with soliton pulse shaping. Opt. Lett., **14** (1989): 1269–1271.
[74] K. Tamura, E. P. Ippen, H. A. Haus, et al. 77-fs pulse generation from a stretched-pulse mode locked all-fiber ring laser. Opt. Lett., **18** (1993): 1080–1082.
[75] B. Ortac, M. Baumgartl, J. Limpert, el al. Approaching microjoule-level pulse energy with mode-locked femtosecond fiber laser. Opt. Lett., **34**(10): 1585–1587.
[76] J. Lhermite, G. Machinet, C. Lecaplain, et. al. High energy femtosecond fiber laser at 976 nm. Optics Letters, **35** (2011): 3459.
[77] D. Ma, Y. Cai, Z. G. Zhang, et al. 37.4 fs pulses generation in an Er: fiber at a 225 MHz repetition rate. Opt. Lett., **35** (2011): 2858–2860.
[78] X. Li, W. W. Zou and J. P. Chen. 41.9 fs hybridly mode-locked Er-doped fiber laser at 212 MHz repetition rate. Opt. Lett., **39** (2014): 1553–1556.
[79] Z. Y. Wei, J. Zhang, J. F. Xia, et al. Highly efficient TW multipass Ti:sapphire laser System. Science in China Series A, **43** (2000): 1083–1087.
[80] M. D. Perry, D. Pennington, B. C. Stuart, et al. Petawatt laser pulses. Opt. Lett., **24** (1999): 160–162.
[81] F. G. Patterson and M. D. Perry. Design and performance of a multi-terawatt, subpicosecond Neodymium glass laser. J. Opt. Soc. of Am, B, **8** (1991): 2384–2391.
[82] E. B. Treacy. Optical pulse compression with diffraction gratings. IEEE J. Quantum Electronics, QE **5** (1969): 454.
[83] O. E. Martinez. Matrix formalism for pulse compressors. IEEE J. Quantum Electronics, **24** (1988): 2530–2536.
[84] A. Offner. Unit power imaging catoptrics anastigmat. United States Patent, 1973, US 3748015
[85] Z. Cheng and W. Zhao. Group-delay Dispersion in double-prism pair and limitation in broadband laser pulses. Chin. J. Laser. B, **11** (2002): 359–363.
[86] V. Pervak, I. Ahmad, J. Fulop, et al. Comparison of dispersive mirrors based on the time-domain and conventional approaches, for sub-5-fs pulses. Opt. Express, **17** (2009): 2207–2217.
[87] A. Sullivan, H. Hamster, H. C. Kapteyn. et al. Multiterawatt, 100-fs Laser. Opt. Lett., **16** (1991): 1406–1408.
[88] S. Sartania, Z. Cheng, M. Lenzner, et al. Generation of 0.1-TW 5-fs optical pulses at a 1-kHz repetition rate. Opt. Lett., **22** (1997): 1562–1564.
[89] D. Herrmann, L. Veisz, R. Tautz, et al. Generation of sub-three-cycle, 16 TW light pulses by using non-collinear optical parametric chirped-pulse amplification. Opt. Lett., **34** (2009): 2459–2461.
[90] Z. G. Zhang, T. Yagi and T. Arisawa. Ray-tracing model for stretcher dispersion calculation. Appl. Opt., **36** (1997): 3393–3399.
[91] J. Itatani, J. Faure, M. Nantel, et al. Suppression of the amplified spontaneous emission in chirped-pulse-amplification lasers by clean high-energy seed-pulse injection. Opt. Commun., **148** (1998):70–74.
[92] J. Kawanaka, K. Yamakawa, K. Ueda, et al. 30-mJ, diode pumped, chirped-pulse Yb:YLF regenerative amplifier. Opt. Lett., **28** (2003): 2121–2123.
[93] K. Ogawa, Y. Akahane and K. Yamakawa. 100-mJ diode-pumped, cryogenically- cooled Yb:YLF chirped-pulse regenerative amplifier. OSA/CLEO, 2011: CMB4.

[94] Q. K. Ai, M. Chen, Y. Xu, et al. Picosecond Nd:YLF five-passes laser amplifier with 20 mJ pulse encrgy. Laser Phys., **22**(7) (2012): 1169–1172.
[95] D. W. E. Noom, S. Witte, J. Morgenweg, et al. High-energy, high-repetition-rare picosecond pulses from a quasi-CW diode-pumped Nd:YAG system. Opt. Lett., **38** (2013): 3021–3023.
[96] A. Mareczko, V. Clet, A. Courjaud, et al. 100 mJ level Yb:CaF_2 regenerative amplifier for multi-pulse fcmtosecond regime. OSA/ASSL2013, ATu3A.
[97] A. Kessler, M. Hornung, S. Keppler, et al. 16.6 J chirped femtosecond laser pulses from a diode-pumped Yb:CaF_2 amplifier. Opt. Lett., **39** (2014): 1333–1336.
[98] K. K. Young, S. P. Chard, N. Hay, et al. 200 W output power at 10 ps from a scalable Z-slab Nd:YAG laser. Proc. Of the SPIE, **8959** (2014): 89590V.
[99] C. J. Saraceno, F. Emaury, O. H. Heckl, et al. 275 W average outpower from a femtosecond thin disk oscillator operated in a vacuum environment. Opt. Express, **20** (2012): 23535–23541.
[100] C. J. Saraceno, F. Emaury, C. Schriber, et al. Ultrafast thin-disk laser with 80 μJ pulse energy and 242 W of average power. Opt Lett., **39** (2014): 9–12.
[101] S. Erhard, K. Contag, A. Giesen, et al. Advanced Solid State Lasers, OSA, Washington, DC., OSA Trends in Optics and Photonics **34** (2000): 137.
[102] C. J. Saraceno, F. Emaury, O. H. Heckl, et al. 275 W average output power from a femtosecond thin disk oscillator operated in a vacuum environment. Opt Express, **20** (2012): 23535
[103] C. J. Saraceno, F. Emaury, C. Schriber, et al. Ultrafast thin-disk laser with 80 μJ pulse energy and 242 W of average power. Opt Letter, **39** (2014): 9–12.
[104] C. Honninger, I. Johannsen, M. Moser, et al. Diode-pumped thin-disk Yb:YAG regenerative amplifier. Appl Physics B, **65** (1997): 423.
[105] D. Müller, S. Erhard and A. Giesen. Advanced Solid-State Lasers, 2001, 319.
[106] D. Müller, S. Erhard, O. Ronsin, et al. Thin disk multi-pass amplifier. Advanced Solid-State Photonics/OSA (2003): 278–284.
[107] T. Metzger, Y. C. Teisset, et al. High-repetition-rate picosecond pump laser based on a YbYAG disk. Optics Letters, **34**(14) (2009): 2123–2125.
[108] J. P. Negel, A. Voss, M. A. Ahmed, et al. 1.1 kW average output power from a thin-disk multi-pass amplifier for ultrashort laser pulses. Opt. Lett., **38** (2013): 5442.
[109] J. Limpert, T. Schreiber, T. Clausnitzer, et al. High-power femtosecond Yb-doped fiber amplifier. Optics Express, **10** (2002): 628.
[110] T. Eidam, S. Hadrich, F. Roser, et al. A 325-W-average-power fiber CPA system delivering sub-400 fs pulses. IEEE J. Select. Top. Quantum Electronics, **15** (2009): 187–190.
[111] C. Jocher, T. Eidam, S. Hädrich, et al. 23 fs pulses at 250 W of average power from a FCPA with solid core nonlinear compression. International Society for Optics and Photonics, **860** (2013): llF-86011F-6.
[112] F. Röser, J. Rothhard, B. Ortac, et al. 131 W 220 fs fiber laser system. Opt. Lett., **30** (2005): 2754–2756.
[113] F. Röser, T. Eidam, J. Rorhhardt, et al. Millijoule pulse energy high repetition ratefemtosecond fiber chirped-pulse amplification system. Opt. Lett., **32** (2007):3495–3497.
[114] T. Eidam, S. Hanf, E. Seise, et al. Femtosecond fiber CPA system emitting 830 Waverage output power. Opt. Lett., **35** (2010): 94–96.
[115] T. Eidam, J. Rothhardt, F. Stutzki, et al. Fiber chirped-pulse amplification systememitting 3.8 GW peak power. Optics express, **19** (2011): 255–260.
[116] S. Hädrich, S. Demmler, J. Rothhardt, et al. High-repetition-rate sub-5-fs pulses with 12 GW peak power from fiber-amplifier-pumped optical parametric chirped-pulse amplification. Opt. Lett., **36** (2011): 313–315.
[117] M. Baumgartl, C. Lecaplain, A. Hideur, et al. 66 W average power from a microjoule class sub-100 fs fiber oscillator. Opt. Lett., **37** (2012): 1640–1642.

[118] G. Mourou, B. Brocklesby, T. Tajima, et al. The future is fiber accelerators. Nature Photonics, **7** (2013): 258–261.
[119] M. Yan, W. Li, K. Yang, et al. High-power Yb-fiber comb with feed-forward control of nonlinear-polarization-rotation mode-locking and large-mode-area fiber amplification. Opt. Lett., **37** (2012): 151–1513.
[120] M. Yan, W. Li, Q. Hao, et al. High-power nanosecond ytterbium-doped fiber laser passively synchronized with a femtosecond Ti:sapphire laser. Opt. Lett., **34** (2009): 3331.
[121] J. Yu, W. Zhao, Y. Wang, et al. Low-repetition rate, nanosecond, high-power pulse amplifier system based on Yb-doped rod-type fiber. Chinese Opt. Lett., **11** (2013): 050601.
[122] B. W. Lin, M. L. Hu, Y. J. Song, et. al. 39 fs, 16 W. Chinese Journal of Lasers, **35** (2009): 811–814.
[123] E. Snitzer. Proposed Fiber Cavities for Optical Masers. J Appl Phys, **32** (1961): 36–39.
[124] J. K. C and E. Snitzer. Amplification in a Fiber Laser. Applied Optics, **3** (1964): 1182–1186.
[125] K. Tamura, E. P. Ippen, H. A. Haus, et al. 77-fs pulse generation from a stretched pulse mode-locked all- fiber ring laser. Opt. Lett., **18** (1998): 1080–1082.
[126] K. Tamura, E. P. Ippen and H. A. Haus. Pulse dynamics in stretched-Pulse Fiber Lasers. Appl Phys Lett, **67** (1995): 158.
[127] F. Ö. Ilday and F. W. Wise. Nonlinearity management: a route to high-energy soliton fiber lasers. J. Opt. Soc. Am. B, **19** (2002): 470–476.
[128] A. Ruehl, V. Kuhn, D. Wandt, et al. Normal dispersion erbium-doped fiber laser with pulse energies above 10 nJ. Opt Express, **16** (2008): 3130–3135.
[129] M. E. Fermann, M. L. Stock, M. J. Andrejco, et al. Passive mode locking by using nonlinear polarization evolution in a polarization-maintaining erbium-doped fiber. Opt. Lett., **18** (1993): 894–896.
[130] B. C. Collings, K. Bergman, S. T. Cundiff, et al. Short cavity erbium/ytterbium fiber lasers mode-locked with a saturable Bragg reflector. IEEE J. Select. Top. Quantum Electronics, **3** (1997): 1065–1075.
[131] N. H. Seang and D. Y. Kim. A new figure-eight fiber laser based on a dispersion-imbalanced nonlinear optical loop mirror with lumped dispersive elements. IEEE Photonics Technology Letters, **14** (2002): 459–461.
[132] D. Y. Tang and L. M. Zhao. Generation of 47-fs pulses directly from an erbium-doped fiber laser. Opt. Lett., **32** (2007): 41–43.
[133] D. Ma, Y. Cai, C. Zhou, et al. 37.4 fs pulse generation in an Er:fiber laser at a 225 MHz repetition rate. Opt. Lett., **35** (2010): 285–2860.
[134] V. Cautaerts, D. J. Richardson, R. Paschotta, et al. Stretched pulse Yb^{3+}:Silica fiber laser. Opt. Lett., **22** (1997): 316–318.
[135] F. Ö. Ilday, J. Buckley, L. Kuznctsova, et al. Generation of 36-femtosecond pulses from a ytterbium fiber laser. Optics Express, **11** (2003): 3550–3554.
[136] M. E. Fermann, V. I. Kruglov, B. C. Thomsen, et al. Self-similar propagation and amplification of parabolic pulses in optical fibers. Phys Rev Lett, **84** (2000): 6010–6013.
[137] J. M. Dudley, C. Finot, D. J. Richardson. et al. Self-similarity in ultrafast nonlinear optics. Nature Physics, **3** (2007): 597–603.
[138] A. Chong, W. H. Renninger and F. W. Wise. Properties of normal-dispersion femtosecond fiber lasers. J Opt Soc Am B, **25** (2008): 140–148.
[139] C. Li, G. Z. Wang, T. X. Jiang, et al. 750 MHz fundamental repetition rate femtosecond Yb:fiber ring laser. Opt. Lett., **38** (2013): 314–316.
[140] H. W. Chen, G. Q. Chang, S. H. Xu. et al. 3 GHz, fundamentally mode-locked, femtosecond Yb-fiber laser. Opt. Lett., **37** (2012): 3522–3524.
[141] K. Kieu, W. H. Renninger, A. Chong, et al. Sub-100 fs pulses at watt-level powers from a dissipative-soliton fiber laser. Opt. Lett., **34** (2009): 593–595.

[142] T. Eidam, S. Hanf, E. Seise, et al. Femtosecond fiber CPA system emitting 830 W average output power. Opt Lett., **35** (2010): 94–96.

[143] M. Baumgartl, C. Lecaplain, A. Hideur, et al. 66 W average power from a microjoule class sub-100 fs fiber oscillator. Opt. Lett., **37** (2012): 1640–1642.

[144] Y. Kobayashi, N. Hirayama, A. Ozawa, et al. 10-MHz, Yb-fiber chirped-pulse amplifier system with large-scale transmission gratings. Optics Express, **21** (2013): 12865–12873.

[145] P. Wan, L.-M. Yang and J. Liu. All fiber-based Yb-doped high energy, high power femtosecond fiber lasers. Optics Express, **21** (2013): 29854–29859.

[146] A. Dubictis, G. Jonušauskas and A. Piskarskas. Powerful femtosecond pulse generation by chirped and stretched pulse parametric amplification in BBO crystal, Optics. Commun, **88** (1992): 437.

[147] I. N. Ross, P. Matousek, M. Towrie. et al. The prospects for ultrashort Pulse duration and ultrahigh intensity using optical parametric chirped pulse amplifiers. Optics Commun, **l 44** (1997): 125.

[148] I. N. Ross, J. L. Collier, P. Matousek, et al. Generation or terawatt pulses by use of optical parametric chirped pulse amplification, Applied Optics, **39** (2000): 242–2427.

[149] F. Tavella, K. Schmid, N. Ishii, et al. High-dynamic range pulse-contrast measurements or a broadband optical parametric chirped-pulse amplifier. Appl. Phys. B, **81** (2005): 753.

[150] C. Dorrer, I. A. Begishev, A. V. Okishev, et al. High-contrast optical-parametric amplifier as a front end or high-power laser systems. Opt. Lett., **32** (2007): 2143.

[151] Y. S. Huang, C. M. Zhang, Y. Xu, et al. Ultrashort pulse temporal contrast enhancement based on noncollinear optical parametric arnplification. Opt. Lett. **36** (2011): 781.

[152] Y. J. Wang and B. Lutherdavies. Optical parametric amplification based prepulse eliminator for a chirped-pulse-amplification Nd:glass laser. J. Opt. Soc. Am. B, **11** (1994): 1531.

[153] A. E. Boguslavskiy, J. Mikosch, et al. The multi-electron ionization dynamics underlying attosecond strong-field spectroscopies. Science **335** (2012): 1336.

[154] F. Krausz and M. Ivanov. Attosecond physics, Reviews of Modern Physics, **81** (2009): 163.

[155] Y. Huismansl, A. Rouzée, A. Gijsbertse, et al. Time-Resolved Holography with Photoelectrons. Science, **331** (2011): 61.

[156] S. T. Cundiff and J. Ye. Colloquium: Femtosecond Optical Frequency Combs. Reviews of Modern Physics, **75** (2003): 325.

[157] A. Baltuska, Z. Wei, M. S. Pshenichnikov, et al. Optical pulse compression to 5 fs at a 1-MHz repetition rare. Opt. Lett., 1997, 22: 102.

[158] M. Nisoli, S. D. Silvestri, O. Svelto, et al. Compression or high-energy laser pulses below 5 fs. Opt. Lett., **22** (1997): 522–524.

[159] A. Shirakawa, I. Sakane, M. Takasaka, et al. Sub-5-fs visible pulse generation by pulse-front-matched noncollinear optical parametric amplification, Appl. Phys. Lett., **74** (1999): 2268.

[160] U. Morgner, F. X. Kartner, S. H. Cho, et al. Sub-two-cycle pulses from a Kerr-lens mode-locked Ti:sapphire laser. Opt. Lett., **24** (1999): 411.

[161] D. H. Sutter, G. Steinmeyer, L. Gallmann, et al. Semiconductor saturable-absorber mirror assisted Kerr-lens mode-locked Ti:sapphire laser producing pulses in the two-cycle regime, Opt Lett., **24** (1999): 631.

[162] G. Steinmeyer, D. H. Sutter, L. Gallmann, et al. Frontiers in Ultrashort Pulse Generation: Pushing the Limits in Linear and Nonlinear Optics. Science, **286** (1999): 1507.

[163] W. Zhang, H. Teng, C. X. Yun, et al. Generation of Sub-2 Cycle Optical Pulses with a Differentially Pumped Hollow Fiber. Chinese Physics Letters, **27** (2010): 054211.

[164] L. Xu, Spielmann Ch., A. Poppe, et al. Route to phase control of ultrashort light pulses. Opt. Lett., **21** (1996): 2008–2010.

[165] H. R. Telle, G. Steinmeyer, A. E. Dunlop, et al. Carrier-envelope offset phase control: A novel concept for absolute optical frequency measurement and ultrashort pulse generation. Applied Physics and Lasers and Optics, **69**(4) (1999): 327.
[166] Fuji Takao, A. Apolonski, F. Krausz, et al. Self-stabilization of carrier-envelope offset phase by use of difference-frequency generation. Opt. Lett., **29**(6) (2004): 632–634.
[167] S. Koke, C. Grebing, H. Frei, et al. Direct frequency comb synthesis with arbitrary offset and shot-noise-limited phase noise. Nature Photonics, **4**(7) (2010): 462.
[168] D. J. Jones. Carrier-envelope phase control of femtosecond mode-locked lasers and direct optical frequency synthesis. Science, **288**(5466) (2000): 635.
[169] R. Holzwarth, T. Udem., T. W. Hänsch et al. Optical Frequency Synthesizer for Precision Spectroscopy. Phys. Rev. Lett., **85**(11) (2000): 2264–2267.
[170] T. Fuji, J. Rauschenberger, A. Apolonski, et al. Monolithic carrier-envelope phase stabilization scheme. Opt. Lett., **30**(3) (2005): 332.
[171] S. Koke, A. Anderson, H. Frei, et al. Noise performance of a feed-forward scheme for carrier-envelope phase stabilization. Applied Physics B-Lasers and Optics, **104**(4) (2011): 799.
[172] G. Sansone, E. Benedetti, F. Calegari, et al. Isolated single-cycle attosecond pulses. Science, **314**(5798) (2006): 443.
[173] G. Paulus, F. Grasbon, H. Walther, et al. Absolute-phase phenomena in photoionization with few-cycle laser pulses. Nature, **414**(6860) (2001): 182.
[174] A. Baltuska, M. Uiberacker, E. Goulielmakis, et al. Phase-controlled amplification of few-cycle laser pulses. IEEE J. Select. Top. Quantum Electronics, **9**(4) (2003): 972.
[175] J. Rauschenberger, T. Fuji, M. Hentschel, et al. Carrier-envelope phase-stabilized amplifier system. Laser Physics Letters, **3**(1) (2006): 37.
[176] H. Wang, M. Chini, E. Moon, et al. Coupling between energy and phase in hollow-core fiber based f-to-2f interferometers. Optics express, **17**(14) (2009): 12082.
[177] C. Li, E. Moon, H. Mashiko, et al. Mechanism of phase-energy coupling in f-to- 2f interferometry. Applied Optics, **48**(7) (2009): 1303.
[178] J. F. Hergott, O. Tchctbakoff, P. M. Paul. et al. Carrier-Envelope phase stabilization of a 20 W, grating based, chirped-pulse amplified laser, using Electro-Optic effect in a $LiNbO_3$ crystal. Optics Express, **19**(21) (2011): 19935.
[179] G. Gademann, Plé F, P. M. Paul. et al. Carrier-envelope phase stabilization of a terawatt level chirped pulse amplifier for generation of intense isolated attosecond pulses. Optics Express, **19**(25) (2011): 24922.
[180] A. M. Sayler, T. Rathje, W. Müller, et al. Precise, real-time, every-single-shot, carrier-envelope phase measurement of ultrashort laser pulses. Opt. Lett., **36**(1) (2011): 1.
[181] M. Möller, A. Sayler, T. Rathje, et al. Precise, real-time, single-shot carrier-envelope phase measurement in the multi-cycle regime. Appl. Phys. Lett., **99** (2011): 121108.
[182] D. Adolph, A. M. Sayler, T. Rathje, et al. Improved carrier-envelope phase locking of intense few-cycle laser pulse: using above-threshold ionization. Opt. Lett., **36**(18) (2011): 3639.
[183] D. D. Yavuz. Toward Synthesis of Arbitrary Optical Waveforms. Science, **331** (2011): 1142.
[184] Chan Han-Sung, Hsieh Zhi-Ming, Liang Wei-Hong, et al. Synthesis and Measurement of Ultrafast Waveforms from Five Discrete Optical Harmonics. Science, **331** (2011): 1165.
[185] M. T. Hassan, A. Wirth, I. Grguraš, et al. Invited Article: Attosecond photonics: Synthesis and control of light transients. Review of Scientific Instruments, **83**(11) (2012): 111301.
[186] A. Wirth, M. T. Hassan, E. Goulielmakis, et al. Synthesized Light Transients. Science, **334** (2011): 195–200.
[187] S. Fang, G. Cirmi, Shih-Hsuan Chia, et al. Multi-mJ Parametric Synthesizer Generating Two-Octave-Wide Optical Waveforms. Conference on Lasers and Electro-Optics Pacific Rim (CLEO-PR), Kyoto, Japan, 2013.

[188] M. Y. Shverdin, D. R. Walker. D. D. Yavuz, et al. Generation of a Single-Cycle Optical Pulse. Phys. Rev. Lett., **94** (2005): 033904.
[189] B. Schenkel, J. Siegert, U. Keller, et al. Generation of 3.8-fs pulses from adaptive compression of a cascaded hollow fiber supercontinuum. Opt. Lett., **28**(20) (2003): 1987–1989.
[190] E. Matsubara, K. Yamane, T. Sekikawa, et al. Generation of 2.6 fs optical pulses using induced-phase modulation in a gas-filled hollow fiber. J. Opt. Soc. Am. B. **24**(4) (2007): 985–989.
[191] L. E. Chipperfield, J. S. Robinson, J. W. G. Tisch, et al. Ideal Waveform to Generate the Maximum Possible Electron Recollision Energy for Any Given Oscillation Period. Phys. Rev. Lett., **102** (2009): 063003.
[192] M. Ramaswamypaye and J. G. Fujimoto. Opt. Lett., **19** (1994): 1756–1758.
[193] A. Stingl, C. Spielmann, R. Szipocs, et al. in Conference on Lasers and Electro-optics (Opt. Soc. Am., 1996): 66.
[194] T. Udem, J. Reichert, R. Holzwarth, et al. Phys. Rev. Lett., **82** (1999): 3568–3571.
[195] D. J. Jones, S. A. Diddams, J. K. Ranka, et al. Science, **288** (2000): 635–639.
[196] S. A. Diddams, D. J. Jones, J. Ye, et al. Phys. Rev. Lett., **84** (2000): 5102–5105.
[197] S. A. Diddams, D. J. Jones, L. S. Ma, et al. Optical frequency measurement across a 104-THz gap with a femtosecond laser frequency comb. Opt. Lett., **25** (2000): 186–188.
[198] Academic K. Femtosecond Optical Frequency Comb: Principle, Operation, and Applications. 2004.
[199] M. Endo, A. Ozawa and Y. Kobayashi. Kerr-lens mode-locked Yb:KYW laser at 4.6 GHz repetition rate. Opt. Express, **20** (2012): 12191.
[200] A. Bartels, T. Dekorsy and H. Kurz. Femtosecond Ti:sapphire ring laser with 2-GHz repetition rate and its application in time-resolved spectroscopy. Opt. Lett., **24** (1999): 996–998.
[201] A. Bartels, D. Heinecke and S. A. Diddams. Passively mode-locked 10 GHz femtosecond Ti:sapphire laser. Opt. Len., **33** (2008): 1905–1907.
[202] J. J. McFerran, L. Nenadovic, W. C. Swann, et al. A passively mode-locked fiber laser at 1.54 um with a fundamental repetition frequency reaching 2 GHz. Opt. Express, **15**(20) (2007): 13155–13166.
[203] A. Martinez and S. Yamashita. Multi-gigahertz repetition rare passively mode locked fiber lasers using carbon nanotubes. Opt. Express, **19**(7) (2011): 6155–6163.
[204] L. Krainer, R. Paschotta, G. J. Spuhler, et al. 29 GHz mode locked miniature Nd:YVO_4 laser. Electron. Lett., **35**(14) (1999).
[205] L. Krainer, R. Paschotta, S. Lecomte, et al. Compact Nd:YVO_4 lasers with pulse repetition rates up to 160 GHz. IEEE J. Quantum Electron. **38**(10) (2002): 1331–1338.
[206] T. C. Schratwieser, C. G. Leburn and D. T. Reid. Opt. Lett., **37** (2012): 1133.
[207] B. R. Washburn, S. A. Diddams, N. R. Newbury, et al. Opt. Lett., **29**(3) (2004): 250–252.
[208] U. Keller. Recent developments in compact ultrafast lasers. Nature **424** (2003): 831.
[209] T. Südmeyer, C. Kränkel, C. R. E. Baer, et al. High-power ultrafast thin disk laser oscillators and their potential for sub-100-femtosecond pulse generation. Appl. Phys. B, **97** (2009): 281.
[210] P. Wasylczyk, P. Wnuk and C. Radzewicz. Passively modelocked, diode-pumped Yb:KYW femtosecond oscillator with 1 GHz repetition rate. Opt. Express. **17**(7) (2009): 5630–5635.
[211] Yang Hee-Won, C. Kim, S. Y. Choi, et al. 1.2-GHz repetition rate, diode-pumped femtoscecond Yb:KYW laser mode-locked by a carbon nanotube saturable absorber Mirror. Opr. Express. **20**(28) (2012): 29518–29523.
[212] M. Endo, A. Ozawa and Y. Kobayashi. Kerr-lens mode-locked Yb:KYW laser at 4.6 GHz repetition rate, Opt. Express. **20**(11) (2012): 12191–12197.
[213] J. V. Rudd, G. Konl, S. Kane, et al. Chirped-pulse amplification of 55-fs pulses at a 1 kHz repetition rate in a Ti:Al_2O_3 regenerative amplifier, Opt. Lett., **18**(23) (1993): 2044–2046.

[214] M. Lenzner, C. Spielmann, E. Wintner, et al. Sub-20-fs kilohertz-repetition-rare Ti:sapphire amplifier, Opt. Lett., **20**(12) (1995): 1397–1399.
[215] S. Backus, J. Peatross, C. P. Huang, et al. Ti:sapphire amplifier producing millijoule-level, 21-fs pulses at 1 kHz. Opt. Lett., **20**(19) (1995): 2000–2002.
[216] S. Santania, Z. Cheng, M. Lenzner, et al. Generation of 0.1-TW 5-fs optical pulses at a 1-kHz repetition rate, Opt. Lett., l997, **22**(20): 1562–1564.
[217] M. Hentschel, Z. Cheng, F. Krausz, et al. Generation of 0.1 TW optical pulses with a single-stage Ti:sapphire amplifier at a 1-kHz repetition rate, Appl. Phys. B, **70**(Suppl.) (2000): 161–164.
[218] E. Zeek, R. Bartels, M. M. Munanc, et al. Adaptive pulse compression for transform-limited 15 fs high-energy pulse generation. Opt. Lett., **25**(8) (2000): 587–589.
[219] Z. Cheng, F. Krausz and C. Spielmann. Compression of 2 mJ kilohertz laser pulses to 17.5 fs by pairing double-prism compressor: analysis and performance, Opt. Commun., **201** (2002): 145–155.
[220] S. Backus, C. G. Durfee III, G. Mourou, et al. 0.2-TW laser system at l kHz. Opt. Lett., **22**(16) (1977): 1256–1258.
[221] Y. Nabekawa, Y. Kuramoto, T. Togashi, et al, Generation of 0.66-TW pulses at l-kHz by a Ti:sapphire laser. Opt. Lett., **23**(17) (1998): 1384–1386.
[222] V. Bagnoud and F. Aalin. Amplifying laser pulses to the terawatt level at a 1-kHz repetition rate, Appl. Phys. B, **70**(Suppl) (2000): 165–170.
[223] J. Seres, A. Muller, E. Seres, et al. Sub-10-fs, terawatt-scale Ti:sapphire laser system. Opt. Lett., **28**(19) (2003): 1832–1834.
[224] S. W. Bahk, P. Rousseau, T. Planchon, et al. The generation and characterization of the highest laser intensity (1022 W/cm^2). Opt. Lett., **29** (2004): 2837.
[225] A. Sullivan, H. Hamster, H. C. Kapteyn, et al. Multiterawatt, 100-fs laser. Opt. Lett., **16** (1991): 1406.
[226] C. P. J. Barty, C. L. Gordon and B. E. Lemoff. Multiterawatt 30-fs Ti:Sapphire Laser System. Opt. Lett., **19** (1994): 1442.
[227] A. Sullivan, J. Bon lie, D. F. Price, et al. 1.1-J, 120-fs laser system based on Nd:glass pumped Ti:sapphire. Opt. Lett., **21** (1996): 603.
[228] C. P. J. Barty, T. Guo, C. L. Blanc, et al. Generation of 18-fs, multiterawatt pulses by regenerative pulse shaping and chirped-pulse amplification. Opt. Lett., **21** (1996): 668.
[229] J. P. Chambaret, C. L. Blanc, G. Cheriaux, et al. Generation of 25-TW, 32-fs pulses at 10 Hz. Opt Lett., **21** (1996): 1921.
[230] K. Yamakawa, M. Aoyama, S. Matsuoka, et al. Generation of 16-fs, 10-TW pulses at a 10-Hz repetition rate with efficient Ti:sapphire amplifiers. Opt. Lett., **23** (1998): 525.
[231] K. Yamakawa, M. Aoyama, S. Matsuoka, et al. 100-TW sub-20-fs Ti:sapphire laser system operating at a 10-Hz repetition rate. Opt. Lett., **23** (1998): 1468.
[232] J. D. Bonlie, F. Patterson, D. Price, et al. Production of > 1021 W/cm^2 from a large-aperture Ti:sapphire laser system. Appl. Phys. B, **70**(Suppl.) (2000): S155.
[233] M. Aoyama, K. Yamakawa, Y. Akahane, et al. 0. 85-PW, 33-fs Ti:sapphire laser. Opt. Let 1., **28** (2003): 1594.
[234] Z. Wang, W. Ling, P. Wang, et al. Amplification Ti:sapphire Laser to 355 TW with a Compact Design. TuB1-2, The 7th Pacific Rim Conference on Laser and Electro-Optics (2007).
[235] X. Liang, Y. Leng, C. Wang, et al. Development of Femtosecond Petawatt Laser Technology. TuBl-l (Invited), The 7th Pacific Rim Conference on Laser and Electro-Optics (2007).
[236] X. Zeng, X. Wei, X. Huang, et al. Experim ental research on amplification of large-aperture Ti:sapphire crystal. High Power Laser and Particle Beams **395** (2005): 17 (in Chinese).

[237] Y. Zhao, P. Wang, W. Zhang, et al. Generation of 7-fs laser pulse directly from a compact Ti:sapphire laser with chirped mirrors. Science in China Series G: Physics, Mechanics and Astronomy, **50** (2007): 261.

[238] P. Wang, Z. H. Wang, Z. Y. Wei, et al. Measurement of spectral phase of femtosecond laser pulse using SPIDER technique. Acta Phys. Sin. **53**(2004): 3004 (in Chinese).

[239] P. Wang, Z. Huan, Y. Y. Zhao, et al. Pulse width measurement of ultra-broad-bandwidth Ti:sapphire oscillator using SPIDER technique. Acta Phys. Sin. **56** (2007): 224 (in Chinese).

[240] M. D. Perry, D. Pennington, B. C. Stuart, et al. Petawatt laser pulses, Opt. Lett., **24** (1999): 160.

[241] Z. H. Wang, C. Liu, Z. W. Shen, et al. High-contrast 1.16 PW Ti:sapphire laser system combined with a doubled chirped-pulse amplification scheme and a femtosecond optical parametric Amplifier. Opt. Lett., **36** (2011): 3194.

[242] D. Umstadter. Review of physics and applications of relativistic plasmas driven by ultra-intense lasers. Phys. of Plasmas B, 2011, 1774.

[243] M. Nantel, J. Itatani, A. C. Tien. et al. Temporal contrast in Ti:sapphire lasers: Characterization and control. IEEE J. Select. Top. Quantum Electronics **4** (1998): 449.

[244] M. Kalshnikov, K. Osvay and W. Sandner. High-power Ti:Sapphire lasers: Temporal contrast and spectral narrowing. Laser and Particle Beams **25**, 2007, 219.

[245] P. Zhang, J. T. He, D. B. Chen, et al. Effects of a prepulse on γ-ray radiation produced by a femtosecond laser with only 5-mJ energy. Phys. Rev., E **57** (1998): R3746.

[246] T. Hosokai, Kenichi Kinoshita, Alexei Zhidkov, et al. Effect of a laser prepulse on a narrow-cone ejection of MeV electrons from a gas jet irradiated by an ultrashort laser pulse. Phys. Rev., E **67** (2003): 036407.

[247] M. Kaluza, J. Schreiber, M. I. K. Santala, et al. Influence of the laser prepulse on proton acceleration in thin-foil experiments. Phys. Rev. Lett., **93** (2004): 045003.

[248] M. H. Xu, Li YT, X. H. Yuan, et al. Effect of shock waves on spatial distribution of proton beams in ultrashort laser-foilinteractions. Phys. Plasmas **13** (2006): 104 507.

[249] S. P. D. Mangles, A. G. R. Thomas, M. C. Kaluza, et al. Effect of laser contrast ratio on electron beam stability in laser wavefield acceleration experiments. Plasma Phys. Controlled Fusion **48** (2006): 1383.

[250] F. Dorchics, T. Caillaudl, F. Blasco, et al. Investigation of laser-irradiated Ar cluster dynamics from K-shell x-ray emission measurements. Phys. Rev., E **71** (2005): 066410.

[251] K-H. Hong, B. Hou, J. A. Nees, et al. Generation and measurement of > 108 intensity contrast ratio in a relativistic kHz chirped-pulse amplified laser. Appl. Phys. B. **81** (2005): 447.

[252] J. Itatani, J. Faure, M. Nantel, et al. Suppression of the amplified spontaneous emission in chirped-pulse-amplification lasers by clean high-energy seed-pulse injection. Opt. Commun. **148** (1998): 70.

[253] A. Jullien, O. Albert, F. Burgy, et al. 10^{-10} temporal contrast for femtosecond ultraintense lasers by cross-polarized wave generation. Opt. Lett., **30** (2005): 920.

[254] L. P. Ramirez, D. N. Papadopoulos, A. Pellegrina, et al. Efficient cross-polarized wave generation for compact, energy-scalable, ultrashort laser sources. Opt. Express, **19** (2011): 93.

[255] A. Dubietis, G. Jonušauskas and A. Piskarskas. Powerful femtosecond pulse generation by chirped and stretched pulse parametric amplification in BBO crystal. Opt. Commun. **88** (1992): 437.

[256] C. Dorrer, I. A. Begishev, A. V. Okishev, et al. High-contrast optical-parametric amplifier as a front end of high-power laser systems. Opt. Lett., **32** (2007): 2143.

[257] C. Liu, Z. H. Wang, W. C. Li, et al. High temporal contrast in a Ti:sapphire CPA laser system with a non-collinear femtosecond optical-parametric amplifier. Opt. Lett., **35** (2010): 3096.

[258] C. Thaury, F. Quere, J. P. Geindre, et al. Plasma mirrors for ultrahigh-intensity optics. Nat. Phys., **3** (2007): 424.

[259] I. B. Földes, A. Barna, D. Csati, et al. Plasma mirror effect with a short-pulse KrF laser. The Sixth International Conference on Inertial Fusion Sciences and Applications **244** (2010):032004.
[260] H. Kiriyama, T. Shimomura, H. Sasao, et al. Temporal contrast enhancement of petawatt-class laser pulses. Opt. Lett., **37** (2012): 3363.
[261] S. Fourmaux, S. Payeur, S. Buffechoux, et al. Pedestal cleaning for high laser pulse contrast ratio with a 100 TW class laser system. Opt. Express, **19** (2011): 8486.
[262] A. Yogo, K. Kondo, M. Mori, et al. Insertable pulse cleaning module with a saturable absorber pair and a compensating amplifier for high-intensity ultrashort- pulse lasers. Opt. Express, **22** (2014): 2060.
[263] V. V. Ivanov, A. Maksimchuk and G. Mourou. Amplified spontaneous emission in a Ti:sapphire regenerative amplifier. Appl. Opt. **42** (2003): 7231.
[264] J. Itatani, J. Faure and M. Nantel. Suppression of the amplified spontaneous emission inchirped-pulse-amplification lasers by clean high-energy seed-pulse injection, Optics Communications, **148** (1998): 70–74.
[265] Y. Yanovsky, C. Felix and G. Mourou. Why ring regenerative amplification (regen)?, Appl. Phys. B, **74** (2002): 5181–5183.
[266] S. Kazamias, Kevin Cassou, David Ros, et al. Laser XUV haute cadence pompe par laser Titane:Saphir, vers la station LASERIX, J. Phys. IV France **138** (2006): 1319.
[267] M. P. Kalashnikov, E. Risse, H. Schnnagel, et al. Characterization of a nonlinear filter for the front-end of a high contrast double-CPA Ti:sapphire laser, Optics Express, **12**(21) (2004): 5088–5097.
[268] D. Homoelle, A. L. Gaeta, V. Yanovsky, et al. Pulse contrast enhancement of high-energy pulses by use of a gas filled hollow waveguide, Opt. Lett., **27**(18) (2002): 16 461648.
[269] A. Jullien, O. Albert, F. Burgy, et al. 10^{-10} temporal contrast for femtosecond ultraintense lasers by cross polarized wave generation, Opt. Lett., **30**(8) (2005): 920922.
[270] M. P. Kalashnikov, E. Risse, H. Schonnagel, et al. Double chirped pulse amplification laser: a way to clean pulses temporally, Opt. Lett., **30**(8) (2005): 923–925,
[271] F. Tavella, K. Schmid, N. Ishii, et al. High-dynamic range pulse-contrast measurements of a broadband optical parametric chirped pulse amplifier. Appl. Phys. B, **81** (2005): 753.
[272] C. Dorrer, A. Begishevl, A. V. Okishev, et al. High-contrast optical parametric amplifier as a frontend of high power laser systems. Opt. Lett., **32** (2007): 2143.
[273] Y. S. Huang, C. M. Zhang, Y. Xu, et al. Ultrashort pulse temporal contrast enhancement based on noncollinear optical parametric amplification. Opt. Lett., **36** (2011): 781.
[274] Y. J. Wang and Luther Davies. Optical parametric amplification based prepulse eliminator for a chirped pulse amplification Nd:glass laser. J. Opt. Soc. Am. B, **11** (1994): 153L.
[275] R. C. Shah, R. P. Johnson, T. Shimada, et al. High temporal contrast using low gain optical parametric amplification. Opt. Lett., **34** (2009): 2273.
[276] C. Rodel, M. Heyer, M. Behmke, et al. High repetition rate plasma mirror for temporal contrast enhancement of terawatt femtosecond laser pulses by three orders of magnitude. Appl. Phys. B, **103** (2011): 295.

Heping Zeng

2 Femtosecond optical frequency combs

2.1 Introduction

In the past decades, the development of optical frequency combs (OFCs) has led to a remarkable revolution in laser technology and significantly advanced research in precision spectroscopy, optical clocks, precision optical frequency metrology, and so on [1–8]. As an important link between radio frequency and optical frequency, this technique opens a door for optical clocks that can provide higher time accuracy and will be a promising tool for improving tests of the fundamental laws of physics. Such an accurate clock is also quite in demand for various interesting applications, such as remote navigation, signal synchronization, and global positioning systems (GPS). Besides optical clocks, many techniques that require high precision, high accuracy, and high stability can benefit from subtle development of precision OFCs [9–12]. For instance, for optical frequency measurements, OFCs function as precise optical rulers for direct and convenient measurements of unknown frequencies in the optical region. An unknown optical frequency can in principle be visualized by direct optical measurements as shown in Figure 2.1.

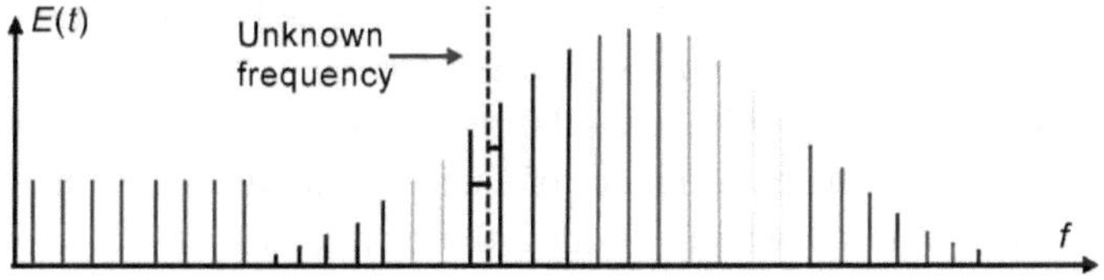

Fig. 2.1: Optical frequency comb in precision frequency measurements.

Basically, an OFC is a set of evenly distributed spectral lines produced by an ultrashort mode-locked laser. For a short timescale, a femtosecond mode-locked laser can be seen as a frequency comb. The space between two successive comb lines (in the frequency domain) is determined by the laser repetition rate, and the frequency comb lines can be extrapolated to the so-called offset frequency, which is actually related to the carrier envelope (CE) phase of the femtosecond laser. However, during long-term operation, these comb lines shift and fluctuate in the frequency domain. The frequency shift and fluctuation originate from variations of the laser itself due to unstable environmental factors like temperature, air pressure, and humidity. Therefore, in or-

Heping Zeng, State Key of Precision Spectroscopy, East China Normal University, Shanghai 20062, China

https://doi.org/10.1515/9783110304558-002

der to get a long-term stable frequency comb, its repetition rate and offset frequency need to be stabilized.

The fundamental conception of the frequency comb was initially developed in the 1970s. Professor T. W. Hänsch and his colleagues first implemented this concept in a pulsed dye laser with clearly narrowed linewidth for exciting the sodium 4d fine structure [13]. The first long-term self-referenced stabilized frequency comb was experimentally achieved on a Ti:sapphire mode-locked laser by Professor John Hall and his research group at JILA in Boulder, Colorado [14]. In fact, Professor T. W. Hänsch first proposed this self-referenced frequency comb project in 1997. But because of the lack of a suitable medium for generating octave spanning white light for the self-reference interference, this project was postponed. Finally, at the beginning of the 21st century, thanks to advances in fiber technology, especially microstructured fibers, long-term stabilized self-referenced combs became possible [15].

Following those breakthroughs of the self-referenced frequency comb, there has been tremendous progress in the knowledge and techniques in this research field during the last decade. In 2002, Ye et al. [16] achieved long-term phase stabilized ultrashort comb pulses with a measured phase noise of 720 mrad by using the self-referenced structure. In their experiment, the noise level was limited by extra phase noises induced by fluctuations of the photonic crystal fiber (PCF) and the self-reference interferometer. To avoid the extra noise from the interferometer, Fuji et al. [17] proposed a collinear difference frequency generation structure in 2005 for detecting CE offset frequency, which largely simplified the self-reference interferometer and reduced unnecessary noises from the phase detection part. As a result, they improved the stability of their laser comb with a reduced CE phase noise of 130 mrad, corresponding to a timing jitter of 62 attoseconds (as). With a similar technique, T. W. Hänsch and his collaborators further improved the performance of a laser comb in 2006, achieving a record CE phase noise of 99 mrad with an ultralow timing jitter of 44 as [18]. For further improvement of the long-term stability and output power upscaling, the frequency comb technique was gradually shifted from solid-state laser systems to fiber lasers. In 2003, Washburn et al. at the National Institute for Standards and Technology (NIST) in Boulder, Colorado, built the first fiber-based self-referenced frequency comb on an erbium-doped fiber mode-locked laser [19].

In this chapter, we present some basic ideas of optical frequency combs and their applications. After this brief background introduction, we first introduce the basic principles of optical frequency combs in Section 2.2. As an important concept for the frequency comb, the CE phase and its measurement, particularly based on self-referenced f-2f interferometers, is introduced in Section 2.3. The impacts on the CE phase of a frequency comb, including pump power, dispersion, and nonlinear polarization rotation, are discussed in Section 2.4. The main part of this review chapter is given in Section 2.5, where we review three different methods to stabilize the CE phase of laser combs. In the last part, Section 2.6, we briefly give some application examples for the frequency combs, including optical clock developments, optical frequency measurements, and accurate laser ranging.

2.2 Basis of optical frequency combs

For a pulse train produced by a femotosecond mode-locked laser, there is a tight relationship between its characteristics in the time domain and in the frequency domain. A train of these ultrafast laser pulses in the time domain turns into a set of equally spaced spectral lines in the frequency domain as a result of the Fourier transform (as shown in Figure 2.2). The pulse train has a vivid metaphor, the optical frequency comb, because of its evenly spaced spectral lines. The space between two successive comb lines is the laser repetition rate, f_r, and the comb can be mathematically extrapolated to the first line closest to zero, which is defined as its offset frequency, f_0, which is determined by the CE phase of the pulse train.

The basic concept of the CE phase is illustrated in Figure 2.2. The time relationship of the electric field with respect to the light pulse can be expressed as a rapidly oscillating sinusoidal function (often called the carrier signal) multiplied by a relatively slow-varying envelope function. When the laser pulse propagates in a medium, the time relationship between the carrier signal and the envelope signal changes. This change is mainly caused by the inconsistent phase and group velocities in the dispersive medium. In other words, the pulse envelope travels with the group velocity, but the carrier travels with the phase velocity. Additionally, nonlinear phase shifts induced by fluctuations of laser intensity may also make carrier-envelope changes.

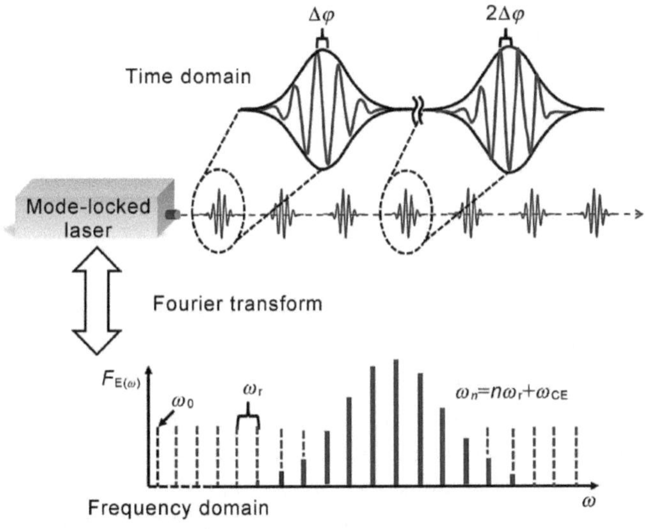

Fig. 2.2: Mode-locked laser pulse train in time and frequency domains.

In order to characterize the behavior of this change, the concept of carrier-envelope phase is defined as the shift of the vertex positions of the carrier signal and the envelope. Due to the difference between the group velocity (v_g) and phase velocity (v_p) of the pulses in the laser cavity, the phase shift per pulse ($\Delta\varphi$) will be produced as:

$$\Delta\varphi = (1/\upsilon_g - 1/\upsilon_p)\omega_c L\,, \tag{2.1}$$

where $\upsilon_g = c/n_g$ and $\upsilon_p = c/n$ are the pulse group velocity and the carrier phase velocity respectively. ω_c is the laser center frequency, n_g is the refractive index for the group velocity, n is the average refractive index, and L is the cavity length corresponding to the period of pulses of $T_r = Ln/c$ and a repetition rate of $f_r = 1/T_r$.

The pulse train can be mathematically expressed in the time domain as

$$\tilde{E}(t) = \tilde{A}(t) \otimes \sum_{n=-\infty}^{+\infty} \delta(t - n/f_r)\exp(-in\Delta\varphi)\,, \tag{2.2}$$

where n represents an arbitrary integer.

The carrier-envelope amplitude is given by

$$\tilde{A}(t) = A_0(t)\exp\left[i(\omega_c t + \varphi_0)\right]\,, \tag{2.3}$$

where φ_0 is the initial phase.

By simply applying a Fourier transform, equation (2.2) becomes

$$F\left[\tilde{E}(t)\right] = F\left\{A_0(t)\exp\left[i(\omega_c t + \varphi_0)\right]\right\} \times F\left\{\delta(t - n/f_r)\exp(-in\Delta\varphi)\right\}\,, \tag{2.4}$$

from which one can understand the pulse train in the frequency domain as

$$E(\omega) = e^{i\varphi_0} A_0(\omega - \omega_c) \sum_{n=-\infty}^{+\infty} \delta\left[\omega - 2\pi(nf_r + f_0)\right]\,, \tag{2.5}$$

where $A_0(\omega - \omega_c)$ is the carrier signal of the pulse train in the frequency domain. The offset frequency in this expression is given by

$$f_0 = \frac{\Delta\varphi}{2\pi} f_r\,. \tag{2.6}$$

For each comb line, its frequency would be $\omega_n = 2\pi v_n = 2\pi(Nf_r + f_0)$.

If the repetition rate f_r and the CE phase or offset frequency f_0 of a laser are both stabilized, all the comb lines of the laser become stable. In practice, a fully stabilized frequency comb can be readily controlled to have a frequency linewidth of less than 1 Hz for each comb line, which may outperform a standard frequency-stabilized continuous-wave (cw) laser. In comparison to a single frequency CW laser, a frequency comb usually provides on the order of millions of comb lines in a single laser cavity, which greatly facilitate applications of lasers in the field of high precision laser spectroscopy and rapid optical imaging.

2.3 Measurements of carrier-envelope phase

2.3.1 Self-referenced f-2f interferometer

As mentioned before, the n-th teeth of a laser comb in a frequency domain can be described as $f_n = nf_r + f_0$, where f_r is the laser repetition rate and f_0 is the offset frequency, also known as the carrier-envelope offset frequency (CEO). Therefore, the frequency stabilization of a laser comb needs to detect and stabilize two signals, f_r and f_0. For detecting the f_r, one can easily use a fast photodiode to obtain the laser repetition rate. However, since it is impossible for any photodetectors to directly sense the phase of optical light, the detection of f_0 becomes more difficult and complicated than that of f_r.

One solution for detecting the CE offset frequency of a laser comb is self-referenced f-to-2f interferometry. Its working principle is illustrated in Figure 2.3. The spectrum of a laser comb is extended to an octave spectral range that includes the low-frequency and high-frequency components $f_n = nf_r + f_0$ and $f_{2n} = 2nf_r + f_0$ respectively. To extract the CEO f_0, one needs to double the frequency of the low-frequency component to generate a new light at optical frequencies $2f_n = 2(nf_r + f_0)$, and then use a photodetector to obtain the beat note (f_{beat}) between the high-frequency components and frequency-doubled low-frequency components. As a result, the beat signal represents the CE offset frequency according to

$$f_{beat} = 2f_n - f_{2n} = 2(nf_r + f_0) - (2nf_r + f_0) = f_0 \,. \tag{2.7}$$

Generally speaking, there are two experimental schemes for self-referenced f-to-2f interferometer, collinear and noncollinear schemes.

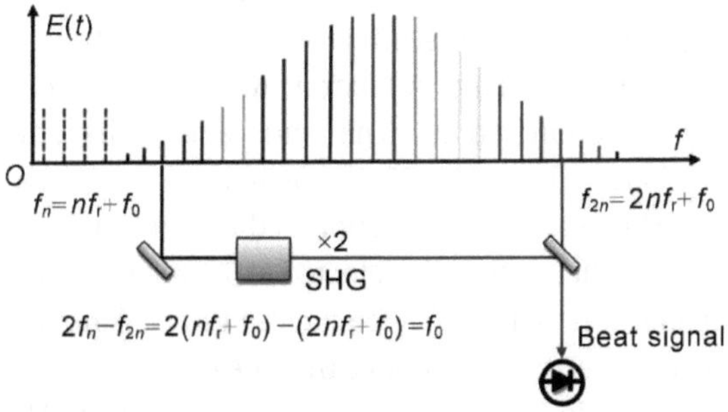

Fig. 2.3: Schematic of self-referenced f-2f interferometer.

2.3.1.1 Noncollinear f-to-2f interferometer

The setup of a typical noncollinear f-to-2f interferometer is shown in Figure 2.4. The laser pulses are delivered from a mode-locked laser and then are injected into a non-

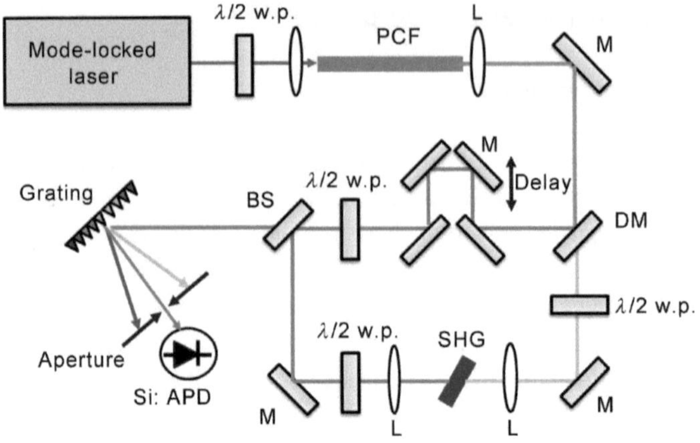

Fig. 2.4: Noncollinear f-to-2f interferometer; PCF: photonic crystal fiber, DM: dichromatic mirror, SHG: sum-frequency generation, BS: beam splitter, Si:APD: avalanched photodiode, and L: Lens.

linear optical fiber (usually photocrystal fiber, or PCF) to generate an octave spectrum. After being spectrally broadened, the laser beam contains short-wavelength and long-wavelength spectral components, which are spectrally separated by a dichroic mirror. The low-wavelength light goes through a nonlinear optical crystal to generate its second harmonic light. The long-wavelength beam passes a time delay line and then combines with the second harmonic light using a beam splitter. By tuning the delay line, one can make the two pulse beams from different interferometer arms temporally overlap. Once the pulse beams temporally and spatially overlap, their beat note signal can be detected with a photodetector.

Figure 2.4 also shows an example of how the noncollinear f-to-2f interferometer works. The laser pulse is generated from a Kerr mode-locked Ti:S laser with the center wavelength at 800 nm. After passing through a short section of PCF, its spectrum is widely broadened from 500 nm to more than 1100 nm. A dichroic mirror is chosen to reflect the low-wavelength part of the beam centered at 515 nm and transmit the long-wavelength part at 1030 nm. Then a β-barium borate (β-BBO) crystal is used to generate the second harmonic of the 1030-nm beam. At the output port of the interferometer, the two spatially overlapped beams are further spectrally selected at 515 nm by a combination of a grating and a small pin hole and then are detected by a Si-APD. Watching the output of the APD with a spectral analyzer, one can notice that the beat signals are symmetrically distributed around the repetition frequency of the laser pulse train. The beat note frequencies can be described successively as f_0, $f_r - f_0$, $2f_r - f_0$, $2f_r + f_0$, and so on.

The noncollinear f-to-2f interferometer offers an easy way to control the time overlap of pulses from its two arms. Meanwhile, the pulse delay between the two beating light pulses can also be compensated for by controlling dispersion with gratings or prisms. In this case, the whole CEO detection system is simplified to a certain extent by using a collinear interferometer structure.

2.3.1.2 Collinear f-to-2f interferometer

In a collinear f-to-2f interferometer, the long-wavelength and short-wavelength light propagate in the same optical arm. The time delay between the two lights is mainly caused by the dispersion of the propagation medium, such as air or fibers. In order to compensate for the delay time, dispersion management components like gratings and prisms can be used. After the time delay compensation, the collinear lights go through a nonlinear optical crystal to generate frequency-doubled light for the long-wavelength part. Then the lights are delivered into a photodetector after a band-pass filter. Such a collinear interferometer is usually used for the CEO detection of broadband femtosecond Ti:S lasers [20–24]. For example, in [24], the femtosecond Ti:S laser has a broadband output spectrum from 570 to 1280 nm, broad enough for CEO detection. As shown in Figure 2.5, the laser beam is compressed by a prism pair and then focused on a β-BBO crystal for the second harmonic generation of the spectral fraction at 1160 nm. In order to improve the signal-to-noise ratio of the beat note, a photomultiplier tube (PMT) is used right after an optical band-pass filter. The beat note is generated at a wavelength of 580 nm.

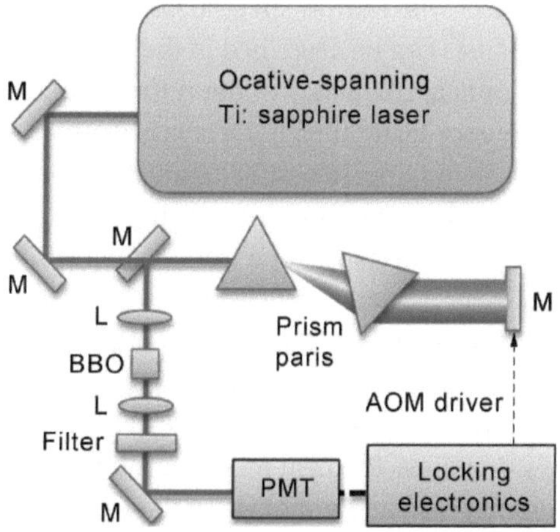

Fig. 2.5: Collinear f-to-2f interferometer for Ti:S laser.

With the rapid development of fiber techniques such as fiber lasers, amplifiers, nonlinear fibers, and so on, the implementation of f-to-2f interferometers for CEO detection has become more convenient [20]. As shown in Figure 2.6, the output of a mode-locked fiber laser is further amplified by a fiber amplifier to gain enough power for spectral broadening in nonlinear fibers. The amplified pulses are spectrally extended to an octave span with a small piece of high-nonlinearity fiber (HNLF). The connections be-

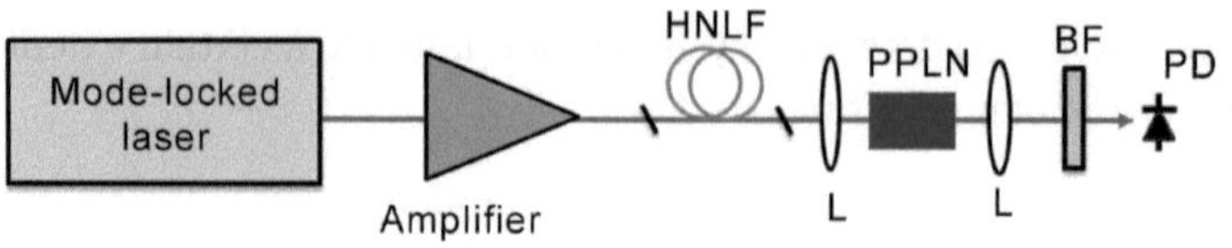

Fig. 2.6: Collinear f-to-2f interferometer for fiber. HNLF: highly nonlinear fiber, L: lens, BF: band-pass filter, PD: photodetector, PLL: phase-lock loop, PPLN: periodically poled lithium niobate.

tween the fiber laser, the fiber amplifier, and the HNLF can be easily achieved with a commercialized fiber fusion splicer, which makes the whole system more compact and easy for operation. Meanwhile, for most femtosecond erbium-doped fiber lasers, the octave span generated by the HNLF can cover a large spectral range from 900 nm to 2200 nm. In this case, nonlinear optical crystals like periodically poled lithium niobate (PPLN) are more favorable for frequency doubling on the long-wavelength side because of the high optical nonlinearity and quasi phase matching of PPLN [20].

2.3.2 Beat note detection between a CW laser and a laser comb

In fact, there is a straightforward way to measure the optical phase noise as well as the frequency stability of a laser comb, which is based on the beat signal of the comb and a stable single longitudinal mode laser (a CW laser with narrow spectral linewidth). As we know, for the laser comb, each of its comb lines can be described in the frequency domain as $f_n + \Delta f_n$, where f_n is the absolute optical frequency of the n-th comb line and Δf_n represents its frequency noise or frequency instability. If the comb beats with a CW laser that works at a frequency of f_{cw} close to f_n, its lowest beat note frequency will be $f_{\mathrm{beat}} = f_n + \Delta f_n - (f_{\mathrm{cw}} + \Delta f_{\mathrm{cw}})$, where Δf_{cw} is the linewidth of the CW laser. Here the frequency noise of the beat note includes two parts, one from the contribution of the comb and the other from the CW laser. Usually, the frequency of the CW laser is stabilized within a linewidth (about kHz or even less) far narrower than that of a free-running comb. Therefore in this case the noise of the beat note (Δf_{beat}) can be used to represent the comb noise. Figure 2.7 shows the beat notes of a CW laser and an optical comb in both the time and frequency domains. The beat notes distribute symmetrically around the comb repetition frequency and its harmonics in the frequency domain. The values of the beat notes are determined by the frequency differences between the CW laser and the comb lines. The beat note linewidth is defined by the convolution of the linewidths of the CW laser and the optical comb. By stabilizing this beat signal, the laser comb can also become stable, which means the laser comb can become relatively stabilized to the CW laser.

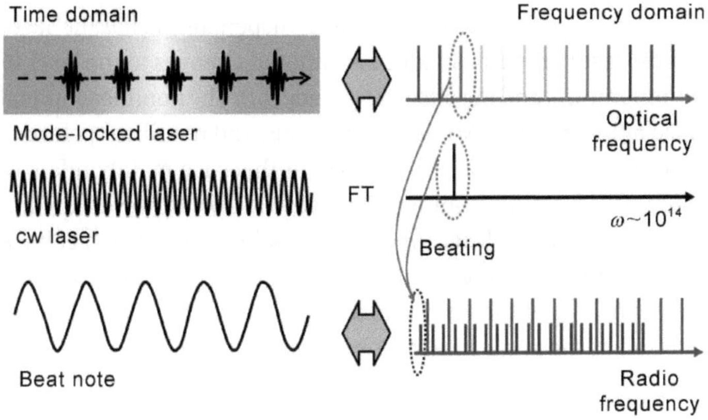

Fig. 2.7: Beat note detection between a CW laser and a laser comb.

2.4 Impacts on the carrier-envelope phase

For a mode-locked laser, the laser pump power (P) and the cavity length (L) are two important factors for controlling the laser optical frequency and suppressing its phase noise. Figure 2.8 shows the impacts of the pump power and the cavity length on the offset frequency (f_0) and repetition rate of the laser (f_r). As illustrated in Figure 2.8, the pump power plays a crucial role in modulating the laser CE phase and the offset frequency. Changes in the pump power can directly influence the gain in the laser, which results in variations of laser properties like pulse peak power, intracavity and output spectra, single pulse energy, and so on.

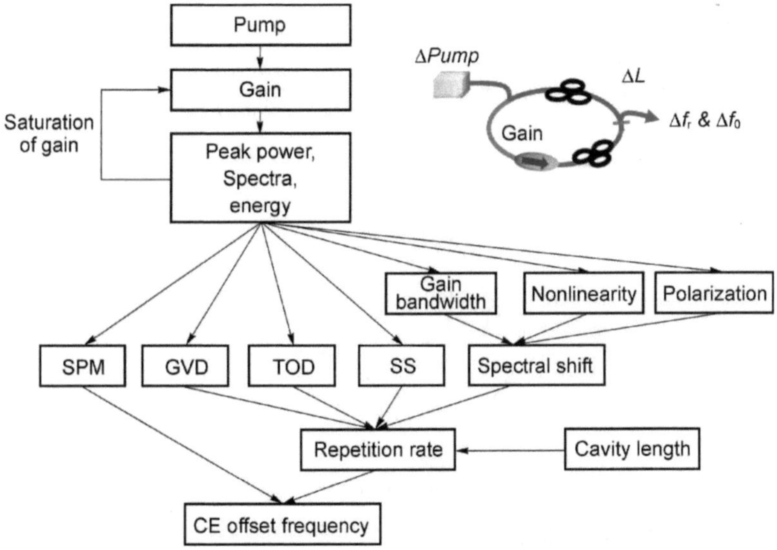

Fig. 2.8: Dependence of CE offset frequency on pump power.

Due to a balance between nonlinear optical effects and dispersion inside the laser cavity, the variations of the laser properties induce additional changes to the laser in terms of the laser repetition rate (f_r) and the offset frequency (f_0). The nonlinear optical effects include self-phase modulation (SPM), self-steepening, and nonlinear polarization rotation for fiber lasers, and so on. The dispersion in this case mainly refers to group velocity dispersion (GVD) and third-order dispersion (TOD). However, the impact of pump power on the gain of the laser system is not a linear positive feedback mode, but in the form of nonlinear negative feedback mechanism (so-called gain saturation), so that the mode locking of the laser is in a stable state, otherwise the laser is unable to be mode-locked or the mode locking is extremely unstable.

2.4.1 Dependence of CE phase on pump power

As we know, the relationship between the CE phase (or offset frequency) and the laser repetition rate can be written as $f_0 = \frac{\Delta\varphi_0}{2\pi} f_r$, where $\Delta\varphi_0$ is the relative phase shift between two successive pulses. In fact, $\Delta\varphi_0$ varies with time t and therefore is labeled as φ_{CE} to distinguish with its time variable $\Delta\varphi_{CE}$ in the following part.

As shown in Figure 2.9, $\Delta\varphi_{CE}$ is influenced not only by the change of pulse periods $\Delta T_r (\Delta\varphi_{CE} = -\omega_c \Delta T_r)$, but also by the carrier light phase shift $\Delta\varphi_{PS} (\Delta\varphi_{CE} = \Delta\varphi_{PS})$ where ω_c is the light angular frequency. Therefore, the dependence of the offset frequency on pump power can be expressed as [25]

$$\frac{df_0}{dP} = \frac{\beta_0}{2\pi}\left(\frac{df_r}{dP}\right) + \frac{f_r}{2\pi}\left(\frac{d\varphi_{PS}}{dP}\right), \tag{2.8}$$

where $\beta_0 = nL\omega_c/c$ is the average fiber propagation constant, L is the laser of length and n is the average refraction index of the fiber cavity. Meanwhile, the repetition rate of the laser f_r can be described as [26]

$$f_r = \frac{1}{T_r} = \left(\beta_1 + \Delta\omega\beta_2 + \frac{1}{2}\omega_{rms}^2\beta_3 + \frac{g}{\Omega_g} + \frac{\mu A^2\delta}{\omega_0}\right)^{-1}, \tag{2.9}$$

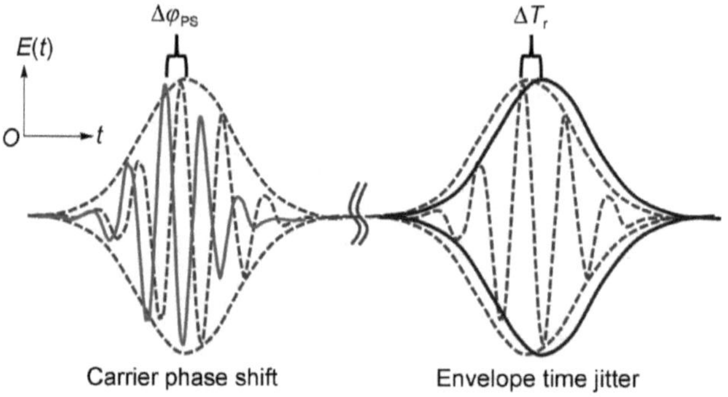

Fig. 2.9: Carrier-phase envelope evolution.

where $\beta_n = d^n\beta/d\omega^n|_{\omega_0}$ are the n-th order frequency derivatives of the linear fiber propagation constant calculated at the gain peak, ω_0. The first term on the right side is the influence from the group velocity inside the cavity. The second term represents the spectral shifts $\Delta\omega = \omega_c - \omega_0$ of the carrier away from the center wavelength ω_0. The third term gives the contribution from third-order dispersion, where ω_{rms} is the root-mean-square pulse spectral width, which is supposed to be $\omega_{rms} \gg \Delta\omega$. The fourth term contains the contribution of the gain assuming a Lorentzian gain shape with a peak value of g and bandwidth Ω_g. The final term is deduced from nonlinear self-steepening effects in fibers, where A^2 is the peak intensity and $\mu = 1.3$ is a minor numerical correction related to the pulse shape [25].

In fact, the parameters β_n in equation (2.9) are independent from pulse energy and pump power, while the other parameters like ω_{rms}, g and A^2 are closely related to the laser pulse energy. The dependence of these parameters on laser pulse energy is determined by dispersion inside the laser cavity. For instance, as a soliton laser with negative intracavity dispersion, these parameters are linearly dependent on pulse energy w [26]. However, the dependence becomes complicated for most stretched-pulse lasers in which the intracavity dispersion contains normal and abnormal parts. According to [25], the relationships among ω_{rms}, w and A^2 can be expressed as

$$\omega_{rms} \propto w^{\frac{1}{2}},\ A^2 \propto w\omega_{rms} = w^{\frac{3}{4}}\ ,$$

and the pulse energy w is dependent on pump power (P) as

$$\frac{dw}{dP} = (1+\eta)^{-1}\left(\frac{w}{P}\right) \approx \frac{w}{P}\,, \quad \text{and} \quad \eta \ll 1\,.$$

Hence the derivatives of ω_{rms}^2 and A^2 with respect to P can be written as

$$\frac{d\omega_{rms}^2}{dP} \approx \frac{\omega_{rms}^2}{P}\,,$$
$$\frac{dA^2}{dP} \approx \frac{3A^2}{2P}\,.$$

If considering $dg/dP = V_{BW}/V_g \times (2P)^{-1}$, one can easily get the dependence of f_r on P by taking a derivative of f_r with respect to P using equation (2.9), yielding [25]

$$\frac{df_r}{dP} = -f_r^2\left(\beta_2\frac{d\Delta\omega}{dP} + \frac{\omega_{rms}^2\beta_3}{2P} + \frac{V_{BW}}{2PV_g\Omega_g} + \frac{3\mu A^2\delta}{2P\omega_0}\right)\,. \tag{2.10}$$

By replacing df_r/dP in equation (2.8) with equation (2.10), one can get the specific relationship between the CE offset frequency and the pump power.

2.4.2 Polarization dependence of CE phase

Because of the nonlinear polarization rotation (NPR) effect inside a fiber laser cavity, the laser pulse properties, including its spectra, pulse width, repetition rate, and CE offset, frequency are closely related to the laser polarization states [27]. It is especially crucial for NPR mode-locked fiber lasers. The polarization-sensitive components inside these fiber lasers, like polarization-sensitive isolators and birefringence single-mode fibers (SMF), make the polarization dependence of fiber lasers even stronger. In fact, the relationship between the polarization and the CE offset frequency is rather understandable for a fiber laser. The influence of the polarization change is directly imposed on the pulse energy through polarization-related gain and loss competition in the laser cavity. Meanwhile, followed by the change of pulse energy, a spectral shift is induced by the SPM effect and further leads to a shift of the CE offset frequency due to dispersion inside the cavity. [27] illustrates the influence of the light polarization state on the CE offset frequency (f_0) of an ytterbium-doped fiber laser. The fiber laser with a typical stretched-pulse ring cavity produces femtosecond pulses at a center wavelength around 1040 nm. In order to control the polarization state, an electrical polarization controller (EPC) is placed inside the fiber cavity, as shown in Figure 2.10 (a). Changing the voltage imposed on the EPC is equivalent to changing the polarization of the light passing through the EPC. As a result, the offset frequency of the fiber laser is studied by adjusting the EPC voltage, as shown in Figure 2.10 (b). Within a certain region, it is observed that the CE offset frequency has an approximately linear dependence on the EPC voltage. Linear fit of the experimental data gives the slope of the CEO change with the EPC voltage as 2.5 MHz/mV. Compared with the pump-controlled CE phase stabilization schemes, this actually provides an alternative way to control the CE offset frequency without inducing detrimental effects of output power variation by directly changing the pump power.

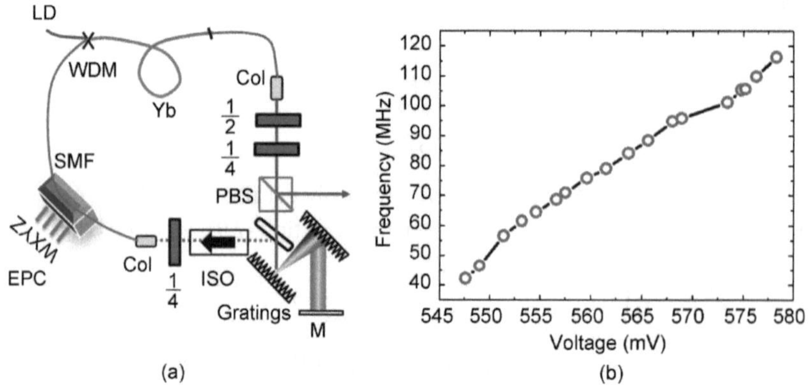

Fig. 2.10: (a) Experimental setup, (b) polarization dependence of f_0.

2.4.3 Impact of intracavity dispersion

Intracavity dispersion is an important factor for generating ultrashort laser pulses and is also fundamental in determining their repetition rate and CE phase. According to equations (1) and (6), the CE offset frequency (f_0) can be expressed as

$$f_0 = \frac{\Delta\varphi}{2\pi} f_r = \frac{f_r}{2\pi}(1/\upsilon_g - 1/\upsilon_p)\omega_c L \,, \tag{2.11}$$

where υ_g is group dispersion velocity, υ_p is carrier phase velocity, and L is the cavity length. Since there is an inverse relationship between the repetition rate and the cavity length, the influences of the two factors on f_0 are cancelled out. Therefore, the offset frequency is mainly dependent on center wavelength and the reciprocal difference between υ_p and υ_g. Many experiments have confirmed that the net intracavity dispersion plays an important role in determining the CE offset frequency. As for a stretched-pulse fiber laser, the frequency shift of f_0 induced by a dispersion change as small as 0.01 ps^2 could be as large as 10 MHz or even more. Meanwhile the f_0 linewidth (Δf_0) is also found to be largely related to intracavity dispersion [28].

In order to investigate the dependence of the f_0 linewidth on intracavity dispersion, a stretched-pulse ytterbium-doped fiber laser is used (as shown in Figure 2.10 (a)). By carefully adjusting the distance between the two intracavity gratings, the net dispersion can be tuned successively. The offset frequency of the fiber laser is monitored by a digital spectral analyzer when the net dispersion is changed. The typical experimental results are shown in Figure 2.11. It is interesting to note that the linewidth of f_0 reaches its lowest value of 100 kHz with zero dispersion inside the fiber cavity. Meanwhile, the linewidth increases in both normal and abnormal dispersion regions, and the linewidth is broadened more seriously in the abnormal dispersion region than in the normal dispersion region. This implies that the laser

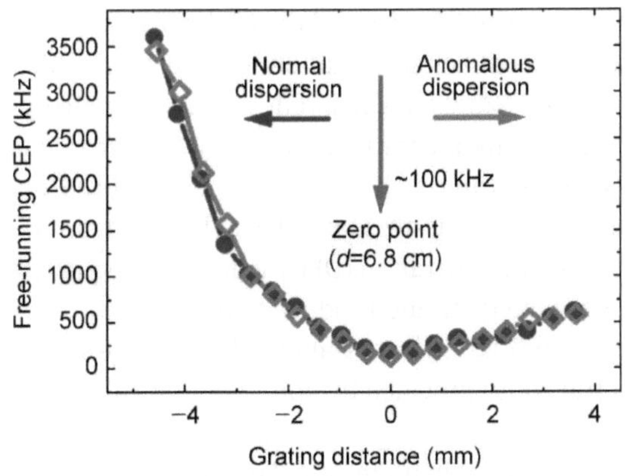

Fig. 2.11: Intracavity dispersion and offset frequency.

can be optimized to get its narrowest f_0 linewidth with lowest phase noise by setting the intracavity dispersion close to zero.

2.5 Optical frequency combs and phase noise suppression

Generally speaking, perturbations from the environment such as air flow disturbance, temperature changes, mechanical or optical platform vibrations, etc., unavoidably give rise to the instability of the mode-locked laser. As a result, the laser pulses exhibit variations in both repetition frequency and the carrier-envelope (CE) phase, which are respectively known as timing jitter and phase noise [26]. For a free-running mode-locked laser, its phase noise contributes to the linewidth broadening of the offset frequency (f_0), which could be readily monitored and actively controlled. Meanwhile, some slowly-varying environmental factors such as temperature are determinative to f_0, resulting in an observable frequency drift to f_0. As monitored with a spectral analyzer, one can see the drifting of the center of f_0. The drifting range varies for different laser systems from less than 1 MHz to more than tens of MHz. In fact, the linewidth broadening and the frequency drifting of f_0 are basically the same thing. Both are caused by kinds of perturbations such as phase noises. The difference between them is the frequency of the perturbations. The slow f_0 drifting represents a kind of noise distributed in a very low radio frequency region, while the linewidth broadening is 'faster' and indicates a kind of noise located in a high frequency region. Therefore, usually when we talk about stabilizing f_0 to achieve effective suppression of phase noises, it means preventing both the frequency drifting and the linewidth broadening in the f_0.

In this section, we will introduce methods for suppressing phase noises of ultrashort pulses, which to a certain extent can be specified as the stabilization of their CE offset frequency f_0. Currently, there are three solutions for f_0 stabilization. The most popular one is using electric servo systems to actively control the f_0 signal against environmental perturbations. This is called a feedback stabilization scheme. Meanwhile, there is also a feed-forward scheme for stabilizing f_0, which relies on an acoustic-optical modulator (AOM) to compensate for the frequency variation of f_0. Both methods are active approaches and dependent on electronic devices. To avoid limitations from the bandwidth of electronics, one can also use a passive method to achieve phase-stabilized lasers, by using a difference frequency generation (DFG) scheme to frequency-mix two spectral components from the same laser and produce a third light with exactly zero offset frequency. However, the DFG-based laser stabilization scheme suffers from $1/f$ noise and the restricted spectral width provided by nonlinear optical materials.

An OFC also needs accurate control of the repetition rate. For precision control, the CE phase stabilization may be affected by the cavity length feedback control in practical fiber lasers. This section will address some useful techniques for the simultaneous stabilization of repetition rate f_r and the offset frequency f_0, as required by the generation of ultrashort frequency combs.

2.5.1 CE phase stabilization based on feedback loops

A phase-locked loop (PLL) was originally used as an electronic control technology that was widely employed in radio, telecommunications, computers, and other electronic applications. The aim of PLL is to maintain a fixed phase relationship between an output signal and an input signal. As illustrated in Figure 2.12, a PLL consists of four parts: the phase detector (PFD), low pass filter (LPF), voltage-controlled oscillator (VCO), and frequency divider (DIV). A periodic signal is generated by a frequency oscillator and is sent into the PLL. The phase detector compares the phase of that signal with the phase of a reference signal and adjusts the VCO to keep the phases matched. Meanwhile, a small fraction of the output of the PLL is used as a reference signal. This forms a feedback loop, which brings the output signal back to the input signal for comparison. The input and output phase locking implies that the input and output are locked at the same frequency. In addition, the divider is used to make the PLL generate a frequency that is a multiple of the input frequency. Therefore, in addition to synchronizing signals, a PLL can track an input frequency or even multiply the input frequency, which makes it more popular in demodulation and frequency synthesis.

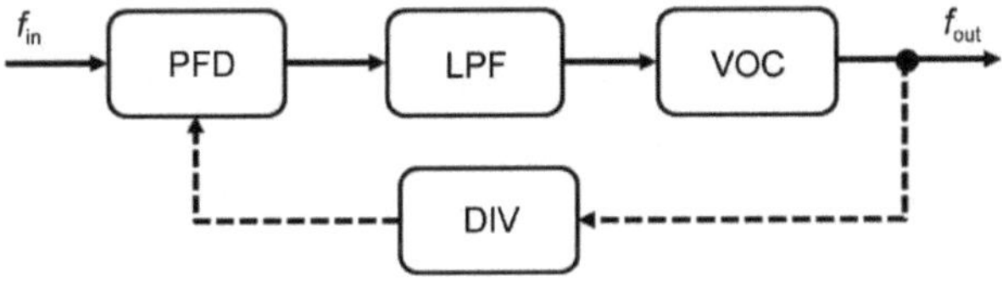

Fig. 2.12: Basic structure of a phase-locking loop.

Based on the principle of PLL, a feedback lock loop was developed for controlling the laser phases. The mechanism of the feedback phase lock loop is similar to that of the electronic PLL. Here, a mode-locked laser itself is the oscillator just like the VCO in the PLL, which generates ultrashort pulses as optical signals. A CE phase detection system is used to obtain the phase information of the laser, which is compared with a reference signal in a phase comparator. After an electronic low-pass filter, the error signal of the phase comparator is fed back to control the laser pump power and finally to modulate the laser output phase. The whole scheme is shown in Figure 2.13. The reference signal usually comes from an RF generator synthesized by an atomic clock so that the frequency accuracy can be transferred from the atomic clock to the pulsed laser through the feedback lock loop.

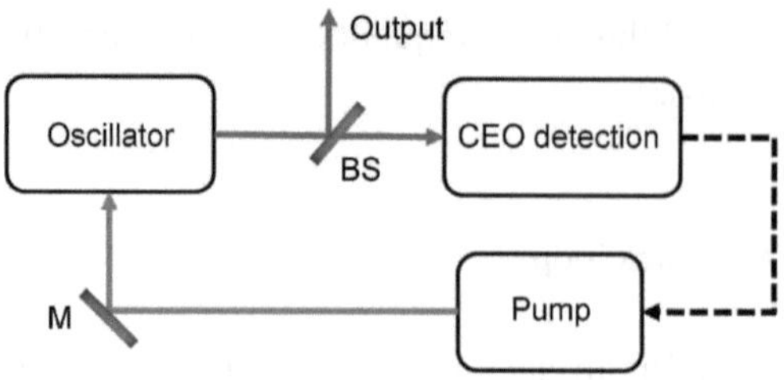

Fig. 2.13: Structure of a feedback controlled comb; M: mirror, BS: beam splitter.

A common way to control the laser pump power as well as the laser CE phase is to adjust the pump current. However, this is challenged by its electronic bandwidth, which is about 10 to 100 kHz. Clearly, for a feedback loop, such a narrow bandwidth will limit the feedback speed and increase the whole response time. This will become a major drawback when the free-running CE phase signal has a broadband noise. This problem can be solved by setting an acousto-optic modulator (AOM) between the pump and the laser oscillator. By adjusting the drive frequency of AOM, the pump power through the first order output port of AOM can be modulated more effectively with a response bandwidth of 1 MHz or even more. In what follows, we give an experimental example of CE phase stabilization with the AOM-based feedback lock loop to further illustrate details of such a controlling scheme.

2.5.1.1 Experimental example I:

The purpose of this experiment is to generate CE-phase-stabilized femtosecond laser pulses. The ultrashort pulses are delivered by a mode-locked Ti:sapphire laser (FEMTOSOURCE rainbow). The center wavelength of the output is at 780 nm with a spectral range from 680 to 900 nm. The pulse width can be further compressed to 7 fs by using chirp mirrors with an output power of 200 mW under a pump power of 3.5 W (Coherent Verdi).

CE offset frequency detection. A collinear DFG-based f-to-2f interferometer is employed in the experiment to obtain a low-noise CE offset beat signal. The details are illustrated in Figure 2.14. The compressed laser pulses are focused into a highly nonlinear periodic poled lithium niobate crystal (PP-MgO:LN). Because of the large optical nonlinearity of this PPLN crystal, the nonlinear SPM and DFG processes can occur in the same crystal. Figure 2.15 shows how the two processes work together to generate the CE offset beat note. Assume that there are two comb lines of the laser labeled m and l. Their frequencies can be expressed as $mf_r + f_0$ and $lf_r + f_0$ ($l < m$). As a result of the difference in frequency generation between the two comb lines, a new component is produced as $(mf_r + f_0) - (lf_r + f_0) = (m - l)f_r = nf_r$ ($m - l = n$). At the same time, the initial laser spectrum can be largely broadened due to the SPM effect that generates

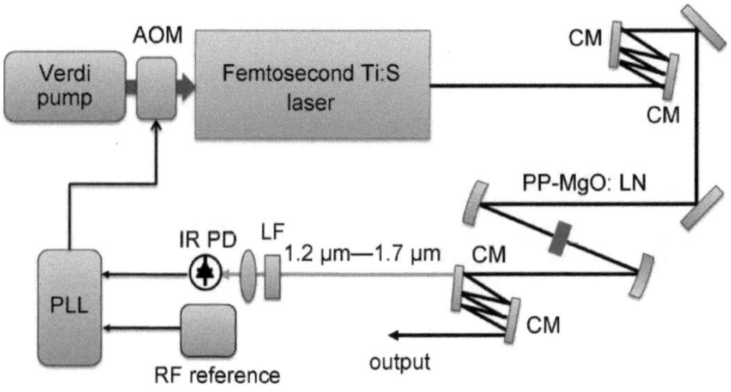

Fig. 2.14: Experimental setup; AOM: acoustic-optical modulator, CM: chirped mirror, M: high-reflection mirror, LF: long wave pass filter.

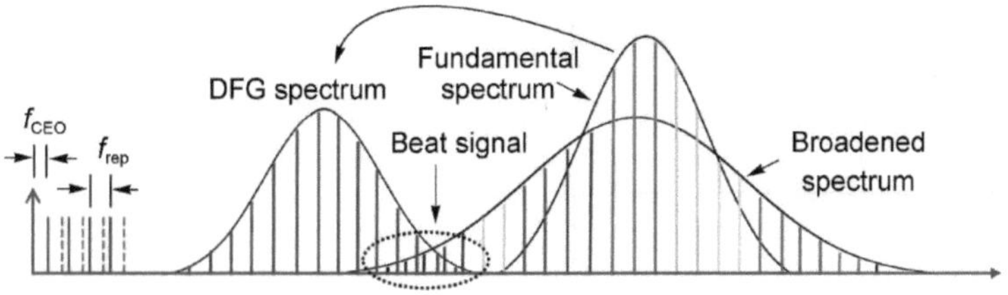

Fig. 2.15: Nonlinear difference frequency generation for CE phase detection.

a spectral overlap with the DFG beam. Correspondingly, in the overlapped region a comb line of the spectrally broadened beam can be recognized as $nf_r + f_0$. In this case, a beat note between the DFG beam and the SPM-broadened beam will be detected with a photo-detector at a radio frequency of $(nf_r + f_0) - nf_r = f_0$, which gives the CE offset frequency of the laser.

In the beat note detection, one needs to avoid the dispersion induced walk-off effect inside the PP-MgO:LN crystal. For this purpose, the crystal length is properly chosen as 2 mm. After the PP-MgO:LN crystal, the laser is separated by a dichromatic chirped mirror (CM) which highly reflects light below 1.2 µm and is transparent for light from 1.2 to 1.7 µm. The reflected beam is used as the output of the whole laser system, while the transmitted light is used for CE phase detection.

Free-running CE offset signals. The free-running CE offset signal f_0 is shown in Figure 2.16 (a). The SNR of f_0 is about 35 dB under a resolution bandwidth of 100 kHz. It is noticed that the f_0 signal exhibits two distinguishable features in the time domain, representing fast variation and slow drifting. Figure 2.16 (b) shows the unlocked f_0 signal with a record bandwidth of 1 kHz measured by a digital spectra analyzer. The fast frequency variation of f_0 can be clearly noticed within a variation range of ~40 kHz. Meanwhile, the center of f_0 can also be observed as a drifting signal by using

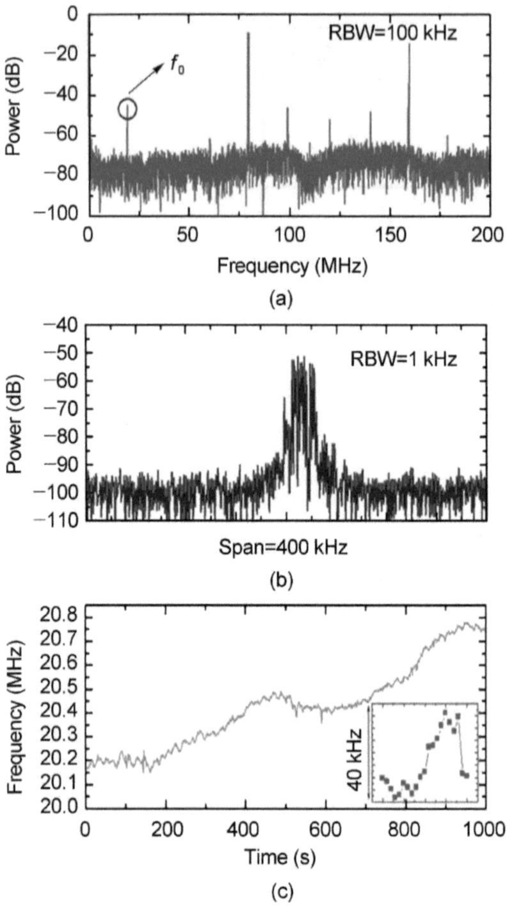

Fig. 2.16: (a) RF spectrum of f_0 beat note, (b) free-running f_0, (c) f_0 recorded by frequency counter.

a frequency counter to record its frequency on a long time scale of 1000 s. The result is shown in Figure 2.16 (c). The signal slowly drifts in a range of 600 kHz within the 1000 s. By calculating the standard deviation of the signal frequency recorded within 20 s (inset picture), a fast variation value of 40 kHz is obtained, which is consistent with the variation value measured by the spectra analyzer.

Stabilized CE offset frequency. For easily and roughly tuning the CE offset frequency, a pair of wedges is placed inside the Ti:S laser cavity. By changing the relative distance between the two wedges, the CE offset frequency is set to 20 MHz, approximately one fourth of the laser repetition rate. A fast-response AOM is inserted between the pump and the laser cavity in order to rapidly modulate the pump power as well as the laser CE offset frequency. With an electronic phase-locking system, the f_0 is locked to a fre-

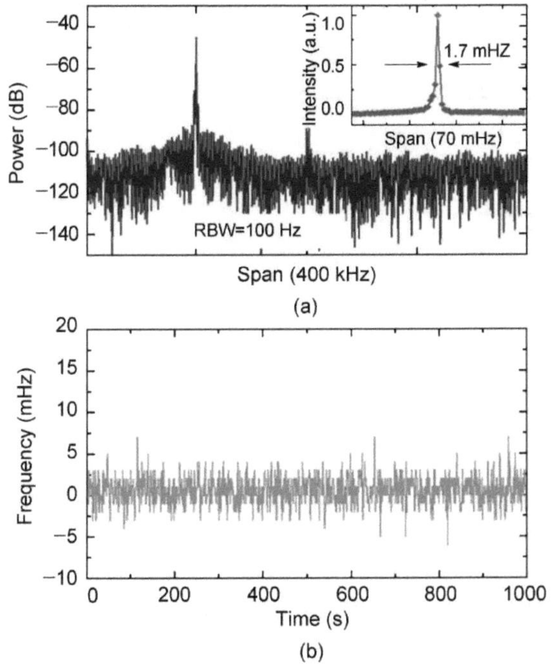

Fig. 2.17: (a) Locked f_0 with a linewidth of 1.7 mHz, (b) locked f_0 measured by a frequency counter.

quency synthesizer. The phase-locking system will be introduced in the next section. The locked f_0 is shown in Figure 2.17 (a) with an observation bandwidth of 100 Hz. As we can see, the signal is quite different from the unlocked one. The center of f_0 is fixed at 20 MHz and its linewidth is reduced from a kHz-level value to 1.7 mHz, which indicates the CE offset signal is well locked. Meanwhile, by using a frequency counter, the frequency variation of the locked f_0 is also recorded, as shown in Figure 2.17 (b). The frequency drifting disappears as noted within 1000 s.

Electronic phase-lock loops. Here we will introduce electronic phase-lock loops for the frequency stabilization of laser combs in Experiment I. Figure 2.18 illustrates the diagram of the lock loop principle. The laser repetition frequency signal (f_r) and its harmonics are detected by a fast photodetector (PD). For the purposes of improving the accuracy of the lock loop, the signal's m-th harmonic is used. Typically, for a laser at a repetition rate of 100 MHz, the m value is ideally larger than 10, in order to improve precision by more than ten times, compared directly to the fundamental repetition frequency. To stabilize the repetition rate, the output of the PD is mixed with a standard reference (f_{SYN2}) in an electronic mixer to generate an error signal. The error signal is the frequency difference ($mf_r - f_{SYN2}$) between the laser repetition signal (mf_r) and the reference (f_{SYN2}). It coexists with other signals including mf_r, f_{SYN2}, and $mf_r + f_{SYN2}$ at

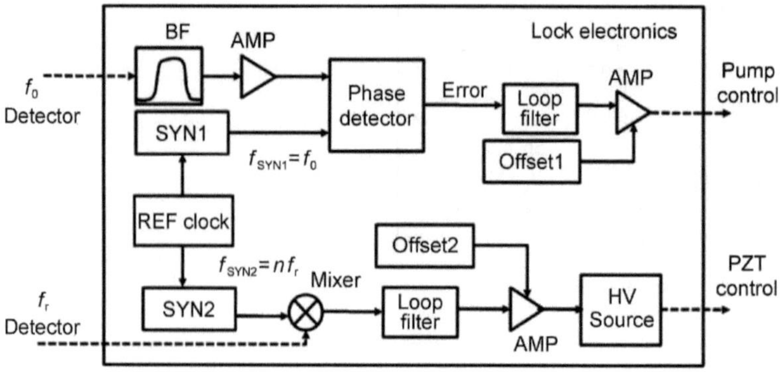

Fig. 2.18: Electronic phase-lock loops for frequency stabilization of laser combs; SYN1, SYN2: reference signals from RF frequency generators, AMP: amplifier.

the output of the mixer. Therefore, a low-pass filter should be used to pick up the error signal only and filter out the others in the meantime. Then the pure error signal is sent to an electronic amplifier with a tunable offset (Offset 2). This offset can provide larger dynamic range for the feedback loop. Finally, the error signal is sent to a high-voltage driver for controlling the piezo-electronic transition (PZT) glued on the backside of a cavity mirror inside the laser. The error signal after the power driver is used to modulate the voltage imposed on the PZT, which influences the laser repetition rate in order to minimize the error signal (both its frequency and amplitude). When the phase-lock loop works, the laser repetition rate, or its harmonics, are identical with the reference frequency.

Basically, the phase-lock loop also works for stabilization of f_0. However, there are some small differences between the stabilization of f_0 and f_r, which should be considered in the electronic loop design. First, the amplitude of the f_0 signal is usually much weaker than the f_r signal and thus a low-noise electric amplifier is helpful for increasing the signal amplitude without inducing unnecessary noise. Second, a phase detector is more suitable than a simple frequency mixer for distinguishing not only the frequency difference but also the phase difference between the f_0 signal and a standard reference signal (SYN1). Third, the output of the phase-lock loop is applied on an acoustic-optic modulator (AOM) placed between the laser oscillator and its pump. Generally, the zero order output of the AOM is used to pump the mode-locked laser. When the phase-lock loop works, the pump power after the AOM will be modulated by the error signal of the lock loop. As a result, the CE phase of the laser can be precisely controlled.

2.5.2 Direct feed-forward scheme for frequency combs

The acoustic-optic frequency shifter (AOFS) is based on acousto-optic effects. In the presence of sound waves in some optical media, their refractive indexes are changed in accordance with the acoustic modulation. If the applied sound waves are controlled to oscillate inside the medium in a form of a standing wave, some kinds of refractive index grating are generated [29]. Accordingly, incident lights propagating through the acoustic grating undergo acousto-optic effects like refraction, diffraction, and interference. For an AOFS, the intensity, frequency, and direction of diffraction light are dependent on the applied sound wave. The mechanism of an acoustic-optic frequency shifter is shown in Figure 2.19.

Assuming the acoustic wave in the crystal oscillates at a wavelength of Λ, the relationship between the output angle θ of the diffraction light and input light wavelength λ of can be expressed as

$$\sin\theta = \left(\frac{m\lambda}{2\Lambda}\right), \tag{2.12}$$

where $m = \ldots, -2, -1, 0, 1, 2, \ldots$ are diffraction orders. Usually, for the thin acoustic optic medium, only the diffraction lights at $m = -1, 0, 1$ can be produced under a single sinusoidal RF driving signal, which is the case for most commercialized AOFS devices. In fact, the process of acousto-optic diffraction of light obeys two basic laws, namely conservation of momentum and energy conservation. The former indicates that phase-matching is crucial for the acousto-optic diffraction light, while the latter can explain the Doppler frequency shift in the diffraction process. Assuming the wavelength of a sound wave is Λ, corresponding to frequency F, and the incident light is at a frequency $\omega = 2\pi/\lambda$, then the mth order diffraction light has a frequency $\Omega = \omega + mF$.

In addition, the efficiency of diffraction light is influenced by the RF power of the applied sound wave. Usually, the efficiency of the zero order output of an AOFS can be adjusted from 10 to 99%, while the first order diffraction efficiency is tunable from

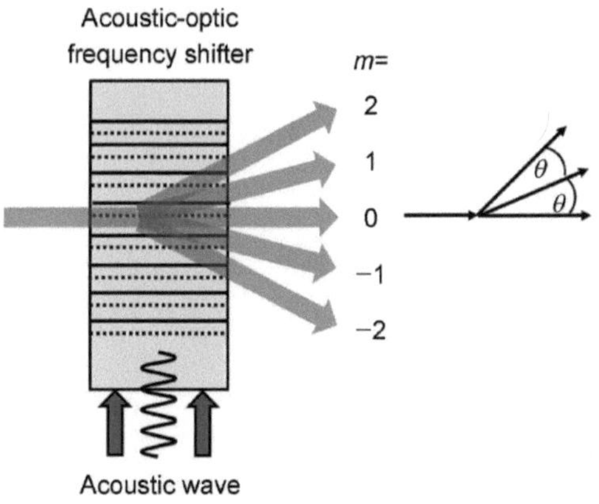

Fig. 2.19: The working mechanism of an acoustic-optic frequency shifter.

0 to 80%. Since the rapid development of acoustic-optic devices, the diffraction efficiency of some acoustic-optic modulators can reach beyond 80%, which is even more promising for highly-efficient modulations of lights.

As we all know, CE phase stabilized laser sources have been widely used in many areas, such as precision spectroscopy, coherent laser ranging, and optical frequency metrology. As promising laser sources for optical frequency combs, mode-locked fiber lasers have attracted much attention for their compactness and their capability for stable, long-term operation. Additionally, developments in large-mode-area ytterbium-doped double-clad photonic crystal fibers (LMA-YDCF) have greatly contributed to pulse amplification up to hundreds of watts and therefore stimulated the advance of high-power fiber-based frequency combs. Usually, for the stabilizing CE phase of the high-power lasers, either a phase servo loop or a difference-frequency-generation scheme can be used. However, each approach has its own distinct drawbacks. The former suffered from its limited feedback bandwidth and negative side effects on laser oscillators, while the latter had a zero offset frequency severely influenced by the increased $1/f$ noise and other low frequency noises. Recently, a self-referenced feed-forward scheme based on an acousto-optic frequency shifter (AOFS) was demonstrated to be able to overcome those shortcomings [30]. Considering that the first-order output of AOFS could be decreased by exactly the driving frequency (f_D), the CE phase variation of an input laser, $f_O(t)$, would be cancelled by independently feeding the unlocked offset signal to drive the AOFS, $f_{\text{drive}} = f_O(t)$, as illustrated in Figure 2.20. Consequently, the measured phase noise was reported to approach the shot-noise-limited level. Moreover, the offset frequency of the first-order beam was tunable by setting $f_{\text{drive}} = f_O(t) + f_A$, where f_A could be a well-controlled radio frequency. To further demonstrate details of this technique, an experiment (Experiment II) will be introduced.

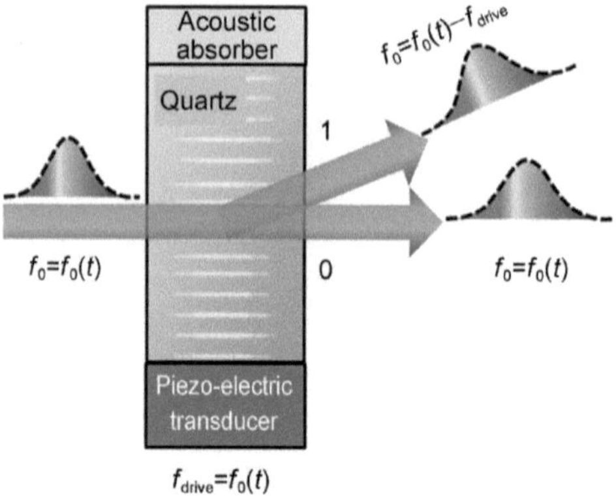

Fig. 2.20: Mechanism of AOFS-based phase stabilization.

2.5.2.1 Experimental example II: AOFS-based phase stabilization experiment

The schematic of the experiment is illustrated in Figure 2.21 [31, 32]. The near-infrared pulse train was generated from an NPR-mode-locked self-started break-free ytterbium-doped fiber laser at a repetition rate of about 64.5 MHz. The initial pulse width was ~1.4 ps with a spectral width of ~50 nm around the center wavelength 1030 nm. The average output power approximated 30 mW and was then amplified by a two-stage amplifier at an average output power > 10 W. The last power amplification stage consisted of 2 m of polarization-maintaining LMA-YDCF. After amplification, we employed two fused-silica transmission gratings to recompress the amplified pulse to ~130 fs at an average output power up to 10 W. For CEO detection, 1 W of the high-power output is delivered into a nonlinear f-2f interferometer and the rest is fed into an AOFS (DTSX-400-1064) with a central drive frequency at 80 ± 5 MHz and a measured first-order diffraction efficiency of > 90% at 1030 nm.

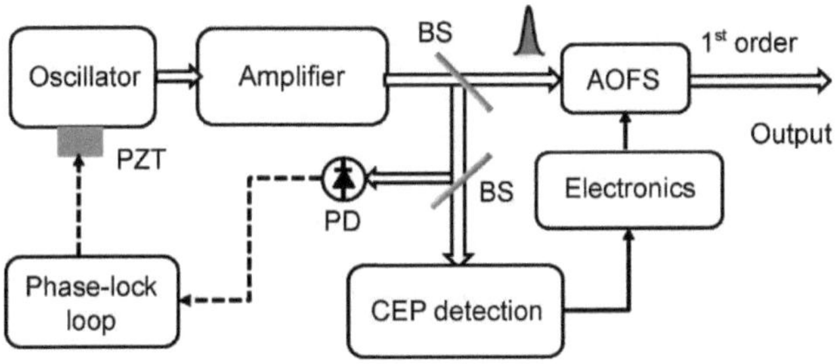

Fig. 2.21: Schematic of the feed-forward CE phase stabilization.

In the f_0 detection part, a 3-cm-long photonic crystal fiber (PCF, 3.3 μm core-diameter) with a small negative dispersion around 1030 nm is used to obtain octave spectral coverage. The f_0 is determined by the heterodyne beat signal in the nonlinear f-2f interferometer (at a wavelength of 610 nm) with an avalanche photodetector (APD 210, Menlosystems). The signal is recorded with a spectral analyzer and is shown in Figure 2.22. Note that not only the center frequency, but also the linewidth of f_0 is changeable by slightly adjusting the grating pair distance inside the fiber laser oscillator. In the experiment, the free-running f_0 was tuned to 17.5 MHz with a linewidth below 100 kHz.

By feeding the free-running f_0 to the AOFS, we obtain a phase stabilized comb light at the output port of the first-order diffraction of the AOFS. The linewidth of the locked offset frequency f_0 is measured and recorded by a FFT spectrum analyzer (SRS, SR760). The result is shown in Figure 2.23, indicating a relative linewidth of 1.4 mHz. The phase-noise power spectral density of the laser comb is also analyzed with the FFT spectrum analyzer, as shown in Figure 2.24. The integrated phase noise was calculated to be 370 mrad, corresponding to a timing jitter of 120 as.

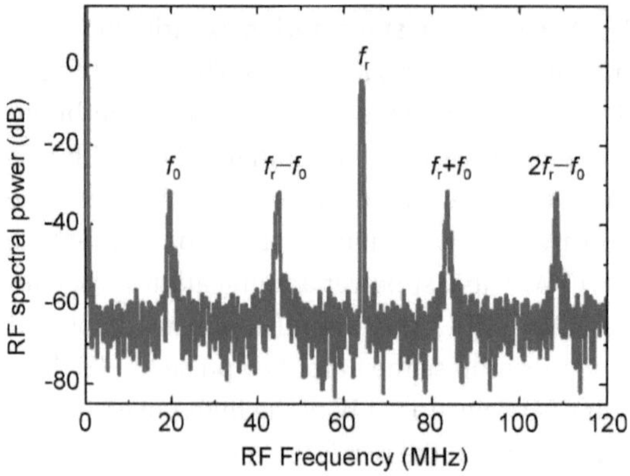

Fig. 2.22: Measured beat note signals of Experiment II.

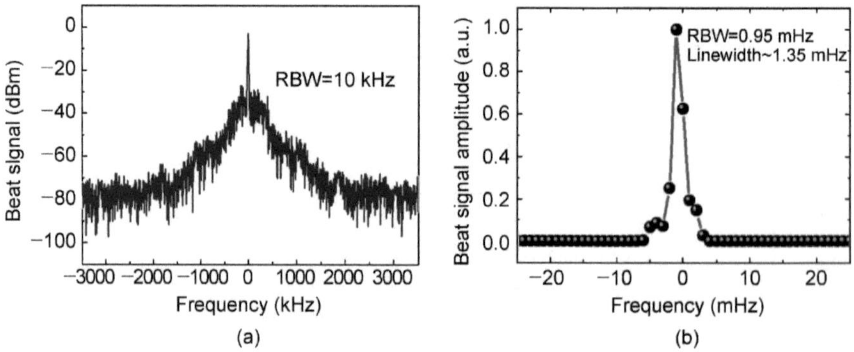

Fig. 2.23: Locked f_0 signal with (a) recorded RBW=10 kHz, (b) RBW=0.95 mHz.

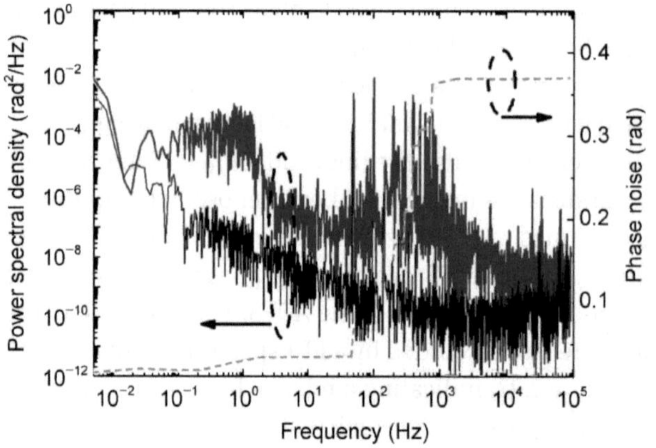

Fig. 2.24: Power spectral density of CE phase noise (black curve: noise background from electronics, gray curve: locked noise power density, dashed curve: calculated phase noise for the phase-locked laser).

The feed-forward CE phase stabilization scheme has many merits. First, AOFS has a response bandwidth of 1~1.5 MHz, much wider than that of a feedback loop (usually at kHz level), which is helpful for suppressing noise in high frequency regions and therefore can improve the performance of the frequency comb. Second, the time response of AOFS is faster and therefore can provide real time control of the CE offset frequency of the laser comb. Third, the AOFS-based CE phase control part is independent from the laser oscillator, which simplifies the comb system to a large extent and protects the oscillator from unexpected influences of the control system. However, the AOFS-based combs do have some obvious shortcomings. For example, in Experiment II, the working range of AOFS is from 75 to 85 MHz. Once the offset frequency signal drifts away from this range, AOFS becomes helpless. Since the oscillator is independent from the control system, the drifting of the signal is inevitable, which definitely reduces the working time of the laser comb. Therefore, to compensate for this shortcoming, a combination of feed-forward and feedback CE phase control systems would be a perfect solution.

2.5.3 Self-stabilized frequency combs

It is well-known that a mode-locked laser contains millions of evenly distributed discrete comb lines in a frequency domain. We can choose two comb lines whose frequencies can be described as $\omega_1 = nf_r + f_0$ and $\omega_2 = mf_r + f_0$, from the laser to generate a new light at $\omega_3 = (n-m)f_r$ through the difference frequency generation (DFG) process (suppose the integer $n > m$). Clearly, the new light has a zero offset frequency, which means the offset frequency as well as the carrier-envelope (CE) phase noise of the laser is cancelled out through the DFG process. The DFG scheme provides CE-phase self-stabilized and self-referenced laser pulses in a very simple configuration, as shown in Figure 2.25. A periodical poled lithium niobate (PPLN) crystal is used to provide a highly efficient DFG process. After optical filters, the CE phase stabilized light can be selected as a source for the laser comb. Since the output power of the DFG process is limited by the PPLN crystal and the nonlinear process itself, optical amplifiers can be used to compensate for this shortcoming. As an example, [33] employed an ytterbium-doped mode-locked fiber laser (centered at 1030 nm) as a light source. Because of a narrow spectral width of the Yb-laser, a 23-cm long photonic crystal fiber was applied to extend the spectrum to 615 nm. The CE phase self-stabilized light around 1530 nm

Fig. 2.25: Self-stabilized frequency combs.

was generated through the DFG process between lights centered at 615 and 1030 nm. Choosing a DFG light around 1.5 um is mainly because of the facility of erbium-doped fiber amplifiers.

Although the DFG-based self-stabilized scheme clearly has many merits like ease of operation, compactness, simplicity, and so on, it suffers from several disadvantages. The biggest drawback for such a scheme would be the CE offset frequency at zero. As we know, there are lots of noise parasites in the low RF frequency region, including $1/f$ noise, mechanical vibration, thermal fluctuation, and so on. As DFG relies on large optical nonlinearity, the nonlinear processes are sensitive to intensity fluctuation, which may produce additional CE offset frequency drifts due to amplitude-to-phase noise. In addition, phase noise may be induced by high-order nonlinear processes, stimulated Raman scattering, and other incoherent scattering processes in the nonlinear optical crystal. Modulational instability may also compete with DFG processes that bring about undesired noise and make output power and the spectrum instable. These noises become a major threat to the stability of the laser comb. Therefore, the application of the self-stabilized comb is limited by its noise level.

2.6 Applications of optical frequency combs

Along with their rapid developments, optical frequency combs have been successfully applied in many research areas, such as precision spectroscopy, optical clocks, light detection and ranging (LIDAR) [34], quantum coherent control [35–38], UV frequency combs, and so forth. In this section, we will list some examples of those well-developed applications, including optical frequency measurement, optical clock applications, and rapid distance measurement.

Optical frequency measurements

One of the most interesting and straightforward applications of OFC is direct frequency measurement for an unknown optical wave [39]. Because of the high precision and wide spectral range of the frequency comb, accurate measurements of an optical frequency within the comb spectral range become easily achievable through beating signals of the unknown wave and the comb. Assuming the unknown wave has a frequency at f_l whose exact value is unknown, the beat note frequency between the light wave and the n-th tooth of the comb would be f_b. f_l can then be expressed by

$$f_l = nf_r \pm f_0 \pm f_b \,. \tag{2.13}$$

Meanwhile, for another comb tooth at n' ($n' = n + m$, n and m are integers), the f_l can also be determined by

$$f_l = (n + m)f_r' \pm f_0' \pm f_b' \,. \tag{2.14}$$

Using equations (2.13) and (2.14), one can deduce the value of n, which is:

$$n = \frac{[\pm f_0' - (\pm f_0) + mf_r' \pm f_b' - (\pm f_b)]}{(f_r - f_r')} \tag{2.15}$$

The value of m can be deduced by slightly tuning the laser repetition rate (f_r) to a different value and comparing the difference to the frequency shift of the beat note before and after tuning f_r. Actually, this is an easy and precise approach for measuring an unknown frequency. To guarantee accuracy of the frequency measurement, a high-accuracy reference is necessary. Usually, atomic clocks with an uncertainty level of 10^{-14} are used as accurate references for stabilizing frequency combs [40, 41]. As we know, the higher the frequency of the oscillator used for a clock, the higher the accuracy of the clock. Therefore, the optical clock, a clock oscillating at optical frequency, appears to be a perfect solution for further raising time and frequency standards.

Optical clocks

As an important part of optical clocks, frequency combs fill in the missing gap between microwaves and optical frequency standards. Before the birth of frequency combs, most accurate time standards at a level of 10^{-15} s came from atomic clocks with cooled or trapped atoms or single ions such as Rb, Ca, CH_4, Hg^+, I_2, Sr^+, Yb^+, and In^+ [43, 44]. While frequency combs serve as a perfect carrier for transferring the time accuracy to optical frequency, the time standard can be further improved. As we know, the uncertainty of an atomic clock can be described as

$$\sigma_y(\tau) = \frac{\delta\nu}{\nu_0} = \frac{1}{\omega_0\sqrt{NT_R\tau}}, \tag{2.16}$$

where $\delta\nu$ is the uncertainty of the atomic resonance frequency of ν_0, N is the number of atoms, T_R is observation time, and τ is average sampling time. From this expression, it can be easily deduced that raising the resonance frequency ν_0 can seriously reduce the uncertainty of an atomic clock. To a certain extent, the uncertainty is dependent on the ratio between the comb line frequency and its linewidth. Since optical frequency is five orders of magnitude higher than the microwave, the uncertainty of an optical clock should be as small as 10^{-18} s, which implies that optical clocks will bring about an unprecedented high standard of time and frequency.

Rapid distance measurement

After a decade of development, frequency combs have become more and more important not only in fundamental scientific research areas but also in many useful applications. Laser ranging and distance measurement is a perfect example. The proposal of dual-comb configuration in distance measurement is especially valuable for greatly increasing laser ranging distance as well as improving measurement accuracy. In this approach, two pulsed laser combs are used as detection and reference pulse generators respectively. The two laser combs have a slight difference in their repetition rates, which can be described as f_1 (transmitting terminal) and $f_2 = f_1 + \Delta f$ (reference), respectively, where Δf is the frequency difference. The pulse that is sent from the transmitter will be reflected by a target and then received by the receiving terminal. It requires a time of t, and therefore the distance between the receiving terminal and the target would be $L = ct/2$ for a time-of-flight distance measurement. However, in standard time-of-flight laser ranging, the measurement range is limited by the repetition rate of the detecting pulses. Moreover, the accuracy of measurement is restricted by the pulse duration and detector response time. In the dual-comb distance measurement, the reference comb (f_2) is used to heterodyne with the received light. Since the relative phase $\Delta\varphi$ between the two combs can linearly increase with the measured distance, one can calculate the distance as

$$L = (m + \Delta\varphi/2\pi)\lambda/2 \,, \tag{2.17}$$

where λ is the wavelength of the two combs and m is an integer representing optical cycles that can be determined by the time-of-flight measurement. Meanwhile, since the reference pulse heterodynes with the detecting light in a period of $\Delta t = 1/\Delta f$, the measurement range can be significantly enlarged by $L'_{\text{max}} = c\Delta t/2$. With such an approach, Dr. I. Coddington and his colleagues measured a section of unknown distance with an accuracy of submicrometers under a long time integration condition in 2010 [45].

Acknowledgment: The experiments reviewed in this chapter are financially supported by the National Key Scientific Instrument Project (2012YQ150092), National Basic Research Program of China (2011CB808105), and National Natural Science Fund (61127014). The author thanks Dr. Ming Yan for offering valuable aid in editing this chapter. Experiments reviewed in this chapter include collaborative contributions from Prof. W. Li, Prof. Q. Hao, Dr. M. Yan, X. Shen, K. Yang, D. Bai, J. Zhao, Q. Ru, Y. Liu, and other graduate students.

Bibliography

[1] K. Kitayama, Highly spectrum efficient OFDM/PDM wireless networks by using optical SSB modulation, J. Lightwave Technol. **16** (1998): 969–976.

[2] M. Kourogi, K. Nakagawa and M. Ohtsu, Wide-span optical frequency comb generator for accurate optical frequency difference measurement, IEEE J. Quantum Electron. **29** (1993): 2693–2701.

[3] J. Ye, L. Ma, T. Daly and J. L. Hall, Highly selective terahertz optical frequency comb generator, Opt. Lett. **22** (1997): 301–303.

[4] N. C. Wong, Optical frequency division using an optical parametric oscillator, Opt. Lett. **15** (1990): 1129–1131.

[5] D. Y. Tang and L. M. Zhao, Generation of 47-fs pulses directly from an erbium-doped fiber laser, Opt. Lett. **32** (2007): 41–43.

[6] X. Zhou, D. Yoshitomi, Y. Kobayashi and K. Torizuka, Generation of 28-fs pulses from a mode-locked ytterbium fiber oscillator, Opt. Express **16** (2008): 7055–7063.
[7] F. Ilday, J. Buckley, L. Kuznetsova and F. Wise, Generation of 36-femtosecond pulses from a ytterbium fiber laser, Opt. Express **11** (2003): 3550–3554.
[8] A. Stolow and D. M. Jonas, Multidimensional Snapshots of Chemical Dynamics, Science **305** (2004): 1575–1577.
[9] Yao Li, Precision control of ultrashort pulses in time and frequency domains, Shanghai: East China Normal University (2010): 13–21.
[10] J. Rauschenberger, T. Fuji, M. Hentschel, et al. Carrier-envelope phase-stabilized amplifier system, Laser Phys. Lett. **3** (2006): 37–42.
[11] T. M. Fortier, A. Bartels and S. A. Diddams, Octave-spanning Ti:sapphire laser with a repetition rate > 1 GHz for optical frequency measurements and comparisons, Opt. Lett. **31** (2006): 1011–1013.
[12] A. Bartels, R. Gebs, M. S. Kirchner and S. A. Diddams, Spectrally resolved optical frequency comb from a self-referenced 5 GHz femtosecond laser, Opt. Lett. **32** (2007): 2553–2555.
[13] R. Teets, J. Eckstein, T. W. Hänsch, et al. Coherent Two-Photon Excitation by Multiple Light Pulses, Physical Review Letters. **38**(14) (1977): 760.
[14] D. J. Jones, S. A. Diddams, J. K. Ranka, et al. Carrier-envelope phase control of femtosecond mode-locked lasers and direct optical frequency synthesis, Science **288** (2000): 635–639.
[15] J. K. Ranka, R. S. Windeler and A. J. Stentz, Visible continuum generation in air-silica microstructure optical fibers with anomalous dispersion at 800 nm, Opt. Lett. **25**(1) (2000): 25–27.
[16] T. M. Fortier, D. J. Jones, J. Ye and S. T. Cundiff, Long-term carrier-envelope phase coherence, Opt. Lett. **27** (2002): 1436–1438.
[17] T. Fuji, J. Rauschenberger, A. Apolonski, et al. Monolithic carrier-envelope phase-stabilization scheme, Opt. Lett. **30** (2005): 332–334.
[18] J. Rauschenberger, T. Fuji, M. Hentschel, A. J. Verhoef, et al. Carrier-envelope phase-stabilized amplifier system, Laser Phys. Lett. **3** (2006): 37–42.
[19] B. R. Washburn, C. G. Jørgensen, J. W. Nicholson, et al. Phase-locked, erbium-fiber-laser-based frequency comb in the near infrared.,Opt. Lett. **29**(3) (2004): 250–252.
[20] A. Apolonski, A. Poppe, G. Tempea, et al. Controlling the phase evolution of few-cycle light pulses, Phys. Rev. Lett. **85** (2000): 740–743.
[21] U. Morgner, R. Ell, G. Metzler, et al. Nonlinear optics with phase-controlled pulses in the sub-two-cycle regime, Phys. Rev. Lett. **86** (2001): 5462–5465.
[22] R. Ell, U. Morgner, F. X. Kärtner, et al. Generation of 5-fs pulses and octave-spanning spectra directly from a Ti:sapphire laser, Opt. Lett. **26** (2001): 373–375.
[23] F. X. Kärtner, U. Morgner, R. Ell, et al. Ultrabroadband double-chirped mirror pairs for generation of octave spectra, J. Opt. Soc. Am. B **18** (2001): 882–885.
[24] 24. T. M. Fortier, D. J. Jones and S. T. Cundiff, Phase stabilization of an octave-spanning Ti:sapphire laser, Opt. Lett. **28** (2003): 2198–2200.
[25] N. R. Newbury and W. C. Swann, Low-noise fiber-laser frequency combs, J. Opt. Soc. Am. B **24** (2007): 1756–1770.
[26] B. R. Washburn, W. C. Swann and N. R. Newbury, Response dynamics of the frequency comb output from a femtosecond fiber laser, Optics Express **26**, 10622–10 (2005): 633.
[27] X. Shen, W. Li, M. Yan and H. Zeng, Electronic control of nonlinear-polarization-rotation mode locking in Yb-doped fiber lasers, Opt. Lett. **37** (2012): 3426.
[28] L. Glandorf, T. A. Johnson, Y. Kobayashi and S. A. Diddams, Impact of dispersion on amplitude and frequency noise in a Yb-fiber laser comb, Opt. Lett. **36** (2011): 1578–1560.
[29] J. Lan. Technology of laser (2nd edition), Beijing: Science Press, 2005: 35–56.

[30] S. Koke, C. Grebing, H. Frei, et al. Direct frequency comb synthesis with arbitrary offset and shot-noise-limited phase noise Nature Photonics **4** (2010): 462–465.

[31] M. Yan, W. Li, K. Yang, et al. High-power Yb-fiber comb with feed-forward control of nonlinear polarization rotation mode-locking and large-mode-area fiber amplification Opt. Lett. **37** (2012): 1511.

[32] Ming Yan, High-power amplification and phase noise suppression of high-repetition rate ultrashort pulses, Shanghai: East China Normal University, 2012: 74–105.

[33] Y. Deng, F. Lu and W. H. Knox, Fiber-laser-based difference frequency generation scheme for carrier-envelope-offset phase stabilization applications, Opt. Express **13** (2005): 4589.

[34] R. T. H. Collis, Lidar Observation of Cloud, Science **149**(3687) (1965): 978–981.

[35] A. Apolonski, A. Poppe, G. Tempea, et al. Controlling the phase evolution of few-cycle light pulses, Phys. Rev. Lett. **85** (2000): 740–743.

[36] A. Baltuska, T. Udem, M. Uiberacker, et al. Attosecond control of electronic processes by intense light fields, Nature **422** (2003): 6928.

[37] A. Baltuska, T. Udern, M. Uiberacher, et al. Attosecond control of electronic processes by intense light fields, Nature **421**(6923) (2003): 611–615.

[38] H. R. Teile, G. Steinmeyer, A. E. Dunlop, et al. Carrier-envelope offset phase control: a novel concept for absolute optical frequency measurement and ultrashort pulse generation, Appl. Phys. B **69** (1999): 327–332.

[39] K. L. Corwin, I. Thomann, T. Dennis, et al. Absolute-frequency measurements with a stabilized near-infrared optical frequency comb from a Cr:forsterite laser, Opt. Lett. **29** (2004): 397–399.

[40] T. Udem, S. A. Diddams, K. R. Vogel, et al. Absolute frequency measurements of the Hg+ and Ca optical clock transitions with a femtosecond laser, Phys. Rev. Lett. **86** (2001): 4996–4999.

[41] J. Stenger, C. Tamm, N. Haverkamp, S. Weyers and H. R. Telle, Absolute frequency measurement of the 435.5-nm 171Yb+-clock transition with a Kerr-lens mode-locked femtosecond laser, Opt. Lett. **26** (2001): 1589–1591.

[42] M. Chwalla, J. Benhelm, K. Kim, et al. Absolute frequency measurement of the 40Ca+ 4s(2)S_(1/2)-3d(2)D_(5/2) clock transition, Phys. Rev. Lett. **106** (2009): 023002.

[43] A. Y. Nevsky, Frequency comparison and absolute frequency measurement of I2 stabilized lasers at 532 nm, Opt. Comm. **263** (2001): 192–272.

[44] M. Boyd, A. D. Ludlow, S. Blatt, et al. 87Sr lattice clock with inaccuracy below 10–15, Phys. Rev. Lett., **98** (2007): 083002.

[45] I. Coddington, W. C. Swann, L. Nenadovic and N. R. Newbury, Rapid and precise absolute distance measurements at long range, Nature Photonics **3** (2009): 351–356.

Fei He, Yang Liao, Jintian Lin, Jielei Ni, Bin Zeng, LingLing Qiao, Ya Cheng

3 Three-dimensional integration of hybrid functionalities in transparent dielectrics by femtosecond laser direct writing

3.1 Introduction

In the past decades, the demand for reducing the volume of samples and reagents used in chemical reactions, biological analysis, and medical inspections has increased significantly due to the need to reduce reagent consumption, side production, analysis time, manufacturing and labor costs, as well as to enhance detection speed and sensitivity. To this end, the use of biomicrochip systems with integration of hybrid functionalities such as microfluidic devices, microreactors, lab-on-a-chip (LOC) devices, micrototal analysis systems (μ-TAS), and optofluidic devices has been proposed [1–5]. Such devices are highly attractive for next generation high throughput biochemical analysis and medical diagnostics, and have already been responsible for introducing Moore's law into the field of life sciences.

Using these devices, conventional processes normally carried out in a lab can be miniaturized and performed on a single chip. This leads to enhanced efficiency and portability, and also reduces the volume of sample and reagent required when performing multilevel assessments involving, for example, chemical, biological, and medical analyses. Due to their high portability and sensitivity, these devices have become powerful detection and analysis tools for a broad range of applications including biomedical research, healthcare, pharmaceuticals, environmental monitoring, and homeland security. Moreover, there is also the possibility of further enhancing the performance by monolithically integrating microfluidic, electronic, mechanical, and optical capabilities.

To date, these microchips have mainly been manufactured using soft lithography, which is carried out using the optically transparent, soft elastomer polydimethylsiloxane (PDMS) [6, 7]. Although soft lithography is a well-established, rapid, and cost-effective technique, and hence is suitable for surface microfabrication, it cannot directly form microfluidic structures inside PDMS. To fabricate 3D microfluidic structures in transparent substrates, additional stacking and bonding processes are

---

Fei He, Yang Liao, Jintian Lin, Jielei Ni, Bin Zeng, LingLing Qiao, Ya Cheng (corresponding author),
Shanghai Institute of Optics and Fine Mechanics Chinese Academy of Sciences, Shanghai 201800, China

https://doi.org/10.1515/9783110304558-003

required. In addition, PDMS is chemically incompatible with many organic solvents, and compositional inhomogeneities frequently cause optical scattering, which is detrimental for optofluidic applications. Other conventional methods for producing microfluidic systems include planar microfabrication techniques such as injection molding [8] and semiconductor processes based on photolithography [9, 10], both of which also require stacking and bonding in order to construct 3D structures. Furthermore, the above techniques have encountered considerable challenges with regard to the monolithic integration of multiple functionalities, for which 3D fabrication approaches are typically needed.

During the past two decades, femtosecond laser microfabrication has been shown to be a highly attractive solution for the 3D manufacturing of transparent materials [11–13]. It exhibits great promise for the fabrication of microfluidic, photonic, micro-optical, microelectronic, and micromechanical components. Its unique capability for the 3D integration of functional microcomponents makes it a powerful state-of-the-art micromachining tool, in particular for 3D integration of hybrid functionalities in a single chip.

Materials processing using ultrafast lasers – lasers that emit light pulses shorter than a few tens of picoseconds – was first reported in 1987 by Srinivasan et al. [14], and Küper and Stuke [15]. They demonstrated the clean ablation of polymethyl methacrylate almost without the formation of a heat-affected zone (HAZ) using femtosecond ultraviolet excimer lasers. The ablation threshold is significantly lower than that of nanosecond laser ablation. These experiments had a significant impact and the research in this field rapidly expanded in the 1990s. In addition, the development of the chirped pulse amplification technique in Ti:sapphire regenerative amplifiers [16], which emit energetic femtosecond pulses without inducing damage or undesirable nonlinear effects in the amplification medium, further accelerated fundamental research on ultrafast laser processing. One important feature of ultrafast laser processing is that it reduces heat diffusion to surrounding regions of the processed area [17]. This feature is well adapted to the high quality microfabrication of soft materials such as biological tissues [18] and hard or brittle materials such as semiconductors and insulators [19] without HAZ formation. Furthermore, suppression of heat diffusion to the surroundings improves the spatial resolution for nanoscale processing [20]. Another important aspect of ultrafast laser processing is that nonlinear absorption (i.e. multiphoton absorption) can induce strong absorption even in materials that are transparent to the ultrafast laser beam [21, 22]. Multiphoton absorption not only permits surface processing [23], but also 3D internal microfabrication of transparent materials such as glass and polymers [24–27]. Davis et al. [24] and Glezer et al. [25] pioneered this field and demonstrated respectively optical waveguide writing and formation of nanovoid arrays inside glass in 1996. Currently, internal microfabrication is widely applied to the fabrication of photonic devices and biochips [11, 28]. In 2001 it was also reported that multiphoton absorption improves spatial resolution to the point of exceeding the diffraction limit, due to the nonlinearity combined with the threshold effect [29]. Careful control of the laser power and scanning speed enabled a minimum fabrication resolution of ~18 nm [30]. One of the major application fields of this feature is two-photon polymerization (TPP) for the fabrication of photonic crystals [31], micromachines [32], and biochips [33]. In the 2000s, it was determined that ultrafast laser irradiation at intensities near the ablation threshold forms nanoripple structures on various materials with periodicities much shorter than the wavelength [34–36]. Regular arrays of conical microstructures are also produced on Si by irradiation with an ultrafast laser beam in a halogen atmosphere (e.g. SF_6 or Cl_2) [37, 38]. The surface structures produced exhibit unique properties of antireflectivity, superhydrophobicity and infrared absorption.

Owing to the above advantages, femtosecond laser processing has attracted significant attention and has promised a wide variety of intriguing applications in not only integrated photonic devices (e.g. waveguide writing [24], polarization sensitive optics [39], quantum circuits [40], etc.) but also integrated microfluidic and optofluidic systems (e.g. LOC devices, μ-TAS systems) [28, 41]. In the meantime, this new interaction of femtosecond lasers with matter, i.e. irradiation of intense ultrafast pulses inside transparent materials such as glass and crystals, has led to many other intriguing phenomena, such as the formation of nanovoids [42] and periodic nanogratings [43], element redistribution [44], nanocrystallization [45], and 'quill' nonreciprocal writing [46]. Several review papers as well as books are available on femtosecond laser micromachining, providing comprehensive overviews on its fundamental physics, the basic processing systems, and many novel applications (Table 3.1).

This chapter mainly focuses on recent advancements in femtosecond laser fabrication of multifunctionalities as well as their hybrid integrations in transparent dielectrics. The remainder of this article is organized as follows. In Section 3.2, the

Tab. 3.1: Review articles and books in the field of femtosecond laser micromachining.

Review Articles	General reviews	Overview of femtosecond laser micromachining [8, 9, 13]
	Fundamentals	Mechanism of ultrafast laser ablation [47–49]
	Techniques	Beam/pulse shaping [50] Special beams [51] Super resolution [52]
	Applications	Photonic devices [53–59] Microfluidics and optofluidics [28, 41, 60–62] Surface and bulk nano-/micro-structuring [63–66] Two-photon polymerization [33, 67–72] Ultrafast laser surgery [73–75]
Books		Misawa et al. (2006) [76] Sugioka et al. (2010, 2013, 2014) [77–79] Osellame et al. (2012) [80]

fundamentals, including principles and characterizations, of femtosecond laser processing are discussed. In Section 3.3, the currently proposed spatiotemporal focusing of femtosecond laser pulses for materials processing is specifically described, including its principles, applications in micromachining, and novel nonlinear effects of this techniques. Sections 3.4, 3.5, and 3.6 introduce femtosecond laser fabrication of microfluidics, micro-optical components and electronic microstructures respectively. Section 3.7 discusses a variety of microchips with hybrid functionalities based on femtosecond laser fabrication, including electro-optical integration in $LiNO_3$, optofluidic sensors, and electrically tunable micro-optical resonators fabricated with FsLDW. Finally, a summary is given in Section 3.8, which also discusses the future outlook for these subjects.

3.2 Fundamentals of femtosecond laser processing

3.2.1 Introduction to femtosecond lasers

Laser pulses with time durations of a few femtoseconds (1 fs = 10^{-15} s) to several hundred femtoseconds are referred to as femtosecond pulses. Such pulses have a broad optical spectrum (e.g. a 40-fs pulse with a center wavelength of 800 nm typically has a spectral width of about 30 nm). Currently, these ultrashort pulses can be generated using mode-locked oscillators. Amplification of ultrashort pulses typically requires a technique referred to as chirped pulse amplification (CPA) in order to avoid damage to the gain medium of the amplifier [81].

Retrospectively, the development of ultrafast optics was triggered by the invention of laser mode locking, one of the most striking interference phenomena in nature [82, 83]. In the 1980s, continued work on the development of solid-state laser materials gave rise to the emergence of a number of new laser media. Various host crystals (YAG, sapphire, forsterite, LiSAF, etc.) doped with transition metal (titanium, chromium) ions now provide laser transitions with enormous bandwidths on the order of 100 THz in the near-infrared wavelength range. The discovery of self-mode-locking [84] or Kerr-lens mode locking (KLM) [85, 86] in a titanium-doped sapphire (Ti:S) laser [87] revolutionized ultrafast laser technology.

Many intriguing applications of ultrashort pulses call for intensities requiring peak powers far exceeding the power levels that can be directly obtained from CW mode-locked oscillators. To this end, the pulses delivered by mode-locked lasers need to be amplified. Femtosecond pulse amplification is first demonstrated by using dye cells or jets pumped by Q-switched Nd:YAG and copper-vapor lasers [88]. Recently, modulation of the multilayer period of dielectric laser mirrors is demonstrated to result in a wavelength-dependent penetration depth of the incident radiation. These devices have been referred to as chirped multilayer mirrors [89], which has paved the way towards the single-cycle regime.

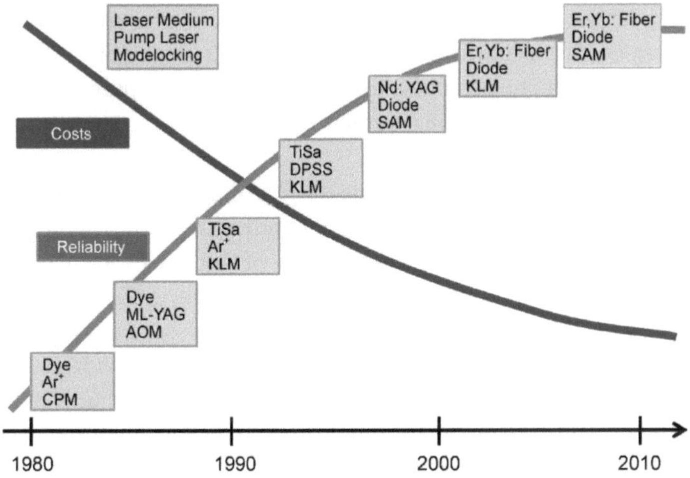

Fig. 3.1: The development with reliability and costs of ultrafast lasers [93].

A robust, stable and very compact fiber chirped pulse amplifier was also developed in the 2000s [90], which facilitated the application of femtosecond lasers. More recently, a rare earth-doped laser medium has been adopted to realize a compact and high-power ultrafast laser system by diode pumping, i.e. core-pumped erbium or ytterbium-doped fibers act as the active laser media, although the pulses are much broader than pulses generated by Ti:sapphire systems [91, 92]. The pump-power originates from long-life pigtailed laser diodes at wavelengths close to 980 nm. Passive mode-locking techniques for ensuring ultrafast pulse generation are based on non-linear polarization rotation or saturable absorber mirrors (SAM). Picosecond lasers based on this system that are suitable for industrial applications have become commercially available. In the 2010s, ultrafast laser processing has thus become a more reliable tool for practical and industrial applications. The development with reliability and costs of ultrafast lasers is shown in Figure 3.1.

Theoretically, an ultrashort pulse is a wave whose field amplitude follows a Gaussian envelope and whose instantaneous phase has a frequency sweep. The real electric field corresponding to an ultrashort pulse is oscillating at an angular frequency ω_0 corresponding to the central wavelength of the pulse [94]. The central angular frequency ω_0 is usually explicitly written in the complex field, which may be separated as a temporal intensity function $I(t)$ and a temporal phase function $\varphi(t)$:

$$E(t) = \sqrt{I(t)}\exp(\mathrm{i}\omega_0 t)\exp(\mathrm{i}\varphi(t)) \tag{3.1}$$

The expression of the complex electric field in the frequency domain is obtained from the Fourier transform of $E(t)$, and can be defined in the frequency domain:

$$E(\omega) = \sqrt{S(\omega)}\exp(\mathrm{i}\varphi(\omega)) = \mathcal{F}\{E(t)\}\ , \tag{3.2}$$

where $S(\omega)$ is the spectrum (or intensity spectral density) of the pulse, and $\varphi(\omega)$ is the spectral phase. When $\varphi(\omega)$ is a constant, the pulse is called a bandwidth-limited pulse, while the pulse is called a chirped pulse for a quadric phase of $\varphi(\omega)$. Such a chirp may be induced by propagating the pulse through dispersive optics, which leads to a temporal broadening of the pulse.

The intensity functions, $I(t)$ and $S(\omega)$, determine the time duration and spectrum bandwidth of the pulse. This minimum value of time-bandwidth product depends on the definition used for the duration, and on the shape of the pulse. For a given spectrum, the minimum time-bandwidth product, the shortest pulse, is obtained by a transform-limited pulse, i.e. for a constant spectral phase $\varphi(\omega)$.

For femtosecond optics, several optical devices have been specifically designed for ultrashort pulses, such as pulse compressors and stretchers, devices that can be used to control the spectral phase of ultrashort pulses [95]. They are composed of a sequence of prisms, or gratings. When properly adjusted, they can alter the spectral phase $\varphi(\omega)$ of the input pulse so that the output pulse is a bandwidth-limited pulse with the shortest possible duration. A pulse shaper can be used to make more complicated alterations on both the phase and the amplitude of ultrashort pulses [96]. To accurately control the pulse, a full characterization of the pulse spectral phase is a must in order to get certain pulse spectral phase (such as the transform-limited). Then, a spatial light modulator (SLM) can be used in the 4-f plane to control the pulse.

Currently, several techniques are available to measure ultrashort optical pulses: (1) intensity autocorrelation: gives the pulse width when a particular pulse shape is assumed; (2) spectral interferometry (SI): a linear technique that can be used when a pre-characterized reference pulse is available, also giving the intensity and phase; (3) spectral phase interferometry for direct electric-field reconstruction (SPIDER): a nonlinear self-referencing technique based on spectral shearing interferometry [97]; (4) frequency-resolved optical gating (FROG): a nonlinear technique that yields the intensity and phase of a pulse, which is a spectrally resolved autocorrelation [98]; and (5) grating-eliminated no-nonsense observation of ultrafast incident laser light e-fields (GRENOUILLE), a simplified version of FROG [99].

Since the development of ultrafast optics, intense femtosecond laser pulses have been increasingly used for materials processing. Currently, femtosecond laser processing is not only a subject of intensive academic research but also a reliable tool for both practical and industrial applications.

3.2.2 Principles of femtosecond laser processing

Fundamental investigations into the laser processing of materials have revealed that the time dependence of ultrafast electron and lattice dynamics is quite different for long and for ultrashort laser pulses. It is established that intense laser pulses with durations of nanoseconds can be used to anneal the lattice in a manner of thermal

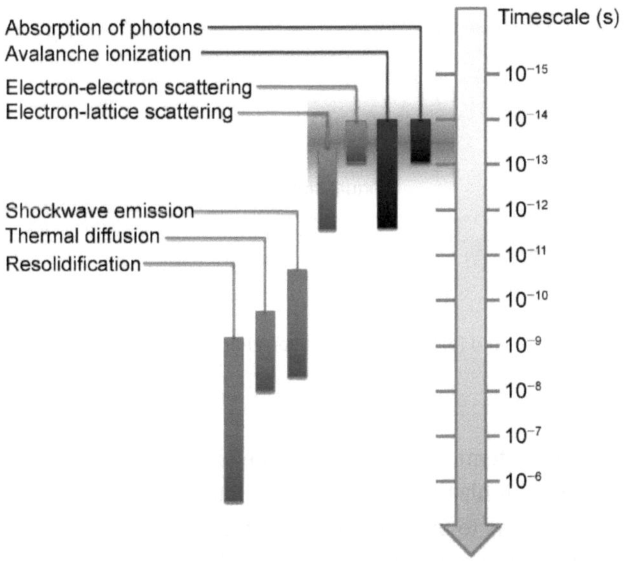

Fig. 3.2: Typical timescales during interaction of ultrashort laser pulses with transparent materials.

processing. For ultrashort laser pulses in the femtosecond regime, however, the electrons and the lattice are driven far out of equilibrium and disordering of the lattice can occur. During the interaction of an ultrashort pulse with a transparent material, the laser energy is first deposited into the electrons and then transferred to the lattice, with several regimes of electron excitation and relaxation [12, 100]. These regimes and the timescales of the corresponding processes are shown in Figure 3.2.

The interaction of femtosecond laser pulses with transparent materials with wide bandgaps (e.g. 800-nm femtosecond pulses interacting with fused silica glass) is generally regarded as a multiphoton absorption (MPA) process. When a single photon has an energy exceeding the bandgap of the material, it is absorbed by exciting an electron from the valence band (VB) to the conduction band (CB) by single-photon absorption, as illustrated in Figure 3.3 (a). However, when an extremely high density of photons (i.e. extremely high-intensity light) is incident on the material, an electron can be excited by multiple photons via virtual states, even if the photon energy is smaller than the bandgap, as shown in Figure 3.3 (b). Due to the nonlinear nature of MPA, tightly focused femtosecond laser pulses can be strongly absorbed even by a transparent medium. This light-matter interaction occurs only near the focal volume where the peak intensity is sufficiently high to induce MPA with a large absorption cross section. This is fundamentally different from the possible avalanche ionization (AI) process that can induce breakdown in the case of interaction of long laser pulses (typically nanoseconds) with transparent materials [101]. For the long pulses, such AI dominated processes are mainly caused by impurities and dislocations in the focal volume. They therefore occur stochastically and are irreproducible. In contrast, the

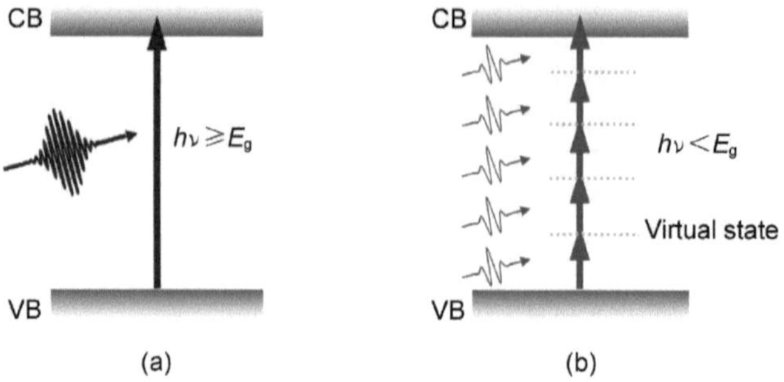

Fig. 3.3: Schematic illustration of (a) single- and (b) multiphoton absorption processes. CB: conduction band, VB: valence band.

MPA dominated process induced by femtosecond lasers is deterministic in that it is only seeded by multiphoton/tunnel ionization. This unique characteristic allows reproducible and controllable processing to be performed using femtosecond lasers.

At high laser intensities and low frequencies, electron excitation is induced by tunneling ionization rather than multiphoton absorption. In tunneling ionization, the potential barrier formed by the valence and conduction band structures is first drastically deformed by the intense electric field of a femtosecond laser before the barrier length is reduced. When this occurs, an electron can tunnel through the barrier and eventually easily escape from the molecule to be excited from the valence band to the conduction band (Figure 3.4).

The probabilities of multiphoton absorption and tunneling ionization in femtosecond laser interaction with transparent materials can be determined by the Keldysh parameter [102],

$$\gamma = \frac{\omega}{e}\sqrt{\frac{m_e c n \varepsilon_0 E_g}{I}}\,, \tag{3.3}$$

where ω is the laser frequency, I is the laser intensity, m_e is the electron effective mass, e is the fundamental electron charge, c is the speed of light, n is the linear refractive index, ε_0 is the permittivity of free space, and E_g is the bandgap of the material.

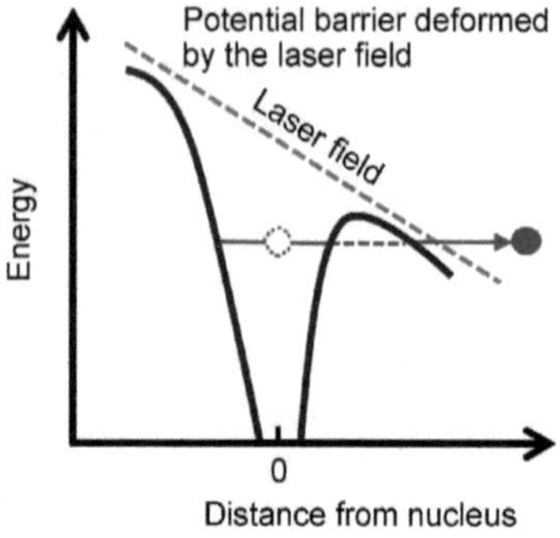

Fig. 3.4: Tunneling ionization induced by femtosecond laser irradiation.

When γ is much greater than 1, multiphoton absorption is dominant and when γ is much smaller than 1, tunneling ionization is dominant. For $\gamma \approx 1$, photoionization is induced by a combination of both processes. Thus, femtosecond laser scans induce strong absorption (electron excitation) even in transparent materials, thereby allowing high-quality microprocessing of glass materials.

3.2.3 Characterization of femtosecond laser processing

Compared with conventional laser fabrication techniques that use continuous wave (CW) lasers or lasers with nanosecond pulses, femtosecond laser processing has several distinct advantages.

(1) Elimination of thermal diffusion

The use of a focused femtosecond laser beam can significantly reduce thermal effects during material processing, due to the fact that most of the energy associated with the pulse can be deposited into electrons before being transferred to the lattice when the pulse duration is shorter than the electron-phonon coupling time (typically a few to a few tens of picoseconds), as indicated by the dark shaded region in Figure 3.5. Fundamental investigations into laser processing of materials have revealed that the time dependence of ultrafast electron and lattice dynamics is quite different for long and for ultrashort laser pulses. It is established that intense laser pulses with durations of nanoseconds can be used to anneal the lattice in a manner of thermal processing. For ultrashort laser pulses in the femtosecond regime, however, the electrons and the lattice are driven far out of equilibrium and disordering of the lattice can occur. During the interaction of an ultrashort pulse with a transparent material, the laser energy is first deposited into the electrons and then transferred to the lattice, with several regimes of electron excitation and relaxation. These regimes and the timescales of the corresponding processes are shown in Figure 3.2.

The free-electron temperature increases rapidly and becomes much higher than the lattice temperature, a process that can be characterized by a two-temperature model [103]. Material in the laser interaction region is then ejected in a very short time in the form of hot dense plasma, leaving the local lattice still 'cold.' This effect is obvious for femtosecond lasers with a low repetition rate (a few kHz or less), and even for a high repetition rate (a few hundred kHz or more) such a process can effectively suppress the formation of HAZ, thus enabling the fabrication of fine structures with micro- or nanoscale features. On the other hand, for laser pulse durations significantly longer than the electron-phonon coupling time, although the radiation energy is first transferred to the electrons, the electrons transfer it to the lattice before the pulse has terminated. This allows the electrons and the lattice to reach an equilibrium state, and the laser simply 'heats' the solid to its melting temperature within the duration of the

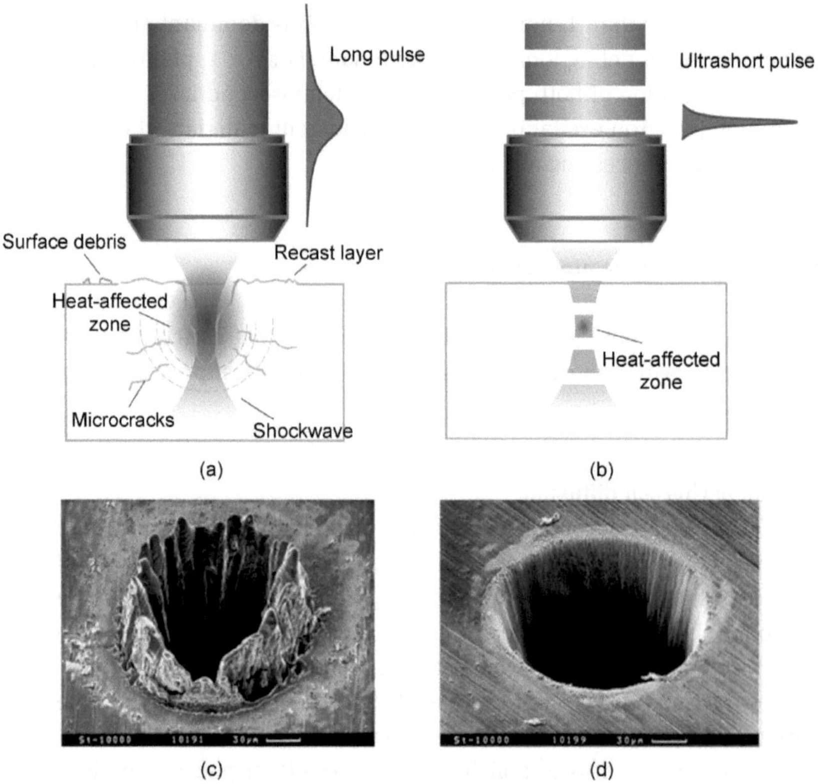

Fig. 3.5: Schematic illustration of the interaction of (a) long and (b) ultrashort laser pulses with transparent material. Scanning electron microscope (SEM) images of holes drilled in 100-μm-thick steel foils by ablation using (c) 3.3 ns-pulses and (d) 200 fs-pulses.

pulse. This gives rise to significant thermal diffusion, and consequently a reduction in fabrication quality and achievable resolution.

When the laser pulse width is shorter than the electron-phonon coupling time in laser-matter interactions, thermal diffusion to the interior of the material can be almost ignored. Most metals have an electron-phonon coupling time on the order of picoseconds [104], which is sufficiently long compared with the pulse width of femtosecond lasers. In this regime, when the material is heated to near the melting point T_{im} by femtosecond laser irradiation, the thermal diffusion length l_{d} is given by

$$l_{\mathrm{d}} = \left[\frac{128}{\pi}\right]^{\frac{1}{8}} \left[\frac{DC_{\mathrm{i}}}{T_{\mathrm{im}}\gamma^2 C_{\mathrm{e}}'}\right]^{\frac{1}{4}}, \tag{3.4}$$

where D is the thermal conductivity, C_{i} is the lattice heat capacity, C_{e}' is given by $C_{\mathrm{e}}' = C_{\mathrm{e}}/T_{\mathrm{e}}$ (where C_{e} is the electron heat capacity and T_{e} is the electron temperature), and γ is the electron-phonon coupling constant [105]. For example, when copper is heated to its melting point of $T_{\text{im}} = 1356\,\text{K}$ by a femtosecond laser, l_{d} is calculated to be 329 nm [106].

On the other hand, when the laser pulse width τ is much longer than the electron-phonon coupling time, l_{d} can be approximately estimated using

$$l_{\text{d}} = \sqrt{\kappa\tau}\,. \tag{3.5}$$

Here, κ is the thermal diffusivity. For copper, l_{d} is estimated to be $1.5\,\mu\text{m}$ for $\tau = 10\,\text{ns}$. Thus, femtosecond laser processing can clearly reduce the thermal diffusion length, reducing HAZ formation in the processed region.

(2) Internal processing

Another important benefit of femtosecond laser microfabrication is that it can be used to machine transparent materials in a 3D space-selective manner, because MPA only occurs in the beam focal volume, where the light intensity is sufficiently high. It is noteworthy that this unique 3D capability cannot be achieved using nanosecond UV lasers since when single photon absorption is involved, modification always begins at the surface and moves inwards, as illustrated in Figure 3.5 (a) and 3.5 (b). Even if a nanosecond laser with a wavelength sufficiently long for the material to be transparent is used, 3D processing still cannot be realized because a radiation intensity sufficiently high to induce MPA will also lead to strong thermal effects such as cracking, due to the long pulse duration. The ability to create arbitrary 3D structures within transparent substrates by the use of femtosecond lasers enables the production of novel device architectures with enhanced functionality and improved compactness.

The probability of nonlinear absorption processes such as multiphoton absorption and tunneling ionization occurring is strongly dependent on the laser intensity. For instance, the absorption cross section for n-photon absorption is proportional to the n-th power of the laser intensity. Therefore, nonlinear absorption can be efficiently induced only at laser intensity above a specific critical value that is dependent on both the material and the pulse width. When a femtosecond laser beam is focused inside a transparent material with adequate pulse energy, nonlinear absorption can be confined to a region near the focal point inside the material where the laser intensity exceeds the critical value. In this way, internal modification of transparent materials and fabrication of structures inside them can be performed, but only using ultrashort pulse lasers. Internal modification is currently widely applied to TPP, photonic device and biochip fabrication, and glass bonding, as discussed below.

(3) Subdiffraction-limited resolution

Again due to the characteristics of MPA, the effective processing profile of a femtosecond laser beam can be much narrower than its Gaussian intensity profile, as indicated in Figure 3.6 (a). This is because the distribution of absorbed energy for N-photon absorption is proportional to the N-th power of the laser intensity (N is the number of

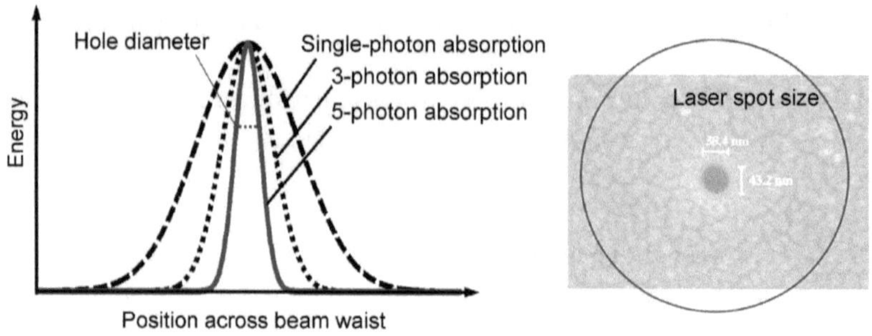

(a) (b)

Fig. 3.6: (a) Profiles of energy absorbed during single-photon absorption, three-photon absorption and five-photon absorption. The hole-diameter is indicated by the inner solid line. (b) SEM micrograph of nanometer-scale hole fabricated in glass. The dark circle indicates the $1/e^2$ focus spot size [107] (Reproduced with permission from PNAS. ©2004 by the National Academy of Sciences, U.S.A.).

photons simultaneously absorbed by the material to produce a transition from the VB to the CB). The effective beam diameter is then given by

$$D_{\text{eff}} = \frac{\omega_0}{N^{\frac{1}{2}}}\,, \tag{3.6}$$

where ω_0 is the actual beam diameter. Meanwhile, suppression of the HAZ by femtosecond laser processing restricts the laser-modified region to within a few nanometers of the volume where the laser fluence is above the threshold for inducing MPA. This makes it easy to achieve a fabrication resolution far beyond the optical diffraction limit by adjusting the intensity at the center of the Gaussian beam profile so that it is as close as possible to the threshold intensity for modification. For example, a resolution of $\lambda/20$ has been realized for both surface modification [107] (Figure 3.6 (b)) and internal modification [108], and a resolution of $\lambda/40$ has been reported for a TPP process [109].

In principle, the threshold effect provides no limit to the fabrication resolution, because the resolution can be significantly improved if the incident energy fluency is set to as near to the intensity at the center of laser beam as possible. Careful stabilization and control of the laser power has resulted in resolution better than 20 nm for TPP using a 780 nm femtosecond laser. However, in practice, fluctuations of the femtosecond laser output power make fabrication of 3D nanostructures with such high fabrication resolution rather difficult. Thus, the typical resolution for the fabrication of 3D nanostructures by TPP is 100–200 nm [29].

(4) Multifunctional integration

The tightly focused femtosecond laser beam can be used to precisely modify or even permanently change the physical and/or chemical properties of the material in a spa-

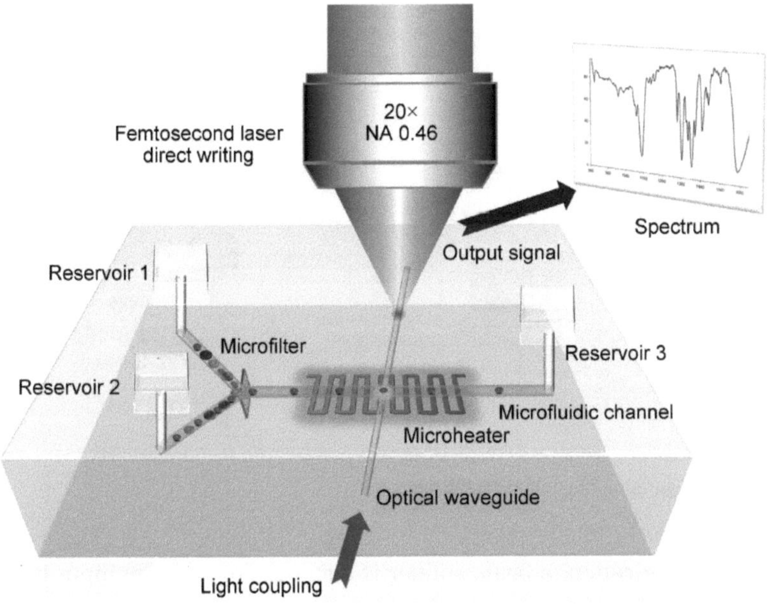

Fig. 3.7: Conceptual illustration of hybrid integration of multifuntionalities using FsLDW.

tially selective manner. A host of exciting novel effects, including surface/bulk microstructuring [63–66], subtractive micromachining [60–62], refractive index modification [53–59], selective metallization [110, 111], and TPP [67–72], can be achieved by FsLDW, which enables multifunctional integration in a single substrate.

Due to the powerful manufacturing capability of FsLDW, these components can be produced in a very similar manner. A typical conceptual illustration of the hybrid integration of multifuntionalities using FsLDW is shown in Figure 3.7. The fabrication of an optical waveguide is one of the first demonstrations of the potential of femtosecond laser micromachining for photonic applications in transparent dielectrics. Preparation of microfluidic channels can be realized by FsLDW with assisted chemical etching. Furthermore, a microheater component can be fabricated using femtosecond-laser-induced electroless plating. Detailed procedures of these processes will be illustrated in the following sections.

3.2.4 Femtosecond laser direct writing systems

Figure 3.8 shows a schematic diagram of a typical FsLDW system. Generally, a femtosecond amplifier system is required to process glass. A $\lambda/2$ plate in combination with a Glan prism is used to coarsely adjust the laser power, while an additional variable neutral density (ND) filter is used to tune the laser power more precisely. The exposure time is controlled by a mechanical shutter. To monitor the stability of the input

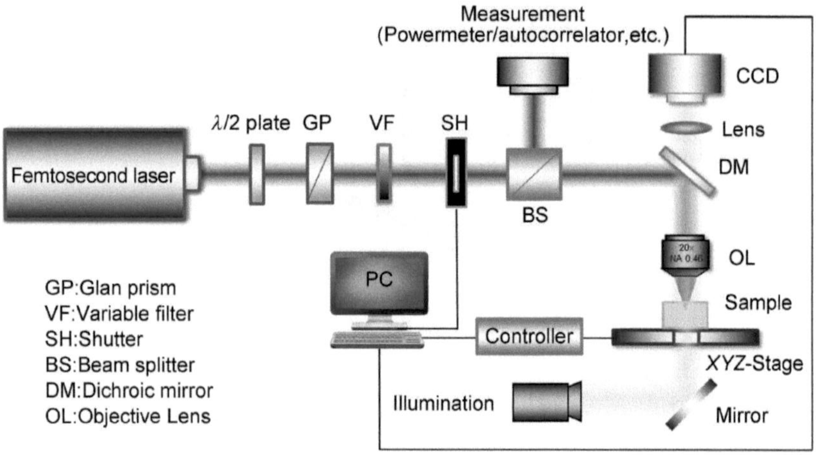

Fig. 3.8: Schematic configuration of general FsLDW system.

laser power during fabrication, a beam splitter is employed to split the input laser beam into two with a fixed intensity ratio. One of the split beams is monitored using a power meter, and an autocorrelatoris is used to evaluate the pulse duration [50]. In the general case, the second split beam is passed through a microscope with the objective of producing a tightly focused spot. For special purposes, however, other types of optical elements can be used, such as a cylindrical lens for line scanning during high-volume processing [112], or an axicon for producing a nondiffractive beam for high-aspect-ratio drilling [51]. The sample is placed on a high-precision *XYZ* translation stage controlled by a computer for 3D translation. A dichroic mirror and a charge-coupled device (CCD) camera connected to the computer are installed above the focusing lens to enable the fabrication process to be monitored in real time.

In femtosecond laser processing of transparent material, laser conditions such as the spatiotemporal beam profile [113], repetition rate [114], scanning speed [115], and polarization [116] can have significant impacts on the fabricated microstructure. These parameters can be precisely tuned using electrical and optical devices. A combination of a wave plate and a polarizer can be used to control the polarization of the laser beam, which is important for some types of applications [116, 117]. A ND filter or an attenuator can be used to produce inhomogeneous illumination, which can lead to a surprising 'quill' writing effect [118]. Using a programmable shutter or an external acousto-optic modulator, a burst of pulses can be produced for single-step writing of fiber Bragg grating (FBG) structures [119]. To compensate for the low throughput of conventional FsLDW, a point-by-point scanning method, a microlens array, or a diffractive optical element can be employed to achieve parallel processing [120, 121].

A key optical component in this system is the focusing lens because it determines the focal spot size, which is directly related to the fabrication resolution. As the femtosecond laser pulses possess a broad spectra, both spherical and chromatic

Tab. 3.2: Parameters of typical microscope objectives.

Objective lens	Numerical aperture	Working distance (mm)	Focal distance f (mm)
MPLN 5×	0.10	12.0	36
MPLN 10×	0.25	6.5	18
MPLN 20×	0.40	1.3	9
MPLN 50×	0.75	0.38	3.6
MPLN 100×	0.90	0.21	1.8

aberrations should be minimized when focusing them. For this reason, commercially available microscope objective lenses are frequently used for femtosecond laser processing, especially when high spatial resolutions are required. The fabrication resolution achievable with these objectives largely depends on their numerical aperture (NA), as well as the lateral direction, where the focal spot diameter is given by $1.22\,\lambda/\mathrm{NA}$, which is inversely proportional to the NA. However, a high-NA objective lens usually has a short working distance, limiting its ability to fabricate 3D microstructures deep in a substrate. Consequently, a tradeoff exists between the fabrication resolution and the working distance for each specific application. Parameters of typical microscope objectives are shown in Table 3.2. As a reference, in previous studies a relatively long working distance objective with an NA of 0.46 and a working distance of ~2 mm is often employed for writing optical waveguides and fabricating microfluidic channels in 2-mm-thick glass substrates. On the other hand, very high NA objectives (e.g. oil-immersion objectives) are often employed for surface nanostructuring or TPP using femtosecond pulses because these applications do not require long working distances as they involve surfaces and thin samples [122, 123].

Another important optical component in the FsLDW systems is a tunable attenuator for externally controlling the femtosecond laser power. To ensure a stable operation (e.g. minimizing fluctuations in the laser parameters, including the pulse width, pulse energy, and pointing direction), the output energy of the femtosecond laser should remain constant during fabrication. The pulse energy is usually varied by rotating a half-wave plate placed in front of a polarizer. If this combination of a half-wave plate and a polarizer does not allow the femtosecond laser pulses to be sufficiently attenuated because of the limited extinction ratio, neutral density filters can be inserted after the polarizer to achieve a wider tunable range.

In addition to the basic optical components mentioned above, a device for characterizing the spatiotemporal profiles of the femtosecond laser pulses is frequently required. To date, various devices have been developed for the characterization of ultrafast laser pulses. In the simplest case, such a device can be an autocorrelator by which the temporal duration or structure of ultrashort pulses can be quantitatively evaluated with a moderate accuracy. The autocorrelator measures the femtosecond pulse duration by splitting the pulse into two replicas, mixing them in a nonlinear optical crystal, and recording the generated nonlinear signals (e.g. second or third harmonics) as a function of the relative delay between the two replicas. As the intensity of the nonlinear signal relies on the overlap between the two replicas, the temporal duration can be estimated from the autocorrelation trace by assuming a certain functional form for the temporal shape of the pulses to be measured. The autocorrelation measurement cannot provide information on the electric field. Extracting the complete information of the electric field is critical for understanding the mechanisms behind the interaction between femtosecond laser pulses and the materials, which can be achieved either by FROG or SPIDER. For an overview of these common pulse characterization techniques, the reader can refer to some review articles [124, 125]. Fortunately, the pulse characterization devices based on the above-mentioned techniques have all been commercialized today, and can be purchased depending on the requirements of the specific applications.

Most current pulse characterization techniques, including autocorrelation, FROG, and SPIDER, can only be used to measure the temporal characteristics of the femtosecond pulses with a uniform temporal profile over the cross section of the laser beam. However, for investigating the microscopic processes involved in the FsLDW in various transparent materials, the direct characterization of the spatiotemporal profiles of the electric field of the tightly focused femtosecond laser pulses is sometimes more desirable. As in the focal volume, the spatiotemporal distortion of the pulses can easily occur due to the imperfection in the optical path alignment and/or the aberrations of the focusing systems. Such distortion can alter the laser-matter interaction at the microscopic scales, as discovered by Kazansky et al. [126]. In ambient air, the characterization of the femtosecond laser pulses at the focus can be realized by replacing the bulk nonlinear crystals used in the traditional techniques such as the FROG and SPIDER with nonlinear nanoprobes [127, 128]. However, techniques enabling the characterization of the femtosecond laser pulses focused on transparent materials are yet to be developed.

One drawback of the FsLDW is that it has a low throughput as it is a serial process. Fortunately, compact, high-power, high-repetition rate ultrafast laser systems have recently been demonstrated, although their pulses are much broader than the pulses generated by the Ti:sapphire systems [91, 92]. With these lasers, the scan speed in direct writing can be dramatically increased by replacing the XYZ motion stage with a galvo scanner [129].

3.2.5 Overview of beam/pulse shaping methods

Most of the FsLDW processes have displayed the manner of transverse direct writing, i.e. the writing direction is perpendicular to the laser propagation direction. Since there is a tradeoff between the fabrication resolution and the working distance in a commercially available objective lens (e.g. an NA of 0.46 corresponding to a working

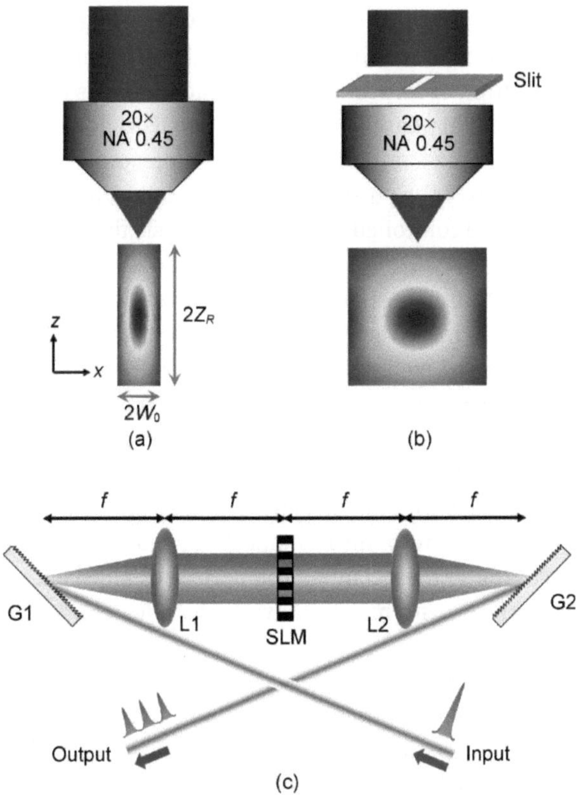

Fig. 3.9: Schematic of spatial and/or temporal beam shaping methods typically used in FsLDW: (a) conventional focusing (CF), (b) slit beam shaping, (c) temporal pulses shaping.

distance of ~2 mm), and a relatively longer distance is required for deep processing of bulk material, the focal spot produced by an objective lens typically has an asymmetrical shape elongated in the direction of propagation, resulting in unbalanced lateral and axial resolutions, i.e. the lateral size of the focal spot is given by $1.22\,\lambda/\mathrm{NA}$ while the axial size is $4\lambda/\mathrm{NA}^2$. Efforts have therefore been made in solving this problem [130–133], of which the slit beam shaping method has been widely employed due to its simplicity. As shown in Figure 3.9 (b), a narrow slit with a width of ~500 µm can be placed right above a 20× objective lens, thus forming microstructures with symmetrical cross sections when the laser direct writing direction is parallel to the orientation of the slit.

The temporal characteristics of ultrashort pulses such as the pulse duration and temporal profile of the pulse also show significant effects in some special processing [113, 134]. The developed pulse shaping methods in ultrafast technology [96] can be borrowed by femtosecond laser processing, and used in improving the fabrication quality as well as high efficiency. As shown in Figure 3.9 (c), a typical optical system for temporal pulse shaping consists of a pair of gratings and a pair of focusing lenses, which are arranged in a 4-f configuration for zero dispersion. A pulse-shaping mask, which can be a SLM, a deformable mirror, or an acoustic-optical modulator, is placed in the Fourier plane for tuning the spectral phase and/or amplitude. This device can generate almost arbitrary pulse profiles with a resolution (e.g. finest temporal feature) that is limited only by the spectral width of the pulses, enabling photoionization and electron-electron impact ionization to be controlled even on an intrinsic time and intensity scale.

Among different laser processing techniques, photoinscription employing ultrashort laser pulses has demonstrated increased potential for three-dimensional optical functionalization of bulk transparent materials. The resulting refractive index changes can be viewed as building blocks for embedded optical functions, used for example in wave guiding applications. The requirements of a desired photonic response involve precise adjustments of the refractive index, which usually depend on the material relaxation paths. Mechanisms with distinct characteristic timescales are involved, such as defect creation or thermomechanical compaction, and by adjusting irradiation conditions, a certain control of their competition may be established, resulting in a better design of the material transformation. Advanced strategies are then necessary to improve the irradiation results.

Recently, beam manipulation concepts have been developed, which allow a modulation of the energy feed according to material transient reactions, thus enabling a synergetic interaction between light and matter and, therefore, optimal results. The emerging idea is that suitable light time delivery can enable the control of the excitation and relaxation processes with consequences for the final modification. The photo writing technique may therefore be conveniently tuned to determine suitable material behaviors, i.e. positive refractive index changes in materials of optical interest, with both fundamental and technological consequences.

This emphasizes a fundamental question; namely, how irradiation results can be improved, taking into account various criteria for quality processing. Considering the complexity of processes occurring between the initial excitation and the final structural transformation, the task of defining the possible improvement factors is rather difficult, requiring more efficient optimization procedures. We consequently discuss structuring approaches based on adaptive spatiotemporal pulse manipulation using automated light modulators capable of influencing ultrafast laser-induced processes inside transparent materials to achieve better processing results. Two strategies for gaining impact on material transformations under light exposure will be presented: adaptive temporal tailoring techniques and spatial beam forming, together with their integration in feedback-driven approaches. These strategies are discussed in the context of optical functionalization for delivering specific designs of the refractive index, accompanied by concepts of efficient processing approaches. This also involves an engineering aspect related to the simultaneous processing of structural modifications in 3D arrangements where a feasible solution is represented by dynamic spatial beam shaping techniques. The approach has a dual aspect and includes corrections for beam propagation errors and spatial intensity distributions in desired forms for parallel processing, mastering both the nonlinearity and the geometry of interaction. Photowritten structures can be arranged in patterns generating complex propagation effects. Offering then multiple opportunities for upgrading laser interactions and responding 'intelligently' to material reactions, adaptive schemes in spatiotemporal domains can have an extended range of action, enabling applications that can largely benefit from light-matter synergies.

More recently, a technique termed spatiotemporal focusing has been proposed as a very attractive new choice [135, 136] in femtosecond laser microfabrication. The hope is that by focusing a spatially dispersed pulse, the duration manifests itself as a function of the distance away from the focus, allowing for a significant reduction of the peak intensity of the pulse out of the focus. Compared with CF, this approach is advantageous in the achievement of 3D isotropic resolution and suppression of nonlinear effects out-of-focus in the materials. Details of this technique will be further described in Section 3.3.

3.3 Spatiotemporal focusing of femtosecond pulses for materials processing

3.3.1 Introduction

In 2005, simultaneous spatial and temporal focusing (SSTF), also named 'spatiotemporal focusing,' 'space-time focusing' or 'temporal focusing (TF),' was first developed by the bio-imaging community for suppressing the background signal in two-photon fluorescence microscopy, thus improving the axial resolution, as well as achieving high speed imaging [137, 138]. The core of this technique involves directing the laser pulse through an optical dispersive element (e.g. prisms, gratings) to produce an angularly dispersive beam. In this situation, the spectral components that comprise the ultrashot pulse are not spatially overlapped, and thereby cannot add together to produce a transform-limited pulse in time. The spatially separated frequencies can be recombined together through an objective lens, producing a shortest pulse only at the focal plane while the out-of-focus pulse is stretched. This focusing mode can localize the peak intensity in both space and time domains, thus achieving an improved axial resolution. In 2010, the spatiotemporal focusing method was implemented into 3D femtosecond laser micromachining, leading to tremendous advances in improving fabrication resolution, eliminating nonlinear self-focusing (Figure 3.10), and increasing fabrication efficiency [135, 139–141].

Compared with CF, this approach is advantageous in that: (1) the axially localized peak intensity of the pulse can substantially suppress self-focusing out of focus, while providing a 3D isotropic fabrication resolution for FsLDW or offering the ability for

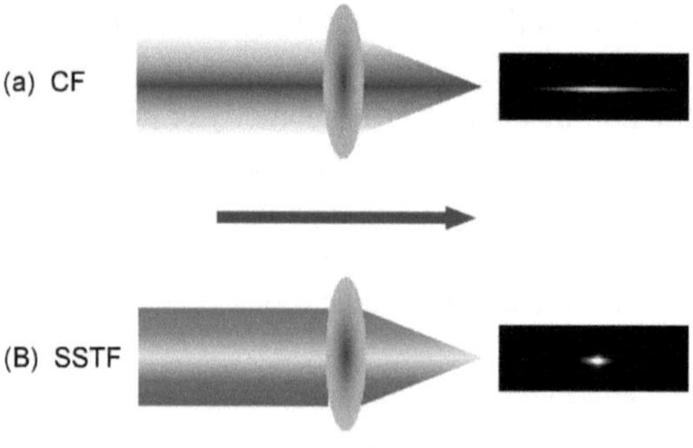

Fig. 3.10: Schematic illustration of femtosecond filamentation in air using (a) CF and (b) simultaneous spatiotemporal focusing (SSTF).

low-NA micromachining of bulk transparent materials, which has unique advantages in femtosecond laser processing of nonlinear crystals and bio-tissues; (2) it allows operation at a wide-field laser exposure scheme, thus enabling high-throughput 3D lithographic microfabrication [136, 142]; (3) the suppression of out-of-focus nonlinear interaction, together with precisely manipulating the spatiotemporal characteristic in a wide range, facilitate unraveling the fundamentals behind a couple of exotic ultrafast optical effects in light-matter interaction, such as the enhancement of clamped intensity in femtosecond filamentation, second harmonic generation (SHG) in centrosymmetric gas, and the intensity plane tilt (IPT) effect at the focal plane [143, 144].

In the field of micromachining, we categorize typical SSTF systems according to their specific configurations and applications into two types: (1) wide-field SSTF, comprising a single grating and a 4-f system, and with the grating arranged right at the focal plane of the objective lens; and (2) focusing-mode SSTF, consisting of a parallel grating pair, an objective lens, and a chirp compensator to compensate the negative chirp introduced by the grating pair. The wide-field SSTF technique was originally proposed for high-speed depth-resolved microscopy [137, 138], and has since been extensively studied in optogenetic control of neuronal activities [145, 146]. By transferring the pattern into the substrate (e.g. photoresist) with this wide-field SSTF illumination, high-throughput 3D lithography microfabrication based on the TPP processing has been demonstrated [136], and is meanwhile capable of optically sectioning the fabricated microstructures in a 3D inspection manner [142]. On the other hand, the focusing-mode SSTF has been widely used for cross-sectional shape control of FsLDW, e.g. by focusing this spatially dispersed laser beam with a 20× (0.46 NA) objective lens an approximately spherical focal spot can be created, enabling FsLDW of 3D optical waveguides or microfluidic channels with isotropic circular cross-sections [136, 139]. It is also reported that by using a low NA parabola mirror (f = 25 mm, NA = 0.05), this technique can be applied to the fabrication of microfluidic devices by ablation through the back surface of 6 mm-thick fused silica substrates [140]. Recently, fabrication of nonlinear material such as biotissues using this very similar setup has been successfully demonstrated, promising some previously inhibited surgical implementations [147].

Another characteristic of SSTF is that it can inherently induce the pulse front tilt (PFT) [141, 148], accompanied with an intensity plane tilt bizarrerie when arranged in a focusing mode [149]. For FsLDW applications, it is also demonstrated that SSTF can be used for 'quill writing,' i.e. the laser-induced modifications in homogeneous materials show a remarkable change when the writing direction is reversed. Moreover, the tilting angle can be precisely controlled in a wide range by adjusting the spatial chirp of the incident beam at the entrance of the focal lens, which may find potential in controlling nanotexture formation such as polarization-sensitive devices fabrication, superhydrophobic/hydrophilic surface processing, and so on. The mechanism underlying this effect has not yet been fully understood, though it has long been considered to be closely linked to the structure existing in the electron plasma. A recently revealed IPT effect brings this exotic phenomenon a step closer to being deciphered [149].

Finally, this temporally focused femtosecond pulse can be considered a special case of the ultrafast laser beam in high field laser physics research, and has hastened a variety of novel ultrafast optical phenomena. When implemented into femtosecond filamentation investigations, significant enhancement of the peak intensity beyond the conventional clamped intensity has been demonstrated [143], which would be of great interest in remote sensing for achieving better spatial (longitudinal) resolution and high signal-to-noise ratio. Meanwhile, it has recently been shown that unexpectedly strong SHG can be produced in a centrosymmetric argon gas by use of temporally focused femtosecond laser pulses, which may originate from the gradient of nonuniform plasma dynamically controlled by such a special laser field [144]. Although these experiments are carried out in air, we expect a similar effect can occur in glass, which can lead to stronger modification inside glass materials. These novel effects may enable fascinating possibilities in the synthesis of new materials by femtosecond laser induced shock wave.

3.3.2 Principles of spatiotemporal focusing methods

Currently, focusing-mode SSTF has been widely adopted in materials processing due to its flexibility for FsLDW. In the following discussion, we refer to spatiotemporal focusing as this scheme. Figure 3.11 shows a schematic configuration of this spatiotemporal focusing, where a single-pass compressor G1-G2 is used to disperse the spectrum components of the input pulses in space, before which the pulse should be prechirped to compensate the negative chirp induced by G1-G2. Recombination of the spectrum components is then achieved by a lens L, and thus the pulses can only reach the shortest duration at the focus of the lens while being broadened out of the focus.

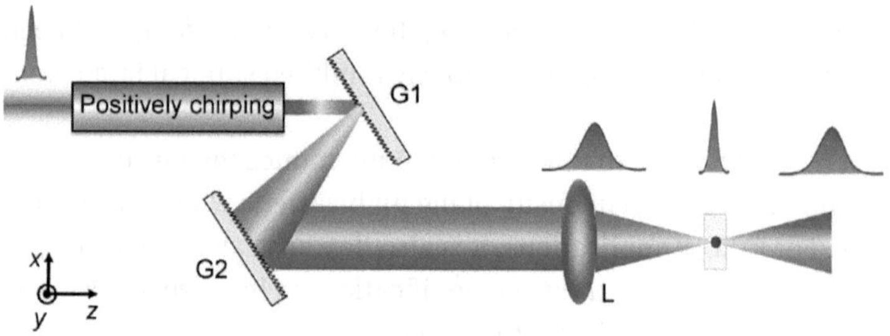

Fig. 3.11: Schematic configuration of spatiotemporal focusing; G1&2: grating, L: lens.

Theoretically, calculation of the spatiotemporally focused femtosecond laser beam propagation can be achieved using Fresnel–Kirchhoff's diffraction theory [135, 139, 149]. Assuming that the temporal chirp of the incident pulse is precompensated, and that the input spectral phase is $\varphi(\omega) = 0$, the normalized light field of a spatially dispersed pulse E_1 at the entrance aperture of the lens thus can be expressed as:

$$E_1(x, y, \omega) = A_0 \exp\left[-\frac{(\omega-\omega_0)^2}{\Delta\omega^2} + \mathrm{i}\frac{1}{2}\varphi_{2\mathrm{in}}(\omega-\omega_0)^2\right] \exp\left\{-\frac{[x-\alpha(\omega-\omega_0)]^2 + y^2}{w_{\mathrm{in}}^2}\right\}, \quad (3.7)$$

where A_0 is the constant field amplitude, ω_0 is the carrier frequency, $\Delta\omega$ is the bandwidth ($1/e^2$ half width) of the pulses, and w_{in} is the initial beam waist ($1/e^2$). φ_{2in} is the initial second-order phase, $\alpha(\omega-\omega_0)$ is the linear shift of each spectral component at the entrance aperture of the lens, with a detailed derivation of α in the Appendix in this section.

After passing through the lens and propagating a distance z from the lens, the light field can be written as:

$$E_2(x, y, z, \omega) = \frac{\exp(\mathrm{i}kz)}{\mathrm{i}\lambda z} \int\int_{-\infty}^{\infty} E_1(\xi, \eta, \omega) \exp\left(-\mathrm{i}k\frac{\xi^2+\eta^2}{2f}\right) \exp\left[\mathrm{i}k\frac{(x-\xi)^2+(y-\eta)^2}{2z}\right] d\xi d\eta\,, \quad (3.8)$$

where k is the wave vector and f is the focal length of the lens. The intensity distribution in the time domain can be obtained by performing an inverse Fourier transform of $E_2(x, y, z, \omega)$ as:

$$I_{\mathrm{SSTF}}(x, y, z, t) = |E_2(x, y, z, t)|^2 = \left|\int_{-\infty}^{\infty} E_2(x, y, z, \omega) \exp(-\mathrm{i}\omega t) d\omega\right|^2. \quad (3.9)$$

The intensity distributions of the focused spatially dispersed beam in both the XZ and YZ planes can be obtained according to equation (3.9). In the numerical simu-

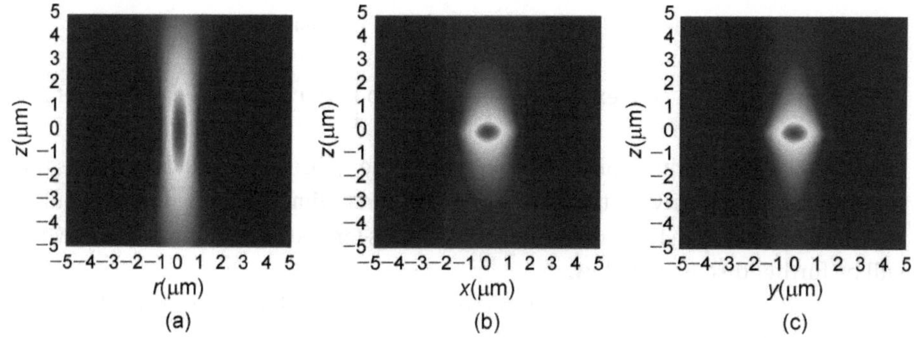

Fig. 3.12: Theoretical calculations of laser intensity distributions at the focus produced by an objective lens with (a) CF and with (b, c) spatiotemporal focusing technique in the *XZ* and *YZ* planes, respectively.

lation, the pulse duration and the spectrum bandwidth are 40 fs, 30 nm respectively. The distance between the gratings (1200 l/mm, blazing at 45°, which is also the incident angle of the laser beam) is set to be ~180 mm. The beam waist of the input laser is $2w_{\text{in}} = 5$ mm. The intensity distribution of the CF laser beam can be obtained by substituting $\alpha = 0$ into equation (3.7). As shown in Figure 3.12, one can clearly see that a nearly spherically shaped intensity distribution has been obtained with the spatiotemporal focusing.

In order to obtain an analytical expression of the peak-intensity distributions near the focus for the SSTF beam, the calculations are simplified by employing the slow varying envelope approximation (SVEA) to equation (3.9), namely, assuming that the wave vector k for each frequency is approximately the center wavelength of the pulse k_0, which is valid for a pulse width of more than 100 fs. The peak intensity can then be expressed as

$$I_{\text{SSTF}}(x,y,z,t) \propto \left| R(z)\exp\left(-\frac{x^2+y^2}{Q}\right)\exp\left[-\frac{\Delta\omega^2}{2(1+\psi)}\left(t+\frac{f\alpha x}{P}\right)^2\right]\right|^2 , \tag{3.10}$$

where

$$\psi = \mathrm{i}\frac{(z-f)\alpha^2\Delta\omega^2}{P} , \tag{3.11}$$

$$R(z) = \frac{1}{1-\frac{z}{f}+\mathrm{i}\frac{z}{kW_{\text{in}}^2}}\frac{1}{\sqrt{1+\psi}} , \tag{3.12}$$

$$P = \frac{zf}{k} + \mathrm{i}(z-f)W_{\text{in}}^2 , \tag{3.13}$$

$$Q = \left(\frac{zf}{k^2W_{\text{in}}^2} + \mathrm{i}\frac{z-f}{k}\right)\Big/\left(1-\frac{\mathrm{i}f}{kW_{\text{in}}^2}\right) . \tag{3.14}$$

At the focal plane, where $z = f$, the spatiotemporal intensity becomes

$$I_{\mathrm{SSTF}}(x, y, z = f, t) \propto \exp\left(-\frac{x^2 + y^2}{f^2/2k^2 W_{\mathrm{in}}^2}\right) \exp\left[-\Delta\omega^2 \left(t + \frac{k\alpha x}{f}\right)^2\right]. \tag{3.15}$$

Obviously, there is a time shift $t_0 = k\alpha x/f$ along the x-direction, which indicates that the shortest pulses arrive at the focal plane at different times in the x-direction but arrive simultaneously in the y-direction. This effect is termed as PFT, the origin of which will be further discussed below.

There is also a growing interest in the mathematical characterization of the spatiotemporally focused laser beam, including both its free-space propagation and the nonlinear interaction with materials. Early theoretical models are mostly based on the Fresnel diffraction integral calculations, in which a paraxial approximation is required, as depicted above. By use of the SVEA (i.e. the spectral bandwidth of the pulse is a small fraction of the carrier frequency for 100-fs pulses typically used in two-photon microscopy applications) or neglecting the high-order chirp, analytical expressions for a 4D spatiotemporal profile of the pulses and even the varying pulse duration can be intuitively described [139, 150]. By taking into account the refractive index of the material, a generalized Huygens–Fresnel integral (i.e. Collins integral) method can be employed by describing each optical element (including the propagation medium) with an ABCD matrix [149]. By considering each spectral component of the spatially dispersed laser pulses as Gaussian beamlets and incorporating the spectral phase information, a double-ABCD method is developed to describe the spatiotemporal structure of this kind of beam, in which analytical expressions for the varying pulse duration, the longitudinal intensity profile, PFT, and IPT can be derived [151]. To precisely characterize temporally focused ultrashort pulses with large bandwidth and/or ones focused with a high NA objectively, a vector TF theory is employed to analyze the nonparaxial case of this technique, revealing that the analytic solution of the paraxial approximation gives an axial resolution almost twice that of the vector case [152]. Meanwhile, a Rayleigh–Sommerfeld diffraction process with low-order Zernike modes considered is performed to analyze the effects of aberrations in spatiotemporal focusing of ultrashort laser pulses [153]. In addition to the above theoretical models on the beam propagation characterization of temporal focusing, several works have been completed on more complicated beam propagation through scattering media [154] or nonlinear interaction with material [155], which can be potentially used for predicting temporal focusing characteristics in practical micromachining applications, and getting a full understanding of this burgeoning technique.

3.3.3 3D isotropic fabrication resolution using spatiotemporal focusing

Intuitively, improving the axial resolution can be used to match the lateral resolution, thus realizing a symmetrical cross sectional shape of microstructure when using fem-

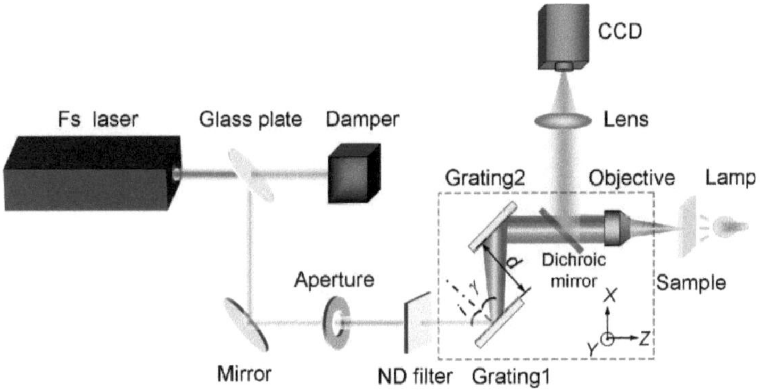

Fig. 3.13: Experimental setup of femtosecond laser micromachining using spatiotemporal focusing technique.

tosecond laser direct writing. A schematic of the experimental setup for spatiotemporally focused laser direct writing is shown in Figure 3.13. The femtosecond lasers (Libra-He/Rega-9000/Legend-Elite; Coherent Inc.) employed for the direct writing are the chirped pulse amplification (CPA) systems, with pulse durations of 40~200 fs and center wavelengths of 800 nm, which have been described in detail elsewhere [135, 139, 149]. In this experiment, the input laser beam size is controlled by an adjustable circular aperture. A pair of parallel gratings (blazing at 800 nm) is used to separate the spectral components of the incident pulse in X coordinates; that is, to produce a spatial dispersion that is different from the grating-lens setup used in spatiotemporal focusing bio-imaging systems.

To precompensate all the chirps through the optics, the input pulse is positively prechirped by decreasing the distance between the grating pair of the CPA pulse compressor. The best precompensation of dispersion can be achieved when the strongest ionization (i.e. the brightest plasma) of the air is observed at the focus of the objective. The power of the spatial dispersed pulse is adjusted by neutral density (ND) filters, before being focused by an objective lens (20×, 0.46 NA, f = 8 mm; BX-51, Olympus Inc.) into a glass sample. The glass can be arbitrarily translated by a PC-controlled XYZ stage (P-545; Physik Instrumente Inc.).

Experimental proofs of the cross sectional shape control by comparing the CF with the SSTF beam are also demonstrated. To do this, 3D microfluidic channels are fabricated in fused silica by FsLDW followed by chemical wet etching. The laser power and scanning speed are chosen to be 3.5 mW and 50 µm/s respectively. Other parameters are chosen according to the above simulation. After the laser irradiation, the sample is subjected to a 150-min wet chemical etching in a solution of 10% HF diluted with water in an ultrasonic bath. The cross sections of the channels are examined by cutting and polishing the sample. The optical micrographs of the cross sections of microflu-

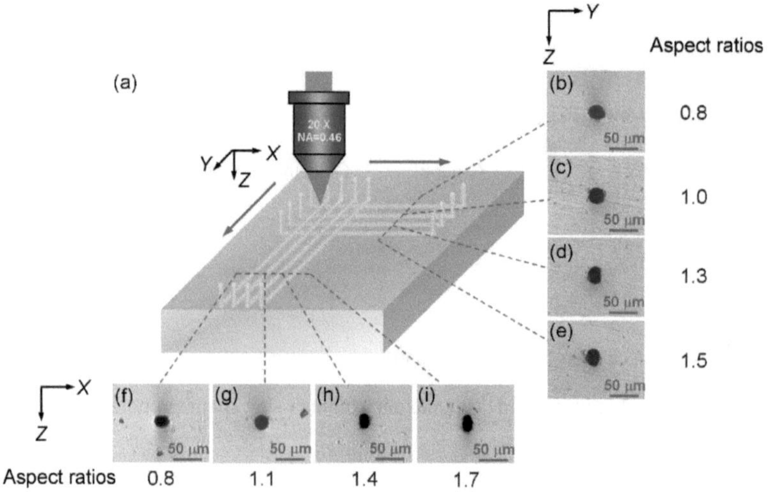

Fig. 3.14: (a) Configuration of 3D microfluidic channels fabricated using the temporal focusing method. The arrows indicate the translation directions of the stage. (b)–(i) Optical micrographs of cross sections of microfluidic channels fabricated with different beam sizes and laser powers.

idic channels, shown in Figure 3.14, are clearly evident of the validity of these shape control methods.

It should be specifically noted that the threshold intensity of femtosecond laser fabrication of the transparent materials not only strongly depends on the peak intensity but also can be affected by the ultrashort laser pulse duration. In addition, the axial resolution would also be affected by the order of optical nonlinearity of the interaction between the femtosecond laser pulse and the transparent material (e.g. the n-th-order optical nonlinearity means the simultaneous absorption of n photons during the light-matter interaction in this situation). Therefore, since the order of optical nonlinearity of the interaction between the femtosecond laser and fused silica is not clearly defined in this experiment, quantitative reproduction of the experimental results by theoretical simulation is difficult.

3.3.4 Spatial and temporal characterization of spatiotemporally focused spots

3.3.4.1 Characterization of pulse front tilt of spatiotemporally focused pulses

The PFT, the dependence of the arrival time of the pulse on the transverse coordinate, is controlled by the combination of angular or lateral spatial chirps and spectral chirps [156]. In the present case of SSTF, the temporal chirp is precompensated and the PFT is introduced by the angular dispersion alone. In the above experimental configuration, the pulse is only spatially chirped before the lens, and the lens converts this spatial chirp to an angular chirp. The Gaussian beam written in term of the amplitude and phase is

$$E(x, y, z, \omega) = A(x, y, z, \omega) \exp\left(i\varphi(x, y, z, \omega)\right) . \tag{3.16}$$

A particular frequency component is then treated as an individual beamlet. After being focused by the lens, the amplitude function for the beamlet propagating at a frequency-dependent angle θ_x can be written as

$$A(x, y, z, \omega) = \frac{A(\omega)w_0}{w(z)} \exp\left[-\frac{(x - z\tan\theta_x)^2 + y^2}{w^2(x)}\right] , \tag{3.17}$$

and the phase function, under the paraxial approximation can be written as

$$\varphi(x, y, z, \omega) = n(\omega)k_0 x \sin\theta_x + n(\omega)k_0 z\left(1 - \frac{1}{2}\sin^2\theta_x\right) - \eta(z) + n(\omega)k_0 \frac{(x - z\sin\theta_x)^2 + y^2}{2R(z)} , \tag{3.18}$$

where $R(z)$ is the radius of the curvature of the beam's wavefront and η is the Gouy phase shift. The frequency dependence of the beamlet angle is defined as

$$\sin\theta_x \approx \theta_x = \frac{\alpha(\omega - \omega_0)}{f} . \tag{3.19}$$

For future reference the spatial chirp rate is related to the angular chirp rate α through

$$\beta = \frac{\alpha\Delta\omega}{w_{in}} = \frac{2\alpha}{\tau_0 w_{in}} . \tag{3.20}$$

Expanding to the first order spectral phase of equation (3.18) with the frequency-dependent angular chirp, we can obtain the position-dependent group delay:

$$\varphi_1(x, z) = \frac{z}{c} + \frac{\alpha\omega_0 x}{cf}\left(1 - \frac{z}{R(z)}\right) + \frac{x^2}{2cR(z)} . \tag{3.21}$$

The first term is the arrival time of the pulse, and the last term, which is present even without the spatial chirp, represents a pulse front curvature that results from the divergence of the beam away from the focal plane. The middle term corresponds to the PFT. When equation (3.21) is evaluated at the focus, $z = 0$, the PFT reduces to what we found earlier in equation (3.15).

Although the spatiotemporal focusing method was originally proposed to improve axial resolution, it can also yield interesting applications in nonreciprocal writing. Recently, Vitek et al. [141] have reported that the spatiotemporally focused laser beam, which naturally creates a focal spot with several orders of magnitude larger PFT, can induce nonreciprocal writing in glass more quickly and efficiently. To gain a deep insight of nonreciprocal writing, quantitative characterization of the PFT of SSTF in the focal spot is necessary.

To measure the tilting angle of front-tilted pulses, an interferometric measurement scheme has recently been proposed, which can be used for the characterization of

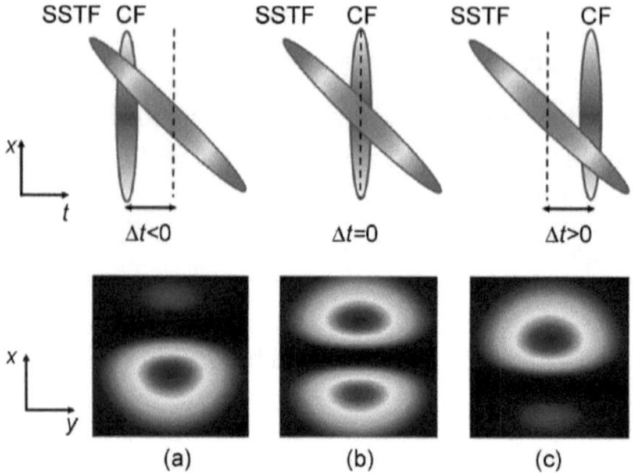

Fig. 3.15: Schematic illustration of the concept of interferometric characterization of PFT of spatiotemporally focused femtosecond laser pulses.

spatiotemporally focused pulses generated in either air or bulk transparent materials [148]. The concept of this technique is illustrated in Figure 3.15. It is known that when focusing a femtosecond laser pulse with a CF system (i.e. an objective lens), no PFT effect will be induced. This pulse will be used as a reference pulse in the experiment. However, spatiotemporal focusing of a femtosecond laser pulse will lead to a sweep of the focal spot across the focal plane at a sweeping velocity, which is inversely proportional to the degree of the PFT. The sweeping direction depends on the direction of the spatial chirp of femtosecond laser pulses before they enter the objective lens. Thus, by intentionally varying the time delay between the reference pulse and the spatiotemporally focused femtosecond pulses, the intersection of two pulses will induce a line-shaped interference pattern, which shifts linearly with the time delay between the two pulses, as illustrated in Figure 3.15. The line-shaped interference pattern formed in the focal plane of the lens can then be directly imaged onto a CCD using a second objective lens. By plotting the shift of the interference fringe as a function of the time delay between the two pulses, the tilting angle of the spatiotemporally focused pulse can be retrieved.

Figure 3.16 schematically illustrates the experimental setup. The femtosecond laser system (Legend-Elite, Coherent Inc.) consists of a Ti:sapphire laser oscillator and amplifier, and a grating-based stretcher and compressor, which delivers ~6 mJ, ~40 fs pulses centered at 800 nm wavelength at a 1 kHz repetition rate. Before the amplified laser beam is recompressed, it is split into two using a 1:1 beam splitter. One beam, which is used to produce the spatiotemporally focused pulse, is spatially dispersed along the x-direction by a pair of 1500 lines/mm gratings (blazing at ~53°), which are arranged to be parallel to each other. The distance between the two gratings is adjusted to be ~780 mm to compensate for the temporal dispersion of pulses. After being dispersed by the grating pair, the laser beam is measured to be ~40 mm ($1/e^2$)

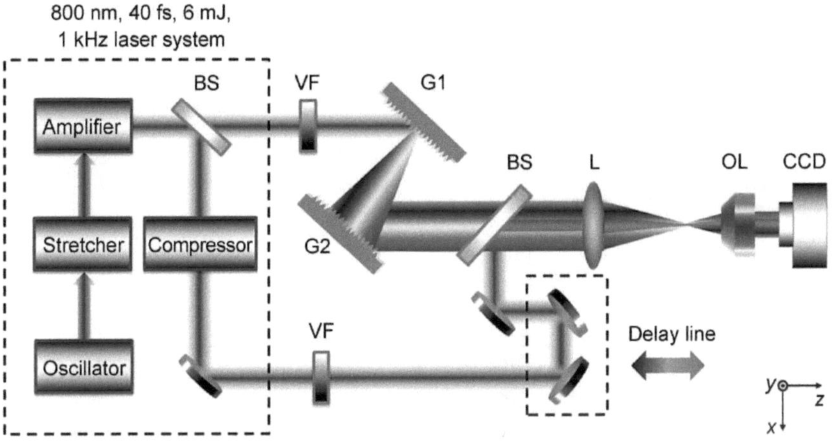

Fig. 3.16: Schematic of the experimental setup; BS: beam splitter, G1-2: gratings (1500 l/mm), L: lens (f = 75 cm), OL: objective lens (20×, NA=0.40), VF: variable neutral density filter.

along the x-axis and ~10 mm($1/e^2$) along the y-axis, as indicated in Figure 3.16. The other beam is compressed by passing through the compressor to generate the reference beam. The spatially dispersed beam and the reference beam are then collinearly recombined using another 1:1 beam splitter. A delay line is inserted into the optical path of the reference beam to adjust the time delay between the two beams. These two beams are then focused using a plane-convex lens with a focal length of $f = 750$ mm and a diameter of 100 mm. The interference patterns are imaged with an objective lens (20×, NA=0.4) on the CCD camera (Wincamd-ucd23) with an integration time of 5 ms. The average power of each beam is individually controlled using a variable neutral density (ND) filter for optimizing the contrast of the interference fringes.

The measurement results are shown in Figure 3.17. The laser pulse energy is sufficiently low to avoid any nonlinear propagation effects under such focusing conditions. Figure 3.17 (a) and 3.17 (b) present the images of the focal spots generated with the CF and SSTF systems on the focal plane captured by the CCD camera, respectively. Both the images show circular focal spots with a Gaussian-like profile. In Figure 3.17 (c)–(g), the images formed by simultaneously focusing the two pulses with both the CF and SSTF systems are presented, in which the time delays between the two pulses are −66.8, −33.4, 0, 33.4, and 66.8 fs, respectively. Here, the positive time delay means that the pulse focused by the SSTF system arrives earlier than the pulse focused by the CF system, and vice versa. Interference fringes have been formed due to the coherent superposition between the two pulses. Due to the fluctuation in the lengths of the optical paths for the two pulses, the interference between the two pulses can be either destructive or constructive. The sweeping velocity of the interference fringe indicated by Figure 3.17 (c)–(g) directly maps out the sweeping velocity of the front-tilted focal spot in the focal plane, i.e. the degree of PFT.

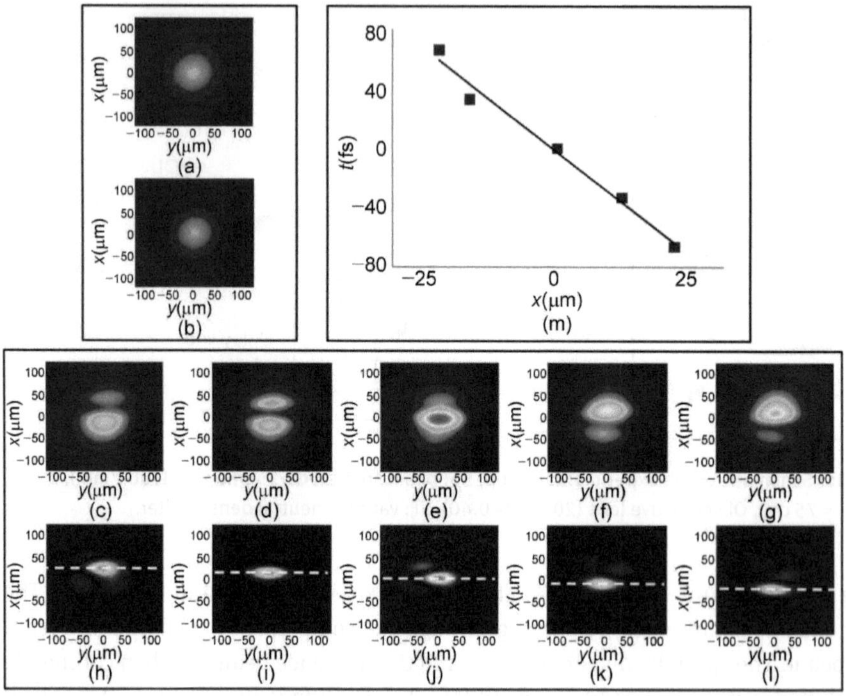

Fig. 3.17: Intensity distribution of the focal spot of the pulses focused with (a) SSTF, and (b) CF systems in the focal plane in air. Intensity distributions of the focal spots formed in air by superposition of the two beams are shown in (c)–(g) with different time delays of the two pulses (h)–(l) being the interference fringes extracted from (c)–(g), respectively. (m) Displays the linear fitting of the position of the interference fringe as a function of the relative time delay.

Interference fringes from the images in Figure 3.17 (c)–(g) by which the center of the fringes can be unambiguously determined by removing the background signals, can be attributed to the incoherent superposition between the two pulses (i.e. the coherent superposition between the two pulses occurs only in the intersection of the two pulses as illustrated in Figure 3.15). The images of Figure 3.17 (a) and 3.17 (b) are subtracted from each image in Figure 3.17 (c)–(g).

The image processing may generate negative values for some pixels particularly when the destructive interference occurs between the two pulses. Therefore, after the subtraction, the value of each pixel is squared. The reproduced images in Figure 3.17 (h)–(l) are generated by performing the above-mentioned image processing on Figure 3.17 (c)–(g), respectively. In Figure 3.17 (h)–(l), the center of interference fringes can be rather precisely determined by looking for the locations of the highest intensities in the fringes, as indicated by the white dots. By plotting the central locations of the interference fringes as a function of time delays between the pulses generated with CF and SSTF systems in Figure 3.17 (m), the data is fitted with a linear

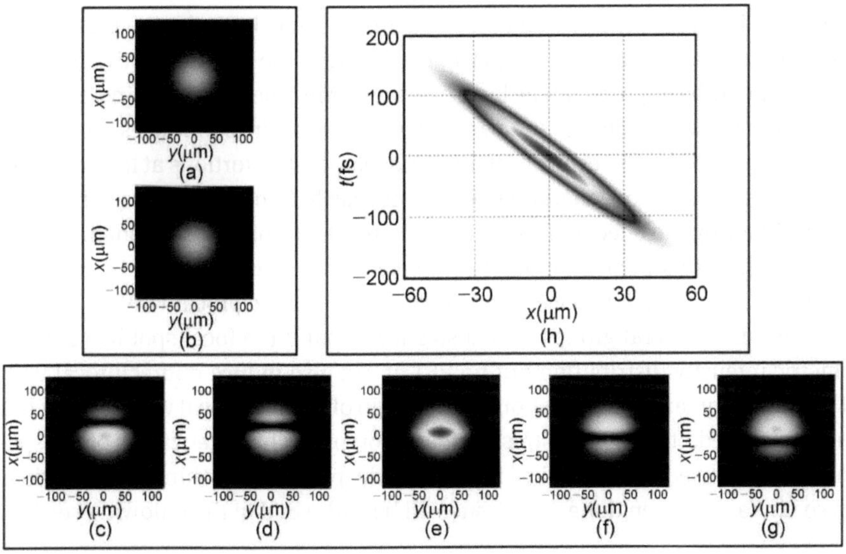

Fig. 3.18: Calculated intensity distribution of the focal spot in the focal plane produced in air with (a) STF and (b) CF systems. Calculated intensity distributions of the focal spots formed by superposition of the two beams are displayed in (c)–(g) with different time delays; (h) displays the calculated PFT of the focal spot produced in air with the STF system.

function whose slope provides the sweep velocity of the focal spot on the focal plane. The measured result of the PFT is inversely proportional to the sweep velocity, which is calculated to be –2.95 fs/μm.

The measured sweeping velocity of focal spot needs to be examined by theoretical simulations using the same parameters as the experimental ones. To obtain the intensity distribution of the focal spot formed by the coherent superposition of the two pulses at various time delays, we can make theoretical calculations based on Fresnel–Kirchhoff's diffraction theory as depicted in Section 3.3.2. The results are presented in Figure 3.18 (c)–(g), where the relative time delays between the two pulses are –66.8, –33.4, 0, 33.4, and 66.8 fs, respectively. It can be seen in Figure 3.18 that the theoretical results reasonably reproduce the experimentally measured results in Figure 3.17 (c)–(g). Again, it can be seen that the interference fringe shifts linearly as a function of the relative time delay, which results from the PFT effect. The simulation result of PFT is presented in Figure 3.18 (h). The slope of the tilt angle of intensity distribution in *x*-*t* domain shows that the degree of PFT is ~3.06 fs/μm, which is in good agreement with the experimental measurement with a slight difference of only ~4%.

The optical nonreciprocal writing, or optical 'quill' writing, is referred to as the effect that the characteristics of the modification change when the scan direction of laser direct writing is reversed, which was initiated by Kazansky et al. [46, 126] and has become another hot topic in femtosecond laser micromachining. They also theorized that the intensity gradient established across the focal spot by the presence of PFT acts to displace electrons in the plasma by virtue of the ponderomotive force. In practice, PFT may be imposed on the laser beam by tuning the laser's grating compressor. However, any adjustment to the parallelism of the gratings produces an angular dispersion that necessarily results in spatial and temporal distortions at focus. Consequently, the conditions for generating PFT are not easily translated between systems, so that deciphering the effect of PFT on nonreciprocal writing is confounded.

At first sight, this phenomenon is completely understandable as PFT has long been regarded as the major cause responsible for the nonreciprocal writing process. However, recently several groups have also shown that if the focal spot has only an asymmetric intensity distribution but no PFT, it can also induce nonreciprocal writing. This discovery raises the question on the origin of nonreciprocal writing, which is still hotly disputed. It should be stressed that a complete characterization of the spatiotemporally focused spot is critical for gaining deeper insight behind this question, which so far has not been done. This issue will be addressed in the following section.

3.3.4.2 Characterization of peak intensity distribution of spatiotemporally focused pulses

In this section, we attempt to characterize the spatiotemporally focused femtosecond laser beam based on the two-photon fluorescence excitation (TPFE) process. In such a manner, instead of examining the fluence distribution in the focal volume, the peak intensity distribution will be examined, since it is more relevant to the nonlinear absorption processes during irradiation of femtosecond laser in glass. A theoretical analysis to reveal the origin of such tilted peak intensity distribution will then be performed. Finally, it is shown that the tilted intensity distribution alone is sufficient for inducing the well-known nonreciprocal writing effect by use of a new direct writing scheme in which the optical axis of the objective lens deviates from the normal incident direction. These results suggest that the role of PFT in inducing the nonreciprocal writing effect observed in glass should be re-examined because of the intrinsically tilted intensity distribution within the spatiotemporally focused spot.

The experiment setup is similar to that employed in Figure 3.13, except that no aperture or slit is used. The incident Gaussian beam is shrunk to a 1 mm diameter, so as to increase the spatial chirp rate. To visualize the spatiotemporally focused spot, a lens with a focal length of ~500 mm is employed to focus the spatially chirp beam into a 0.12 mg/mL sodium fluorescein diluted with water. The total volume of the fluorescein is ~1200 mL, which is contained in a 240 mm ×160 mm × 160 mm cuvette. The laser power measured after the ND filter is ~20 mW. The TPFE image can be captured by a digital camera from both the top view and side view.

Figure 3.19 (a) and 3.19 (c) show digital camera captured images of TPFE using the spatiotemporal focusing scheme, observed from the *XZ* and *YZ* planes, respectively. It

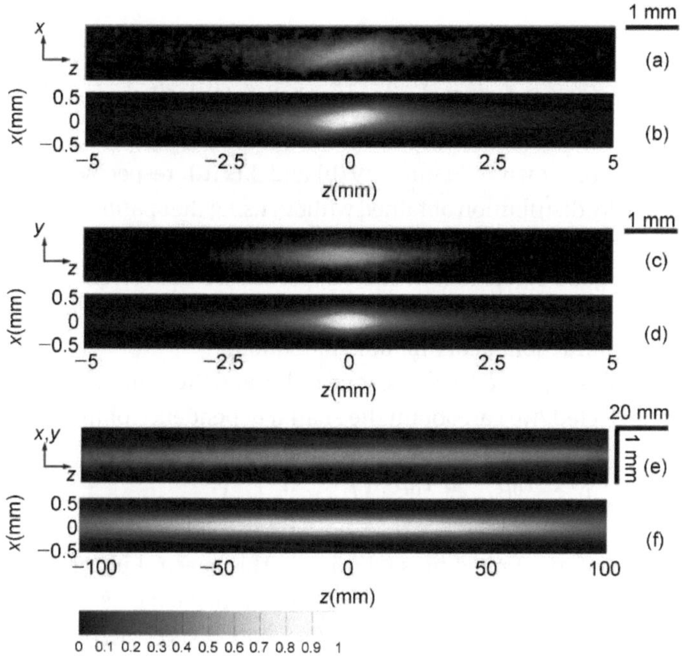

Fig. 3.19: TPFE images (a)–(d) with temporal focusing and (e)–(f) without temporal focusing (lens $f = 500$ mm). (a) and (c) are TPFE images captured in the *XZ* and *YZ* planes respectively, and (b), (d), (f) are corresponding numerical simulation TPFE signals. The gradient bar is shown at the bottom. Notice that the horizontal and vertical scale bars in (e) and (f) are different.

can be seen that the fluorescence intensity distribution in the *XZ* plane is tilted while in the *YZ* plane it appears symmetrical. Such tilted intensity distribution in the *XZ* plane has never been reported before, with characteristics fundamentally different from those of the well-known PFT.

For comparison, the femtosecond laser beam with a CF system, namely without use of the spatiotemporal focusing to perform the similar TPFE experiment, is employed. In this situation, the incident beam has a diameter of 12 mm while the laser power and the focal lens are the same as that used above. The TPFE image is shown in Figure 3.19 (e), which demonstrates a much longer focal depth than that obtained with the spatiotemporal focusing. The result is consistent with the observation of a highly localized focal spot in a thick glass sample enabled by use of spatiotemporal focusing with a low numerical aperture lens.

Theoretically, simulation of the two-photon excitation of dye with a spatiotemporally focused femtosecond laser beam can be achieved using Fresnel–Kirchhoff's diffraction theory, as equation (3.9). The TPFE intensity distribution at the focus of the spatiotemporally focused beam, in both the *XZ* and *YZ* planes, can be obtained with

$$I_{2P}(x,y,z) \propto \int_{-\infty}^{\infty} I_{SSTF}^2(x,y,z,t)\,dt = \int_{-\infty}^{\infty} \left|\mathcal{F}^{-1}\{E_2(x,y,z,\omega)\}\right|^4 dt\,. \tag{3.22}$$

The numerical simulations are shown in Figure 3.19 (b) and 3.19 (d), respectively. On the other hand, TPFE intensity distribution obtained without using the spatiotemporal focusing can be calculated by substituting into equation (3.7). One can clearly see that the theoretical results are in good agreement with the experiment results. It should be noted that the long focal length of the lens used in the experiment makes it reasonable to use Fresnel–Kirchhoff's diffraction theory for our simulation.

When the spectral phase is expanded to the second order, and the material dispersion of air and water is neglected, we can obtain the spatial dependence of the pulse chirp:

$$\varphi_2(x,z) = \left(\frac{x}{w_0}\frac{\tau_0\beta}{\omega_0} - \frac{z}{z_R}\frac{\tau_0^2\beta^2}{4}\right)\left(\frac{n}{1+z^2/z_R^2}\right), \tag{3.23}$$

where $\tau_0 = 2/\Delta\omega$ is the transform limited pulse duration. This spectral phase originates from the evolution of the wavefront curvature of each of the beamlets. It is evident in this expression that there is a line in the XZ plane where this geometric phase is zero. Solving for where $\varphi_2 = 0$ we can get

$$z(x) = \frac{2}{\beta}\frac{x}{w_0}\frac{\Delta\omega}{\omega_0}z_R = \frac{2nf}{\alpha\omega_0}x\,, \tag{3.24}$$

where the focal spot size is $w_0 = \lambda_0 f/(\pi w_{in})$, and the beamlet Rayleigh range is $z_R = \pi n w_0^2/\lambda_0$. Thus the value of the IPT angle θ_t, defined as the angle between the tilted intensity plane and the positive z-axis (indicated in Figure 3.20), can be evaluated as:

$$\tan\theta_t = \frac{\alpha\omega_0}{2nf}\,. \tag{3.25}$$

For a position close to $z = 0$, the denominator of the second parenthesis in the right side of equation (3.28) can be neglected. Working for this linear approximation again, we can obtain the GVD coefficient of this spatiotemporally focused pulse, as $k_{2x} = \partial\phi_2/\partial z$ and $k_{2z} = \partial\phi_2/\partial z$ are GVD coefficients in the x and z directions respectively. Hence, the pulse duration of this spatiotemporally focused beam can be analytically expressed as:

$$\tau(x,z) = \tau_0\sqrt{1+\left(\frac{k_{2x}x + k_{2z}z}{\tau_0^2}\right)^2} \approx \tau_0\sqrt{1+\frac{1}{\tau_0^4}\frac{\alpha^2}{c^2f^2}(x - z\tan\theta_t)^2}\,. \tag{3.26}$$

It is obvious that the pulse duration in this situation is dependent on both the propagation distance z and the lateral distance x away from the geometrical focus.

Intuitively, the slope of IPT is proportional to the angular chirp rate and inversely proportional to the focal length of L3. In the experiment, the IPT angle is calculated to

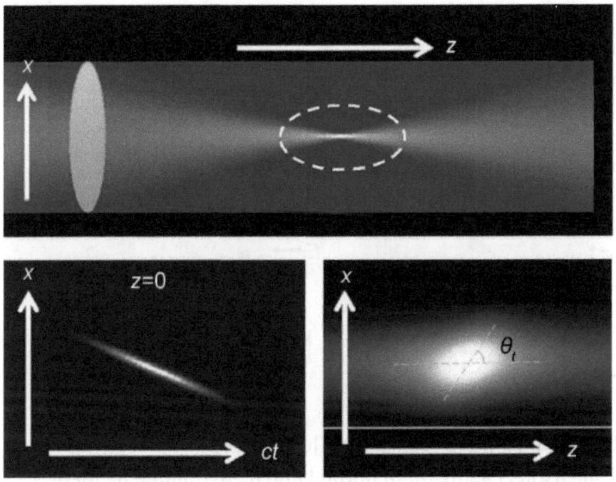

Fig. 3.20: Illustration of the relative orientations of the spatial chirp (top), PFT (lower left), and intensity plane tilt with the IPT angle indicated (lower right). The PFT is represented as a snapshot so that the leading edge of the pulse is larger at $z = ct$.

be ~26° by using equation (3.25). Obviously, for CF ($\alpha = 0$), it should result in $\theta_t = 0$ since no IPT effect will occur. It should be specifically noticed that in the tight focusing case, the IPT angle will be approximated to be ~90° by using equation (3.25). Such a large tilt angle will also mask this IPT effect.

It is instructive to compare the IPT to the PFT. The PFT, i.e. the tilt in x-t in the focal plane, arises from the calculation of the variation of the group delay with respect to position. At $z = 0$, we can write $\varphi_1(x) = \alpha\omega_0 x/(cnf) = \beta\tau_0 x/w_0$, whereas in [151] $\tau_0\Delta\omega = 2$, and $w_0 = 2cnf/(\omega_0 w_{in})$ are used to simplify the expression. The temporal shift of the pulse at the spot radius is $\beta\tau_0$. For positions $x > 0$ in the focal region, the pulse arrives later (larger group delay) and the z position of maximum intensity is farther from the lens ($z > 0$). Figure 3.20 illustrates the relative orientations of the tilt in the peak intensity plane in the XZ plane.

To add to the evidence about the mechanisms behind the nonreciprocal writing with femtosecond laser pulses, we performed an experiment to test for nonreciprocal writing where the beam is tilted with respect to the sample, but with CF, the beam is inspected at the focus to ensure that it does not have a spatial chirp. This is easily accomplished by simultaneously rotating the incident beam and the objective (20×, NA = 0.46) at an angle of $\theta = 15°$ in the YZ plane, as illustrated in Figure 3.21. In this experiment, another femtosecond laser (Libra-He, Coherent Inc.) is used that emits 800 nm, 50 fs pulses with a maximum pulse energy of 4.5 mJ and a repetition rate of 1 kHz. The laser beam is linearly polarized along the Y axis. The material used in the experiment is commercially available fused silica (JGS1), cut to small coupons with dimensions of 10 mm × 5 mm × 2 mm, and with six sides polished. The sample is placed on a XYZ stage with a resolution of 1 μm. The laser power and the scanning speed are

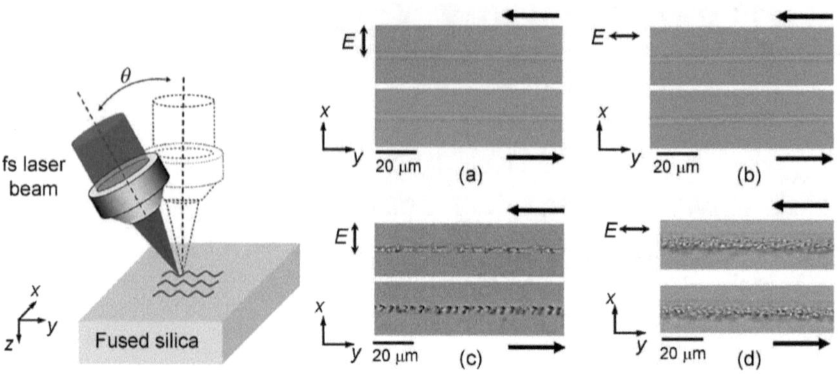

Fig. 3.21: Schematic of femtosecond LDW using a tilted focal plane (left) and optical micrographs of fs direct-writing structures in fused silica glass using (a, b) normal and (c, d) oblique incident beams. The polarization is along *x* and *y* directions in (a, c) and (b, d) respectively. The arrows denote the laser writing directions.

1.5 mW and 20 µm/s respectively. The location of the focal plane is ~300 µm beneath the surface of the sample.

When the beam is at normal incidence, no nonreciprocal writing effect is observed, as shown in Figure 3.21 (a) and 3.21 (b). While in the oblique incident scheme as depicted above, nonreciprocal writing structures are observed when the orientation of the electric field *E* is arranged perpendicular to the laser scanning direction (S polarization), as shown in Figure 3.21 (c). In contrast, when the electric field *E* is arranged to be parallel to the laser scanning direction (P polarization), no reciprocal writing effect is observed, as shown in Figure 3.21 (d). It is noted that for P-polarization there is a component of the electric field that is perpendicular to the surface, allowing for resonance absorption in the plasma gradient. The more effective energy absorption in this case may play a role in the difference between Figure 3.21 (c) and 3.21 (d). If we consider the intensity distribution tilting in a spatiotemporally focused laser beam, the results in Figure 3.21 (c) and 3.21 (d) are consistent with that reported in [143]. In that experiment, the change in incident angle is accomplished by displacing the beam from the center of the lens, so there is the potential for spatial chirp owing to the prismatic effect of the lens. More importantly, this simple experiment shows that a PFT relative to the surface may be sufficient for inducing the nonreciprocal writing effect. Thus, the role of PFT in inducing the nonreciprocal writing will require re-examination with further experimental and theoretical investigations.

This finding is a big step in the understanding of this exotic phenomenon, and suggests that for achieving a complete understanding on the nonreciprocal writing effect in glass observed with spatiotemporally focused laser pulses, the roles of the PFT and peak-intensity distribution tilting should be separately investigated and weighed, which requires the development of more sophisticated beam manipulation techniques in both space and time domains.

3.3.4.3 Control of the intensity plane tilt by tuning the temporal chirp of the pulses

Since the IPT gives rise to an asymmetric focal spot shape, it is generally undesirable in many applications, such as ultrafast laser material processing, nonlinear fluorescence microscopy, two-photon optogenetic applications, and so on. As shown above, the IPT results from the intrinsically varying chirp of the local light field in the focal plane, and thus one may expect that manipulation of IPT could be achieved by tuning the temporal chirp of the output laser pulses from the amplifier. Unfortunately, as we will show below, this idea is not completely correct. Nevertheless, the following results also show that by taking into account the fanning out of the beamlets over the depth of focus, the IPT can be reduced with sufficiently large initial chirps of the output laser pulses from the amplifier.

To vary the chirp of the output laser beam, two flipping mirrors FM1 and FM2 are switched to the positions that enable the amplified pulses to pass through the compressor. The compressor is composed of a single grating and a mirror is currently employed to adjust the temporal chirp of the initiated pulses, which can be achieved by translating the grating mounted on a motor. For the grating compressor/stretcher system, the amount of the second-order phase is given in [95] as

$$\phi_2 = -\frac{\lambda^3 d\sigma^2}{\pi c^2}\left[1-(\lambda\sigma-\sin i)^2\right]^{-3/2}, \tag{3.27}$$

where L is the distance of the beam propagation between gratings. According to the specification of the compressor from Coherent Inc., the estimated value of SOP can be obtained as

$$\phi_{2\text{in}} \approx 5.3\Delta d \times 10^6\ \text{fs}^2, \tag{3.28}$$

where Δd is the distance of the grating away from its chirp-free position in this setup. The corresponding theoretical analysis involving SOP can be developed by varying the parameter $\varphi_{2\text{in}}$ in equation (3.7).

By increasing the temporal chirp of the incident pulses in the experiment, it is noticed that the TPFE signals become weak, so the exposure time of the digital camera is increased to 1 s, which is longer than that used in former cases. Figure 3.22 shows TPFE images using SSTF with different input temporal chirp values. It is observed that the temporally focused spot shifted its position away from the geometrical focal point as the SOP increased, which was proven earlier in the two-photon bio-imaging experiment with the temporally focused beam [157, 158]. When the initial temporal chirp of the incident laser beam becomes sufficiently large, i.e. $\phi_{2\text{in}} > 4000\ \text{fs}^2$, the tilt angle of the focal intensity plane starts to decrease with the increasing SOP. This trend is also evidenced by the corresponding simulation results, as shown in Figure 3.22.

The experimental observations in Figure 3.22 can be understood following the above analysis. With the SOP, the second order spectral phase can be expressed as:

$$\varphi_2(x,z) = \left(\frac{x}{w_0}\frac{\tau_0\beta}{\omega_0} - \frac{z}{z_R}\frac{\tau_0^2\beta^2}{4}\right)\left(\frac{n}{1+z^2/z_R^2}\right) + \varphi_{2\text{in}}, \tag{3.29}$$

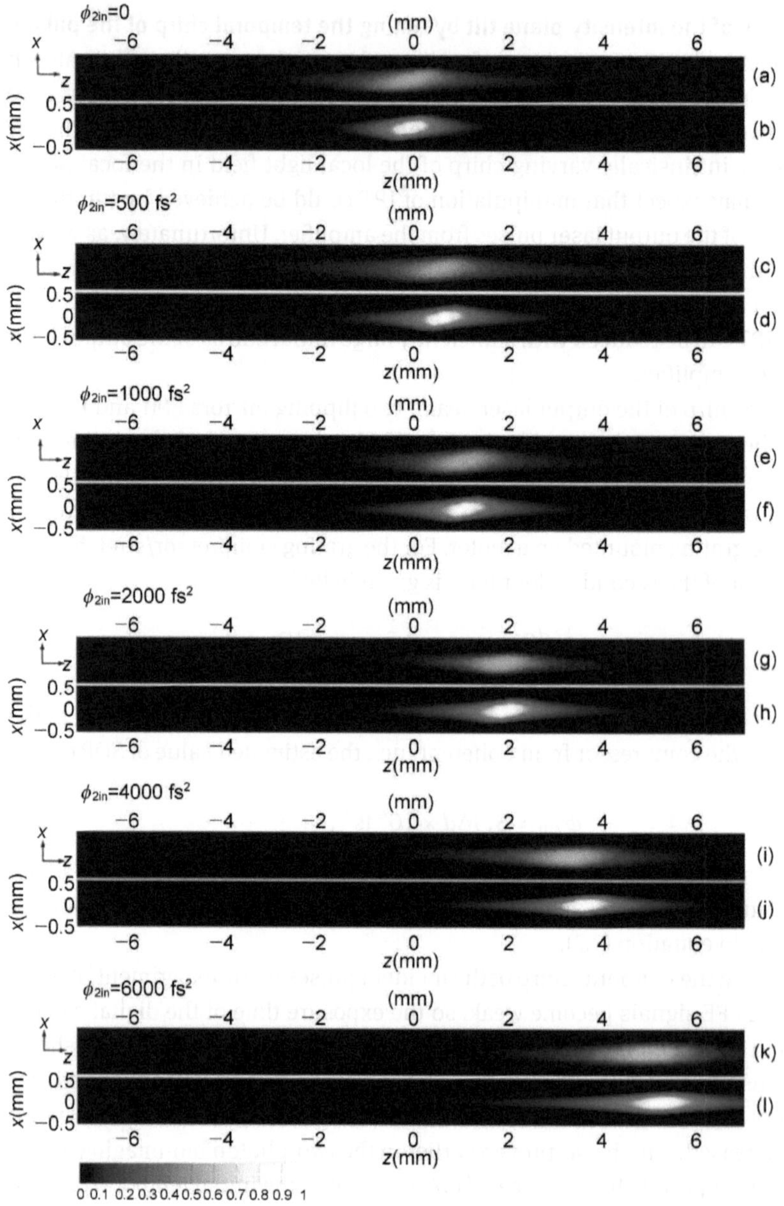

Fig. 3.22: TPFE images using temporal focusing with input SOP values of (a) $\phi_{2in} = 0$, (c) $\phi_{2in} = 500\,fs^2$, (e) $\phi_{2in} = 1000\,fs^2$, (g) $\phi_{2in} = 2000\,fs^2$, (i) $\phi_{2in} = 4000\,fs^2$, and (k) $\phi_{2in} = 6000\,fs^2$. (b), (d), (f), (g), (j) and (l) are corresponding numerical simulation TPFE signals. The gradient bar is shown at the bottom.

where $z = 0$ is set to be the axial position of the geometric focal plane. The input second-order phase $\varphi_{2\text{in}}$ can balance the geometric phase at an axial position z_d away from the geometric focal plane. Working with equation (3.29), we can solve for on-axis positions with the zero second-order phase: $\varphi_2(x = 0, z = z_\text{d}) = 0$. There are two solutions for z_d; the position closer to $z = 0$ results in a higher intensity because there is a larger bandwidth and smaller focal spot size. Neglecting the denominator term in the second parentheses in equation (3.29), and making use of $\beta\tau_0 = 2\alpha/w_\text{in}$ we find:

$$z_\text{d} \approx \varphi_{2\text{in}} \frac{w_\text{in}^2}{n\alpha^2} z_\text{R} = \varphi_{2\text{in}} \frac{\Delta\omega^2}{n\beta^2} z_\text{R} , \tag{3.30}$$

where in the second form we make use of the time-bandwidth relation $\tau_0\Delta\omega = 2$. In the present investigation, we can use equation (3.30) to evaluate the results in Figure 3.22. For comparison, we also estimate the values of z_d with respect to $\varphi_{2\text{in}}$ from the experimental and numerical simulation in Figure 3.22, respectively. The experimental values of z_d can be obtained by directly measuring the distance shift of the peak position of the TPFE signals away from this peak position with zero-SOP input (Figure 3.22 (a)).

One can find that for relatively small values of SOP (i.e. $\varphi_{2\text{in}} < 2000\,\text{fs}^2$), the simple expression equation (3.30) reproduces both the experimental and the numerical simulation results. This calculation precisely describes the relation between the axial shift of the temporally focused spot and the chirp of the incident laser beam. When $\varphi_{2\text{in}} > 2000\,\text{fs}^2$, the temporal focus will shift farther away from the geometric focal point of the lens along z-axis and be located out of the depth of focus (1.3 mm), thus the paraxial approximation fails and equation (3.30) becomes invalid. For $\varphi_{2\text{in}} = 0$, the temporal focused spot and the geometrical focus overlap.

The change of the focal plane tilting angle with the varying SOP can be obtained by calculating the slope of the zero-SOP curve $\varphi_2(x, z) = 0$. In such circumstances, we have

$$z'(x)\,|_{x=0} \approx \frac{1}{\tan\theta_t} \cdot \frac{1}{1 - 2z_\text{d}^2/z_\text{R}^2} . \tag{3.31}$$

The focal plane tilting angle as a function of the SOP is obtained as

$$\tan\theta_T = \left(1 - 2z_\text{d}^2/z_\text{R}^2\right) \tan\theta_t . \tag{3.32}$$

Again, equation (3.32) is only valid when $z_\text{d} \ll z_\text{R}$. Under this condition, for small input chirps, the change in the focal plane tilting angle will be negligible using equation (3.32), which is evidenced in Figure 3.22. To increase the value of z_d, the incident laser pulses should be strongly temporally chirped before they reach the external grating pair, whereas for very large chirp values the approximation involved in the derivation of equations (3.30)–(3.32) would be inappropriate. In such a case, the experimental results and calculations based on the paraxial Fresnel diffraction model show that the IPT can be reduced with a sufficiently large initial chirp of the output laser pulses, which may benefit some applications such as femtosecond laser microprocessing and two-photon fluorescence bio-imaging.

3.3.5 Novel nonlinear effects induced by spatiotemporal focusing

3.3.5.1 Enhanced femtosecond laser filamentation using spatiotemporally focused beams

We extend the application of spatiotemporally focused ultrafast laser pulses from 3D micromachining to femtosecond laser filamentation. It is well known that the peak intensity in the filament core is inherently limited by the intensity clamping effect during femtosecond laser filamentation. We experimentally demonstrate the significant enhancement of the peak intensity beyond the clamped intensity using spatiotemporally focused femtosecond laser pulses. The experimental setup is similar to Figure 3.14. Laser pulses at a center wavelength of 800 nm with a spectral bandwidth of ~30 nm are derived directly from the amplifier without compression. A cylindrical convex lens and a cylindrical concave lens are employed to reduce the beam size by five times in the horizontal direction. Then the beam is dispersed by a parallel pair of gratings in the horizontal direction. The distance between the gratings is optimized to compensate for the temporal dispersion of different spectral components. It is then focused with a 1000 mm focal length lens into the air. The fluorescence from the filament is imaged to the entrance slit of the spectrometer in a 4-*f* geometry configuration. For comparison, the fluorescence spectrum of the filamentation generated without using the spatiotemporal focusing is also measured.

Figures 3.23 (a) and 3.23 (b) show the filament profiles generated with and without the spatiotemporal focusing technique captured by a digital camera from the side of the filaments. When using the 1000 mm focal lens to focus the 40 fs laser pulse without spatiotemporal focusing directly, the length of the filament is ~100 mm. However, when the spatialtemporal focusing technique is employed, the length of the filament

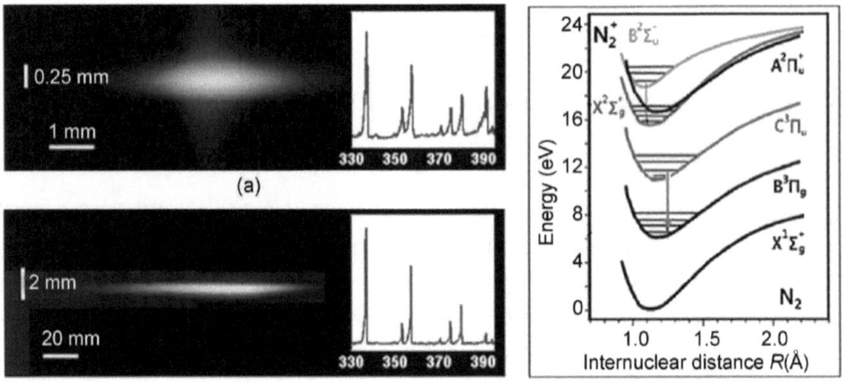

(b) (c)

Fig. 3.23: Filamentation profiles captured by a digital camera (a) with and (b) without using spatiotemporal focusing. Please note the scale change from (a) to (b). Insets show the corresponding N_2 fluorescence spectra. (c) Indicates the energy diagram of a nitrogen molecule.

decreases to less than 4 mm. The shortened filament implies that the peak intensity of the pulse decays quickly away from the focus. The rapid decrease of the peak intensity beyond the focus is due to two reasons. First, as different spectral components of the beam are separated spatially, only at the focus will all the spectral components converge to form an ultrashort intense pulse. Meanwhile, before or after the focus different spectral components of the beam diverge spatially, resulting in a spatially chirped narrow spectral bandwidth, thus a longer pulse. Secondly, as different spectral components are spatially dispersed, only in the geometric focus is the pulse chirp free, while before the focus the pulse is highly positively chirped and after the focus the pulse is highly negatively chirped. The large chirp elongates the temporal pulse duration. Both these two effects ensure that the shortest pulse duration can only be achieved in the focus, while away from the focus, the pulse duration grows quickly.

To investigate the clamped intensity in the filament, the fluorescence of nitrogen molecules and ions generated by focusing the laser beam into the air with a 100 cm focal lens (both with and without a spatiotemporal focusing configuration) is measured. Two representative spectra are demonstrated in the insets of Figure 3.23 (a) and 3.23 (b). Both spectra in the insets feature the 'continuum free' quality as reported previously. In particular, the 337 and 391 nm lines in the spectra can be assigned to the second positive band of $N_2(C^3\Pi_u \rightarrow B^3\Pi_g)$ and the first negative band system of $N_2^+(B^2\Sigma_u^+ \rightarrow X^2\Sigma_g^+)$, respectively. The strength ratio between the two nitrogen fluorescence lines is given by the following:

$$R \equiv \frac{S_{391}}{S_{337}} = \frac{aI^{n1}}{k(aI^{n1} + bI^{n2})} \propto \frac{1}{1 + (b/a)I^{n2-n1}} \,. \tag{3.33}$$

From equation (3.33), it is known that higher laser intensity will give rise to a larger ratio, since $n_2 < n_1$ in view of higher ionization potential of the inner electrons.

The experimentally measured strength ratios of the 391 nm and the 337 nm lines is plotted in Figure 3.24 as a function of the laser energy with and without spatiotemporal

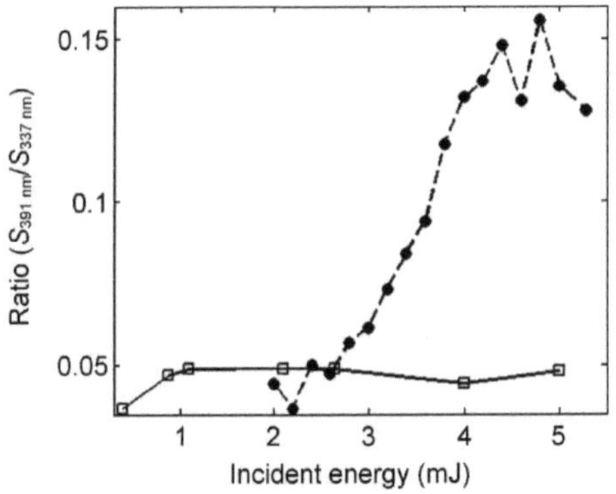

Fig. 3.24: Experimentally measured intensity ratio of the 391 and 337 nm lines as a function of the laser energy with (solid cycles) and without (open squares) using spatiotemporal focusing.

focusing. It can be seen that in the case without using spatiotemporal focusing, the ratio shows a first rapid rise with the increase of the input laser energy. While the energy is higher than 1 mJ, the ratio remains almost constant since the peak intensity does not increase due to the intensity clamping. However, when the spatialtemporal focusing technique is employed, the intensity ratio of the 391 nm to the 337 nm first grows with the increase of the incident energy, and then reaches a plateau at ~0.15 when the laser energy is larger than 4.5 mJ, which indicates that the intensity clamping occurs at a higher energy level with higher intensity.

The peak intensity can then be calculated by the empirical formula [143]

$$I_0 = 79 \times (2.6/R - 1)^{-0.34} \times 10^{12}\ \mathrm{W/cm^2}\,. \tag{3.34}$$

According to equation (3.34), the peak clamped intensity in the filament is estimated to be enhanced by ~50% using spatiotemporal pulses as compared to compressed pulses when using the 100 cm focal length lens. Although this experiment was carried out in air, we expect a similar effect can occur in glass, which can lead to stronger modifications inside glass materials. This effect may enable new applications such as the synthesis of new materials by femtosecond laser induced shock wave.

3.3.5.2 Second harmonic generation in air using spatiotemporally focused pulses

When an electron in a potential energy environment interacts with a laser field, the electron will be pushed side to side by the force of the laser field. In a harmonic potential energy environment, the motion of the electron will be sinusoidal, and there will be no SHG. But in a harmonic potential energy environment, the motion of the election will not be sinusoidal. Therefore a second harmonic can be generated because the symmetry is broken.

When the laser field is strong enough, the coulomb potential is distorted, generating a barrier. The electron in a bound state can escape easily and become a free electron. The free electron exposed in the optical field will quiver by the force of the laser field. Supposing a driver laser field with frequency ω, the quivering velocity v of the electron can be calculated by

$$-\,\mathrm{i}\omega m v = -\frac{e}{m}E\,. \tag{3.35}$$

At the same time, electron density n will fluctuate at the same frequency of the laser field,

$$n = -\frac{1}{4\pi e}\nabla \cdot E\,. \tag{3.36}$$

The generated electric current J will oscillate with double the frequency of the laser field,

$$J = nev \propto E(\nabla \cdot E) \propto \exp\left(-\mathrm{i}2\omega t\right)\,. \tag{3.37}$$

This electric current will be the source of SHG [144]

$$P(2\omega) = \frac{J}{-i2\omega} = \frac{e}{8\pi m\omega^2}E(\nabla \cdot E)\,. \tag{3.38}$$

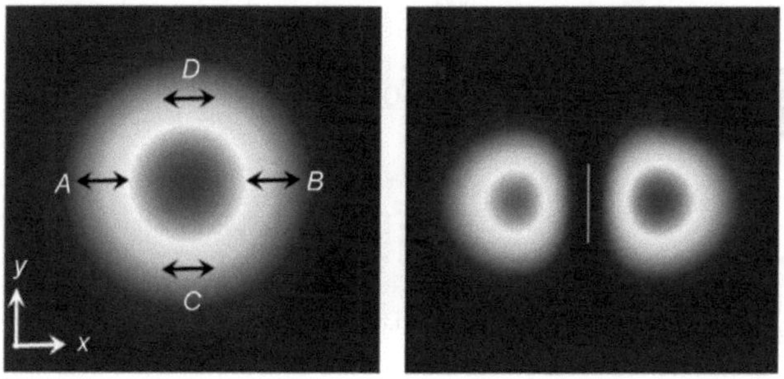

(a) (b)

Fig. 3.25: (a) Calculated Gaussian electron density profile generated by *x*-polarized laser field in a traditional focusing scheme; for electrons at positions A and B, the electron density it feels is different when it quivers, while for those at positions C and D, the electron density is still symmetric when it quivers. (b) The generated second harmonic profile.

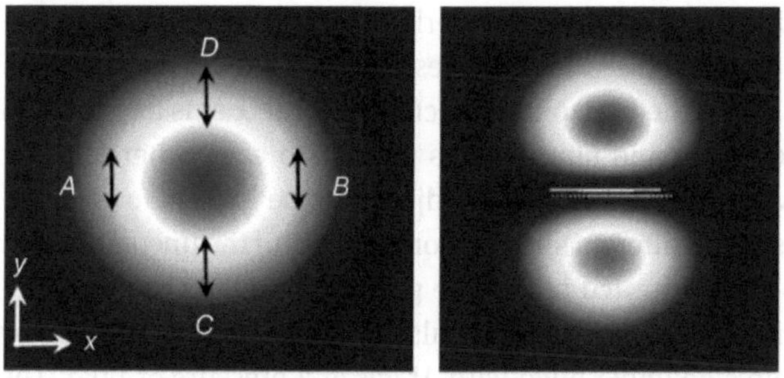

(a) (b)

Fig. 3.26: (a) Calculated Gaussian electron density profile generated by a *y*-polarized laser field in a traditional focusing scheme; for electrons at positions C and D, the electron density it feels is different when it quivers, while for those at positions A and B, the electron density is still symmetric when it quivers. (b) The calculated second harmonic profile.

In traditional focusing, a Gaussian laser spot at the focus will generate a Gaussian electron density profile. If the polarization of the laser field is along the x axis, the electron will quiver along the x direction. For electrons at the gradients of the profile on the x axis, the electron density it feels will not be the same when it quivers from one side to the other side. In this sense, symmetry is broken and a second harmonic can be generated. However, for those at the gradients on the y axis, a second harmonic cannot be generated due to asymmetry. Consequently, the spot of the second harmonic generated due to broken symmetry will have two lobes located along the x direction, as illustrated in Figure 3.25.

Similarly, when the polarization of the laser field is along the y direction, we will have a second harmonic spot with two lobes in the y direction, as demonstrated in Figure 3.26.

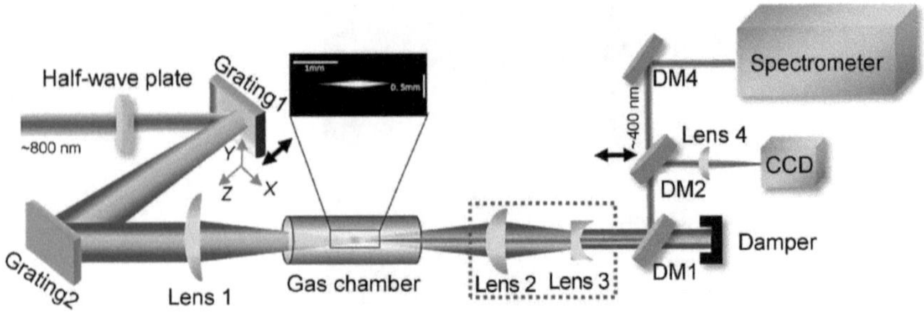

Fig. 3.27: Schematic of the experimental setup.

But what happens when it comes to SSTF? To investigate this, an experiment is carried out with a commercial Ti:sapphire laser system (Legend EliteDuo, Coherent, Inc.), which delivers ~6 mJ pulses with a center wavelength at ~800 nm and a spectral bandwidth of ~30 nm at a repetition rate of 1 kHz. The experiment setup is illustrated in Figure 3.27. The amplified but uncompressed laser pulses emitted from the laser system are first spatially dispersed along the horizontal direction (i.e. the x axis as shown in Figure 3.27) by a pair of 1500 lines and mm gratings parallel to each other (blazing at ~50°). The distance between the two gratings is adjusted to be ~80 cm to compensate for the temporal dispersion of different spectral components of the uncompressed laser pulses. A half-wave plate is inserted before the gratings to adjust the polarization direction of the fundamental beam. The laser pulse of ~1.0 mJ is then focused by a 75-cm-focal-length lens into a chamber filled with Ar gas at a pressure of 1 bar. The generated second harmonic radiation is recorded by a spectrometer (Shamrock 303i, Andor Corp.) after its beam size is approximately reduced by half with a combination of a convex lens and a concave lens (Lens 2 [$f = 25$ cm] and Lens 3 [$f = -12$ cm]). The spatial profile of the second harmonic beam in the focal plane is imaged by a convex lens (Lens 4: $f = 20$ cm) on an area CCD camera (Wincamd-ucd23) with an integration time of 2 ms. Three dichroic mirrors (DM1, DM2, and DM3) with high reflectivity at 400 nm and high transmission at 800 nm, are used to separate the fundamental beam from the second harmonic radiation. A Glan–Taylor prism before the spectrometer is used to measure the polarization of SHG. For comparison, a SHG experiment with a CF scheme is also carried out by removing the grating pair while keeping all the other conditions unchanged.

Figure 3.28 shows various properties of the second harmonic beam from Ar gas by use of SSTF femtosecond laser pulses. When the polarization direction of a fundamental laser is set along the y axis (SSTF–PY), the focus of the SHG beam, which is obtained from the CCD in Figure 3.28, shows two lobes separated from each other along the y axis, which is coincident with previous investigations in traditional focusing schemes. Nevertheless, when the polarization direction of the fundamental laser is set along the x axis (SSTF–PX), i.e. the spatial dispersion direction of the incident

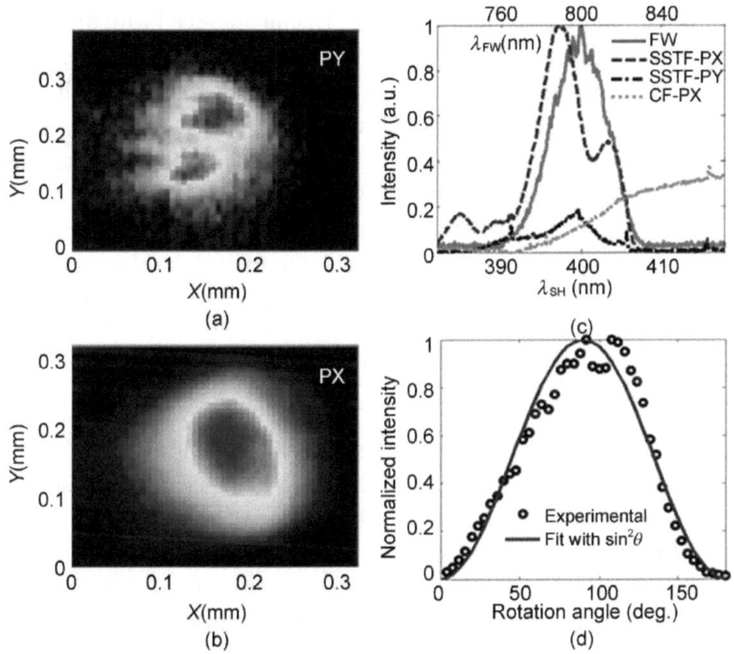

Fig. 3.28: Spatial patterns of SHG by the SSTF scheme with the fundamental laser linearly polarized (a) along the y axis (SF–PY) and (b) along the x axis (SF–PX). (c) The second harmonic spectra obtained in SF–PY (dashed–dotted line), SF–PX (dashed line) and the CF (dotted line) schemes. The fundamental wave spectrum is shown by the solid curve. (d) The measured polarization property of SH for the SSTF–PX case (circle) and the fit (line) with $\sin^2\theta$.

beam before the focal lens, the second harmonic beam profile becomes a Gaussian-like distribution without any splitting, as shown in Figure 3.28 (b).

The second harmonic spectra of Figure 3.28 (a) and 3.28 (b) are also illustrated in Figure 3.28 (c) by the dash-dotted and dashed curves respectively. For comparison, the spectrum generated by the CF scheme with the fundamental laser polarization along the x axis (CF–PX) is also shown in Figure 3.28 (c) by the dotted line. It is found that for the experimental conditions in Figure 3.28 (c), the measured spectra are centered at ~400 nm, which confirms that the second harmonic of the fundamental laser pulse is indeed generated. Note that the narrow peaks appearing in Figure 3.28 (c) are fluorescence emissions from Ar and a very small amount of N_2 in the chamber due to the leakage [144]. Furthermore, it is found that by only changing the fundamental laser polarization, as shown in Figure 3.28 (c), the efficiency of SHG is significantly enhanced by about five times of that obtained with the pump laser polarized along the y axis. Meanwhile, it can be seen from Figure 3.28 (c) that the SHG with the CF scheme (dotted line) is too weak to be distinguished from the tail of a strong supercontinuum white light, which has been generated in the filamentation process, due to self-steepening and self-phase modulation [159]. Since the white light develops progressively along the plasma filament, the much shorter filament length of ~1.5 mm produced with the SSTF scheme results in the dramatically reduced (almost unobservable) white light spectra as shown by the dashed and dashed-dotted curves in Figure 3.28 (c). As a result, the SSTF laser pulses can generate a second harmonic efficiently with an extremely high signal-to-noise ratio, especially when the pump laser is linearly polarized along the direction of the spatial chirp (i.e. along the x-axis direction).

To clarify the origin of difference in second harmonic properties in traditional focusing schemes and spatiotemporal focusing schemes, a theoretical simulation is performed. In spatiotemporal focusing, the laser field in the focal plane can be described by [139]:

$$E(z = f, t) \propto \exp\left[-\Omega^2 \left(t + \frac{\alpha x k}{f}\right)^2 /4\right] , \tag{3.39}$$

where α represents the spatial chirp and $\sqrt{2}\Omega$ is the spectral bandwidth at $1/e^2$.

Equation (3.39) shows that the laser field with SSTF schemes can give rise to a significant PFT along the x-axis direction (i.e. shortest pulses arrive at different times with a time shift $t_0 = \frac{\alpha x k}{f}$ along the x direction).

The influence of the ponderomotive force on the plasma dynamics is greatly enhanced by the PFT, forcing the major part of electrons to be distributed at the edge of the spatiotemporally focused spot. Therefore, in the calculations, we simply assume that the initial electron density produced by photoionization is uniform ($n_0 \approx 2.7 \times 10^{19}$ cm^{-3} estimated from the experimental condition) [160, 161], and the ponderomotive force is responsible for the inhomogeneity of the plasma density by taking the electron dynamics in account as:

$$\frac{d\vec{v}}{dt} = k\nabla I(r, t) , \tag{3.40}$$

where k is a constant. Then, we can obtain the electron density variation δn based on the continuity equation [162]

$$\frac{d\delta n}{dt} = -n_0 \nabla \cdot \vec{v} . \tag{3.41}$$

Since the light field in the femtosecond laser pulses changes with time, the calculated δn is also a function of time. Here, to provide a clear physical picture, we only show the calculated δn in the focal plane at the end of the driving pulse, i.e. at $t = 100$ fs. It can be clearly seen that in Figure 3.29 (b), the calculated δn is centrosymmetric and the minimum is at the central part, which can be easily understood by the ponderomotive force driven by the CF Gaussian laser field. In the central area of the CF laser beam, the laser intensity is nearly uniformly distributed; thus, there is almost no gradient in the plasma density. Surprisingly, with the SSTF scheme, electrons are strongly pushed to one side along the x axis by the ponderomotive force, as evidenced in Figure 3.29 (a).

With the calculated electron density variation δn, the source of SHG, i.e. the current density $\vec{J}_{2\omega}$ of all the quivering electrons can be expressed as $\vec{J}_{2\omega} = \delta n e \vec{v}$, where

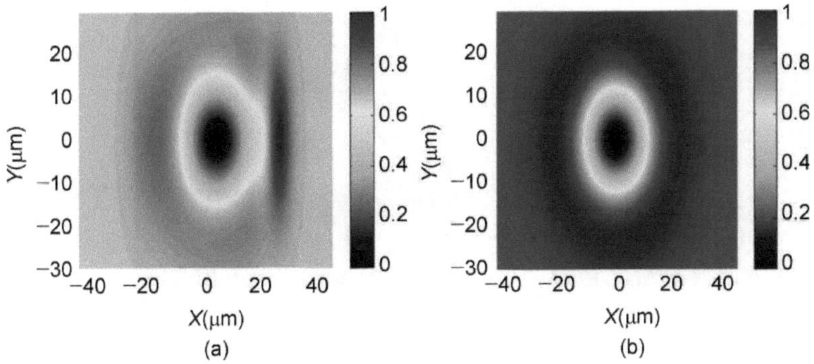

Fig. 3.29: Calculated normalized electron density variation driven by linearly polarized laser with (a) SSTF scheme and (b) CF scheme.

$\vec{v}$ is the velocity of the quivering electrons and δn is the charge perturbation by the quivering electrons. The SHG can be calculated using the following expression:

$$\vec{P}(2\omega) = \frac{\vec{J}_{2\omega}}{-2i\omega} = \frac{e^3}{2im^2\omega^3}\left(\nabla\delta n \cdot \vec{E}\right)\vec{E}\,. \tag{3.42}$$

Based on equation (3.42), we can theoretically calculate the SHG in the focal plane using the experimental parameters, and the results are shown in Figure 3.30. It is noteworthy that the SHG results are averaged temporally over the whole pulse duration. It can be seen in Figure 3.30 that the simulations successfully reproduce all the major features of the experimentally measured second harmonic patterns (Figure 3.30 (a) and 3.30 (b)); that is, the SSTF laser polarized along the y axis produces a double-lobe second harmonic beam separated in the y direction (Figure 3.30 (a)) and the SSTF laser polarized along the x axis generates a main lobe of a Gaussian-like second harmonic beam accompanied by a much weaker satellite second harmonic beam (Figure 3.30 (b)). The satellite second harmonic signal is unobservable in the measurement due to the limited sensitivity of the CCD detector.

With this calculation, the SHG efficiency obtained with the two-pump laser polarization directions in the SSTF scheme are also compared, as illustrated in Figure 3.30. It can be observed that the SHG efficiency with the SSTF laser polarized along the x-axis is more than two times higher than that with the SSTF laser polarized along y axis. In comparison, the SH signals obtained with the CF scheme employing different polarizations of pump laser are much weaker. This qualitatively agrees with the experimental observations as shown in Figure 3.28 (c). The quantitative difference should result from the fact that in the calculation, only the SHG at the focal plane is considered, i.e. all the propagation effects have been ignored. To include these effects, a complete modeling of the laser propagation in the plasma should be carried out, which will be considered in future work.

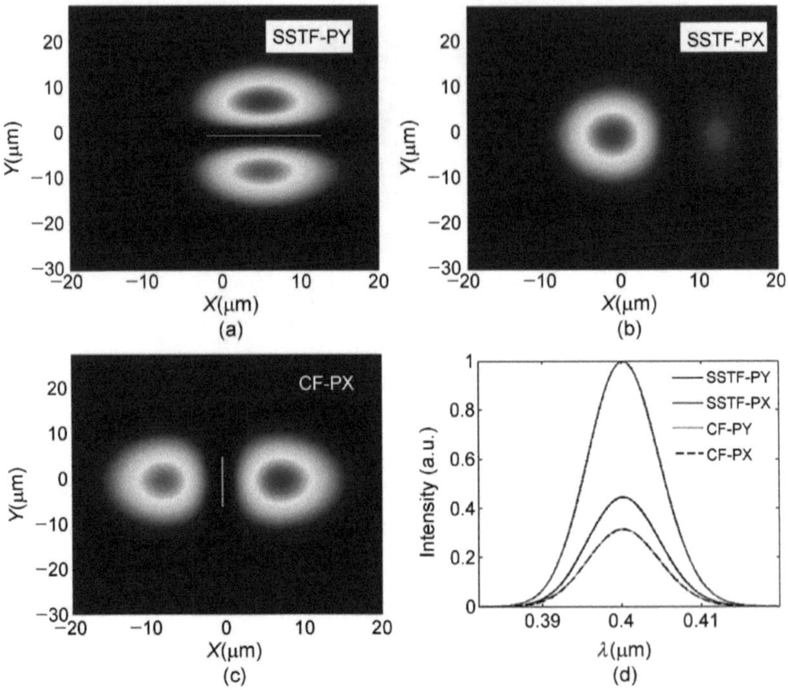

Fig. 3.30: Spatial SH patterns simulated (a) in SSTF–PY, (b) SSTF–PX, (c) CF-PX, and (d) the calculated second harmonic spectra with SSTF by the fundamental light polarized along the x axis (solid line) and along the y axis (dashed line and simulated SH spectra obtained with the CF scheme for different polarizations).

3.3.6 Conclusions

To summarize, this work focuses on 3D femtosecond laser materials processing with spatiotemporally focused laser pulses. Previously, in the femtosecond laser materials processing, major efforts were made in the development of spatial and/or temporal beam shaping methods. With the introduction of simultaneous spatial and temporal focusing of femtosecond laser pulses, we show that several long-standing difficulties in femtosecond laser micromachining can be resolved, including achieving true 3D isotropic resolutions, avoiding nonlinear self-focusing in thick samples, and enhancing the peak intensity at the focus spot by overcoming intensity clamping. Specifically, we further reveal an unexpected tilted peak intensity distribution at the focus of a spatially chirped femtosecond laser beam, which will play important roles in applications such as femtosecond laser micromachining and bioimaging.

The implementation of the spatiotemporal focusing technique in the field of femtosecond laser micromachining provides us with new and exciting opportunities to manipulate the ultrashort pulse for 3D laser material processing. It will offer great opportunities for further fundamental research as well as technological applications in

3D laser-material processing. On the other hand, enhancement of the peak intensity of femtosecond filamentation and second-harmonic generation (SHG) in air can be also realized using this spatiotemporal focusing scheme. Future work will lie in developing methods towards measuring the spatiotemporal characteristics as well as optimizing the intensity profile of these temporally focused laser pulses.

Appendix

(1) Derivation of the spatial chirp rate α

The configuration of the gratings pair is shown in Figure 3.31, where i is the incident angle, γ the −first order diffractive angle, d the distance between the gratings and σ the groove density of the gratings. The −first order diffraction must satisfy the following equation:

$$\sin\gamma = \sin i - \lambda\sigma \,. \tag{3.43}$$

The left and right sides of equation (3.43) can be written using the Taylor expansion to the first order as

$$\sin\gamma = \sin\gamma_0 + \Delta\gamma\cos\gamma_0 + \cdots \,, \tag{3.44}$$

$$\begin{aligned}\sin i - \lambda\sigma &= \sin i - \lambda_0\sigma + \Delta\lambda\sigma + \cdots \\ &= \sin i - \lambda_0\sigma + \frac{\lambda_0\sigma}{\omega_0}\Delta\omega + \cdots \,,\end{aligned} \tag{3.45}$$

where λ_0 and ω_0 are the center wavelength and center frequency respectively. It is noted that in equations (3.44)–(3.45) the approximation $\Delta\lambda \approx -\lambda\Delta\omega/\omega$ is used. Combining equations (3.43)–(3.45), it can be derived that

$$\Delta\gamma \approx \frac{\lambda_0\sigma}{\omega_0\cos\gamma_0}\Delta\omega \,. \tag{3.46}$$

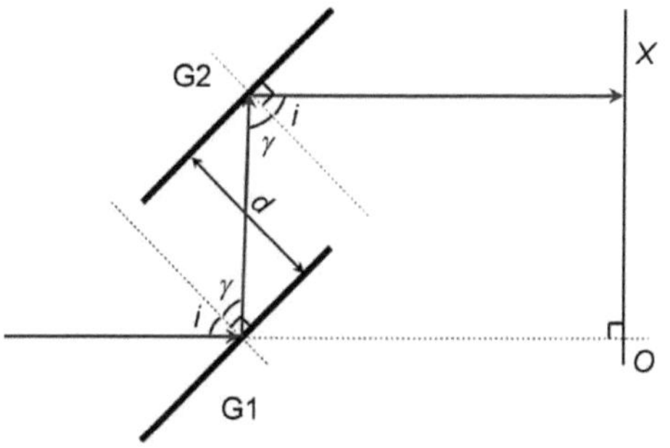

Fig. 3.31: Configuration of the gratings pair used in the spatiotemporal focusing scheme for dispersing and collimating different spectral components of the laser pulses.

The position of different spectral components along the x direction is given as

$$x = \frac{d\sin(i+\gamma)}{\cos\gamma}\,. \tag{3.47}$$

Both sides of equation (3.47) can be written using the Taylor expansion as

$$x = x_0 + \Delta x + \cdots\,, \tag{3.48}$$

$$\begin{aligned}\frac{d\sin(i+\gamma)}{\cos\gamma} &= d(\sin i + \cos i \tan\gamma)\\ &= d(\sin i + \cos i\tan\gamma_0) - \frac{d\cos i}{\cos^2\gamma_0}\Delta\gamma + \cdots\,.\end{aligned} \tag{3.49}$$

Substituting equation (3.48) into equation (3.49), we obtain

$$\Delta x = -\frac{d\lambda_0\sigma\cos i}{\omega_0\cos^3\gamma_0}\Delta\omega + o(\Delta\omega) + \cdots\,. \tag{3.50}$$

Hence the coefficient

$$\alpha = -\frac{d\lambda_0\cos i}{\sigma\omega_0\cos^3\gamma}\,. \tag{3.51}$$

(2) Derivation of the expression of the tilt angle of the temporal focus θ_t for equation (3.25)

Instituting $\beta\tau_0 = 2\alpha/w_{\text{in}}$ into $\phi_2(x,z) = 0$ for equation (3.29), we can obtain

$$\frac{2\alpha}{\omega_0 w_{\text{in}} w_0}x - \frac{\alpha^2}{z_R w_{\text{in}}^2}z + \frac{\phi_{2\text{in}}}{n z_R^2}z^2 + \frac{\phi_{2\text{in}}}{n} = 0\,. \tag{3.52}$$

Taking the first-order derivation of z with respect to x for equation (3.52), we have

$$z'\big|_x = \frac{2w_{\text{in}}z_R}{\alpha\omega_0 w_0}\cdot\frac{1}{1 - 2\phi_{2in}\cdot\frac{z}{nz_R}\cdot\frac{w_{\text{in}}^2}{\alpha^2}} = \frac{1}{\tan\theta_t}\cdot\frac{1}{1 - 2z_d z/z_R^2}\,. \tag{3.53}$$

Considering the derivation at $x = 0$ and from equation (3.29), we have $z(x)|_{x=0} \approx z_d$. Thus, the first order derivation of $z(x)$ at $x = 0$ can be written as

$$z'(x)\big|_{x=0} \approx \frac{1}{\tan\theta_t}\cdot\frac{1}{1 - 2z_d^2/z_R^2}\,. \tag{3.54}$$

3.4 Femtosecond laser fabrication of microfluidics

3.4.1 Introduction

Microfluidics is a rapidly emerging technology that enables miniaturization and integration for biological, chemical, and medical applications. Integration of fluidic functions such as valving, metering, mixing, transport, and separation on a single substrate has enabled construction of microfluidic systems that can control and manipulate tiny volumes of liquids with high precision, leading to downsizing of both chemical and biological analysis [163, 164]. Most current microfluidic systems are primarily two-dimensional. Extension of these devices to three dimensions would allow for full integration of a broad range of different types of components and enable potential high-value applications. However, it is still a challenge to realize complex 3D microfluidic structures with current planar photolithography technologies, which require multilayer and multistep processing procedures to form 3D structures, including stacking, bonding, sealing, and so on.

FsLDW is the main technique currently used to modify the interior of transparent materials in a spatially selective manner [165, 166]. This is possible due to the nonlinear interaction between the tightly focused femtosecond laser beam and the transparent material, since the interaction is effectively confined to the vicinity of a focal point where the laser intensity exceeds the threshold for multiphoton absorption. In combination with wet chemical etching, FsLDW has also been used to fabricate microfluidic structures, including microchannels and chambers [60, 167], microvalves [168], and micropumps [169].

Currently, there are two main strategies for fabricating 3D microfluidic channels embedded in glasses using femtosecond lasers. The first strategy employs femtosecond laser modification inside glass materials followed by chemical wet etching [167]. Unfortunately, with this technique the length of the microfluidic channel that can be fabricated is usually limited to a few millimeters, due to the limited etching ratio between the areas with and without femtosecond irradiation. Another strategy is to perform femtosecond laser 3D drilling from the rear surface of the glass in contact with distilled water, in which the water introduced into the microchannel can help to remove the ablated debris [170]. However, when the length of the channel increases, the debris can no longer be ejected from the microchannel, which in turn causes clogging and the termination of the microchannel. So far, the microchannels fabricated by these methods often suffer from their insufficient lengths for practical microfluidic applications.

Another challenge of fabricating 3D fluidic structures by FsLDW is the limited fabrication resolution mainly caused by the optical diffraction limit. So far, by inducing the TPP of photosensitive materials with femtosecond laser pulses, 3D fabrication with a resolution of sub-100 nm has already been achieved by choosing a laser intensity slightly above a threshold value [69]. However, the TPP technology is intrinsically limited to polymer materials, in which direct formation of thin microfluidic channels cannot be easily achieved due to the difficulty in removing the nonphotopolymerized materials contained in the thin channels. Currently, the thinnest microfluidic channels are obtained by liquid-assisted femtosecond laser 3D drilling, resulting in channel diameters ranging from 500 to 700 nm [171].

Glass is considered to be a good candidate of substrates for microfluidics fabrication by FsLDW because of its wide optical transparency, low thermal expansion, and

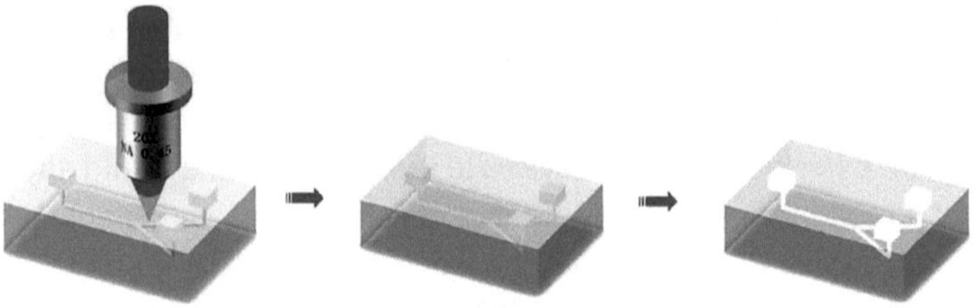

Fig. 3.32: Femtosecond laser modification inside glass materials followed by heat treatment and chemical wet etching.

good chemical inertness and biocompatibility. Although a wide variety of glass has been explored for femtosecond laser processing (particularly for optical waveguide writing), fused silica and a kind of photosensitive glass termed Foturan are the most common materials used for microfluidic sensors, since they exhibit a selective chemical etching ratio of 30:1 or greater in HF acid and other solutions following laser processing.

The mechanism of femtosecond laser processingq of fused silica will be discussed below, while Foturan glass exhibits similar selective etching characteristics but with a mechanism quite different from that of fused silica. The selective etching ratio enhancement in Foturan glass is the result of a photochemical reaction, while in fused silica, it is caused by a photophysical process. More specifically, femtosecond laser irradiation of Foturan glass reduces Ag dopant ions to Ag atoms due to the generation of free electrons by MPA, and during subsequent annealing, Ag nanoclusters are formed by diffusing Ag atoms. These nanoclusters then act as seeds for the formation of crystalline lithium metasilicate, which has an etching rate in HF that is about 50 times higher than that of the unirradiated glass matrix. Thus, the typical method for fabricating microfluidic structures in Foturan glass involves FsLDW, heat treatment in a programmable oven, and wet chemical etching in an ultrasonic bath. Figure 3.32 shows the typical processing procedures of microfluidics fabrication in Foturan glass using FsLDW.

Recently, a novel porous glass has been proposed as an exciting new substrate material for the fabrication of micro- and nanofluidic structures [172–175]. It is produced by removing the borate phase from phase-separated alkali-borosilicate glass in a hot acid solution. The composition of the porous glass is approximately $95.5SiO_2 - 4B_2O_3 - 0.5Na_2O$ (wt.%). The pores have a mean diameter of about 10 nm, and are uniformly distributed in the glass, occupying about 40% of its volume. These pores form a 3D connective network, which allows liquid to flow through the entire glass volume. Interestingly, by annealing the glass substrate at about 1150 °C, the mesoporous glass can be consolidated and the pores sealed. Compared with fused silica and Foturan, this glass offers the advantage of enabling fabrication of 3D microflu-

Tab. 3.3: Typical properties and applications of fused silica, Foturan glass and porous glass [166].

Glass	Properties related to FsLDW	Applications
Fused silica	Composition: amorphous SiO_2 Refractive index: 1.46 (550 nm) Melting temperature: 1665 °C Annealing temperature (for surface smoothing): 1200 °C [207] High optical transparency, low background fluorescence, biocompatible. Selective etching ratio: ~1:30 (HF), ~1:200 (KOH) [182]	Photonic devices [53–59] Microfluidics [41] Optofluidic integration [41] Micro-optics Micromechanics [176, 177] Microelectronics Surface/bulk nanostructuring [63–66]
Foturan	Composition: lithium-potassium glass dotted with small amounts of silver and cerium oxides Refractive index: 1.52 (550 nm) Annealing temperature (for surface smoothing): 570 °C Melting temperature: 660 °C Selective etching ratio: ~1:50 (HF)	Microfluidics Micro-optics Optofluidic integration Micromechanics Microelectrodes Electrofluidics
Porous glass	Composition: $95.5SiO_2 - 4\,B_2O_3 - 0.5\,Na_2O$ (wt.%). Mean pore size: ~10 nm Relative pore volume: 40% Annealing temperature (for sealing): 1120 °C Microfluidic channels with arbitrary 3D geometries and unlimited length Feature size approaching ~$\lambda/20$	Micro and nanofluidics Bulk nanostructuring

idic channels with arbitrary geometries and unlimited lengths, with feature sizes far beyond the optical diffraction limit. In addition, the fabrication procedure is simple and contamination-free, involving only femtosecond laser ablation in water and subsequent annealing to seal the pores.

Typical properties and applications of the three kinds of glass are listed in Table 3.3. This section focuses primarily on recent efforts to tackle the two issues as mentioned above. By use of FsLDW in fused silica, followed by chemical etching as well as glass drawing, centimeter microfluidic channels with an aspect ratio as high as 1000:1 is demonstrated, with an inner wall roughness lower than 1 nm. By using FsLDW in porous glass immersed in water followed by postannealing, microfluidic channels with nearly unlimited lengths and arbitrary 3D geometries are demonstrated [173]. By controlling the laser peak intensity and polarization, a single nanocrack with a sub-50-nm feature size could be achieved inside porous glass [108]. Based on these strategies, several functional devices and their applications are demonstrated, including large-volume hollow chambers [178], 3D passive microfluidic mixers [174], and an integrated micronanofluidic system for single DNA analysis [175]. Finally, future opportunities and current challenges for this novel approach are highlighted.

3.4.2 Microfluidics fabrication in fused silica

Fused silica is conventionally regarded as a nonphotosensitive material in the visible region. It is an ideal substrate material due to its excellent physical and chemical properties, such as chemical inertness, low thermal expansion coefficient, low autofluorescence, exceptional transmittance over a wide spectral range, and so on. Misawa's group pioneered the fabrication of 3D micro-fluidics in fused silica using a femtosecond laser in 2001, although they obtained a low etch selectivity [167]. Later in 2004, Bellouard et al. formed high-aspect ratio microchannels with arbitrary lengths that are open on the top surface of fused silica by femtosecond-laser-assisted wet chemical etching [179].

The fabrication procedure for microfluidic structures in fused silica generally involves FsLDW followed by wet chemical etching [167]. The laser-modified regions exhibit an enhanced etching rate, which is thought to be due to a reduction in the Si-O-Si bond angle, although this is still a subject for debate. A substantial breakthrough in fabrication in fused silica was achieved in 2005 when Hnatovsky et al. discovered that the etch rate and selectivity strongly depend on the polarization of the writing pulses [180]. This is due to the formation of periodic nanograting-like structures that are always oriented perpendicular to the polarization of the laser beam [43, 181]. As shown in Figure 3.33, these nanogratings consist of alternating regions with high and low etching rates. If the nanogratings are oriented perpendicular to the microchannel (Figure 3.33 (c)), the alternating regions prevent the HF solution from entering the microchannel, resulting in the lowest etch rate. Therefore, to achieve the highest etch

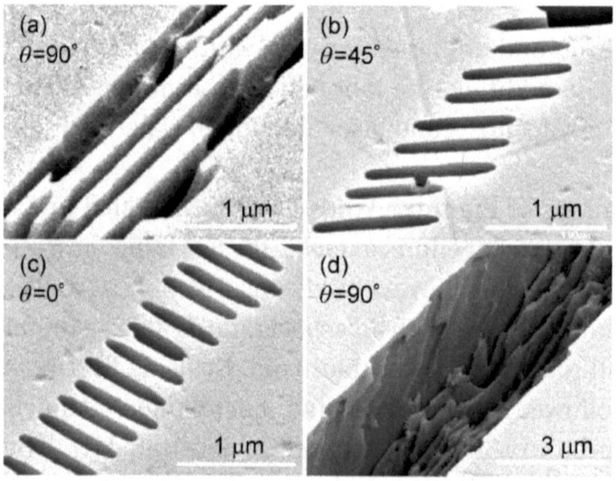

Fig. 3.33: (a)–(c) Top-view (*XY*-plane) SEM images of long-range periodic nanostructures formed along the writing direction for different polarizations for $E_p = 300$ nJ (the polarization is parallel to the writing direction when $\theta = 0°$) and (d) $E_p = 900$ nJ. The structures are revealed after etching for 20 min in a 0.5% aqueous solution of HF. (Reproduced with permission from OSA. ©2005 by the Optical Society of America).

rate, the nanogratings must be oriented parallel to the microchannel; this has the potential to enhance the etch rate by nearly two orders of magnitude relative to that obtained when the nanogratings are oriented perpendicular to the microchannel.

The most commonly used etchant has been HF, which offers a selective etching ratio of up to about 30:1. However, it was recently found that very highly selective etching could be achieved using KOH, allowing for the fabrication of microfluidic channels with an aspect ratio of about 200:1, although the etching rate is extraordinarily low compared to that for HF [182, 183]. In Figure 3.34 [182], it is clearly shown that the channels formed using KOH have much more uniform diameters. Kiyama et al. recently used a KOH etchant to fabricate a 1-cm-long through channel that had a diameter of less than 60 μm near its ends and a much smaller diameter at its middle. Since it was produced without using an ultrasonic bath, it may be possible to further improve the microchannel homogeneity.

Using the above technique, microfluidic channels can be easily fabricated as shown in Figure 3.35 (a). However, the fabricated channels typically are strongly tapered due to the limited etching selectivity, and their cross-sectional shapes are highly elliptical, as shown in Figure 3.35 (c) and 3.35 (d). Furthermore, the surface roughness of the sidewall of the microchannel can reach a level of ~500 nm, which is too high for many fluidic applications. To solve these problems, He et al. performed an additional glass drawing step after the etching of the sample. Namely, the glass sample is heated with an oxyhydrogen (OH) flame until it is softened. Then the glass sample is slightly drawn in the direction parallel to the channel. After the drawing process, a homogeneous channel with perfectly circular cross sectional shapes throughout is obtained, as shown in Figure 3.35 (b), 3.35 (e), and 3.35 (f). In addition, an inner surface roughness of ~0.2 nm was achieved. These improvements were achieved owing to the surface tension in the molten glass during the glass drawing. The glass drawing technique also allows for creating centimeter-long microfluidic channels in complex fluidic networks with a diameter of only ~10 mm, as shown in Figure 3.35 (g)–(j).

3.4.3 Microfluidics fabrication in porous glass

Figure 3.36 schematically shows a typical experimental setup used for FsLDWin porous glass, which consists of a femtosecond laser amplifier system (Coherent, Inc., Santa Clara, CA, USA, center wavelength: ~800 nm, pulse width: ~50 fs, repetition rate: 1~250 kHz), a charge-coupled device (CCD) imaging system, a computer-controlled *XYZ* translation stage, beam control devices, and delivery optics. The Gaussian laser beam with an initial 8.8 mm diameter is passed through a ~3 mm-diameter aperture for ensuring a high beam quality (i.e. only the central part of the beam is used because of its high homogeneity). A long-working-distance water-immersed objective is employed for focusing the beam into the porous glass sample, which is fixed in a

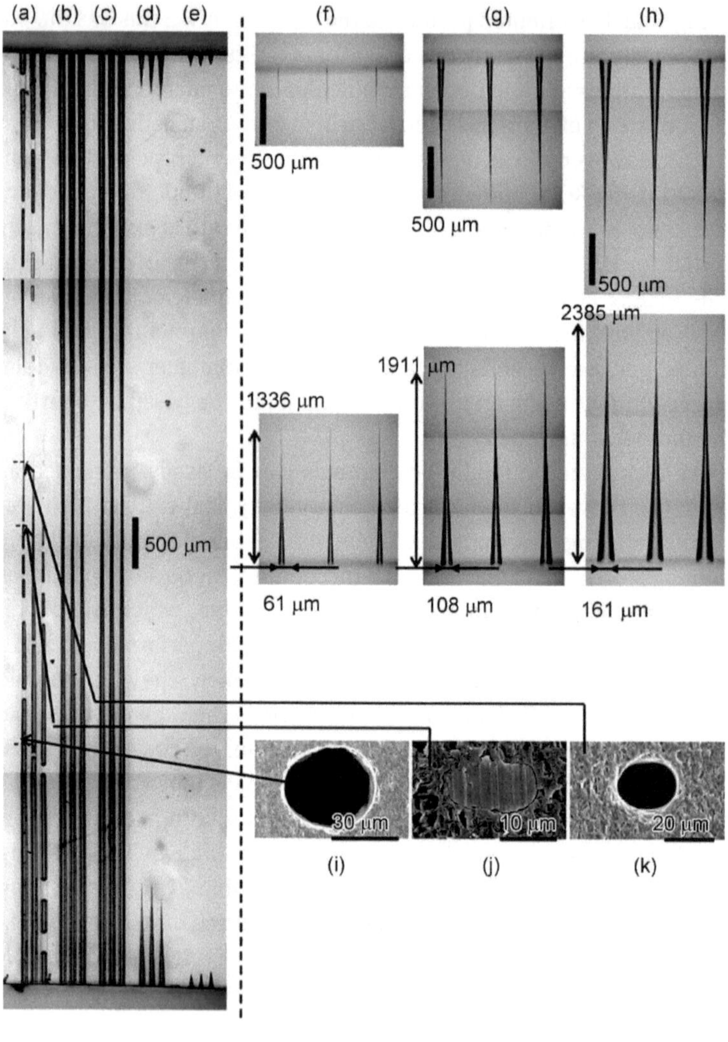

Fig. 3.34: Optical micrographs for comparison of etching profiles in a 9.2-mm-long silica substrate produced by etching by 10 M (35.8%) aqueous KOH (20 mL) and 2.0% aqueous HF (20 mL). (a)–(e) Channels fabricated after soaking for 60 h in KOH at 80 °C after FsLDW focused by a 40× objective lens (NA: 0.65) for pulse energies of (a) 500, (b) 400, (c) 300, (d) 200, and (e) 100 nJ. (f)–(h) Two images of each set of etching fronts, which represent the time evolution of channel formation in aqueous HF solution at ambient temperature. (i)–(k) SEM pictures of cross-sections of the microfluidic channels (Reprinted with permission from ACS. Copyright 2009 American Chemical Society).

petri dish filled with distilled water. The petri dish is placed on the translation stage, as shown in Figure 3.36. Since the high-repetition-rate laser pulses could also lead to temporary evacuation of water in the laser action zone by semicontinuous irradiation, a chopper is used to modulate the femtosecond laser pulse train during the ablation

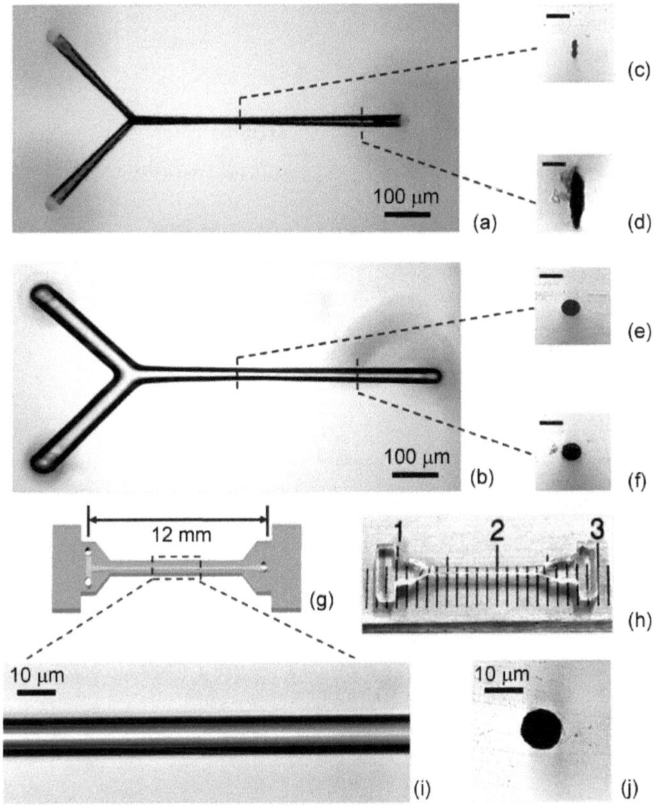

Fig. 3.35: Optical micrographs of the Y-branched channels (a) before and (b) after the glass drawing process; (c)–(f) cross sections of the channels at the locations indicated by the cutting lines; (g) schematic illustration and (h) digital-camera-captured image of a 12-mm-long microchannel; (i) close-up view and (j) the cross section of the microchannel in (h). Scale bars in (c)–(f): 50 μm.

process, so that the water can always fill the ablation zone by infiltration through the porous network.

Figure 3.37 (a) and 3.37 (b) show the schematic view of the experimental setup and the flow diagram for the fabrication process respectively. The main fabrication process includes two steps: (1) direct formation of 3D microchannels from one side of a porous glass substrate immersed in water by FsLDW and (2) postannealing of the glass substrate at ~1150 °C by which the porous glass can be consolidated. The inset SEM image in Figure 3.37 (a) clearly shows that the size distributions of nanopores are homogeneous. After the consolidation process, all the nanopores in the porous glass disappeared; however, the fabricated microchannels could survive due to their large diameter. Moreover, the porous structure is consolidated into a clear impervious glass known as Vycor brand 96% SiO_2 glass, which is similar to fused silica in its thermal properties and light transmittance [172].

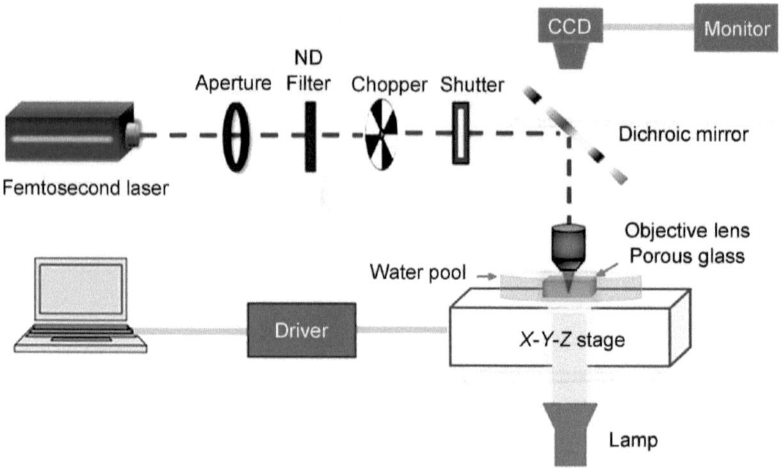

Fig. 3.36: Schematic of the setup used for FsLDW in porous glass.

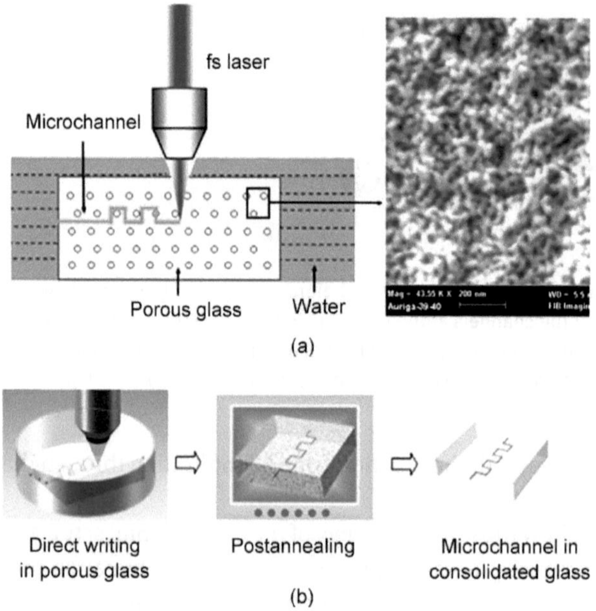

Fig. 3.37: (a) Schematic view of experimental setup; (b) flow diagram for the fabrication process.

3.4.3.1 Fabrication of multicentimeter-long microchannels

For laser direct writing of microchannels in porous glass, relatively high laser intensity is required for inducing laser ablation in microchannel. Typically, a 50× objective (NA 0.80) is employed for focusing the 1 kHz femtosecond laser pulses at a depth of 300 μm below the sample surface. The laser pulse energy is chosen to be 60 μJ, and the

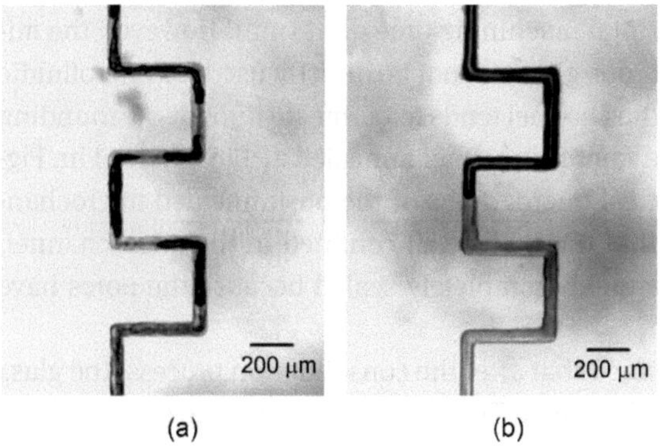

Fig. 3.38: Optical micrographs of microchannels in porous glass fabricated by (a) single scan and (b) multiple scans.

translating speed is set to be 20 μm/s. Figure 3.38 (a) shows a section of microchannel inside porous glass immersed in water, wherein the dark segments consisted of gas bubbles produced when focusing femtosecond laser pulses on the silica/water interface. The discontinuous bubble segments indicate that some parts of the channel are blocked by debris created by laser ablation. By repeatedly scanning, the water could disperse and flush away clogged segments, until a homogeneous and continuous hollow channel is obtained. The continuous bubble section shown in Figure 3.38 (b) indicates that the channel is homogeneous and debris-free, after one slow scan at 20 μm/s followed by repeatedly scanning the whole microfluidic channel six times at a higher translation speed of 1 mm/s.

Figure 3.39 (a) shows a top view micrograph of the homogeneous square-wave-like microchannel embedded in porous glass with a total length of ~1.6 cm, which is

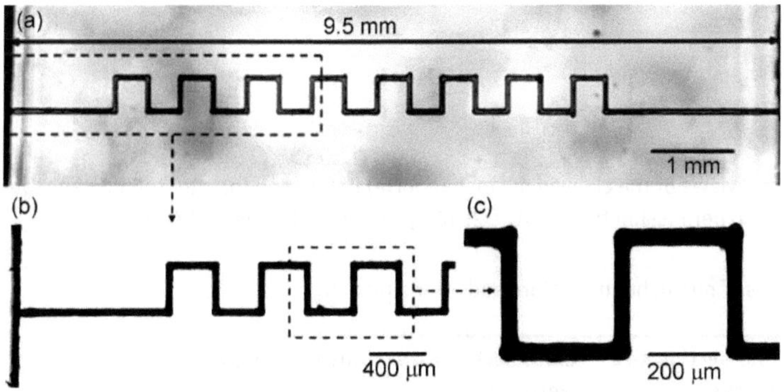

Fig. 3.39: (a) Optical micrograph of a 1.6-cm-long microchannel embedded in porous glass before postannealing; (b) optical micrograph of postannealed microchannel filled with red ink; (c) close-up view of (b) (Reproduced with permission from OSA. ©2010 by the Optical Society of America).

obtained by multiple scans with a total machining time of ~15 min. However, the microfluidic channel embedded in porous glass cannot be directly used for microfluidic application, because the liquid in the channel tends to permeate into the surrounding pores. To collapse all the pores, the sample is further annealed at 1150 °C for 1 hr. Figures 3.39 (b) and 3.39 (c) show optical micrographs of the postannealed microchannel filled with red ink. It is clear that the ink is well confined in the microchannel, which indicates that the microchannel is completely sealed because the pores have been closed after postannealing.

It should be specifically mentioned that after the consolidation process, the glass substrate shrinks in all the three directions. The initial length of ~1.6 cm of the microfluidic channel decreases to ~1.4 cm after the postannealing. Furthermore, we compare the cross sections of microchannels before and after the postannealing by breaking up the microfluidic channels, as shown in Figure 3.40 (a), (b). Table 3.4 gives geometric sizes of the microchannel before and after annealing, which indicate the shrinkage factors in three directions are almost equal. The corresponding volume shrinkage factor is ~39%, approximately equal to the 40 vol% porosity of porous glass. Note that there are some cracks along the laser propagating direction around the microchannels as shown in Figure 3.40 (a),(b), which could have resulted from inhomogeneous axial intensity distribution of the incident laser due to both the self-focusing effect and the spherical aberration effect [184].

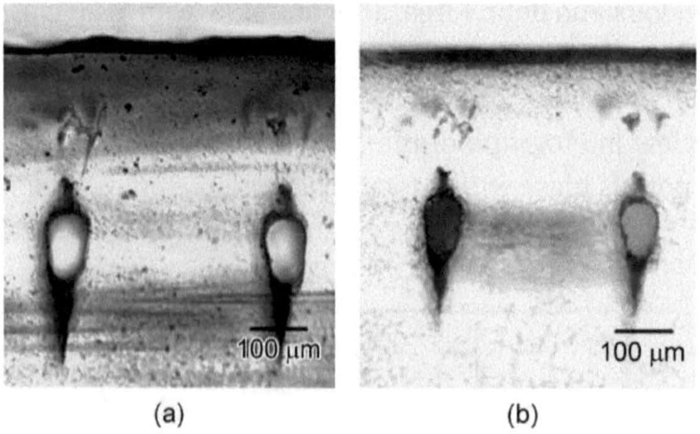

Fig. 3.40: Cross-sectional views of the cleaved microchannel (a) before and (b) after the postannealing [173] (Reproduced with permission from OSA. ©2010 by the Optical Society of America).

Tab. 3.4: Geometric sizes of microchannel before and after annealing

Description	Cross section size (transverse)	Cross section size (vertical)	Channel length
Before annealing	74 µm	114 µm	9.50 mm
After annealing	63 µm	97 µm	8.00 mm
Shrinkage factor	14.9%	14.9%	15.8%

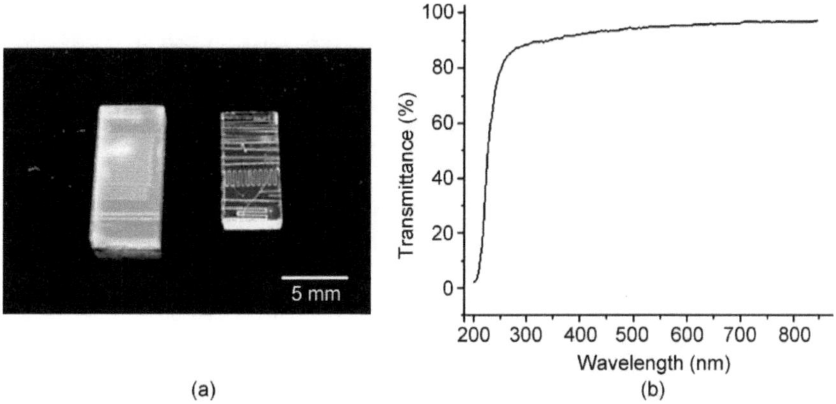

Fig. 3.41: (a) The porous glass before (left) and after (right) postannealing; (b) transmission spectrum of the consolidated glass.

Moreover, the porous glass substrate, which is opaque due to the scattering of nanopores, becomes highly transparent after the consolidation process, as shown in Figure 3.41 (a). Figure 3.41 (b) gives a transmission spectrum of the consolidated glass, which shows superior transmission ranging from ultraviolet radiation (UV) to near infrared radiation (NIR). This facilitates the future incorporation of optical functions in the glass for optofluidic applications.

3.4.3.2 Mechanism for efficient removal of debris

In this section, we discuss the dominant mechanism that is responsible for the removal of debris from the microchannel during the FsLDW in porous glass. Actually, in previous studies of fabrication of microfluidic channels in fused silica by water-assisted femtosecond laser drilling [185], it has been observed that the drilling process relies heavily on water circulation in the microchannel, which is a bidirectional mass transfer process including both the water flowing into the laser ablation zone from the inlet of the microchannel driven by capillary force and bubbles ejecting through the inlet driven by the shock wave produced by the femtosecond laser, as schematically shown in Figure 3.42 (a). The water plays two key roles for producing hollow structures: (1) dispersing the debris produced by laser ablation; and (2) producing bubbles to expel the debris from the channel through the inlet, which in turn promotes the water circulation. Unfortunately, as the length of the channel increases in solid glass, once the bubble and debris build up to the extent that water inflow is blocked, the drilling process will terminate. However, for porous glass, the water can continuously infiltrate into the ablation zone through the porous network, as schematically shown in Figure 3.42 (b), even if the microchannel is blocked. Therefore, the repeated back and forth writing can always disperse debris and produce bubbles independent of the length of the microchannel.

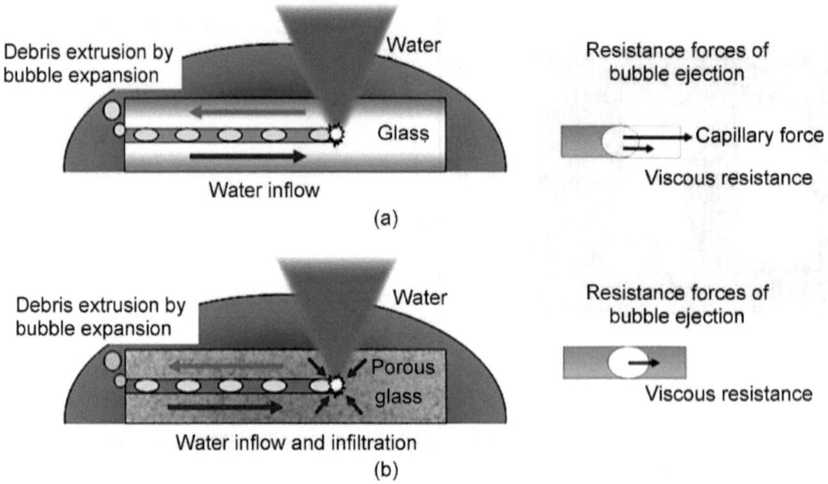

Fig. 3.42: Schematic illustration of water-assisted direct-writing process in (a) solid glass and (b) porous glass.

In the right panels of Figure 3.42, the resistance forces of bubble ejection when the microchannel is blocked by bubbles are illustrated for a solid glass case and porous glass case. For the porous glass case, it is much easier to drive the bubbles out from the channel because the capillary force exerted by the water surrounding the bubbles is balanced out. In the experiment of FsLDW in porous glass, the microchannel could be temporarily blocked somewhere if some of the bubbles in the channel do not have enough time to escape from the channel before the new debris is produced. However, we can easily expel the bubbles from the channel by scanning the femtosecond laser beam along the channel repeatedly.

In addition, due to the porosity of the sample, the amount of debris produced by the femtosecond laser ablation would be less than that of traditional water-assisted drilling inside dense glass. Furthermore, the porosity of the sample will also allow for the direct creation of hollow space by localized solidification and the collapse of the nanopores, which has been verified by the formation of hollow nanogratings in porous glass.

3.4.3.3 High-aspect-ratio 3D microchannels

For the direct writing of high-aspect-ratio (length-to-diameter) microchannels, both high-repetition-rate femtosecond laser pulses and tight focusing are desirable. Here, the 250-kHz femtosecond laser pulses with a pulse energy of ~2 μJ are tightly focused by a water-immersed objective (NA = 1.0). Under such conditions, a peak laser intensity of $\sim 8 \times 10^{15}\,\mathrm{W/cm^2}$ can be reached in a tiny focal volume with a diameter of ~0.8 μm, which is sufficiently strong for inducing nonlinear optical absorption and ablation of the glass material. The significant amount of heat and strong shockwaves generated by the femtosecond laser ablation can facilitate water circulation and help to remove the debris from the outlet of the microchannel. Since the high-repetition-rate laser pulses could also lead to the temporary evacuation of water in the laser action zone by semicontinuous irradiation, a chopper operated at 1 kHz with a duty of 50/50 is used (Figure 3.36) to modulate the femtosecond laser pulse train during the ablation process, so that the water can always fill the ablation zone by infiltration through the mesoporous network [186].

The 3D nature of the FsLDW offers unique flexibility for constructing complex microfluidic networks [187]. Figure 3.43 (a) shows a 3D helical microchannel with a total length of ~1 cm and a diameter of ~16 μm (i.e. corresponding to an aspect ratio of > 600). The coil radius and the pitch of the helical channel are 100 μm and 50 μm respectively. The homogeneous and debris-free helical channel is produced inside the porous glass by performing one slow scan (10 μm/s) followed by three fast scans (100 μm/s) over the whole microfluidic channel. The total fabrication time is less than 30 min. Figure 3.43 (b) shows several overpass structures consisting of spiral-like bends along microfluidic channels, by which one microchannel can cross over the others. The fluorescence micrograph presented in the inset of Figure 3.43 (b) indicates that indeed, the two microchannels do not intersect each other. The insets in Figure 3.43 (a),(b) present fluorescence microscopy images of the fabricated microchan-

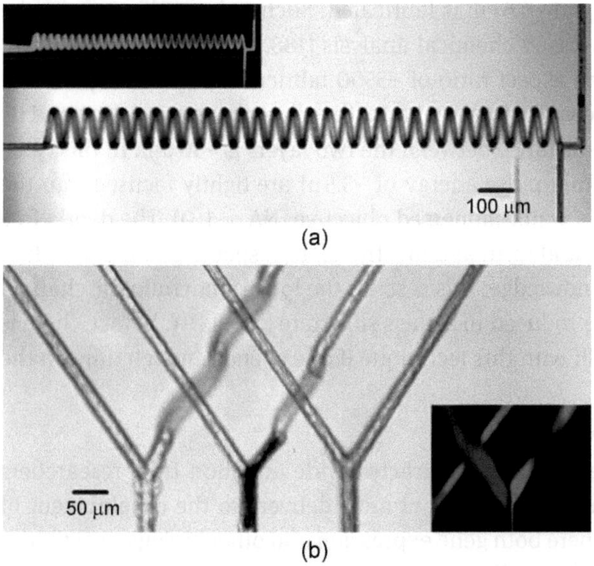

Fig. 3.43: Optical micrographs of (a) a 3D helical microchannel and (b) 3D microfluidic channels crossing over each other embedded in consolidated glass; insets: fluorescence microscopy image of the microchannel filled with fluorescein solutions [174] (Reproduced with permission from RSC. ©2012 by the Royal Society of Chemistry).

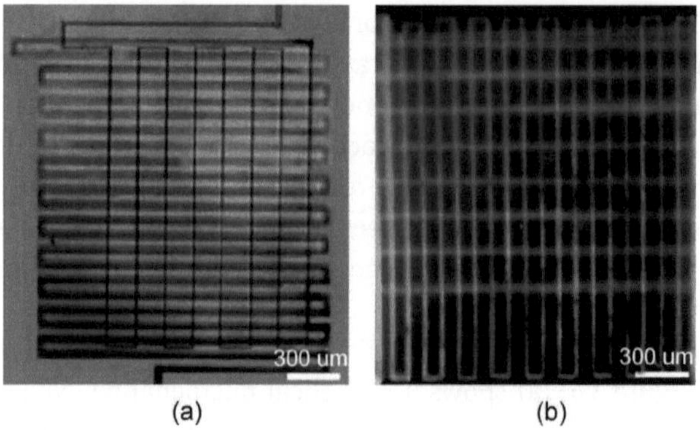

Fig. 3.44: (a) Optical micrograph of a two-layer microfluidic channel in porous glass with a length up to 6.6 cm (b) Fluorescence microscopy image of the postannealed microchannel filled with a solution of fluorescein.

nels filled with fluorescein solutions. The confined fluorescent solution provides proof that the nanopores have all collapsed to form the consolidated substrate.

Many microfluidic applications such as electrophoresis require long channels compactly deployed in a tiny space. Compact microfluidic systems not only reduce the sizes of micro-analysis systems but also promote device performance qualities such as sensitivity. Toward this goal, a 3D two-layer microfluidic channel structure in porous glass with a length up to ~6.6 cm is fabricated. Such a channel length can be sufficient for electrophoresis-based chemical analysis [188, 189]. Figures 3.44 (a),(b) shows microchannels with an aspect ratio of ~5500 fabricated in porous glass and consolidated glass, respectively. The total length of the microchannel is 6.6 cm and its diameter is ~12μm, while the distance between the two layers is ~30 μm. In this case, femtosecond laser pulses with a pulse energy of ~3.5 μJ are tightly focused into the porous glass substrate using a water-immersed objective (NA = 1.0). The debris-free microchannel is written back and forth at a low translation speed, and the total fabrication time is ~3 h. To our knowledge, this is so far the longest microfluidic channel with the highest aspect ratio produced in a glass substrate by FsLDW. In fact, there is no limit on the channel length with this technique if one spends enough time on the fabrication of the structure.

3.4.3.4 Large-volume hollow chamber

Recently, the 3D cell culture system has attracted wide attention from researchers ranging from the engineering of tissues for clinical delivery to the development of models for drug screening, where both gene expression and other biological activities more closely mirror what happen in living organisms [190]. To build 3D microfluidic devices for cell culture, hollow microfluidic chambers with relatively large volume and arbitrary 3D configurations are desirable.

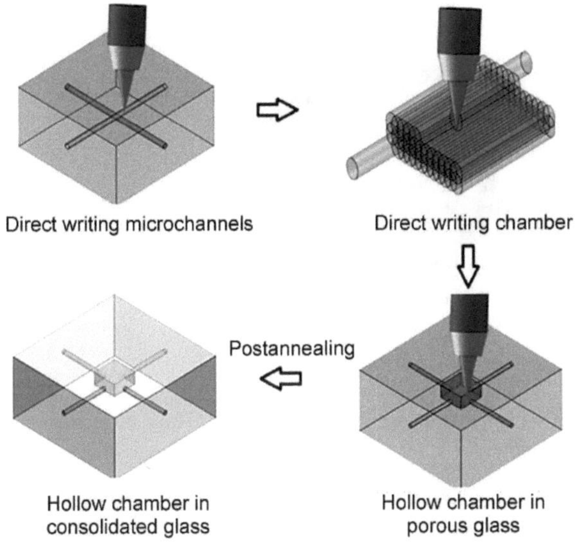

Fig. 3.45: Schematic illustration of the fabrication process for a large-volume hollow chamber.

Figure 3.45 shows the fabrication process of a large-volume chamber inside glass by femtosecond laser direct writing. Two straight channels are first fabricated inside the porous sample for the inflow of water and ejection of debris. Along with the straight channels, a chamber is fabricated by translating the sample in a line-by-line and layer-by-layer manner. The scan of the chamber started from its bottom in order to avoid the extra aberration created by the previously formed hollow layers. After the laser direct writing, a postannealing process is applied to the sample foremoving all the nanopores in the glass. After postannealing, geometric sizes of the annealed hollow chamber decrease to 85% of the original one, while its shapes remain unchanged (Table 3.4).

Figure 3.46 (a) shows a large-volume chamber connected to four microchannels embedded in the porous glass. Here, the 250-kHz femtosecond laser pulses with a pulse energy of ~6.5 μJ are tightly focused by an objective lens (NA = 0.90). First, the microchannels are fabricated with one slow scan at a translation speed of 20 μm/s. The diameter of the microchannels is approximately 20 μm. Secondly, the microchamber is fabricated in a line-by-line and layer-by-layer manner at a higher translation speed of 100 μm/s. The interval between the adjacent lines in the horizontal plane is 3 μm, and the vertical interval between the adjunct layers is 5 μm. The size of the microchamber is 1 mm × 1 mm × 100 μm defined by the number of adjacent lines, and the total scanning process for fabricating the chamber took ~20 h because of its large volume. The thick black curve in Figure 3.46 (b) shows the boundary of air and water in the chamber, providing clear evidence of the hollow structure. The technique allows for the fabrication of arbitrary 3D hollow structures embedded in glass.

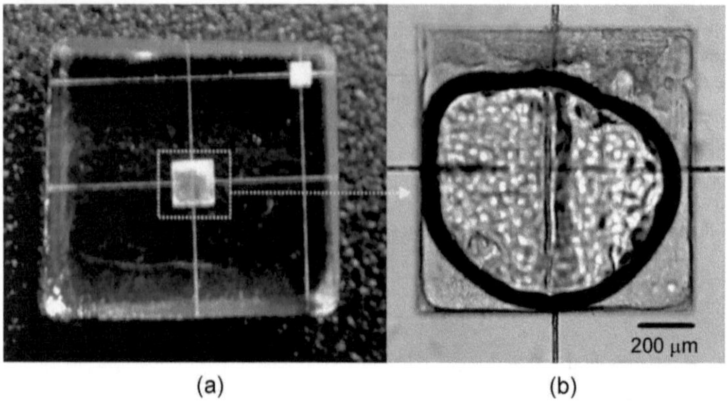

Fig. 3.46: (a) Overview image of a 1 mm × 1 mm × 100 μm chamber connected to four microchannels. (b) Enlarged view of the large volume chamber.

3.4.3.5 3D microfluidic mixers

Fluid mixing is an essential function required by most microfluidic systems. However, quick and efficient fluid mixing inside microchannels is usually difficult to achieve due to the laminar nature of microflows, characterized by low Reynolds numbers. Recently, various passive mixers have been developed to achieve efficient mixing by utilizing 3D geometric structures to induce disturbance in the fluids [191–194]. Nevertheless, the fabrication of 3D microfluidic mixers with arbitrary geometries is still challenging. Here, we demonstrate the rapid fabrication of a passive microfluidic mixer consisting of geometrically complex 3D microchannels by FsLDW in porous glass.

According to the chaotic theories, a passive micromixer based on the baker's transformation concept could provide an ideal mixing performance [195]. However, such kinds of mixers usually require complicated 3D geometries, thus they are difficult to fabricate through the conventional planar fabrication process. By using FsLDW, a passive micromixer consisting of symmetrical 3D units as theoretically proposed by Carrière [196] is fabricated. Figure 3.47 (a),(b) presents the schematic illustrations of the designed 3D mixer which is composed of a Y-shape microchannel embedded 400 μm below the surface of chip and a string of mixing units, connected to two opening inlets and one outlet. The inlets and outlet are all on the backside of the chip. Three circular slots surrounding the inlets and the outlet are also fabricated by femtosecond laser ablation, which are used as connection interfaces with plastic pipes of a diameter of 0.9 mm. Figure 3.47 (c) presents an overview of a 3D micromixer consisted of six mixing units. The length of all the horizontal and vertical channels in each mixing unit is 150 μm, as shown in Figure 3.47 (d) (top view) and (e) (side view). It also can be observed from Figure 3.47 (e) that the cross section of the channel is elliptical with a width of ~50 μm and a depth of ~75 μm. Figure 3.47 (f) shows the inner surface of the microchannel for evaluation of the surface roughness.

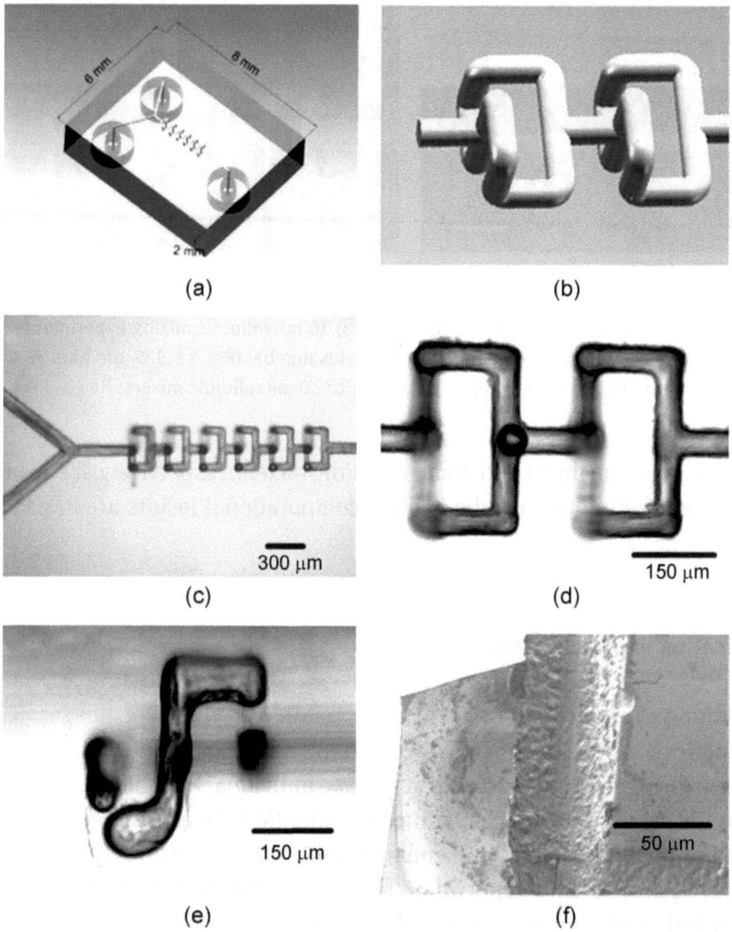

Fig. 3.47: Schematic diagrams of the 3D passive microfluidic mixer: (a) overview and (b) close-up view images; (c) overview optical micrograph of the microfluidic mixer, (d) top view optical micrograph of two mixing units, and (e) cross sectional optical micrograph of the 3D microchannels; (f) SEM images of the inner wall of the fabricated microchannel. The dark sections in (e) are caused by the bubbles. (Reproduced with permission from RSC. ©2012 by the Royal Society of Chemistry).

Figure 3.48 (a),(b) shows the experimental results of mixing the two fluorescent dye solutions (fluorescein sodium and Rhodamine B) in the 1D and 3D microfluidic mixers fabricated under the same direct writing conditions, respectively. Clearly, the mixing behaviors in these two structures agree well with the simultion results. For the 3D mixer, after passing through three mixing units (i.e. a length of 0.9 mm), the two fluids are well mixed, corresponding to a mixing time of ~10 ms. For comparison, Figure 3.48 (a) shows that in the 1D microfluidic channel, efficient mixing is not achieved after a propagation distance of ~1300 μm. The insets in Figure 3.48 (a),(b) show the nu-

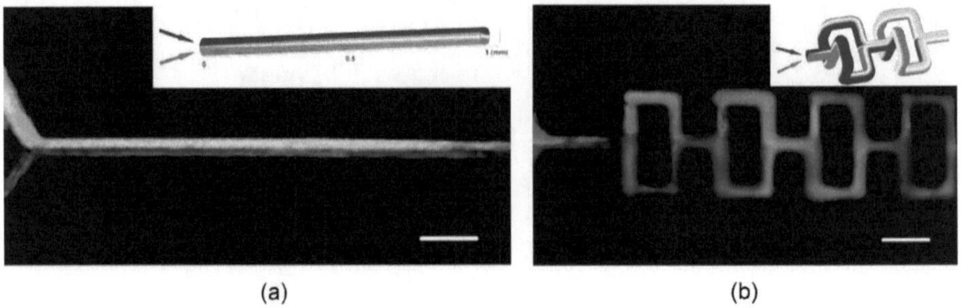

Fig. 3.48: Fluorescence microscopy images of the (a) 1D and (b) 3D microfluidic mixing experiments, where two fluids (light gray and dark gray) are mixed at Reynolds number (Re) = 5.3. Scale bars = 150 µm. Insets: numerical simulations of mixing in (a) 1D and (b) 3D microfluidic mixers; Re= 5.3.

merical simulations of mixing in 1D and 3D microfluidic mixers, respectively. It can be clearly seen that the experimental results and the computational results are in good agreement.

3.4.4 Nanofabrication in porous glass

3.4.4.1 Direct writing of nanochannels

Recently, nanofluidics has attracted increasing attention due to its unique fluid transport properties and great potential to handle and probe individual biomolecules [197]. The incorporation of nanofluidic elements into microfluidic devices enables further downsizing of micro-analysis systems and enhancement of their functionalities, both of which are highly desired for LOC applications [198]. Both fundamental investigation and practical application of nanofluidics heavily rely on fast and simple nanofabrication techniques [199–201]. Here, we show that by uniquely employing the porous glass as the substrate material, nanogratings consisting of an array of elliptical-shaped hollow nanovoids elongated in the direction of laser propagation can be formed in glass using a linearly polarized writing beam. By reducing the peak intensity of the laser pulses close to a threshold value, only a single hollow nanovoid in the central area of the focal spot can survive, serving as the basic element for the construction of 3D nanofluidics inside glass.

Previously, it has been reported that nanogratings can be formed in glass with femtosecond laser irradiation, which consist of alternating regions of high and low etch rates periodically distributed in a direction perpendicular to the laser polarization [180]. We found that the nanograting-like structures can also be induced in the porous glass by femtosecond laser irradiation, whereas the gratings are composed of an array of hollow nanovoids, as schematically shown in Figure 3.49.

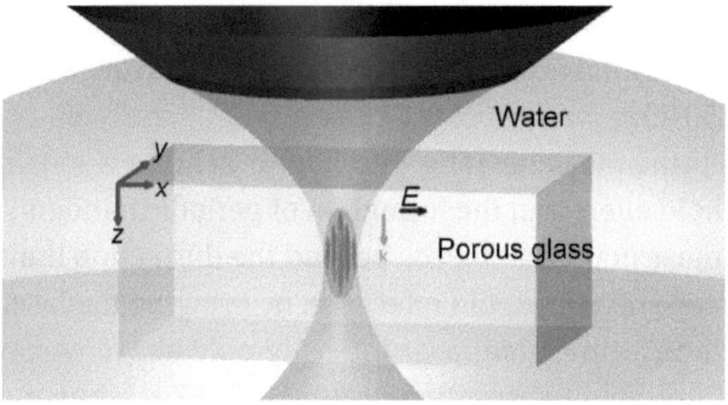

Fig. 3.49: Schematic diagram of inducing nanograting structures inside water-immersed porous glass by femtosecond laser pulses. The sample is translated in the *x*-*y* plane, the writing polarization is aligned to the *x*-axis, and the laser incident direction is along the *z*-axis. (Reproduced with permission from OSA. ©2013 by the Optical Society of America).

Amazingly, we found that a single nanochannel could still be induced in the central area of the focal volume at a very low laser intensity, while other nanograting-like channels vanish. The cross sections of the nanogratings induced by FsLDW at different pulse energies (with a same writing speed of 10 μm/s) are shown in Figure 3.50 (a)–(c). The direction of laser polarization is perpendicular to the translating direction. At a relatively high pulse energy of 90 nJ, the number of the elliptical nanovoids in the grating is large, as shown in Figure 3.50 (a). By reducing the pulse energy to 70 nJ, the number of nanovoids decreases to only four. It is noteworthy that under this condition, the periodicity of the nanograting apparently increases as evidenced in Figure 3.50 (b),

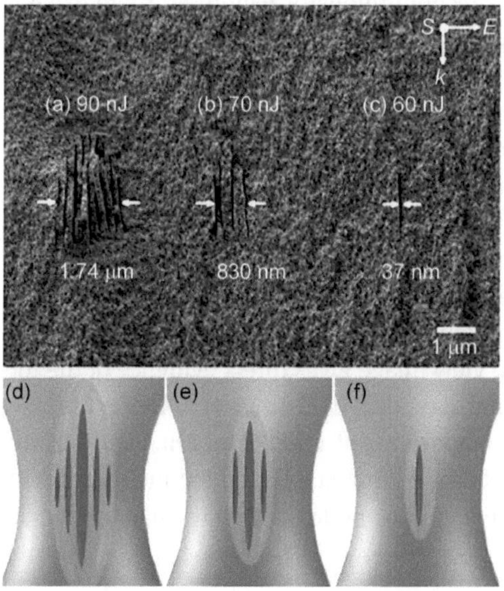

Fig. 3.50: (a)–(c) Evolution from nanovoid array to single nanovoid with decreasing laser intensity. The laser incident direction (*k*), polarization (*E*), and writing direction (*S*) are indicated. (Reproduced with permission from OSA. ©2013 by the Optical Society of America.) (d)–(f) Principle schematic diagrams to achieve a single nanocrack. (Reproduced with permission from Springer. ©2014 by Springer.)

which also helps reduce the number of nanovoids. Finally, by reducing the pulse energy to 60 nJ, only the central nanovoid survives, showing an extremely narrow width of ~37 nm as presented in Figure 3.50 (c).

Figure 3.50 (d)–(f) presents a principle schematic diagram to achieve a single nanocrack by combining the threshold effect and the formation of periodic nanograting. Conventionally, the surface nanostructuring of glass beyond the diffraction limit could be achieved by taking advantage of a threshold effect, i.e. by bringing the laser intensity down to a level approaching a threshold laser intensity only above which the ablation of glass induced by multiphoton absorption occurs [107, 202]. In theory, there is no limit on the fabrication resolution when the threshold effect takes place, because the ablation linewidth can always be reduced if one keeps on reducing the difference between the chosen laser intensity and the threshold intensity. However, in practice, due to the fluctuation of the output power of the femtosecond laser, the fabrication process will become extremely unstable when the laser intensity is near the threshold intensity [52]. In fact, reliable sub-100 nm fabrication linewidths are extremely difficult to achieve for surface ablation. Moreover, a sub-100 nm fabrication linewidth has never been demonstrated for 3D nanostructuring in bulk transparent materials so far due to the different physical mechanism involved in bulk machining [203]. In this case, the formation of a single nanocrack is much less sensitive to the fluctuation of the laser intensity by combining the threshold effect and the formation of a periodic nanograting. As shown in Figure 3.50 (f), when the femtosecond laser intensity is intentionally reduced to a level at which only the intensity in the light gray oval region is higher than the threshold intensity, one will be able to select only one nanocrack in the central area of the focal volume. Note that the single nanocrack is much narrower than the above-threshold-intensity region. So there exists a window (a range of laser intensity) for inducing a single nanocrack, which is between the threshold laser intensity of emergence of a single nanocrack and the minimum laser intensity for inducing two nanocracks in porous glass.

The induced single nanovoids can be connected into a continuous structure for construction of 3D nanochannels in glass. This is achieved by slowly scanning the focal spot of the femtosecond laser at a translation speed of 5–10 μm/s in the porous glass immersed in water. The nanochannel is connected to a prefabricated microchannel through which the debris can be driven out from the nanochannel by bubble generation. Figure 3.51 (a) presents a zoomed-in view of the cross section of the nanochannel surrounded by small nanopores. The top view image gives direct evidence of the formation of a single continuous nanochannel, as shown in Figure 3.51 (b). After the laser writing, the sample is annealed at a high temperature of 1120 °C for 2 hr. Figure 3.51 (c) shows a fluorescence microscope image of the postannealed nanochannels filled with fluorescent dyes, indicating that there are no clogs and leakage in the nanochannels.

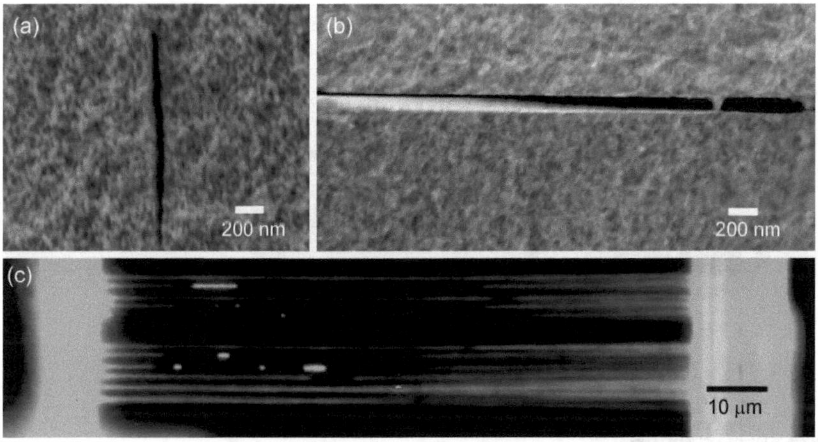

Fig. 3.51: (a) Cross section view and (b) top view SEM images of a continuous nanocrack written in the porous glass; (c) fluorescence microscope image of the postannealed nanofluidic channels filled with fluorescent dyes as an evidence that these nanochannels are through structures without clogging and leakage (Reproduced with permission from OSA. ©2013 by the Optical Society of America).

3.4.4.2 Rapid prototyping of 3D nanofluidic chips

The unique 3D capability of FsLDW allows fabrication of nanochannels at arbitrary depths inside the glass substrate with a high axial resolution. Moreover, the dimensions of the channel can easily be controlled by adjusting the machining parameters, such as the laser pulse energy, the translation speed and the number of repeated scans. In combination with 3D microfluidics we have demonstrated in the previous sections, it is now possible to produce integrated micro-nanofluidic systems of complex 3D geometries and configurations in glass in a single continuous step.

Figure 3.52 shows a schematic diagram of the process flow chart for fabricating integrated 3D micronanofluidic structures by FsLDW in porous glass, showing four major steps: (1) fabrication of reservoirs by femtosecond laser direct writing ablation of porous glass immersed in water with a low-NA objective (NA = 0.45) at a pulse energy of ~10 μJ, as shown in Figure 3.52 (a); (2) direct writing of microchannels connecting the reservoirs with a long-working-distance water-immersed objective (NA = 1.10) at a pulse energy of 0.5–3 μJ, as shown in Figure 3.52 (b); (3) direct writing of a nanochannel array bridged between two microchannels with the same objective used in (2) but at a reduced pulse energy of 125–200 nJ, as shown in Figure 3.52 (c); and (4) postannealing of the glass substrate at ~1120°C for 120 min, by which all the nanopores in porous glass can be collapsed to form a consolidated glass substrate, while the fabricated nanochannels survive due to their larger diameter than the pores.

Figure 3.53 (a) schematically illustrates an integrated micronanofluidic structure, which contains an array of two-layer nanochannels connected to two microfluidic channels. Figure 3.53 (b) shows the optical micrograph of the fabricated nanochan-

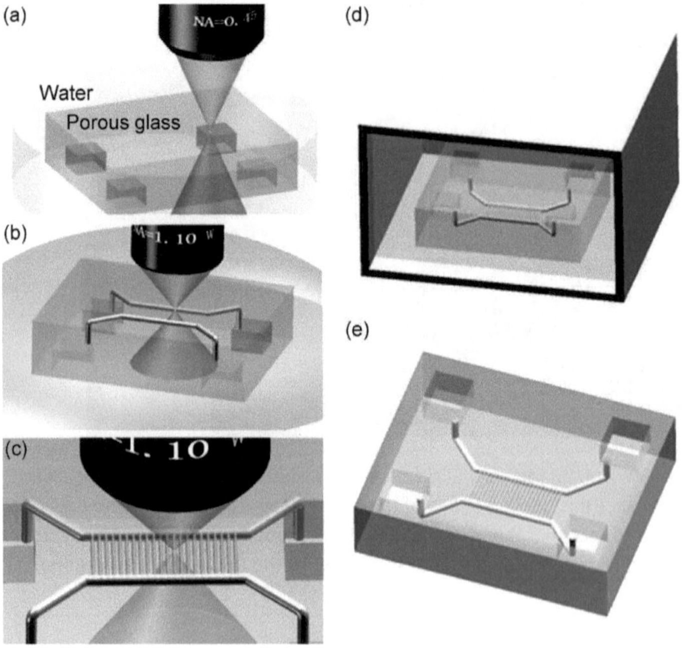

Fig. 3.52: Schematic diagram of process flow chart for fabricating 3D integrated micro-nanochannels by FsLDW.

nel arrays. The nanochannels are directly written at the depths of about 200 μm below the porous glass surface, and the difference in the depths of the two layers of the nanochannels is only ~1.5 μm. All the nanochannels are directly written by a single scan at a translating speed of 10 μm/s with pulse energy of 128 nJ. Note that as we have examined prior to the writing process, the threshold pulse energy for inducing optical breakdown in porous glass immersed in water is ~108 nJ. Therefore, the peak laser intensity chosen for writing the nanochannel is approximately ~20% higher than the threshold intensity. The direction of laser polarization is perpendicular to the translating direction. During the fabrication of the nanochannel in porous glass immersed in water, it is observed that the bubbles are continuously ejected from the inlet of the fabricated nanochannel into the thick microchannel, which can efficiently remove the debris produced by laser ablation. The continuous bubble flow in turn indicates that an unclogged nanochannel has been formed with a single scan (it was observed previously that once clogging occurred, there would be no more bubbles coming out from the outlet). After collapsing the nanopores by postannealing, the size of each nanochannel is reduced by ~14% isotropically in all three dimensions, resulting a total channel length of ~40 μm, a width in the range of 30–50 nm, and a height in the range of 1–1.5 μm, as shown in Figure 3.53 (b),(c). The cross section of the nanochannel appears highly elliptical, showing a typical depth-to-width ratio up to 25, as evidenced

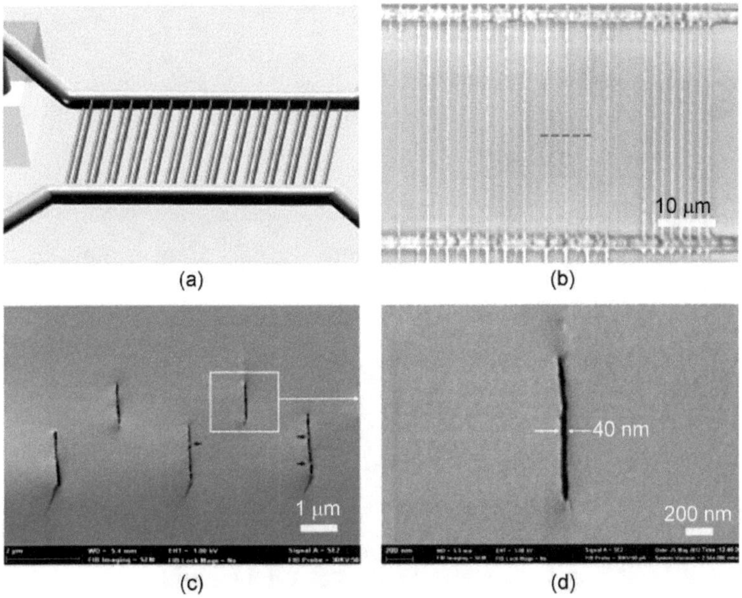

Fig. 3.53: (a) Schematic diagram of an array of double-layer nanochannels to bridge two microchannels; (b) top-view optical micrograph of double-layer nanochannels after postannealing; (c) cross-sectional SEM micrograph of the nanochannels cleaved along the dashed line in (b). The laser is incident from bottom of the figure, and the arrows indicate some tiny occasional protrusions on the edges of the nanochannels, which are formed due to the cleaving process. (d) Close-up SEM micrograph of cross section of a nanochannel. (Reproduced with permission from RSC. ©2013 by the Royal Society of Chemistry).

in Figure 3.53 (d). Such highly elliptical shapes resemble the cross section of an individual nanoplane in a nanograting, indicating a relatively low fabrication resolution in the axial resolution.

To demonstrate the applications of the fabricated nanochannels in investigating single molecular behaviors, a complex 3D micro-nanofluidic device is fabricated as illustrated in Figure 3.54 (a), performing an experiment of DNA stretching using the fabricated device. It is noteworthy that recently, nanochannel-based DNA stretching has attracted enormous attention because of its potential use in DNA analysis, whereas until now direct fabricating nanochannels and realizing 3D micro- and nanofluidic integration within bulk glass have not yet been achieved [204]. The device consists of two nanochannel arrays with different dimensions (i.e. widths and lengths) at different depths embedded in a single glass substrate. The nanochannels are connected to the four common reservoirs through two microchannels with a diameter of ~50 μm. Such a device allows for simultaneously observation of the stretching of DNA molecules in the two nanochannel arrays of different dimensions.

Figure 3.54 (b) presents a top view of the fabricated two nanochannel arrays, in which the short nanochannels have a length of ~30 μm and a cross section of

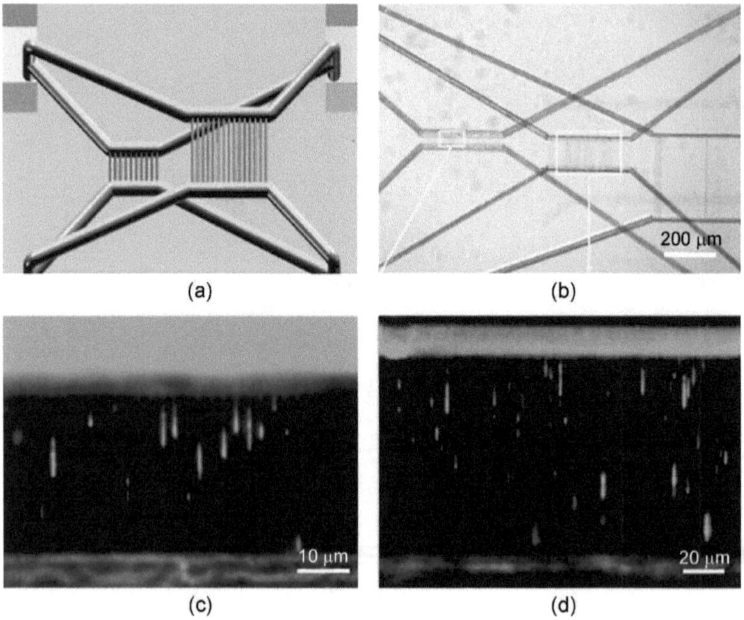

Fig. 3.54: (a) Schematic diagram and (b) top-view optical micrograph of the fabricated 3D nanofluidic device for DNA analysis, and fluorescent images showing the stretching of λ DNA in two arrays of nanochannels with different widths: (c) 50 nm and (d) 200 nm. (Reproduced with permission from RSC. ©2013 by the Royal Society of Chemistry).

~50 nm × 1 μm, whereas the long nanochannels have a length of ~100 μm and a cross section of ~200 nm × 1.2 μm. After an aqueous buffer solution of stained λ DNA is pipetted into a reservoir, the DNA molecules can be readily transported into the nanochannels by capillary force. Figure 3.54 (c),(d) is a typical image of the stretching of λ DNA in the nanochannels. In the thin nanochannels with a cross section of 50 nm × 1 μm, the average stretched length of DNA is ~6.4 ± 1.0 μm (~30% of the dye-adjusted contour length). In the thicker nanochannels with a cross section of ~200 nm × 1.2 μm, the average stretched length of DNA is ~2.8 ± 0.6 μm (~13% of the dye-adjusted contour length). The apparent lengths of some stretched DNA molecules in Figure 3.54 (d) can reach as long as ~15 μm, which can be attributed to the clustering of the DNA molecules. As shown in these images, the continuous flow and uniform stretching of the DNA molecules in the nanochannel arrays also verify the continuity of the nanochannels and the repeatability of the direct writing process.

3.4.5 Conclusions

In this section, we mainly demonstrated the strategies of the fabrication of microfluidic structures in several glass materials using FsLDW. In particular, we have shown that the fabrication of 3D homogeneous microchannels with large sizes and complex geometries can be achieved using water-assisted FsLDW in porous glass followed by a postannealing for the consolidation of the porous glass. By further manipulating the laser pulse energy and polarization, nanofluidic channels with a width of ~30 nm, which is less than $\lambda/25$ (here $\lambda = 800$ nm, which is the wavelength of the writing beam), and an aspect ratio above ~1000 can be directly written in the porous glass, enabling the straightforward integration of microfluidics and nanofluidics in a single platform. Interestingly, two extreme nonlinear effects, namely the femtosecond-laser-induced nanograting formation and the threshold effect in femtosecond laser induced optical breakdown in glass, play important roles in fabrication of the nanochannels. Based on this technique, several functional devices built in glass that cannot be produced by any previous laser direct writing techniques, including a 3D passive microfluidic mixer, an integrated DNA analysis chip, etc. have been demonstrated. This technique will open a broad spectrum of opportunities for micro- and nanofluidic applications.

Currently, however, this emerging field is still in its infancy from both fundamental and applied points of view. First of all, the underlying physical mechanisms behind the formation of nanogratings as well as water-assisted FsLDW in porous glass are quite complicated and not yet fully understood. Secondly, the roughness of microfluidic structures fabricated by laser direct writing needs to be further improved for in vivo and optofluidic applications. Thirdly, the fabrication efficiencies for large-volume and high-aspect-ratio microfluidic devices are still quite low. Lastly, the cross section of the nanochannel appears highly elliptical and asymmetrical, indicating a relatively low fabrication resolution in the axial resolution. It is expected that future advances in 3D micro-/nano-fluidic applied research will strongly promote the further development and optimization of this technique.

3.5 Femtosecond laser fabrication of micro-optical components

3.5.1 Introduction

Fabrication of optofluidic and optoelectronic systems always requires synergetic incorporation of micro-optical components in microfluidic networks. It is not easy to incorporate optical functions in microfluidic systems and electronic chips. The main difficulty is that most microfluidic/electronic systems are currently fabricated using lithography-based planar technologies, which lack the flexibility and/or capability to form three-dimensional (3D) micro-optical elements. Although some optical elements have been successfully incorporated in microfluidic/electronic chips by postassembling, this increases the fabrication complexity and cost. Ideally, a technology will be realized that can simultaneously fabricate 3D microfluidic/electronic and 3D micro-optical structures in a single substrate. FsLDW is the only method that has demonstrated the compatibility and high flexibility to fulfill this requirement.

As essential building blocks for optofluidic/optoelectronic devices, optical waveguides and free-space micro-optical components are frequently implemented for signal generation, propagation, and detection. Their fabrication technique as well as their 3D assembling in a single chip have become big challenges. It is a straightforward process to fabricate optical waveguides by use of FsLDW, as femtosecond laser irradiation can change the refractive index of glass through multiphoton absorption in the focal volume. This technique has been developed since 1996, with details reviewed elsewhere [53–59]. Free-space micro-optical components such as micromirrors and microlenses can be fabricated using femtosecond-laser-assisted wet chemical etching to form hollow structures with planar or curved surfaces in glass that serve as optical interfaces. However, the etched structures exhibit a surface roughness of a few tens of nanometers, thus hampering their usability for optical applications. To solve this problem, several postprocessing heat treatments, including oxygen/hydrogen flame polishing [205, 206], annealing in an oven [207], and CO_2 laser reflow [208] have been proposed. Using these methods, the surface roughness can be substantially reduced to a few nanometers, opening up new possibilities for the design of useful devices. A host of on-chip optical elements such as microlenses, hollow optical waveguides, and optical microresonators, have been developed in this way [205–208].

Procedures for fabricating micro-optical components in Foturan glass mainly involve the following four steps: (1) direct writing of a latent image in the sample by a tightly focused femtosecond laser beam, (2) baking the sample in a programmable furnace to form modified regions, (3) etching the sample in 10% hydrofluoric acid in an ultrasonic bath to selectively remove the modified regions, and (4) postannealing the sample at 570°C for 5 h to smooth the etched surface. Step (4) is very important for fabricating high-performance optics, as discussed below. Details of this technique can be found in some other books and chapters [60–62].

In this section, we discuss the technical details of fabricating several key optical components in fused silica and crystals using FsLDW, including microlenses, hollow optical waveguides and optical micro resonators. Beyond conventional FsLDW technique, water assisted back side femtosecond laser etching and several postprocessing techniques, including heat treatment and FIB polishing, are also employed. Thus, constructing integrated optofluidic sensing devices will be straightforward, as discussed in Section 3.7.

3.5.2 Fabrication of micro-optical components in glass

3.5.2.1 Fabrication of microlenses in fused silica

As discussed in Section 3.4, 3D hollow microfluidic structures can be formed in glass by femtosecond-laser-assisted wet chemical etching. This approach can be extended to fabricate 3D micro-optical components such as optical micromirrors, microbeam splitters [209], and cylindrical and spherical microlenses in Foturan glass [210, 211], as well as spherical microlenses in fused silica that can realize nearly diffraction-limited focusing [206, 212].

The latest progress in this area is the use of fused silica instead of Foturan. This is advantageous since fused silica has superior optical properties to Foturan (e.g. a wider transmission range and lower autofluorescence), which are highly desirable for chip-based optical sensing applications. The excellent homogeneity of fused silica also results in the formation of smooth, regular surfaces over a large area, leading to a nearly diffraction-limited optical performance.

Figure 3.55 shows the process for fabricating micro-optical lenses in fused silica. A layer-by-layer annular scanning method is used to inscribe the contour of the microlens, as shown in Figure 3.55. To impart the hemispherical lens with a high surface homogeneity, dynamic delta Z scanning is performed from bottom to top [213]. After laser irradiation, the sample is etched for ~80 min in an ultrasonic bath containing 10% HF (Figure 3.55 (b)), until a microlens is formed. The sample is then polished by

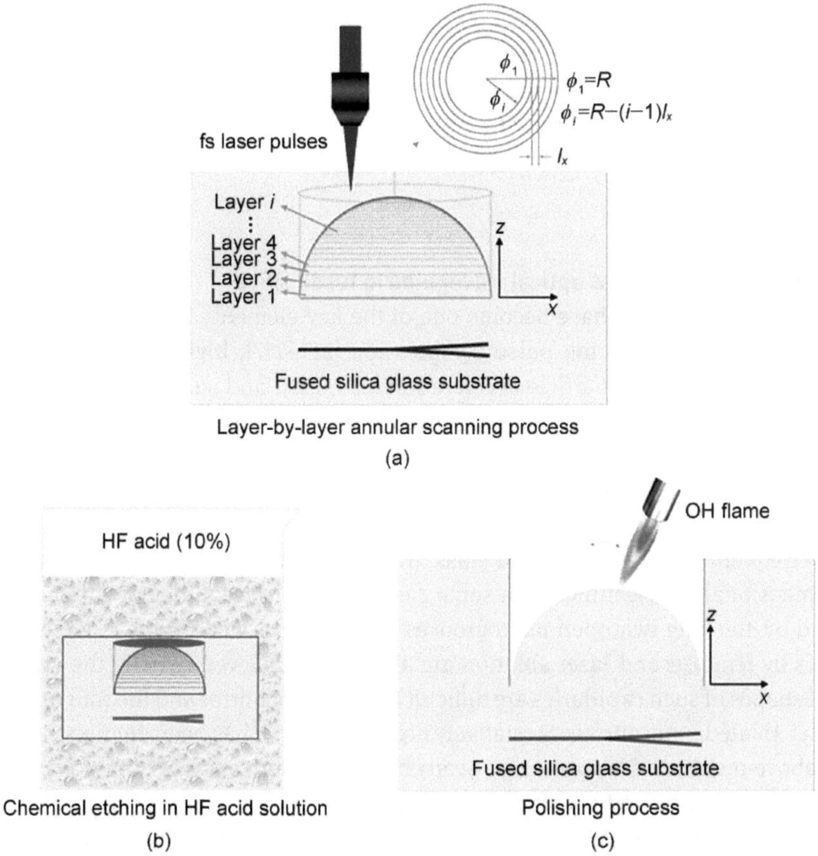

Fig. 3.55: Schematic showing process for fabricating a microlens in fused silica.

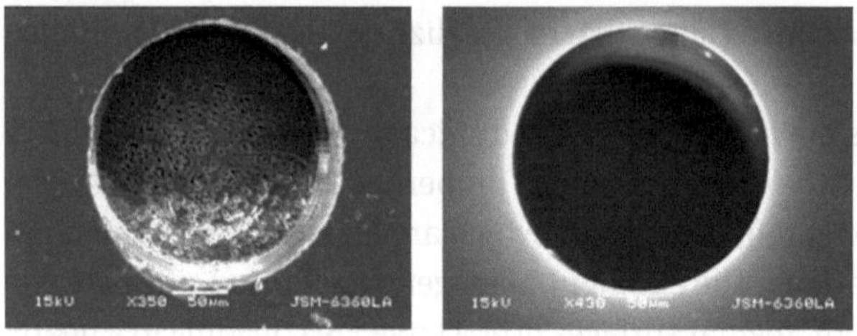

(a) (b)

Fig. 3.56: SEM images of fabricated micro-optical lens (a) before and (b) after flame polishing.

surface reflow with an OH flame (JD90, Zhengzhou, China), as shown in Figure 3.55 (c). By adjusting the flame flow, the temperature of the glass sample could be controlled to be close to the melting point of fused silica (1730°C). At such a high temperature, the glass surface melts slightly and forms a thin layer of a liquid-like phase. The surface tension of this thin liquid layer causes the microlens surface to become smooth.

Figures 3.56 (a) and 3.56 (b) show the fabricated micro-optical lens before and after flame polishing respectively. They show that the surface roughness is greatly reduced by the flame polishing. The microlens is thus capable of producing two-photon fluorescence images of biotissues with a quality comparable to that obtained by a commercially available objective lens [214], which will be discussed in the following section.

3.5.2.2 Fabrication of hollow optical waveguide in fused silica

Hollow optical waveguides have become one of the key elements in high-field laser physics experiments, including pulse compression [215–217], high-order harmonic generation [218, 219], laser particle acceleration [220, 221], and so on. With the hollow waveguides, the confinement of high intensity optical beams can be maintained over a distance significantly longer than the Rayleigh length, thereby boosting the efficiency of nonlinear interactions in gas media through the extension of interaction length and/or improved phase matching conditions. Currently, hollow waveguides are frequently fabricated by the glass drawing technique, which is inherently a one-dimensional (1D) technique. In some cases, hollow waveguides or capillaries fabricated by stacking two open microgrooves separately formed on the surfaces of substrates by femtosecond laser ablation are also used; however, usually the cross-sectional shapes of such capillaries are difficult to precisely control and the roughness of the laser ablated inner surface is relatively high. Recently, it has been demonstrated that the above-mentioned issues can be overcome by the fabrication of hollow waveguides using 3D femtosecond laser microfabrication.

The fabrication of hollow structures buried in glass materials by femtosecond laser essentially consists of two key steps: (1) spatial-selective modification of the etching property of glass by scanning a tightly focused femtosecond laser beam inside glass; and (2) preferential removal of the material in the laser modified regions by chemical etching for creating the hollow structures. For obtaining a smooth surface, an OH flame is employed as described above. However, the OH flame polish technique can only be used for an open structure with its surface exposed. Therefore, postannealing in an oven at a predetermined temperature is currently the only solution to achieve the sufficient inner-surface smoothness for optical applications.

Fabrication of the hollow waveguides involves three steps: (1) inscription of the waveguide inside fused silica chips with temporally focused femtosecond laser beam; (2) etching of the samples in a solution of 20 M/L KOH diluted with water in ultrasonic bath for selective removal of the modified areas; and (3) baking of the sample in a programmable furnace for smoothening of the surface. To achieve a circular cross-section for the hollow waveguide, a spatiotemporal focusing method is used. With this technique, we can fabricate not only one-dimensional straight optical waveguides but also curved waveguides with arbitrary geometries. More details of the experiment setup with spatiotemporally focused femtosecond laser beams can be found elsewhere. The waveguide pattern is inscribed ~500 µm beneath the surface, and the total length is 100 mm. For achieving a higher etching ratio, the waveguide is scanned three times with a speed of 100 µm/s. Since the contrast ratio of etching selectivity for a KOH solution between the laser exposed and unexposed regions is typically limited to 100~200, 21 short vertical channels with an equal interval of 5 mm between each other along the waveguide are fabricated, as illustrated by the schematic in Figure 3.57 (a). The reason for fabricating the 21 vertical outlets is to ensure a homogenous diameter of the hollow channel by reducing the taper caused by the limited selectivity of the etch ratio. On the other hand, increasing the number of vertical outlets along the channels will give rise to stronger light scattering for guiding the optical beam.

After femtosecond laser irradiation, the sample is subjected to wet chemical etching in a solution of 20 M/L KOH diluted with water in an ultrasonic bath. The temperature of the solution is kept to 95 °C constantly during the etching process. The hollow

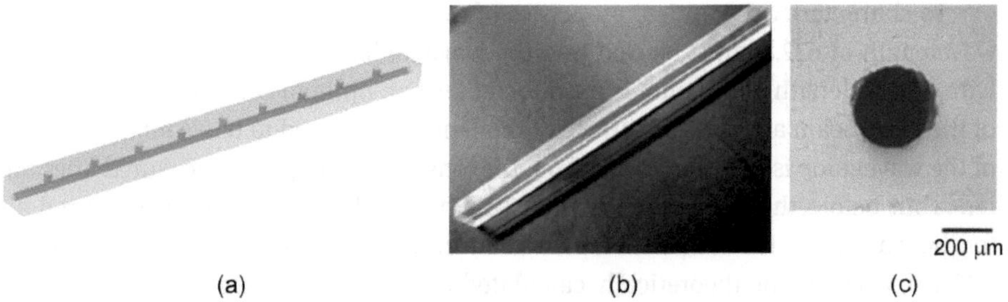

Fig. 3.57: (a) Schematic of the hollow waveguide; (b) overview image of a 62 mm-long hollow waveguide captured by digital camera and (c) optical micrograph of the cross section of hollow waveguide.

channel is formed after ~20 h, while its average diameter is merely ~30 μm. To achieve a hollow channel with a diameter of ~250 μm, which is a typical size of the hollow fiber used for many high-field physics applications, the sample is further etched for another ~150 hrs. Since some reaction products may contaminate the channel during the etching process, the etched sample is ultrasonically cleaned in distilled water for ~ 15 min. At this time, the roughness of the inner surface is very poor, which is unfavorable for optical applications.

To improve the surface smoothness of the sidewall of channel, the sample is annealed in a box-type furnace (HMF 1700-20, Shanghai, China). The furnace can provide an annealing temperature as high as 1700 °C. The postannealing protocol for smoothening the inner-surface of fused silica is as follows: first, the temperature is ramped to 600 °C at 5°C/min and then held at this temperature for 1 h, then the temperature is ramped from 600 to 1200 °C at 5°C/min and held at 1200 °C for 5 h, and finally the sample is cooled down to room temperature at 5°C/min. Figure 3.57 (b) and 3.57 (c) show overview images captured by digital camera and an optical micrograph of the cross section of the channel respectively. The circular cross section should be attributed to the spatiotemporal control of ultrashort pulses during the laser irradiation process.

To quantitatively investigate the roughness improvement, two hollow microstructures with a rectangular cross section under the same experimental conditions are fabricated. One of the samples undergoes the above-mentioned postannealing for inner-surface smoothening, while the other does not. The samples are then broken up to have the laser-fabricated inner surfaces exposed, so that the surface roughness values can be measured with optical microscopy and scanning–probe microscope (SPM). Even from the optical micrographs of the surfaces of these two samples (Figure 3.58 (a) and 3.58 (b)), a dramatic improvement of inner-surface smoothness can be observed. Scanning–probe micrographs of the inner surfaces fabricated without and with the postannealing process are compared in Figure 3.58 (c) and 3.58 (d), respectively. The root-mean-square (RMS) roughness (Rq) of the inner surface is greatly reduced from 120.2 to 5.2 nm with the annealing process.

To characterize the optical properties of the waveguide, a He-Ne laser beam at a wavelength of 632.8 nm is ejected into the waveguide. In this study, an optical lens with a focal length of 400 mm is used to achieve an optimal coupling. The diameter of the laser beam at the front surface of the lens is measured to be ~4 mm. The input of the waveguide is placed 490 mm from the lens. The laser power measured after the lens (but before the waveguide) is ~18.53 μW, while the output power measured after the waveguide is 16.56 μW. Thus the total transmission of the waveguide reaches ~89.4%, whereas the theoretically calculated maximum transmission is ~95%. Both the near-field and far-field modes of the waveguide using a CCD camera are shown in Figure 3.59. One can clearly see that the output beam has a single-mode profile.

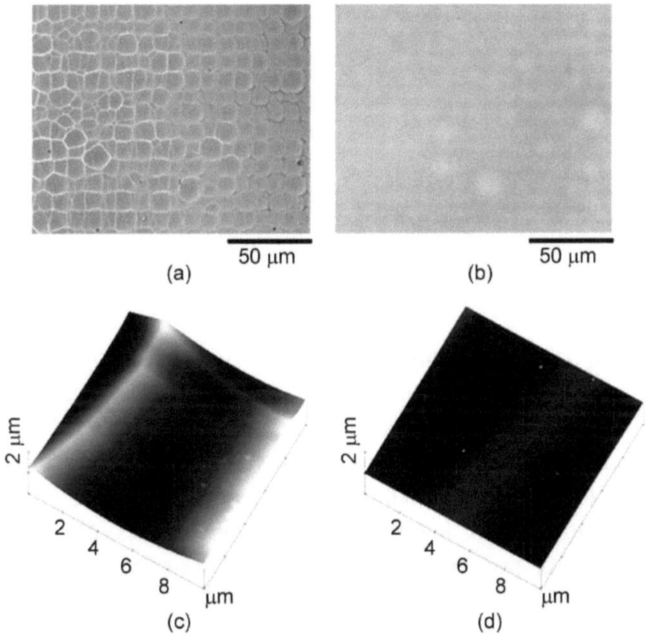

Fig. 3.58: Optical micrographs of inner surface (a) before and (b) after annealing, and SPM images of inner surfaces (c) before and (d) postannealing.

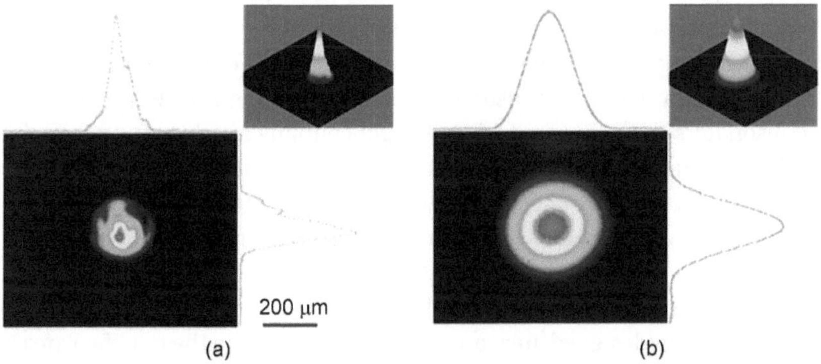

Fig. 3.59: (a) Near- and (b) far-field profiles of the He-Ne laser beam exiting from the hollow waveguide.

Furthermore, various kinds of wavy waveguides with the same diameter of ~250 μm using the above-mentioned process are also demonstrated. Figures 3.60 (a) and 3.60 (b) show digital camera-captured pictures of two sinusoidal hollow fibers. The amplitude and period of the waveguide in Figure 3.60 (a) are 125 μm and 10 mm respectively; whereas for the waveguide in Figure 3.60 (b), its amplitude and period are 125 μm and 50 mm respectively. The total length of the two waveguides is 50 mm.

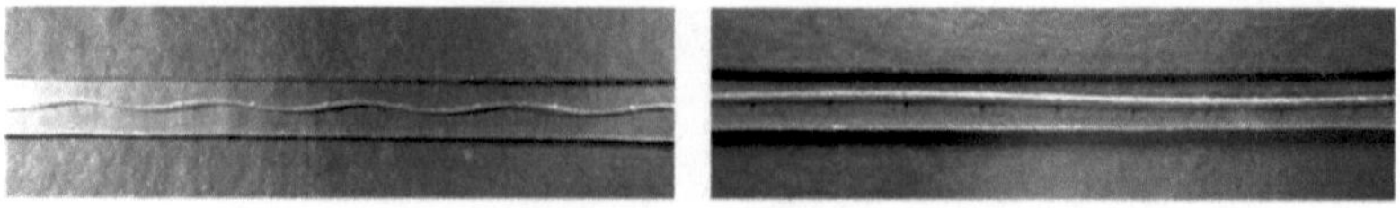

(a) (b)

Fig. 3.60: Overview images of two wavy hollow waveguides corresponding to an amplitude of 125 μm and periods of (a) 10 mm and (b) 50 mm.

3.5.3 Fabrication of optical micro-resonators in various dielectrics

3.5.3.1 Fabrication of high-Q microresonators in glasses

High-quality (high-Q) factor whispering-gallery-mode (WGM) microresonators can efficiently confine light in small spaces by total internal reflection. The combination of low absorption loss in the volume of microresonators and low scattering loss at their smooth peripheries leads to long photon lifetimes, which makes the WGM microresonators excellent candidates for low threshold nonlinear optics [222–224], quantum physics [225–227], and biosensing [228, 229]. Today, most on-chip WGM microresonators, such as microdisks [230], microtoroids [231], and deformed microcavities [232] are fabricated based on planar lithographic approaches. The lithographic techniques frequently rely on selective material removal by chemical wet etching, which limit the choices of materials that constitute the WGM microresonators. Since the 2010s, femtosecond laser micromachining has been employed for fabricating high-Q WGM microresonators in various dielectric materials.

First, the fabrication of 3D high-Q microresonators with either arbitrary tilting angles or non-uniform heights on fused silica substrate using FsLDW is described. Such microresonators can enable out-of-plane light coupling at arbitrary angles with respect to the substrates [208]. The process flow of fabrication mainly consists of: (1) FsLDW followed by selective chemical etching of the irradiated areas to create free-standing microdisk structures, and (2) reflow of the silica microdisk by CO_2 laser annealing to improve the Q factors, as shown in Figure 3.61 (a). In the laser direct writing, the fused silica glass is scanned by a tightly focused femtosecond laser pulses for selectively modifying a predesigned area, and then glass material in the modified area is preferentially removed by wet chemical etching. The femtosecond laser pulses with a pulse energy of ~0.2 μJ are tightly focused by a 100× objective with a NA of 0.9 into the glass sample, which is fixed on the *XYZ* translation stage with a resolution of 1 μm. A layer-by-layer annular scanning method with the lateral scanning step of 1 μm is adopted to form the 3D microdisk supported by a thin pillar. After the laser writing, the sample is subjected to a ~20 min etching in a solution of 5% HF diluted with water, leaving behind an on-chip microdisk, as shown in Figure 3.61 (b). Lastly, a CO_2 laser reflow step is adopted by which the surface tension of the melted glass creates a smooth surface (Figure 3.61 (c)). The 3D capability of the FsLDW offers flexibility for fabricating 3D microresonators with controllable tilting angles and variable heights, as shown in Figure 3.61 (d)–(e).

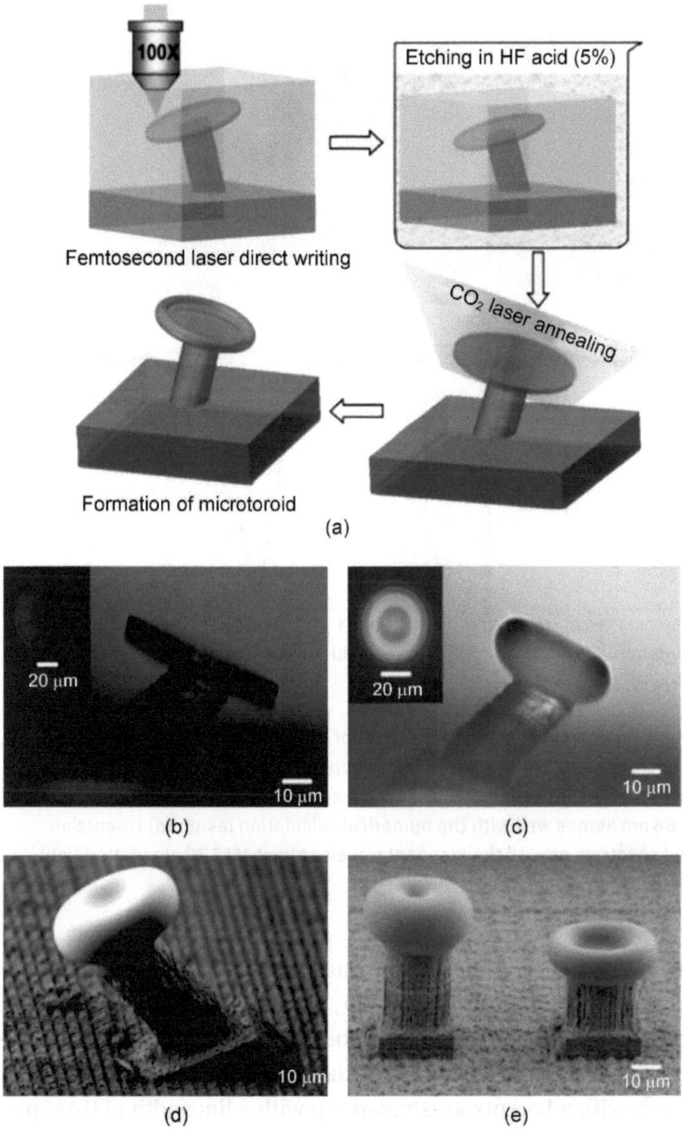

Fig. 3.61: (a) Procedures of fabrication of 3D microresonator by FsLDW. Optical microscope images of a tilted fused silica microdisk fabricated by femtosecond laser micromachining and HF wet etching (b) before and (c) after CO_2 laser annealing. Insets in (b) and (c): top views of the microresonator. SEM images of (d) the tilted microresonator, and (e) two microtoroidal resonators with different heights after CO_2 laser annealing.

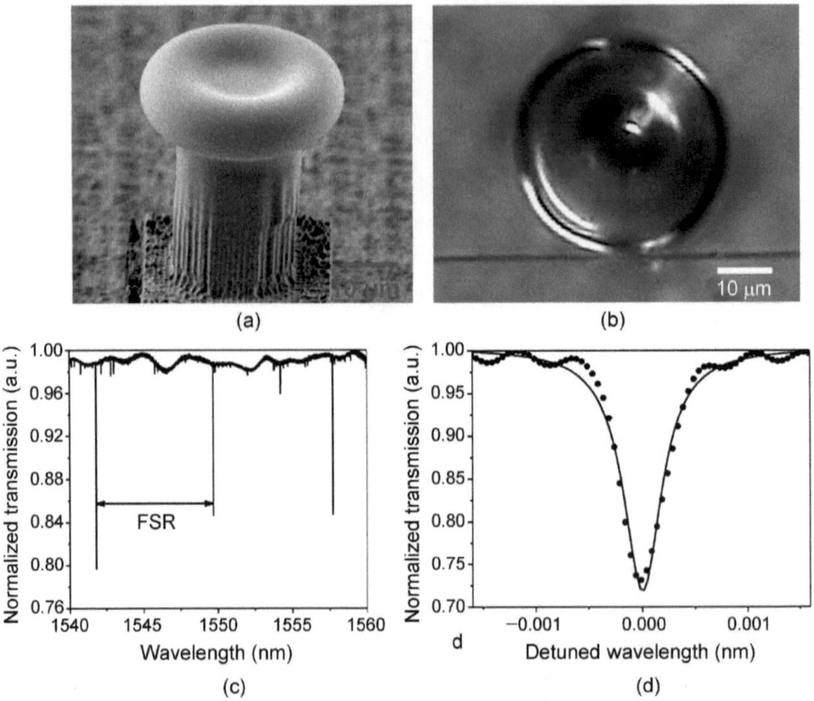

Fig. 3.62: (a) SEM image of a microresonator with a diameter of ~39 μm fabricated on fused silica substrate. (b) An optical micrograph of the microresonator coupled with a fiber taper. (c) Transmission spectrum of a microresonator with a diameter of 66.9 μm coupled with the fiber taper. The free spectral range (FSR) of 7.88 nm agrees well with the numerical calculation result. (d) Lorentzian fit (solid line) of measured spectrum around the resonant wavelength at 1557.70 nm (dotted line), showing a Q factor of 3.58×10^6.

A fiber taper coupling method [233] is chosen to measure the Q factor and characterize the transmission spectrum of the microresonator, as shown in Figure 3.62 (a) and 3.62b. The resonance transmission spectrum of the fiber taper coupled to the microresonator is presented in Figure 3.62 (c). Figure 3.62 (d) shows an individual WGM located at 1557.70 nm wavelength with a Lorentzian-shaped dip with a linewidth of 0.435 pm. The Q factor for the microresonator is then calculated to be 3.58×10^6, which can no doubt be improved in the future by replacing the current motion stage with a stage of higher translation resolution and by replacing the wavelength tunable laser with a laser source of narrower linewidth. The result indicates that femtosecond laser 3D micromachining on glass substrates enables the fabrication of high-Q WGM microresonators.

The above-mentioned method for the fabrication of high-Q microresonators in fused silica relies on chemical wet etching and thus is intrinsically limited to a few kinds of materials that can be modified by femtosecond laser irradiation for selec-

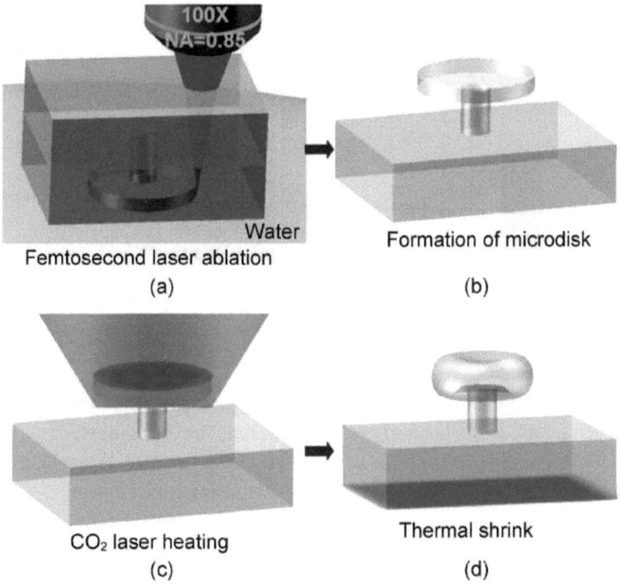

Fig. 3.63: Procedures for fabrication of a Nd:glass microresonator by water-assisted femtosecond laser ablation, followed by CO_2 laser reflow.

tively promoting the etch rate in the irradiated region. To extend the range of substrate materials, water-assisted FsLDW is employed for fabricating 3D microresonators, as schematically illustrated in Figure 3.63. In principle, the technique can be used for fabricating microresonators in all kinds of amorphous transparent materials whose surfaces can be smoothed using thermal reflow.

The fabricated Nd:glass microlaser shows a pump threshold as low as 69 µW under free space CW excitation [234]. For generating lasers at 1064 nm wavelength, the microresonator is optically excited by a CW laser diode operating at 780 nm, as illustrated in the upper inset of Figure 3.64 (a). The emission spectra collected from the microresonator at different pump powers are presented in Figure 3.64 (b). Two bright spots located in both ends of the equatorial ring are observed, as indicated by the arrows in the lower inset of Figure 3.64 (a). This indicates the nature of WGM emission, since light stored in the resonator could only escape tangentially from the edge of the microresonator. At the lowest excitation power, only a broad-spectrum photoluminescence (PL) emission is detected. As the pump laser power is increased, discrete peaks start to appear in the recorded spectra. The intensity of these periodic peaks increases dramatically when the pump laser intensity exceeds a threshold. Figure 3.64 (a) shows the measured output laser power as a function of pump power, where the lasing action occurred at a pump power of only 69 µW. Considering the mismatch between the pump wavelength and absorption peak of the gain medium (808 nm), the threshold could be further reduced by changing the pump laser source.

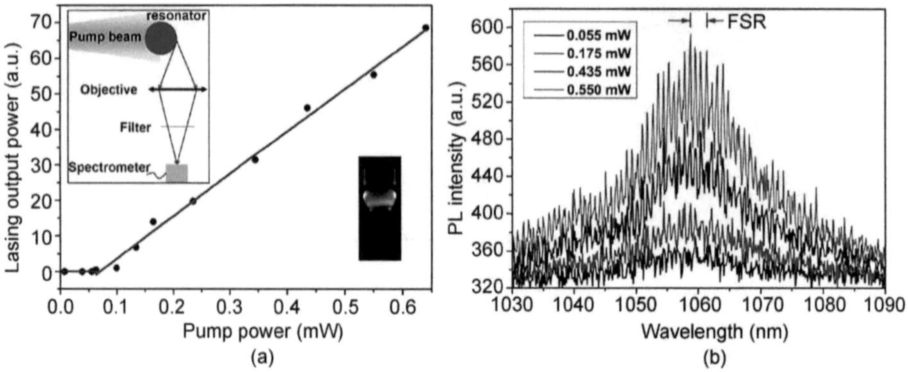

Fig. 3.64: (a) Laser output power as a function of the pump power, showing a lasing threshold of 69 μW. The upper inset in (a) is a schematic of the laser experimental setup. The lower inset in (a) shows laser spots on the edges of microresonator captured with the 2D CCD array detector mounted on the spectrometer. (b) Evolution of the emission spectrum with the increasing pump power.

3.5.3.2 Fabrication of high-Q crystalline microresonators

WGM microresonators fabricated on crystalline substrates based on semiconductor lithography approaches have been frequently employed in nonlinear optical applications, in which either the second ($\chi^{(2)}$) or third order ($\chi^{(3)}$) nonlinear optical process has been demonstrated with pronounced conversion efficiency [223, 235]. The dielectric crystalline WGM resonators have shown great promise as the next generation nonlinear sources of both classical and nonclassical light, owing to their unique characteristics including high nonlinear optical coefficients, low intrinsic absorption loss, and large transparent windows [236, 237]. In particular, as an important ferroelectric nonlinear crystalline material, lithium niobate (LN) crystal has received significant attention because of its high second order nonlinear and electro-optic coefficients [238]. Unfortunately, owing to the technical difficulties related to the material growth and the lithographic fabrication on LN substrate, high-Q LN resonators are now typically realized using mechanical polishing whose sizes are limited to the millimeter-scale [239]. Realization of high-Q on-chip submillimeter LN resonators remains a challenge [240].

Here, we develop a novel technique to fabricate high-Q on-chip submillimeter LN microdisks based on femtosecond laser micromachining, followed by focused ion beam (FIB) polishing and thermal treatment to reduce the scattering loss at the boundary [241].

In the experiment, commercially available ion-sliced LN thin films bonded by silica on a LN substrate are used to fabricate freestanding microresonators, where the thicknesses of the z-cut single crystal LN thin film and silica layer are 0.7 and 2 μm respectively. The procedure of fabrication consists of (1) a femtosecond laser ablation of the sample, which is immersed in water to form cylindrical posts with a height of ~15 μm, (2) smoothing the sidewalls of the cylindrical post by focused ion beam

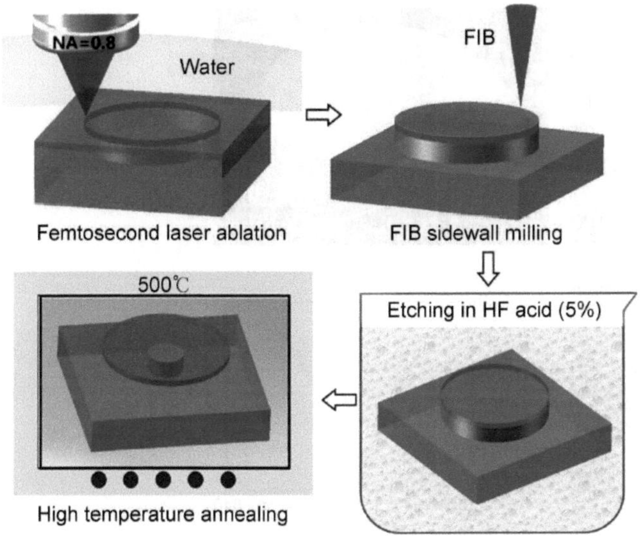

Fig. 3.65: Procedures of fabrication of a LN microresonator by water-assisted femtosecond laser ablation, followed by FIB milling, chemical etching, and annealing.

(FIB) milling, (3) chemical etching to form the freestanding LN microdisk on a silica pedestal by underetching the silica layer, and (4) high temperature annealing of the sample to reduce defects generated by FIB, as schematically illustrated in Figure 3.65. The microstructures obtained in the different stages of fabrication of the LN microdisk resonator are shown in Figure 3.66 (a)–(c). In particular, the inset of Figure 3.66 (c) shows the smooth edge of the freestanding LN microdisk, which ensures the high-Q factor of the microresonator. The Q-factor of a LN microresonator with a diameter of 82 μm is measured as 2.5×10^5, as shown in Figure 3.66 (c).

To demonstrate the nonlinear optical functionality of the LN WGM microresonator, the SHG experiment is carried out in the LN microresonators with two kinds of pump laser sources. The first pump source used in the experiment is a broad bandwidth femtosecond laser at a central wavelength ~800 nm, a repetition rate of 88 MHz, and with a linewidth (full width half maximum) of ~40 nm and a pulse width of ~30 fs. The laser is coupled into the fiber taper with a waist of ~ 1 μm, which is in direct contact with (i.e. over coupling condition) the sidewall of the microresonator to increase the efficiency and stability of the coupling of the pump light into the microresonator. An online polarization controller is used to adjust the polarization of the pump light. The second harmonic generated in the microresonator is collected and directed into the grating spectrometer (TriVista, Princeton Instruments Inc.) by the same fiber taper. The width of the entrance slit of the spectrometer is set to be ~50 μm. The conversion efficiency of SHG is measured with a 55 μm microresonator.

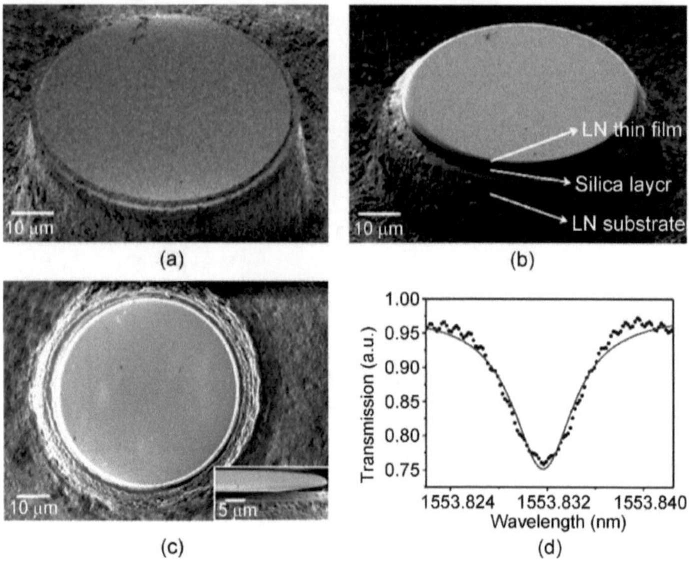

Fig. 3.66: (a) SEM image of a cylindrical post formed after femtosecond laser ablation; (b) SEM image of the cylindrical post after the FIB milling; (c) SEM image (top view) of the microresonator after the chemical etching and postannealing. Insets in (c): side view of the microresonator. (d) Lorentzian fit (solid line) of the measured transmission spectrum of microresonator (diameter ~82 μm) after annealing around the resonant wavelength at 1553.83 nm (dotted line), showing a Q factor of 2.5×10^5.

Figures 3.67 (a) and 3.67 (b) plot the transmission spectra of the fundamental light and SHG respectively. During the frequency conversion, the energy conservation condition should be fulfilled, requiring that $\lambda_P = 2\lambda_S$, where λ_P and λ_s are the pump and signal wavelengths measured in a vacuum. Additionally, the phase-matching condition should also be satisfied to gain efficient conversion. Thanks to the relatively high mode density supported by the resonator and broad excitation spectrum, a strong second harmonic signal without any tuning can be observed. The peak wavelength of the second harmonic wave is measured to be $\lambda_S = 397.14$ nm (Figure 3.67 (b)), and the corresponding fundamental wavelength at $\lambda_P = 794.28$ nm is on resonance, as indicated in Figure 3.67 (a). The side-view optical microscope image of the microresonator is shown in Figure 3.67 (c), where the second harmonic signal is clearly visible. Figure 3.67 (d) plots the measured second harmonic conversion efficiency ($\eta = P_{SH}/P_f$, where P_{SH} and P_f are the powers of the second harmonic and the pump laser respectively) as a function of the power of pump laser coupled into the fiber taper. The fitting slope in Figure 3.67 (d) shows that the power of the second harmonic is proportional to the square of the pump laser power, i.e. $P_{SH} \propto P_f^2$. The normalized conversion efficiency ($\eta_N = \eta/P_f = P_{SH}/P_f^2$) is determined to be 2.3×10^{-6}/mW. Further increase

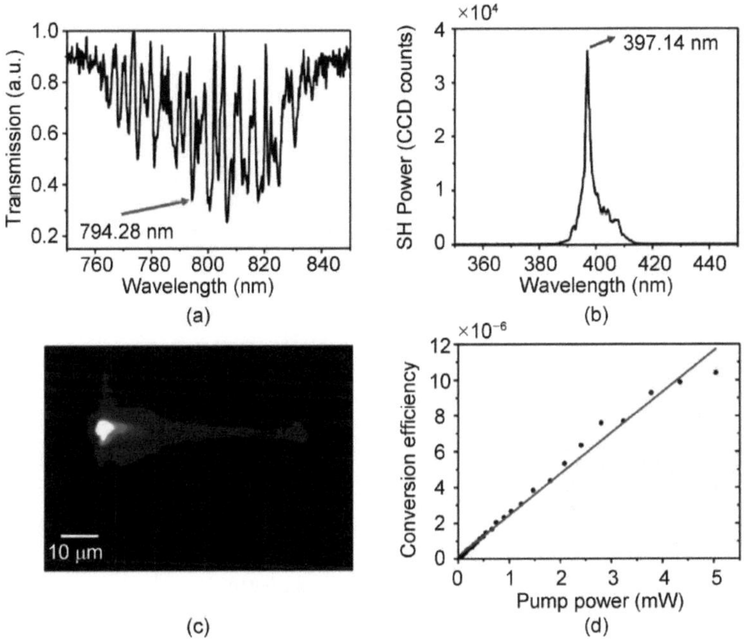

Fig. 3.67: (a) The transmission spectrum of broad-bandwidth femtosecond pump lasers collected from the output port of the fiber taper; (b) second harmonic (SH) spectrum collected from the output port of the fiber taper; (c) optical microscope side view image of the 55 μm microresonator pumped with the femtosecond laser, showing the second harmonic beam scattering out from the microresonator; (d) conversion efficiency of SHG as a function of the pump power. The normalized conversion efficiency is determined to be 2.3×10^{-6}/mW.

of the pump laser power would lead to a decrease of conversion efficiency, because of the saturation effect [236].

SHG experiment is carried out using a CW tunable single-longitudinal mode Ti-sapphire ring laser (Mantisse TX-light, Spectra Physics, Inc.) as the pump source, which has a spectral linewidth of ~60 kHz. By tuning the pump laser wavelength, it is found that the optimum conversion efficiency is observed at a pump wavelength of 799.884 nm. Figures 3.68 (a) and 3.68 (b) plot the spectra of both the pump light (λ_P = 799.884 nm) and second harmonic signal (λ_S = 399.922 nm) collected from the output port of fiber taper respectively. The side-view optical microscope image of the microresonator is shown in Figure 3.68 (c), where the second harmonic is clearly visible after removing the pump light with the filters. Figure 3.68 (d) plots the measured SHG conversion efficiency as a function of the power of the pump laser coupled into the fiber taper. The normalized conversion efficiency, which can be obtained from the slope of the fitting line in Figure 3.68 (d) (i.e. the solid line in Figure 3.68 (d)), reaches 1.35×10^{-5} /mW. The linear fit also indicates that the power of the second harmonic signal is proportional to the square of the pump laser power. The result is

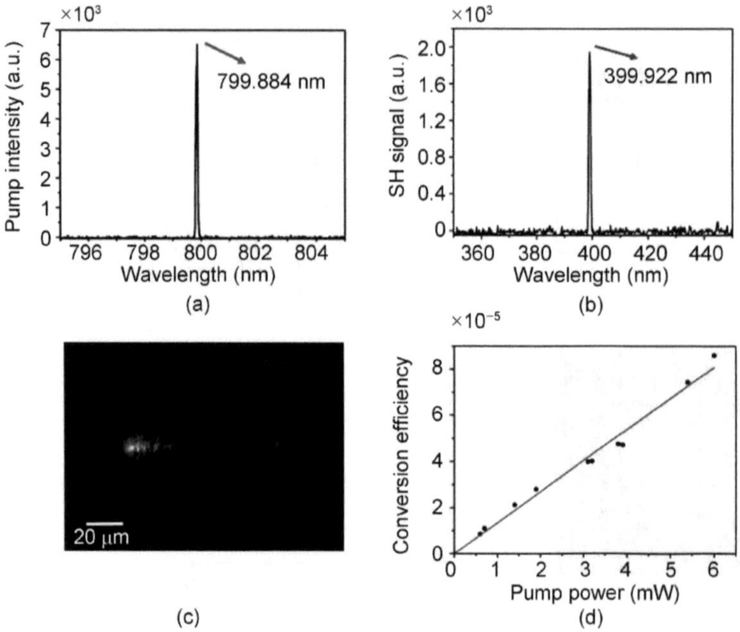

Fig. 3.68: (a) The typical spectrum of CW pump laser for efficiently generating second harmonic; (b) the spectrum of the generated SH signal; (c) side-view image of the 82 μm microresonator under the irradiation of pump laser, showing the second harmonic beam scattering out from the microresonator; (d) SHG conversion efficiency as a function of pump power. The slope indicates a normalized conversion efficiency of 1.35×10^{-5}/mW.

consistent with the theoretical prediction, and provides direct evidence that the SHG signal is not contaminated by other noise signals.

It is observed that the normalized conversion efficiency measured with a broad-bandwidth femtosecond pump laser is lower than that of the CW tunable single-longitudinal-mode pump laser. This is reasonable because the majority of the bandwidth of the femtosecond pump laser cannot fulfill the resonance conditions required by SHG, which causes a significant loss of the pump energy and leads to the reduction of conversion efficiency.

FsLDW can also be used for fabricating high-Q microresonators on other crystalline substrates [242]. Figure 3.69 (a)–(b) are the top-view and side-view images of a CaF_2 microdisk resonator with a diameter of ~88 μm. Figure 3.69 (c) shows the transmission spectrum of the CaF_2 microresonator. Figure 3.69 (d) shows an individual WGM located at 1546.51 nm wavelength with a Lorentzian-shaped dip. The linewidth obtained by a Lorentzian fitting is 36.74 pm, as shown by the solid curve in Figure 3.69 (d). The Q-factor for the microdisk is then calculated to be 4.2×10^4.

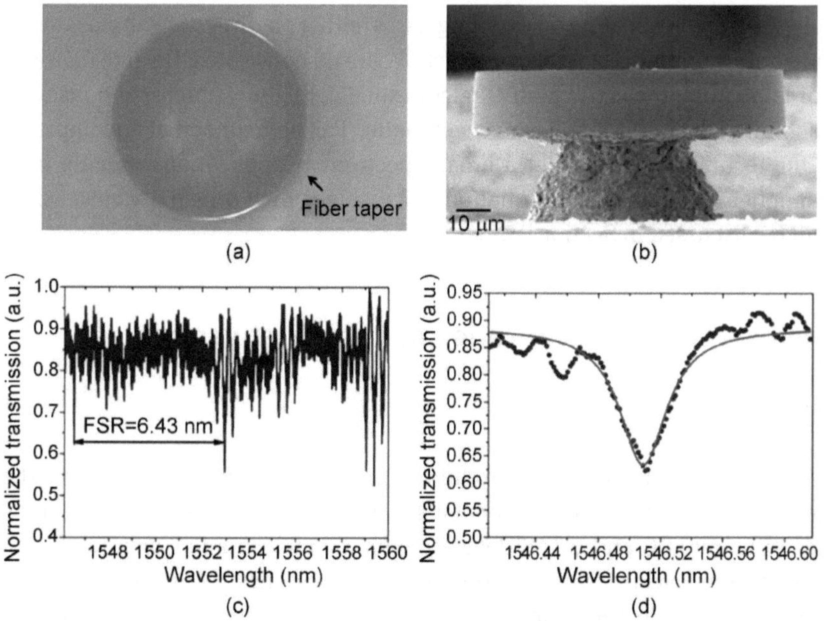

Fig. 3.69: (a) Optical micrograph of the microdisk coupled with a fiber taper (top view); (b) SEM image of the fabricated microdisk (side view), showing the tilting of the sidewall; (c) transmission spectrum of the fiber taper coupled with the microdisk; (d) Lorentzian fit (solid line) of measured spectrum around the resonant wavelength at 1546.51 nm (dotted line), showing a Q factor of 4.2×10^4.

3.5.4 Conclusions

To summarize, FsLDW has been used to fabricate various free-space optical components, such as high quality microlenses, hollow optical waveguides, and optical microresonators, in fused silica glass and several crystals. In these processes, femtosecond-laser-assisted wet chemical etching and subsequent heat treatment (to smooth the surfaces to optical quality through surface reflow) are employed. Typically, micro-optical lenses fabricated by femtosecond-laser-assisted wet chemical etching can produce nearly diffraction-limited focal spots and realize high-quality two-photon fluorescence imaging of biotissues.

The unique flexibility offered by laser direct writing allows us to precisely tailor the geometry of the hollow waveguide, as evidenced by the wavy waveguide. In addition, it is also possible to control the other parameters of the waveguides in a spatial-selective manner, namely each portion of the waveguide can be individually tailored by tuning the laser parameters during laser direct writing. With the above technique, several hollow waveguides can be simultaneously fabricated and fused to form complex photonics networks, such as beam-splitters and couplers, etc. These possibilities make this technique attractive for high-field laser physics research.

Finally, 3D microresonators on various glasses (either passive or active glass materials) and nonlinear optical crystals using FsLDW are demonstrated. Despite the short history of fabricating dielectric high-Q microresonators using femtosecond laser 3D micromachining, the initial results are encouraging. Further refinement and improvement of the technique will benefit a broad spectrum of applications ranging from nonlinear optics and quantum information processing to high-sensitivity biosensing, which will be discussed in Section 3.7.

3.6 Femtosecond laser fabrication of metallic structures

3.6.1 Introduction

In addition to the fabrication of microfluidics and micro-optical components, the incorporation of microelectronic components is further desired to enhance the functionalities of many LOC and μ-TAS devices. Such microelectronic components can be realized by the selective metallization of glass substrates. Conventional metallization process for dielectric materials largely depends on metal deposition by vacuum evaporation [243], sputtering [244], and chemical vapor deposition [245], which are widely utilized in the microelectronic industry today. Due to the inherently planar nature of the lithographic process, this technology is limited in its capability to produce 3D structures. Laser direct writing is a reliable alternative technique for directly patterning thin metal films without using photolithography. For example, laser-induced plasma-assisted ablation (LIPAA) [246, 247] and laser-induced forward transfer (LIFT) [248, 249] using a nanosecond laser are attractive techniques for the selective metallization of transparent materials. However, they can treat only the surfaces of glass substrates and cannot metallize the interiors of microfluidic systems.

Currently, FsLDW has proven to be a powerful tool even for the selective metallization of insulators, and brings several unique advantages compared with nanosecond laser metallization. First, when the femtosecond laser intensity is well-controlled near the threshold of the ablation, selective metallization of insulators can be realized at nanoscale spatial resolution, since tiny microstructures (i.e. nanometer-sized holes and grooves) have been reliably produced in dielectrics using a femtosecond laser [250, 251]. Second, one can fabricate the metallized microstructure deeply embedded in the substrate by combining femtosecond laser ablation and selective metallization, which is a desirable characteristic for many novel microdevices requiring 3D configuration. Lastly, this technique offers the ability to directly incorporate microelectric elements into micro-optic and/or microfluidic circuits in a single glass chip. Further development of this technique has great potential for rapid and cost-effective fabrication of monolithic 3D microelectro-optofluidic devices.

In this section, an approach to integrate electric functions using selective metallization of insulators by femtosecond laser modification combined with electroless chemical plating is proposed [111]. The mechanism of selective metal deposition and its applications in the integration of electronics and photonics are investigated [252]. Additionally, embedded microelectrodes with controllable cross section by exploring the 3D-structuring ability of FsLDW are demonstrated, allowing for the control of the electric field inside a transparent substrate [253]. Moreover, these metallization techniques can be utilized to manufacture surface-enhanced Raman scattering (SERS) platforms potentially used for highly sensitive analysis of biochemical samples [254, 255].

Thus, the selective metallization of glass can be used to incorporate microelectronic components in microfluidic systems, making it an important technique for further enhancing the functions of biochips. Both femtosecond-laser-assisted electroless plating and femtosecond laser surface modification combined with electroless plating can be used to selectively deposit thin metal films only on laser irradiated regions, even on the internal walls of microfluidic structures, which is desirable for fabricating highly functional biochips.

3.6.2 Selective metallization on glass surfaces and its mechanism

The laser system consists of a Ti:sapphire oscillator (Coherent, Inc.) and a regenerative amplifier, which emits 800 nm, 40 fs pulses with a maximum pulse energy of ~2.5 mJ at a 1-kHz repetition. To ensure a high beam quality, the 8 mm-diameter output laser beam is reduced to 5 mm by an aperture in front of the focusing system. In most of our experiments, the focusing system is a 20× microscope objective (NA 0.45). The pulse energy of the beam is controlled by a half-wave plate and a polarizer. The samples can be arbitrarily translated three dimensionally by a PC-controlled *XYZ* stage with a resolution of 1 µm. A CCD camera connected to PC is used to monitor the fabrication in real time.

Commercially available microscope slides (borosilicate glass) and MgO-doped x-cut lithium niobate ($LiNbO_3$) crystal are used in this work. The substrates are polished on four sides and cleaned with acetone in an ultrasonic bath before use. The composition of the electroless copper plating solution is $CuSO_4 \cdot 5H_2O$(5.0 g/L), $C_{10}H_{16}N_2O_8$(EDTA; 14.0 g/L), with HCHO (5.0 g/L) as a reducing agent, 2,2'-dipyridine (2 mg/L) as a stabilizer, and polyethylene glycol (4000 Mw 50 mg/L) as surface activators. The pH of the plating bath is adjusted to approximately 12.5 using tetramethylammonium hydroxide (TMAH) and the bath temperature is maintained at 40 °C.

A field-emission scanning electron microscope (FESEM, Hitachi S4700) equipped with an energy-dispersive X-ray detector (EDX, Oxford INCA300) is used to investigate the morphology and composition in the femtosecond laser irradiated areas on the glass surfaces. Before FE-SEM observation, the samples are sputter-coated with gold to avoid charge-up during the observation. An X-ray photoelectron spectrometer (XPS, Kratos AXIS UltraDLD) is employed to determine the composition in the laser ir-

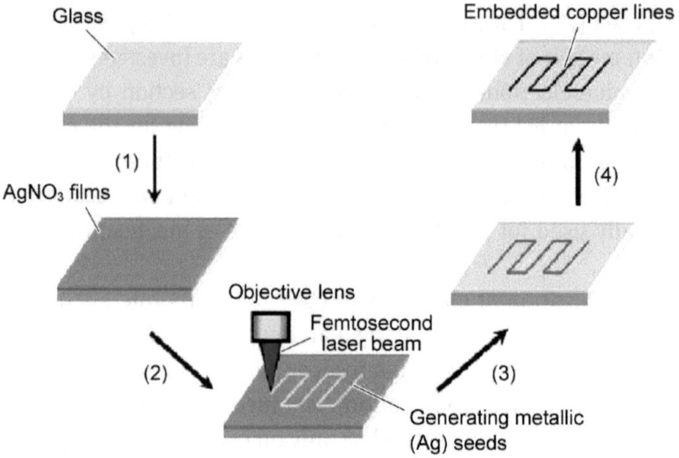

Fig. 3.70: Schematic illustration of the fabrication process for the selective metallization of insulators: (1) formation of silver nitrate thin films on insulator substrates, (2) modification of insulator surfaces by FsLDW, (3) removal of unirradiated silver nitrate films by acetone, and (4) copper coating by selective electroless plating.

radiated areas on the glass surfaces. A reflection optical microscope is used to capture the images of copper microstructures.

For the purpose of realizing selective metallization on the surfaces of dielectric materials, a modified FsLDW is developed. The fabrication process mainly consists of four steps illustrated in Figure 3.70, including (1) formation of silver nitrate thin films on insulator substrates; (2) modification of insulator surfaces by FsLDW; (3) removal of unirradiated silver nitrate ($AgNO_3$) films by acetone; and (4) selective copper coating by electroless plating. As a photosensitive layer, the $AgNO_3$ films are prepared by immersing the substrates into a 0.5 mol/l $AgNO_3$ solution and withdrawing the substrates at an approximate speed of 1 mm/s. Subsequently, the substrates are naturally dried in the dark at room temperature for 12 h. The thickness of the $AgNO_3$ films could be controlled by adjusting the withdraw speed. When the femtosecond laser beam is focused on the insulator surfaces coated with silver nitrate films, Ag particles can be formed from the decomposition of silver nitrate films in the irradiated area. The FsLDW has high spatial resolution and a small heat affected area, facilitating the selective deposition of the Ag particles in the irradiation area. In the successive electroless plating process, these particles can serve as seeds for in-situ selective copper deposition. Before electroless chemical plating, the substrates are cleaned with distilled water in an ultrasonic bath for the removal of unirradiated $AgNO_3$ films.

Figure 3.71 show optical micrographs of the copper microstructures selectively deposited on the glass surface with a laser scanning speed of 60 μm/s. One can see that continuous metal lines with smooth edges are deposited on the glass substrates, and the linewidths of deposited Cu films are approximately 6 μm and 5 μm when the

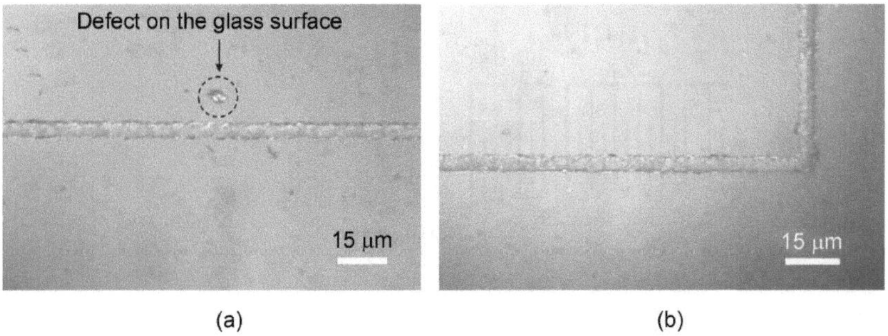

Fig. 3.71: Optical micrographs of selectively deposited copper microstructures on glass surfaces at the scanning speed of 60 μm/s: (a) pulse energy: 5 μJ, (b) pulse energy: 3 μJ.

laser powers are set at 5 and 3 mW respectively. In particular, it should be noted in Figure 3.71 (a) that there is a defect (as indicated in the dashed circle) right above the metal line fabricated by FsLDW that is not metallized, providing clear evidence of the good selectivity. In fact, even when the laser power is reduced to as low as 1 mW, the Cu line can still be fabricated by this technique. Basically, the features of the fabricated microstructures can be controlled by FsLDW parameters such as the laser power, scanning speed, focal conditions, and so on. The thickness of copper microstructures mainly relies on the electroless plating process, which has been described elsewhere [256]. In addition, the thickness of the coated silver nitrate films before laser irradiation is also important for selective copper deposition.

To examine the electrical properties of copper microstructures, microcircuits are further integrated on the glass substrates as shown in Figure 3.72 (a). The resistances of pattern (I) and pattern (II) are approximately 15 and 6 Ω respectively. These results confirm that the copper microstructures formed on the glass substrates have good electrical conductivity. In addition, these copper lines selectively deposited on the glass surface show good adhesion to the glass substrates, as they can remain intact after a cleaning process with distilled water in an ultrasonic bath. The good adhesion is further confirmed by a peel test using scotch tape.

The mechanism of femtosecond laser selective metallization on glass surfaces is investigated by XPS and EDX analyses. A glass substrate coated with $AgNO_3$ film is modified by femtosecond laser irradiation and then ultrasonically cleaned in distilled water to remove unirradiated $AgNO_3$ film. When the femtosecond laser beam is focused on the surface, Ag particles can be formed from the decomposition of $AgNO_3$ films in the irradiated area [257, 258]. The decomposition reaction formula of silver nitrate is as follows:

$$2AgNO_3 \rightarrow 2Ag + 2NO_2\uparrow + O_2\uparrow \ . \tag{3.55}$$

As a result of the multiphoton absorption process, the FsLDW has high spatial resolution and a small heat affected area, facilitating the selective deposition of the Ag parti-

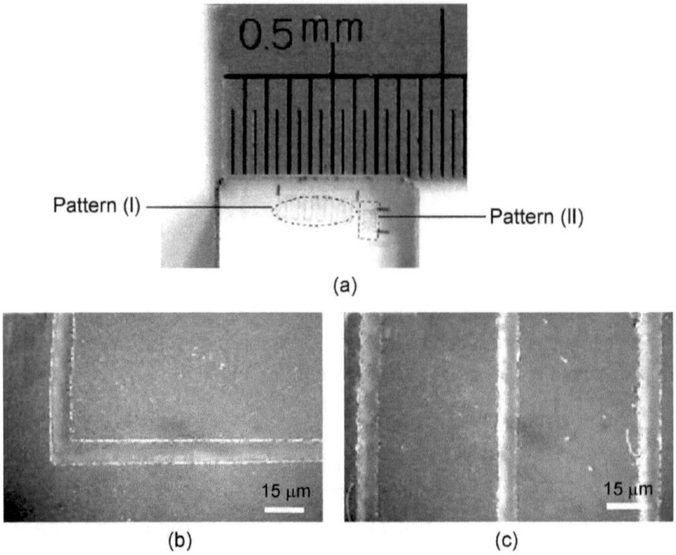

Fig. 3.72: Photographs of metal patterning on the glass when laser pulse energy is 8 μJ: (a) Electric circuits fabricated on glass surface; (b) Cu lines in pattern (I); (c) Cu lines in pattern (II).

cles in the irradiation area. Moreover, after the laser irradiation, although the sample is ultrasonically cleaned in distilled water for 10 min for the removal of unirradiated $AgNO_3$ films, selective electroless copper plating still can be achieved. It is speculated that the femtosecond laser may have played a role in localized melting and quasiwelding at the interface of Ag particles and insulator surfaces [259, 260], which promotes the adhesion between reduced Ag particles and substrates. In the subsequent electroless plating process, these particles can serve as seeds for in-situ selective copper deposition [261, 262].

Figures 3.73 (a) and (b) show the optical micrographs of the irradiated areas on the glass surface. The size of the laser irradiated area of 1.5 mm × 1.1 mm is formed by a raster scan of the laser beam focused by a 20× microscope objective with a line interval of 7 μm, and the pulse energy and the scanning speed of the laser beam are set as 8 μJ and 180 μm/s respectively. XPS is used to analyze the composition and chemical states in the irradiated area, as shown in Figures 3.73 (c) and (d). Figure 3.73 (d) is magnified from Figure 3.73 (c) to analyze Ag 3d signals in more detail. The peaks of Ag $3d_{5/2}$ and $3d_{3/2}$ can be clearly seen at the binding energies at 368.3 and 374.2 eV respectively, which exactly correspond to signals from the atomic silver. We conclude that the irradiation of femtosecond laser beam on the glass surfaces coated with the silver nitrate films can result in the decomposition of silver nitrate, and consequently result in the formation of silver atoms or even silver nanoparticles, which become the seeds for the subsequent electroless plating process.

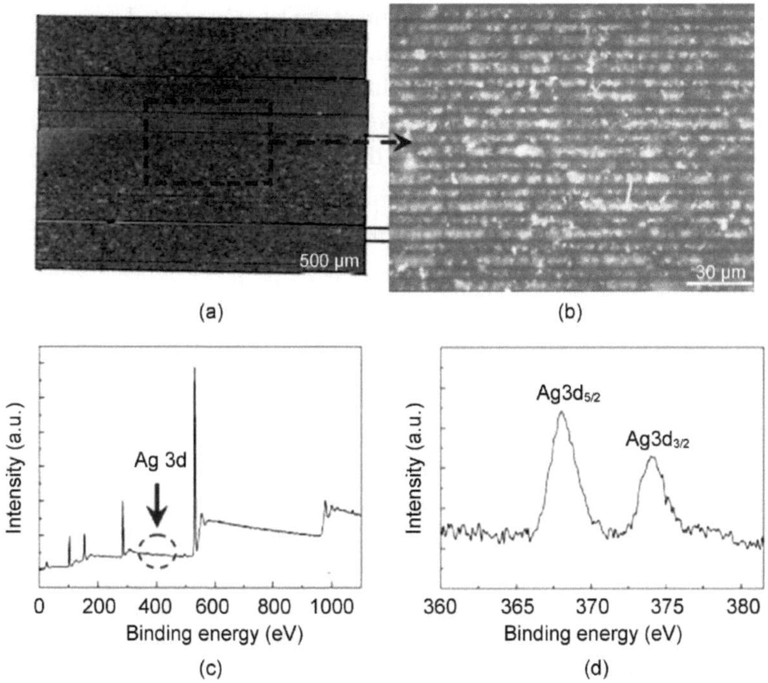

Fig. 3.73: (a) Optical micrograph of glass irradiated by femtosecond laser, (b) magnified image of the area indicated by the square in (a), (c) full XPS spectrum of the irradiation region, and (d) high resolution XPS spectrum highlighting the peaks of Ag 3d.

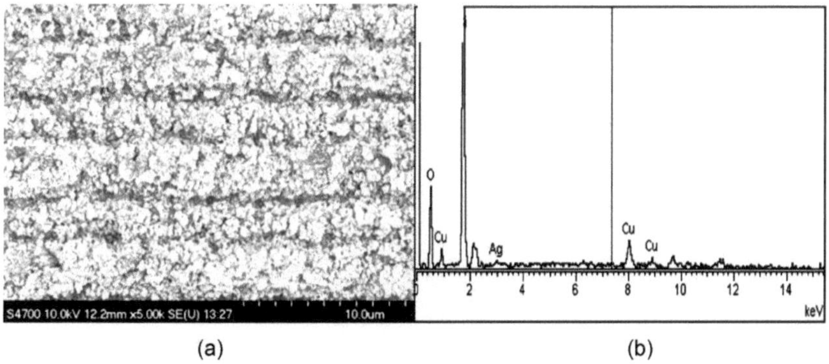

Fig. 3.74: (a) SEM image of femtosecond laser irradiated area on the glass sample electrolessly plated for 5 min with average power of 5 mW; (b) EDX result for the sample in (a). The unlabeled peaks correspond to gold sputtered on the sample surface before EDX analysis, and the ratio of O:Si:Cu:Ag is about 58.40:33.12:8.22:0.26.

Figure 3.74 shows the morphology and composition of the laser irradiated area on the glass surface after the electroless plating for 5 min. The EDX analysis in Figure 3.74 (b) shows that copper element has been detected in the femtosecond laser irradiated area, indicating that deposition of copper has been successfully realized even with a trace amount of silver atoms serving as seeds for the electroless plating process.

3.6.3 Fabrication of embedded microelectrodes

$LiNbO_3$ is a ferroelectric material with large optical nonlinearity and Pockels effects. Electro-optic components based on $LiNbO_3$ waveguiding structures have gained significant importance, such as electro-optic switches, modulators, and electro-optic tuned quasi-phase-matched (QPM) devices [263, 264]. In these components, it is crucial to design and fabricate electrodes, which have been realized commonly by way of lithographic methods. The 3D-structuring ability of femtosecond laser microfabrication allows us to fabricate microelectrodes deeply embedded in $LiNbO_3$ substrate by combining femtosecond laser ablation and selective metallization. The fabrication process for embedded microelectrodes in $LiNbO_3$ mainly consists of the four steps illustrated in Figure 3.75, including (1) micromachining of microgrooves on the surface of $LiNbO_3$ by femtosecond laser ablation; (2) formation of $AgNO_3$ films on substrates; (3) scanning the femtosecond laser beam in the fabricated microgrooves for modification of the inner surfaces; and (4) electroless copper plating.

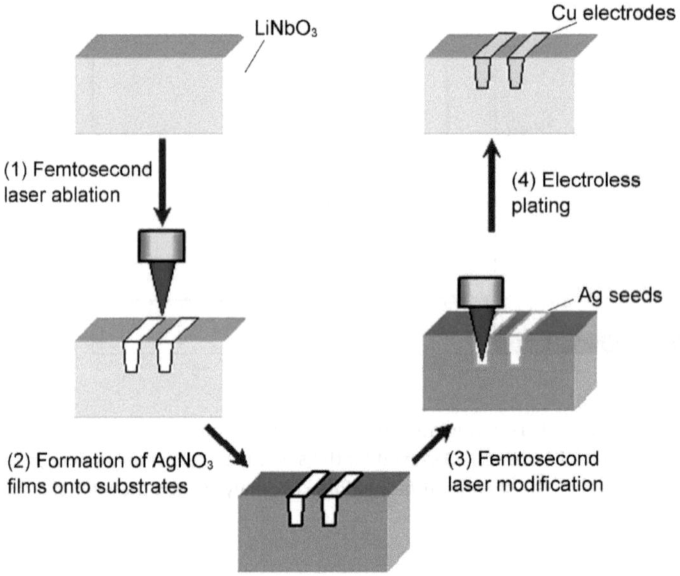

Fig. 3.75: Schematic illustration of the fabrication process for embedded microelectrodes in $LiNbO_3$.

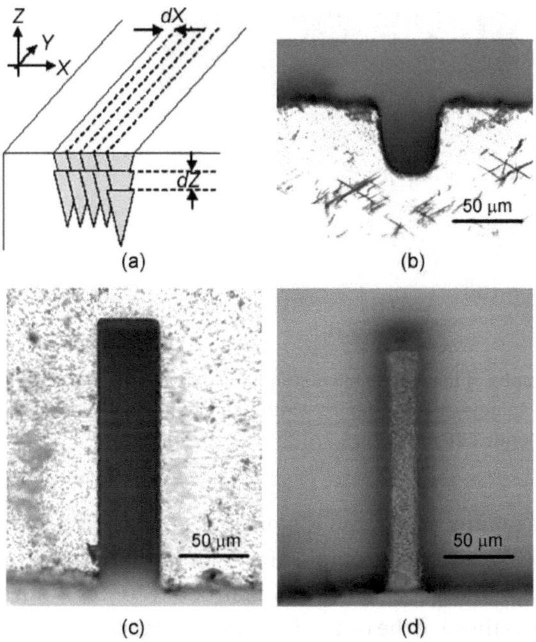

Fig. 3.76: (a) Illustration of fabrication of grooves by femtosecond laser matrix ablation; (b)–(d) optical micrographs of a fabricated groove: (b) end view, (c) top view and (d) bottom view.

Figure 3.76 (a) shows an illustration of the matrix ablation and an example of an obtained U-groove. The desired section pattern of the groove is applied to the substrate by performing laser direct writing line-by-line and layer-by-layer, which define a matrix in the transverse (X, Z) plane. The U-groove with length 200 μm, width 40 μm and depth 50 μm as shown in Figure 3.76 (b)–(d) is made by inscribing a volume consisting of a matrix of (20×10) lines using a laser beam focused by a 20× microscope objective at a speed of 1 mm/s and pulse energies of 2 μJ, and matrix density parameters (dX, dZ) are chosen to be respectively 2 μm and 1 μm. After femtosecond laser ablation, chippings in the groove can be removed by ultrasonic cleaning in distilled water. It is observed that the groove is clean and homogeneous, except that the end of the groove on the substrate edge is broader than other portions as a result of edge effects. The geometries of the cross section of the grooves could be controlled by changing the parameters of the matrix ablation. For instance, a groove with smooth and vertical inner walls could be fabricated by slightly increasing the power of laser irradiation layer by layer in order to compensate for the loss of power associated with additional depth [176].

Figure 3.77 shows optical micrographs of copper microstructures with different cross-section geometries deposited in femtosecond-laser-ablated grooves. As shown in Figure 3.77 (a), two grooves with spindle cross section embedded in the $LiNbO_3$ substrate are fabricated by repetitively inscribing 20 times using a laser beam focused by

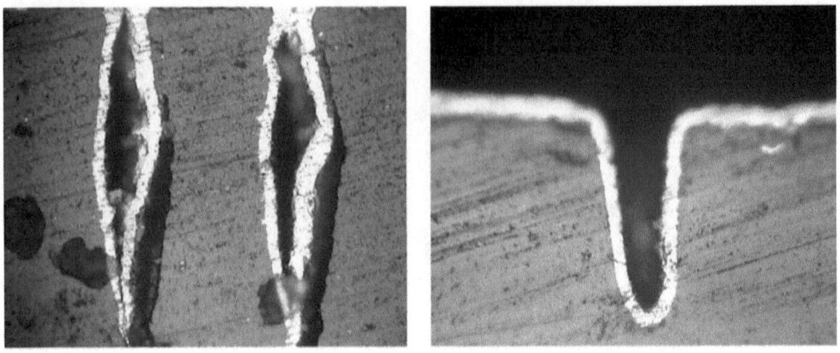

(a) (b)

Fig. 3.77: Cross-sectional optical micrographs of the copper microstructures deposited in the grooves with different cross sections, (a) spindle-shape, opening width: 20μm, and depth: 150μm; (b) U-shape, opening width: 40μm, and depth: 100μm.

a 20× microscope objective. The scanning speed and the pulse energy are 200 μm/s and 60 μJ respectively. Figure 3.77 (b) shows an example of a U-groove fabricated by matrix ablation, whose inscribing volumes consisting of a matrix of (20 × 40) lines and other parameters are the same as those of the example shown in Figure 3.76. After the depositing of $AgNO_3$ films, the inner surfaces of the grooves are modified by femtosecond laser direct writing on the bottom surface of grooves at a speed of 200 μm/s and pulse energy of 2 μJ. In this way, metallic Ag particles can be generated from the decomposition of $AgNO_3$ films on the irradiation area, which can serve as seeds for in-situ selective copper depositing in the following electroless process. After plating for three hours, one can observe that a uniform electroless copper deposition is initiated on the sidewalls of the grooves. Therefore, it can be deduced that Ag particles are deposited uniformly on the entire surface inside grooves by femtosecond laser direct writing, which can be interpreted by taking account of the reflection of the laser beam on the sidewalls of grooves [265]. At the same time, the 'necking' at the top of the spindle-shape grooves due to the narrow openings can be observed, which can result in seams and voids in the microelectrodes. However, the increasing openings of U-grooves enable the disappearance of 'necking,' as shown in Figure 3.77 (b). The thickness of three-hour deposited films is about 6 μm, equivalent to a deposition rate of 2 μm/h. In addition, due to the residual debris of laser ablation around the grooves, the lateral growth of copper beside the grooves is also observed.

The dependence of groove-filling characteristics on plating time is shown in Figure 3.78. One can see that copper deposition rate on the bottom of the grooves is higher than that on the surface, which produces void-free and seamless fill for high-aspect-ratio microgrooves. These micrographs clearly indicate almost no voids and good uniform filling for these grooves, which could be attributed to the surface activators added into the electroless plating solution and the continuous Ag seeds layer [266]. Due to the overgrowth on the grooves, as shown in Figure 3.78 (c), the mechanical polishing of the substrate surface is performed after electroless plating.

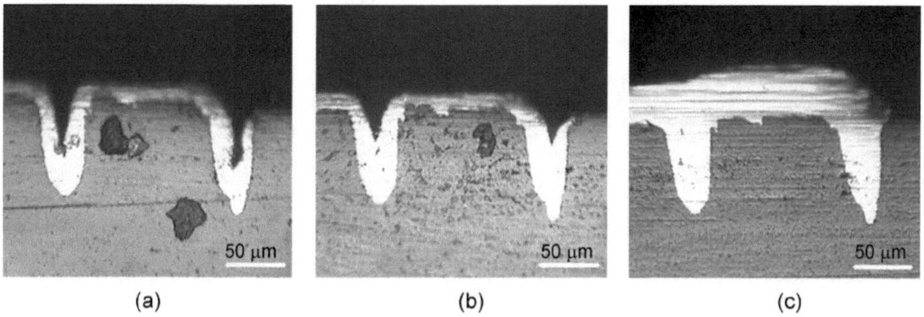

Fig. 3.78: Cross-sectional optical micrographs of the copper microstructures deposited in U-grooves with plating time: (a) 0 h, (b) 5 h and (c) 30 h, opening width: 40 μm; depth: 75 μm.

3.6.4 Fabrication of SERS structures

Since its discovery in the 1970s [267–269], SERS has provided a glimpse into the future of high-throughput single-molecule detection and analysis [270, 271]. The high sensitivity of SERS spectroscopy and its inherent molecular specificity result in several distinct advantages over current state-of-the-art bio-analytical techniques, such as the ability to perform label-free analysis and to reduce photon-damage effects. Much more information about samples can be obtained from Raman spectroscopy than from fluorescence spectroscopy, since all molecules have their own fingerprint Raman spectrum. Electrochemically reduced electrodes, colloidal clusters, and silver, gold, or copper island films with nanoscale roughness can serve as SERS substrates.

The key issue with SERS lies in the fabrication of stable and reproducible substrates with a controllable enhancement factor (EF), which is by no means easy because of the difficulty in controlling the nanoscale behavior of materials. In order to achieve a high EF, a variety of fabrication techniques have been developed for SERS substrates, many of which consist of roughening the surface by some means and then depositing noble metal nanoparticles. Recently, FsLDW has been introduced as a powerful tool for fabricating SERS substrates [272, 273].

It was first demonstrated by Zhou et al. that it is possible to fabricate SERS substrates with a controllable EF using FsLDW on Ag-doped phosphate glass followed by electroless plating. In their investigation, silver seeds are first photon-reduced using femtosecond laser irradiation, and are then transformed into silver nanoparticles with sizes suitable for SERS in the subsequent electroless plating process. Rhodamine 6G is then used as a probing molecule to investigate the enhancement of Raman signals from the substrate. The Raman signal was found to be enhanced almost uniformly over the entire substrate surface. In addition, using FsLDW, a microelectronic circuit and a micro-SERS unit were integrated into a microfluidic chamber, thus forming a prototype Raman chip for bio-sensing applications. Furthermore, Lan et al. fabricated a SERS substrate on the tip of an optical fiber. FsLDW was first employed to etch uniform

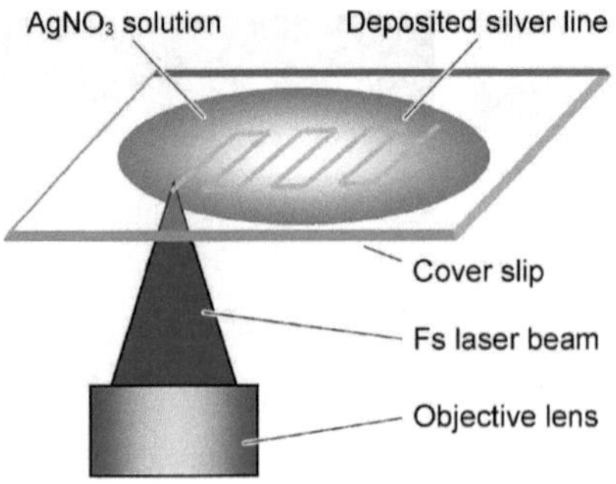

Fig. 3.79: Schematic illustration of the fabrication process.

patterns on the end face of an optical fiber, which was then coated with a thin layer of silver by thermal evaporation [274].

A small volume of $AgNO_3$ solution was dropped on the glass slip. The beam from a Ti:sapphire laser oscillator operating at a wavelength of 800 nm, a pulse duration of 40 fs, and a repetition rate of 80 MHz was focused on the interface between the glass slip and $AgNO_3$ solution using an inverted microscope by an objective lens (10×, NA is 0.3). The silver ions in the solution were reduced by fs laser irradiation and deposited on the slip mounted on a computer-controlled *XYZ* stage to form the silver microstructures. The scheme of the experiment is shown in Figure 3.79. Ag is easily oxidized in the atmosphere, especially for silver microstructures when an electric current is passing through it. In order to form an antioxidation protection layer, electroless copper plating is carried out after the laser scanning process. However, in the case of fabricating the SERS substrate, the additional electroless copper plating process remains unused.

Metallization of glass substrates by femtosecond laser photon reduction using silver nitrate ($AgNO_3$) solution is applied to highly sensitive sensing of molecules based on SERS. The femtosecond laser beam is focused on the glass cover slip surface that is in contact with the $AgNO_3$ solution. Silver ions in the $AgNO_3$ solution are then reduced so that silver thin films are deposited on the glass surface, as shown in Figure 3.80 (a). This shows that silver particles with diameters ranging from tens of nanometers to a few micrometers are partially created. To demonstrate sensing based on SERS, 10 ml of rhodamine 6G (R6G) solution diluted with methanol (10^{-7} M) is dropped on the silver film as a sample. The solvent is evaporated and a watermark area of less than 1/4 of the original area remains on the cover slip. After 3 min, the cover slip is rinsed with distilled water and allowed to dry naturally. Figure 3.80 (b) shows a SERS spectrum of R6G obtained using an integration time of 1 s after the background has been subtracted (the inset shows the original spectrum before subtracting the background). Weak but distinct Raman bands, especially for the aromatic C–C stretching vibrations in the spectral range 1300–1650 cm^{-1}, are observed with an intense fluorescence background. The Raman signal originates from R6G molecules attaching to the SERS-active spots, while the fluorescence comes from dissociative molecules. This result indicates that

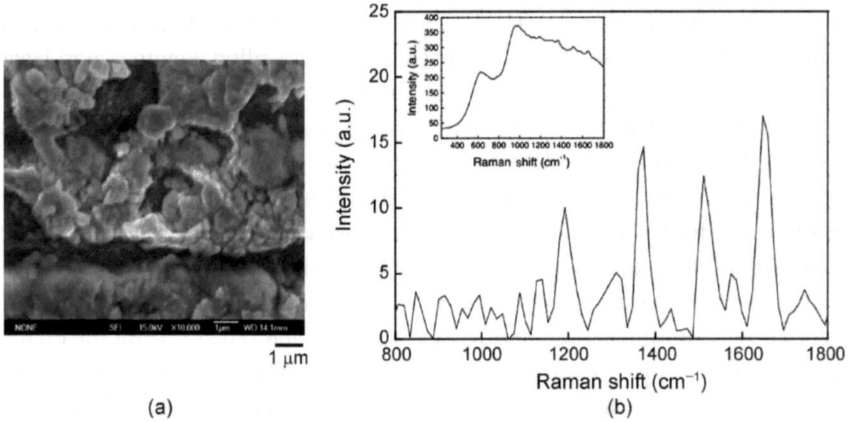

Fig. 3.80: (a) SEM image of silver film deposited on glass chip by irradiation in $AgNO_3$ solution; (b) SERS spectrum of R6G molecules (10–7 M) absorbed on silver nanoparticles after background subtraction (integration time: 1 s). The inset shows the original SERS spectrum.

the deposited silver nanostructures give rise to a SERS effect that permits sensitive detection. However, the Raman enhancement factor (EF) is relatively low because only a few R6G molecules are effectively adsorbed on the SERS active spots by diffusion in the present scheme. Unlike silver colloidal substrates, the lack of electrostatic interactions between the molecules and the silver film prevents them from getting closer and hence limits the total EF.

To enhance the EF, R6G molecules are embedded by mixing R6G with $AgNO_3$ solution and the mixed solution is irradiated by a femtosecond laser beam. Organic R6G molecules are then codeposited on the slip surface together with the silver nanoparticles. Consequently, the EF of a sensor with embedded molecules had an EF that is 40× higher than that without embedded molecules. It could detect R6G molecules with a concentration as low as 10^{-8} M.

3.6.5 Conclusions

In conclusion, the integration of electrical functions into insulators using selective metallization by femtosecond laser modification combined with electroless chemical plating has been demonstrated. This technique involves irradiating a femtosecond laser beam on glass immersed in an electroless plating solution, and can selectively deposit thin metal films on laser-irradiated regions by a single step, even on the internal walls of microfluidic channels, thus providing great flexibility to incorporate potential electrical functions for many biochip applications, such as electro-optics, electrophoresis, electrocapillary and electrowetting, and so on.

One drawback of this technique is the weak adhesion of metal films deposited on glass, although the adhesion can be improved by postannealing. Femtosecond laser modification using a $AgNO_3$ film followed by electroless plating results in higher adhesion of metal films. This technique can be used to deposit metal films on various kinds of materials. Femtosecond laser ablation followed by electroless plating also allows selective deposition of metal films with high adhesion. However, this technique is only applicable to certain materials such as photosensitive glass, since the surface roughness produced by laser ablation is responsible for metal deposition. These selective metallization techniques are demonstrated by fabricating microheaters for spatial control of the temperature in microfluidic devices and a SERS substrate for the highly sensitive analysis of molecules. 3D metal microstructures with high electrical conductivities can be created by the two-photon-induced metal ion reduction of a liquid or a polymer containing metal ions using FsLDW. This technique has not been used to fabricate biochips, but the fabrication of such structures inside the microfluidic systems will enable more functionalities to be added to biochips.

3.7 Integration of hybrid functionalities by femtosecond laser direct writing

3.7.1 Introduction

As in the above descriptions, FsLDW is a promising fabrication technique that can be used to modify the interior of transparent dielectrics (e.g. glass and crystals) in a spatially selective manner. In particular, femtosecond laser pulses modify the chemical properties of the fused silica substrate in the laser-irradiated regions, which can be selectively removed by successive wet etching using HF or KOH, realizing direct fabrication of 3D microfluidics. This two-step process (i.e. FsLDW followed by wet chemical etching) can also be used to integrate micromechanical components such as microvalves and micropumps. It can be further extended to fabricate free-space optics such as micro-optical lenses and optical microresonators inside glass or crystals. In addition, in combination with electroless plating, FsLDW can be used to fabricate microelectrodes in glass and crystals.

Thus, femtosecond lasers allow us to fabricate various functional microcomponents in transparent dielectrics, such as 3D microfluidics, micro-optical components, microelectronic structures, and so on. Using only a single femtosecond laser micromachining system, it is possible to integrate all fluidic, optical, and electrical functions in a single chip without assembly and packaging. Such versatility cannot be realized by any other existing technology. Such integration provides many advantages over existing technologies such as soft lithography and semiconductor processing based on photolithography for fabricating biochips such as microfluidic and optofluidic systems, LOC, and μ-TAS devices.

Using these devices, conventional processes normally carried out in a lab can be miniaturized and performed on a single chip. This leads to enhanced efficiency and portability, and also reduces the volume of samples and reagents required when performing multilevel assessments involving, for example, chemical, biological, and medical analyses. The most significant benefits of these devices are a scaling down of the size, minimal consumption of reagents, reduced manufacturing costs, and enhanced detection speed and sensitivity. Due to their high portability and sensitivity, these devices have become powerful detection and analysis tools for a broad range of applications including biomedical research, healthcare, pharmaceuticals, environmental monitoring, and homeland security. Moreover, there is also the possibility of further enhancing the performance of microfluidic devices by monolithically integrating electronic, mechanical, or optical capabilities.

This section demonstrates typical examples of microchips with hybrid functionalities integrated by using FsLDW, involving optofluidic chips for biosensing, electro-optical integration in $LiNbO_3$, on-chip SERS devices, optical microresonator based sensors, and so on.

3.7.2 Optofluidics fabricated by femtosecond laser direct writing

Optofluidics refer to a class of optical systems that are synthesized by the marriage of microfluidics and optics/photonics. They exhibit tunable optical properties and physical adaptability, and show high potential for LOC and bio-photonics applications [275]. This field emerged in the 2000s when both microfluidics and nanophotonics were maturing, and their synergy became feasible. Although the primary fabrication techniques for optofluidics are based on lithographic methods, FsLDW has been shown to be a promising alternative for the flexible fabrication of true 3D multifunctional structures. Currently, fabrication of optofluidic structures by FsLDW can be achieved by the incorporation of optical waveguides and/or micro-optical components into microfluidic systems. Using a unified process, a wide variety of sensing devices such as a Mach–Zehnder interferometer (MZI) [276], single-cell detection and manipulation systems [277–279], and photonic microcomponents such as microfluidic lasers [280] and waveguides [281] can be fabricated in both fused silica and Forturan glass substrates. It is noteworthy that the integration of microfluidic structures with optical waveguides in porous glass has not yet been demonstrated. However, this is now being attempted by writing waveguides in consolidated porous glass in which the embedded microfluidic structures have been formed beforehand.

Micro-optical components can be fabricated in a slightly different manner, namely by FsLDW in fused silica followed by chemical wet etching in HF acid. In this process, the internal areas modified by femtosecond laser irradiation will gain a significantly higher etch rate than those unmodified areas, so that hollow structures embedded in fused silica can be produced by preferentially removing the materials in laser-scanned

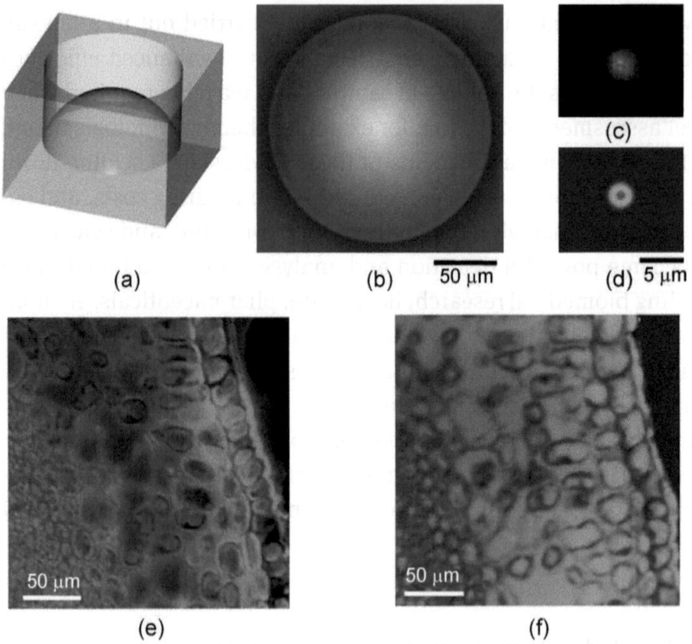

Fig. 3.81: (a) Schematic and (b) optical micrograph of the microlens fabricated by femtosecond laser micromachining; (c) the measured focal spot and (d) the calculated point-spread function of the microlens in (b); two-photon fluorescence images of the leaf tissue acquired using (e) the microlens and (f) an objective lens (5× magnification).

areas. Using this technique, micro-optical lenses with nearly diffraction-limited focusing performance on fused silica substrates can be produced. To create the curved surface as illustrated in Figure 3.81 (a), the sample is first scanned with the tightly focused femtosecond laser beam and then subjected to the wet etching as mentioned above. Afterwards, an additional OH flame polishing is used to smooth the curved surface. The fabricated microlens is shown in Figure 3.81 (b). In comparison to the simulation result (Figure 3.81 (d)) for a model lens with the same geometry, the measured focal spot produced by the fabricated microlens has a comparable size, as shown in Figure 3.81 (c). Recently, the microlens has been used in two-photon fluorescence imaging. The two-photon images of a plant leaf tissue acquired with the micro-optical lens and with a 5× objective lens are compared in Figure 3.81 (e) and Figure 3.81 (f), showing little difference in their imaging performances.

By employing the above-mentioned techniques, a microlens and microfluidic channel can be integrated in a single fused silica chip [212]. To inscribe the contour of the microlens, a similar layer-by-layer annular scanning method is adopted. The curvature radius and aperture diameter of the microlens are 0.6 and 1.2 mm respectively, as depicted in Figure 3.82. To inscribe the microfluidic channels, the laser power and the scanning speed are changed so that the etching duration for fabricating the tiny channel could match that of fabricating the hemispherical microlens. As shown in Figure 3.82, the channels form a Y-shaped structure, and the length of each channel is 600 μm. After laser irradiation, the sample is subjected to an ultrasonic bath in HF solution, until the microlens and channels are formed. Afterwards, the sample is polished with OH flame.

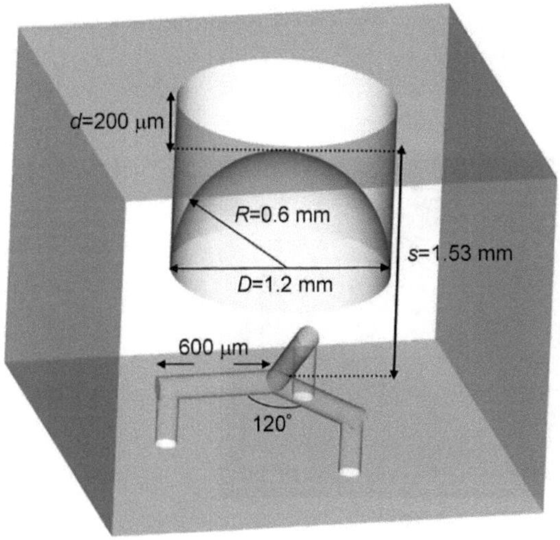

Fig. 3.82: Schematic of the microfluidic chip integrated with a micro-optical lens.

By comparing the measured diameters of channels in Figure 3.83 (a) and 3.83 (b), the magnification of the fabricated microlens is ~4.7, which is in good agreement with the theoretical value. To further verify the imaging performance of the microlens, the fluorescein dye solution (0.12 mg/ml, pH = 13.0, excitation wavelength 490 nm, emission wavelength 520 nm) is fed into the Y-shaped channel. At first, the sample is placed on the stage with the microfluidic channels above the microlens, as shown in Figure 3.83 (a). Then the fluorescent channels are imaged on the CCD camera through the objective lens directly. To investigate the imaging performance of the fabricated microlens again, the sample is inversed, as shown in Figure 3.83 (c) and 3.83 (d). Figure 3.83 (d) shows the magnified image of the microchannel structure formed by the microlens. The integrated chip can therefore be used as an optical imaging system to observe micro-organisms tagged with fluorescent dye and to create an image of the dynamic process in the interested area of the microfluidic channels. Compared with the conventional optical microscope composed of bulky optics and mechanics, this microchip is inexpensive, portable, and disposable.

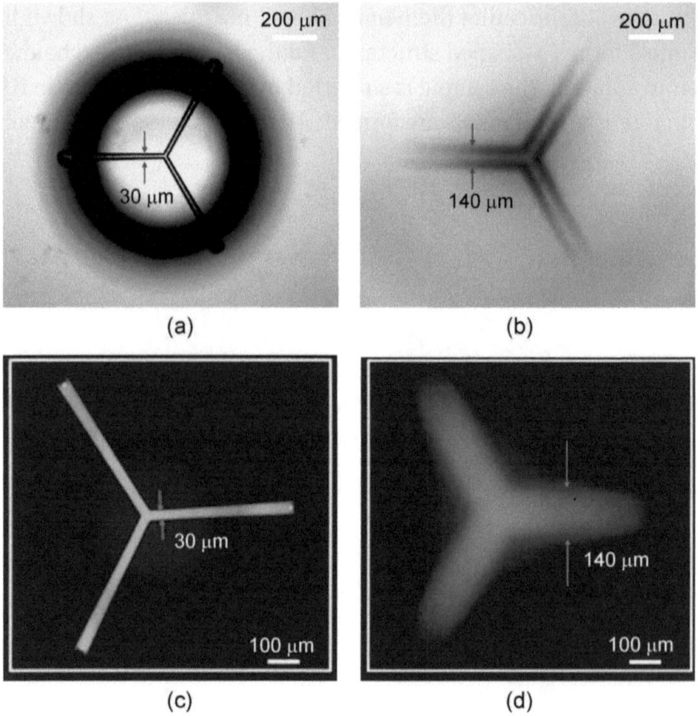

Fig. 3.83: Optical micrograph of (b) Y-shaped microfluidic channel and (c) the enlarged image of the channel formed by the microlens. Micrographs of the fluorescent channels formed (a) without and (d) with the microlens.

3.7.3 Electro-optical integration in $LiNbO_3$ by femtosecond lasers

$LiNbO_3$ has long been used in integrated optics due to its large nonlinear optical coefficients, and excellent electro-optic and piezoelectric properties. It is possible to integrate a substantial number of multifunctional devices on a single $LiNbO_3$ substrate, such as surface-acoustic-wave generation, signal processing, EO modulation, sensing, and other tasks for a wide variety of applications. Recently, it has been demonstrated that buried optical waveguides in $LiNbO_3$ can be fabricated by femtosecond laser inscription [282], which opens the possibility of writing 3D optical circuits in the crystal. Unfortunately, because the optical waveguides fabricated by femtosecond laser inscription are usually deeply buried in the crystal, the conventional surface electrodes gives rise to weak EO interactions. The embedded electrodes could give rise to homogeneous electric fields across the buried waveguides sandwiched between the microelectrodes, and enable effective EO interaction.

A schematic diagram of the Y splitter modulator fabricated by femtosecond laser is presented in Figure 3.84 (a). The angle between the two branches of the Y splitter is 1.2°, and the distance of the two branches is 80 μm. In order to produce thermally

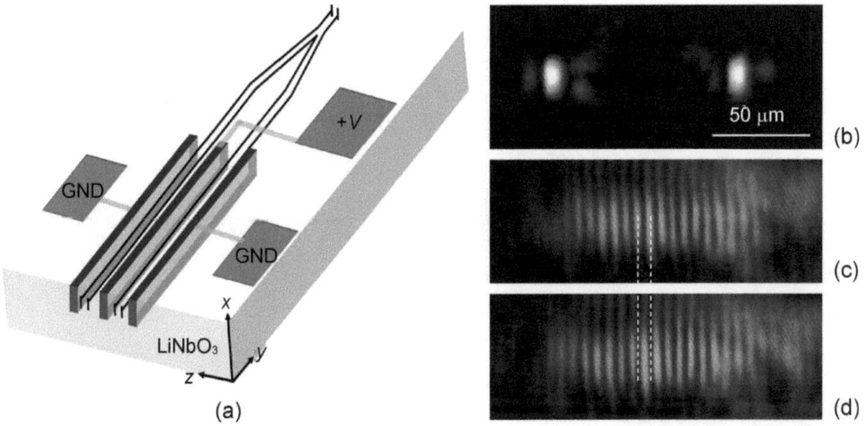

Fig. 3.84: (a) Schematic layout of the Y-splitter modulator; (b) near-field intensity distribution measured at the exit of modulator, and interference pattern of two branches of the Y splitter at different dc voltages of (c) 0 V and (d) 20 V. The attached white lines show the shift of the interference fringe.

stable waveguides in the low repetition rate regime, an approach of writing two parallel lines in close separation is used, which produces a guiding region between the double lines [283]. In comparison with the waveguides guiding light in the irradiated region (type I), the waveguides guiding light between irradiated regions (type II) can preserve the nonlinearity of the bulk crystal [284]. The laser beam is focused with a 100× microscope objective (NA = 0.90), and is incident along the x axis of the crystal and polarized linearly along the y axis (Figure 3.84). The waveguides are fabricated by consecutively writing two parallel lines separated by 10 μm with pulse energies of 0.3 μJ and a translating velocity of 100 μm/s. The focus is located at a depth of 50 μm. After fabricating the Y splitter, three embedded microelectrodes are integrated into the $LiNbO_3$ crystal, as shown in Figure 3.84 (a).

When a 633 nm He:Ne laser beam is coupled into the waveguide splitter with a 20× microscope objective, light is emitted from both branches of the splitter. The near-field beam profiles at the ends of the exit are present in Figure 3.84 (b), showing a horizontal size of 6 μm and a vertical size of 10 μm ($1/e^2$ intensity). By measuring the cone angle of the emerging beam, the estimated NA and refractive index change are approximately 0.03 and 2×10^{-4} respectively. The propagation loss of waveguide is measured to be ~1 dB/cm by comparing the transmission through waveguides with different lengths in order to subtract the coupling losses.

After passing through the splitter, the two exiting beams propagated in the free space with a certain divergence angle, and therefore an interference pattern can be observed in the overlapping area of the two beams in the far field. The device is examined using a He-Ne laser polarized in the extraordinary (z) direction by applying a varying DC voltage to the electrodes, and in the meantime, we collected the interference pattern of two branches of the Y splitter with a CCD camera. It can be seen that the

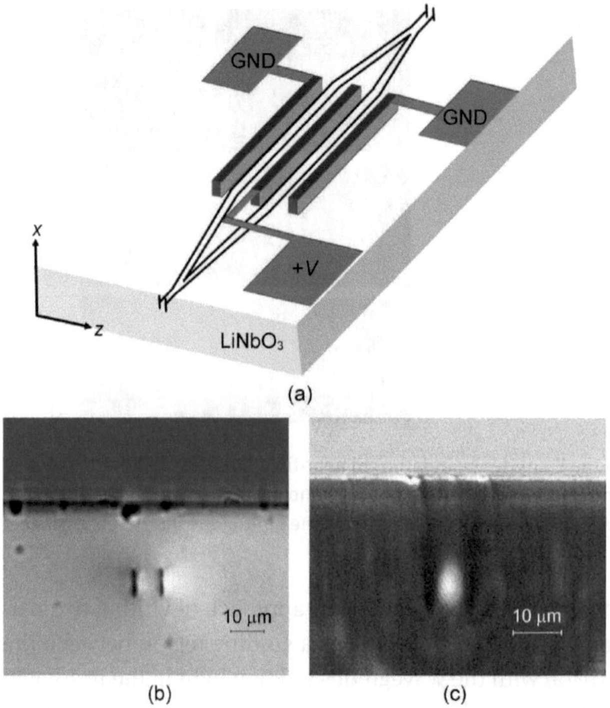

Fig. 3.85: (a) Schematic layout of a Mach–Zehnder EO modulator; (b) a microscope image of the end facet of the MZI; and (c) near-field intensity distribution measured at the exit of MZI.

interference fringes shift horizontally with the varying voltage. The results are shown in Figure 3.84 (c) and 3.84 (d). The interference fringes will experience a Pi-shift if the voltage is changed by 20 V.

Furthermore, a Mach–Zehnder interferometer EO modulator can be fabricated as illustrated in Figure 3.85 (a) [111]. The fabrication procedures are the same as the above-mentioned Y-splitter modulator except for changing the Y-splitter into a Mach–Zehnder interferometer structure. Figure 3.85 (b) shows a microscope image of the end facet of the MZI. A 633 nm He-Ne laser beam is coupled into the entrance of the MZI with a 20× microscope objective, where only the guiding of polarization along the Z-axis is observed. The near-field beam profile at the exit is present in Figure 3.85 (c), showing a single-mode distribution with a horizontal size of ~4 μm and a vertical size of ~6 μm ($1/e^2$ intensity).

The geometries of embedded electrodes are shown in Figure 3.86 (a) and 3.86 (b). A numerical simulation based on finite element method [285] is used to analyze the electric field between the embedded electrodes as shown in Figure 3.86 (b), and the corresponding potential distribution is shown in Figure 3.86 (c). It can be seen that the electric field across the waveguide is nearly uniform and almost confined in the horizontal (z) direction, leading to an improved EO overlap in comparison with the conventional planar electrodes.

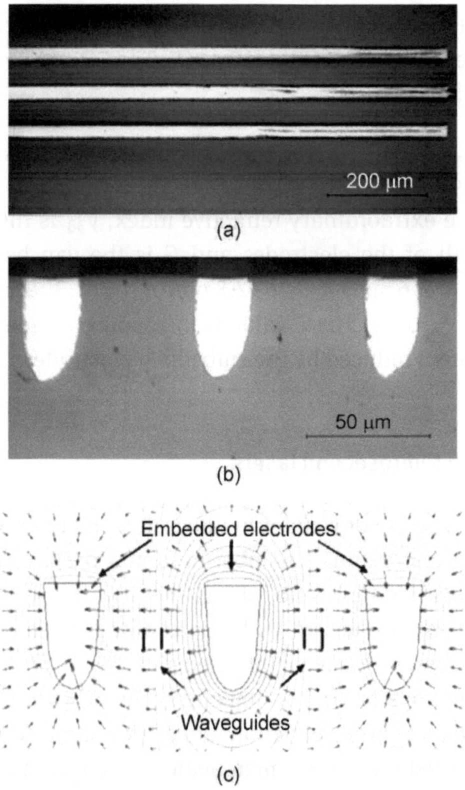

Fig. 3.86: (a) Top view and (b) end view of optical micrographs of the embedded electrodes and optical waveguides; (c) contour plots of the equipotential contour of embedded electrodes.

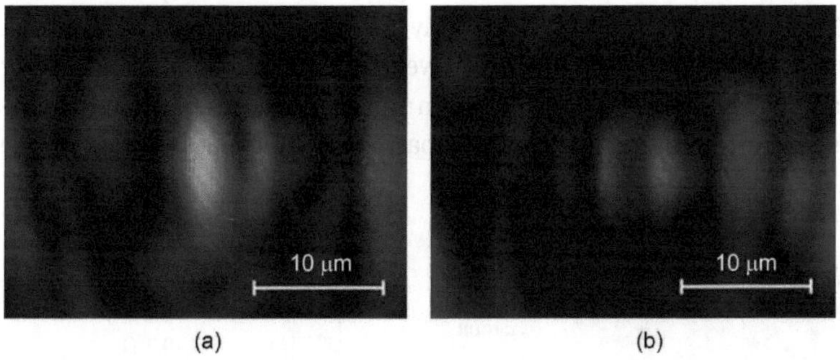

Fig. 3.87: Near-field intensity distribution at the exit of EO modulator at different DC voltages of (a) 0 V and (b) 19 V.

The device is examined using a 633 nm He-Ne laser polarized in the Z direction by applying a varying DC voltage to the electrodes, and in the meantime, the near-field mode at the output end of MZI is collected. The results are shown in Figure 3.87. The measured extinction ratio is only ~9.2 dB, mainly due to the fact that the two arms of MZI cannot be fabricated completely symmetrically because of the limited translation precision of the XYZ stage. The voltage difference between the adjacent maximum and minimum of optical intensity output is measured to be $U_\pi = (19 \pm 1)$ V. The EO overlap integral of modulator Γ can be calculated by [264]

$$\Gamma = \frac{\lambda G}{2 n_e^3 \gamma_{33} V_\pi L}, \quad (3.56)$$

where λ is the laser wavelength, n_e is the extraordinary refractive index, γ_{33} is the EO coefficient of $LiNbO_3$, L is the length of the electrodes and G is the gap between electrodes. With $G = 46\,\mu m$, $L = 2.6\,mm$, $\lambda = 632.8\,nm$, $n_e = 2.2$, and $\gamma_{33} = 2.9 \times 10^{-11}\,m/V$, we can obtain $\Gamma = 0.95$. The Γ value is in good agreement with the simulation result of the electric field induced by the embedded electrodes.

3.7.4 On-chip SERS devices fabricated by femtosecond lasers

Owing to the above fabrication techniques, it is convenient to integrate a microelectronic circuit and a SERS structure into a microfluidic chamber to form a prototype Raman biochip for biosensing applications. Figure 3.88 (a) shows the SEM image of silver microstructures deposited on the glass surface with a laser scanning speed of 1 mm/s and a laser power of 80 mW. One can see that a complicated surface morphology is formed with features at both the micrometer and nanometer scales. Shown in Figure 3.88 (b) is a microheater fabricated by fs laser reduction, deposition, and subsequent electroless Cu plating. The total resistance of the microheater is measured to be ~3.4 Ω, mainly from the resistance of the heating circuit indicated by the solid circle in Figure 3.88 (b). The microheater can be stably operated with a maximum current of 200 mA at which the corresponding heater power is 0.14 W. This is high enough for most microbiosensor applications. The reduction of silver ions from the $AgNO_3$ solution and subsequent deposition of silver nanoparticles on the glass surface can be

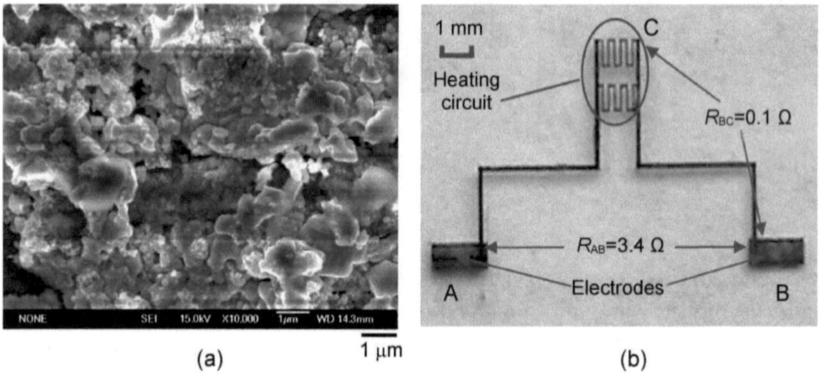

Fig. 3.88: (a) SEM image of morphology on glass surface after fs laser modification; (b) a microheater fabricated by fs laser direct writing in $AgNO_3$ solution followed by electroless Cu plating.

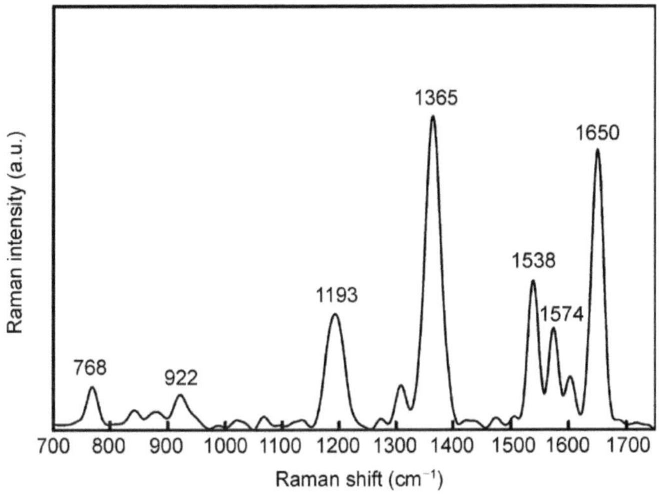

Fig. 3.89: SERS spectrum of R6G molecules absorbed on the reduced silver nanoparticles.

attributed mainly to the high peak power offered by the fs laser pulse and the thermal effect originating from the high repetition rate of the fs oscillator.

The microstructure directly fabricated by the fs laser reduction and deposition also exhibits strong enhancement of the Raman signal. A 10 μL of R6G/methanol solution with a concentration of 10–7 M is dropped on the silver microstructures reduced and deposited by the fs laser oscillator. After three minutes the slip is flushed by distilled water and dried naturally. Raman scattering of the R6G molecules is excited by a 532-nm continuous laser with a power of 10 mW and collected on the same microscope in reflection mode but with a different objective lens (40×, numerical aperture is 0.65). Collected SERS spectrum of R6G is shown in Figure 3.89 with an integration time of 1 s. The aromatic C-C stretch vibrations in the spectral region from 1300 to 1650 cm^{-1} can be clearly observed from Figure 3.89, which comes from the R6G molecules absorbed at the SERS-active sites on the substrate. The low detection concentration and short integration time proved that the nanoparticles reduced by fs laser can lead to a great enhancement of the Raman signal. By the code position of reduced silver nanoparticles and R6G molecules, a SERS substrate with embedded sample molecules is obtained with higher Raman enhancement efficiency for the detection of the fluorescent sample.

Since we can now fabricate both the microelectronic circuit and SERS substrate on the same platform, it is natural to integrate both these functions into a microfluidic channel to form a LOC system. The scheme of the Raman biochip sample is illustrated in Figure 3.90. It is fabricated on a fused silica glass with three steps. First, a microfluidic cavity is created by fs laser ablation using a fs laser amplifier. Secondly, a microheater and two electrodes are built in the microfluidic cavity using the fs laser oscillator. Lastly, the SERS substrate is fabricated in the middle area between two heating

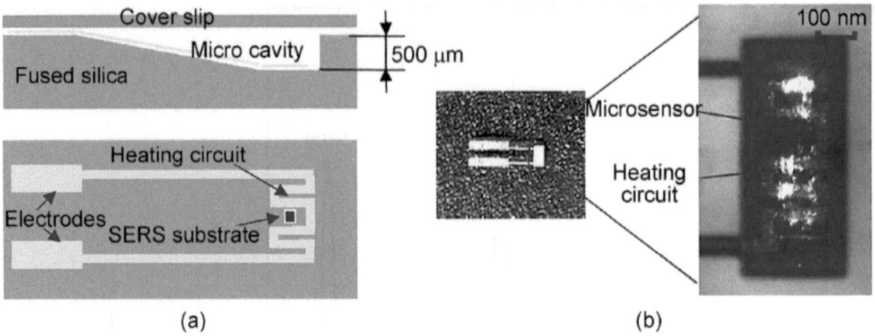

Fig. 3.90: (a) Schematic illustration of side view (upper panel) and top view (lower panel) and (b) digital camera captured image of the integrated Raman sensor.

circuits in the cavity, as shown in the lower panel of Figure 3.90 (a). After the absorption of R6G molecules on the substrate for three minutes, the microcavity is flushed and then filled with distilled water. Then the substrate is excited by a 532-nm laser for a few minutes in order to bleach the fluorescence of the dissociated R6G molecules. When the measured Raman signal become stable, the microelectrodes of the heater are connected to a constant-current power supply and an electric current is applied on the microheater to heat the water in the microcavity. The measured Raman intensity exponentially decreased with increasing electric current. The decrease of Raman signal with the increasing temperature in the heated microfluidic cavity could be explained by the dissociation of the R6G molecules absorbed on the nanoparticles with increasing temperature.

3.7.5 Sensing applications of optical microresonators

High-Q-factor WGM microresonators can efficiently confine light in a small space by total internal reflection. The combination of low absorption loss in the volume of microresonators and low scattering loss at their smooth peripheries leads to a long photon lifetime, which makes the WGM microresonators excellent candidates for many important applications. Integration of the on-chip microresonators with other functional microstructures such as photonics and microfluidics is also challenging, because in most cases their processing techniques are incompatible. The difficulties hamper the widespread use of microresonators in applications ranging from nonlinear optics to high-sensitivity biological/chemical sensing.

3.7.5.1 High-sensitivity optofluidic sensors based on optical microresonators

For many optofluidic sensing applications, the efficient enhancement of the sensing sensitivity can be achieved by use of a high-Q-factor microresonator [286, 287], owing to its unique capability of confining light in a tiny volume for long periods of time by total internal reflection. In most cases, the coupling of light into and out of the WGM microresonators is realized with a fiber taper carefully positioned near the microresonator with precision motion stages to achieve the critical coupling condition [288, 289]. Since the fiber taper and the microresonators are not integrated, the widespread use of such systems in optofluidic and biosensing applications has been hampered because of their poor portability. Further chip-level integration of the microresonators in a microfluidic system is also difficult, because of the incompatibility between the processing techniques of microfluidic systems and optical microresonators.

We fabricated a fully integrated optofluidic sensor that consists of a high-Q microresonator and a microfluidic mixer on fused silica glass substrates (UV grade fused silica JGS1 whose upper and bottom surfaces are polished to optical grade) with a thickness of 1 mm [290]. Both the microresonator and the microfluidic mixer are simultaneously fabricated in a fused silica substrate based on FsLDW. The fabricated optofluidic sensor offers a detection limit of 1.2×10^{-4} refractive index unit (RIU) for bulk refractive index sensing of liquids.

The process flow for fabricating the fully integrated sensor mainly consists of three steps: (1) femtosecond laser exposure followed by selective wet etch of the irradiated areas to create 3D microfluidic channels embedded in glass and the microdisk structure at the outlet of the microchannel; (2) selective reflow of the silica disk by CO_2 laser irradiation to improve the Q-factors; and (3) assembly of the fiber taper with the microresonator by CO_2 laser welding.

The key innovation of this experiment is to assemble the fiber taper onto the sidewall of the microtoroid resonator to produce a fully integrated, portable optofluidic sensor. Figure 3.91 illustrates the procedures of assembling of the optical fiber taper with the resonator. As illustrated in Figure 3.91 (a), the fiber taper is first brought to the vicinity of the toroid using the 3D nanopositioning stage to achieve critical coupling. The highest Q-factor we achieved under the critical coupling condition was 3.24×10^6. Then fiber taper was slightly moved to get it in direct contact with the sidewall of the microtoroid, and we irradiated the microtoroid with the CO_2 laser beam again, as illustrated in Figure 3.91 (b). We notice here that the focusing conditions of the CO_2 laser beam used for welding the fiber onto the microtoroid are the same as that used for the reflow of the microdisk to form the microtoroid, namely no modification on the optical setup that is established for surface reflow of the microresonator is applied. However, to achieve a high Q-factor of the assembled microtoroid-fiber system, we indeed had optimized the duty ratio of the CO_2 laser. In the experiment, eventually a duty ratio of 2.5% (i.e. 50% lower than that used for reflowing the microdisk) is chosen, leading to both a strong welding and a decent Q-factor in air of 3.21×10^5.

Figure 3.91 (c) shows the optical micrograph of the microtoroid in contact with the fiber taper, before and after the fiber taper was welded onto its sidewall with the CO_2 laser irradiation. Specifically, in Figure 3.91 (d), one can see that the CO_2 laser irradi-

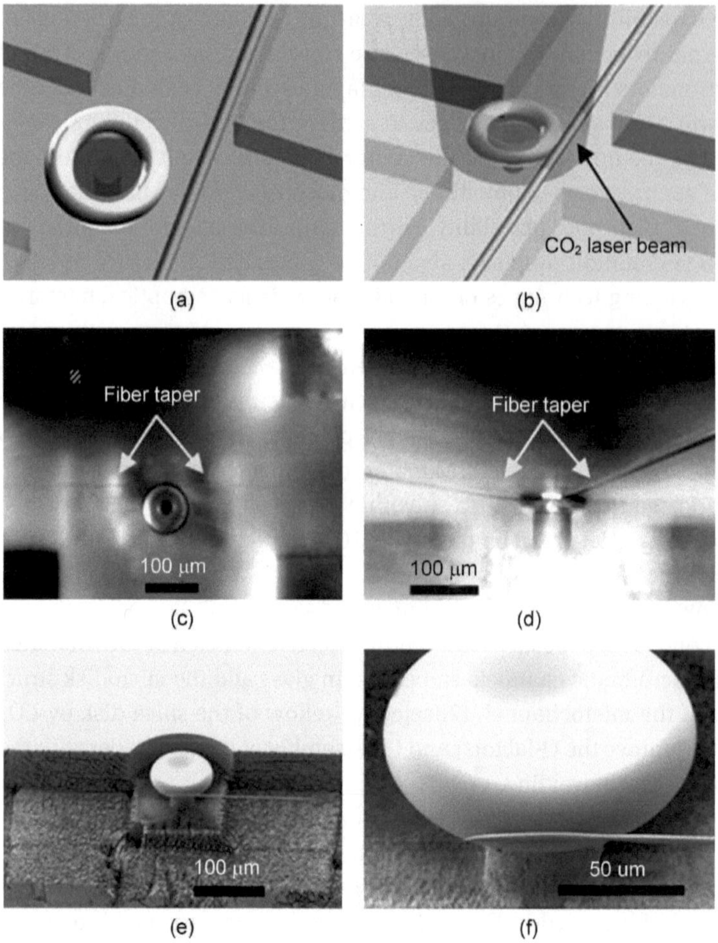

Fig. 3.91: (a) Bring the fiber taper close to the microresonator to achieve a high coupling efficiency. (b) Bring the fiber taper in contact with the microresonator and perform CO_2 laser welding. (c) Top-view optical micrograph of the fiber taper coupled to the microresonator before welding. (d) Side-view optical micrograph of the fiber taper welded onto the sidewall. Fiber can remain assembled on the microtoroid with bending. Top-view (e) and close-up view (f) SEM images of the fiber taper welded to the side wall of microtoroid.

ation creates a strong welding in the contacting area between the fiber taper and the microtoroid, and thus the fiber can be significantly bent without falling away from the microtoroid. The high mechanical strength ensures a reliable operation of the sensor in field-portable uses. Figures 3.91 (e) and 3.91 (f) further provide detailed information on how the fiber taper is welded to the side wall with the respective overview and close-up view SEM images of the integrated fiber-microtoroid system. It can be seen that after the welding, the cross section of the fiber taper is significantly deformed, i.e.

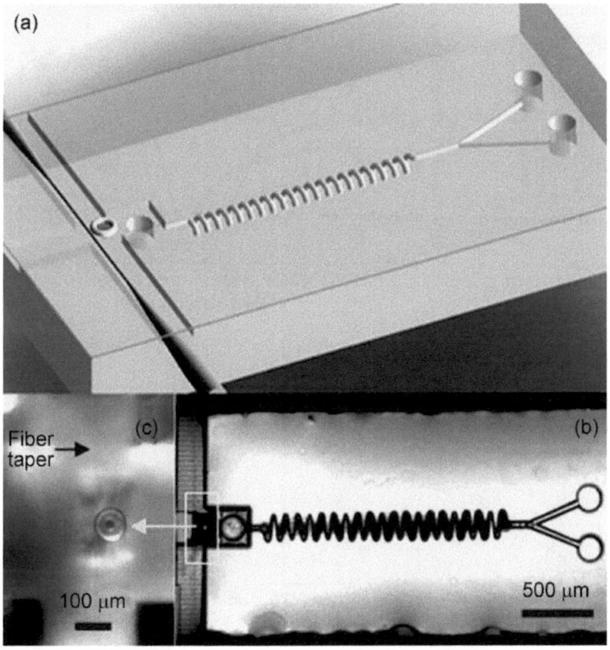

Fig. 3.92: (a) Illustration of layout of the integrated optofluidic sensor; (b) the optical micrograph of the fabricated sensor. The fiber is indicated with the dashed line, and the microtoroid resonator assembled with the fiber taper is sitting in the area indicated by the square. (c) The close-up view of the assembled microtoroid-fiber system fabricated in the microfluidic channel using a femtosecond laser.

the original perfectly round-shaped cross section becomes a compressed flat oblate. Further theoretical analysis is required to understand how the deformation of the tapered fiber influences the Q factor in a quantitative way.

The layout of the designed sensor is shown in Figure 3.92(a). The microfluidic system is embedded 300 µm beneath the glass surface, which is composed of two Y-branched-channels with a length of 600 µm connected with two inlets, and a long spiral channel with a total length of 3 mm and a diameter of ~35 µm connected to an outlet for the mixing of liquid. Near the outlet, a reservoir is fabricated and the microtoroid resonator is fabricated in the center area of the reservoir. Therefore, the mixed liquid coming out from the outlet can be sensed with the microresonator. The optical microscope image of the fabricated sensor is presented in Figure 3.92(b), showing that all the components have been integrated in a single glass chip. Figure 3.92(c) illustrates the close-up view image of the integrated toroid-fiber system. The fiber taper is too thin and can hardly be seen in such images of limited resolution (Figure 3.92(b) and Figure 3.92(c)), which provide a sufficiently large field-of-view for the sensor. Therefore, the fiber taper is indicated with the dashed line in Figure 3.92(b), and with an arrow in Figure 3.92(c).

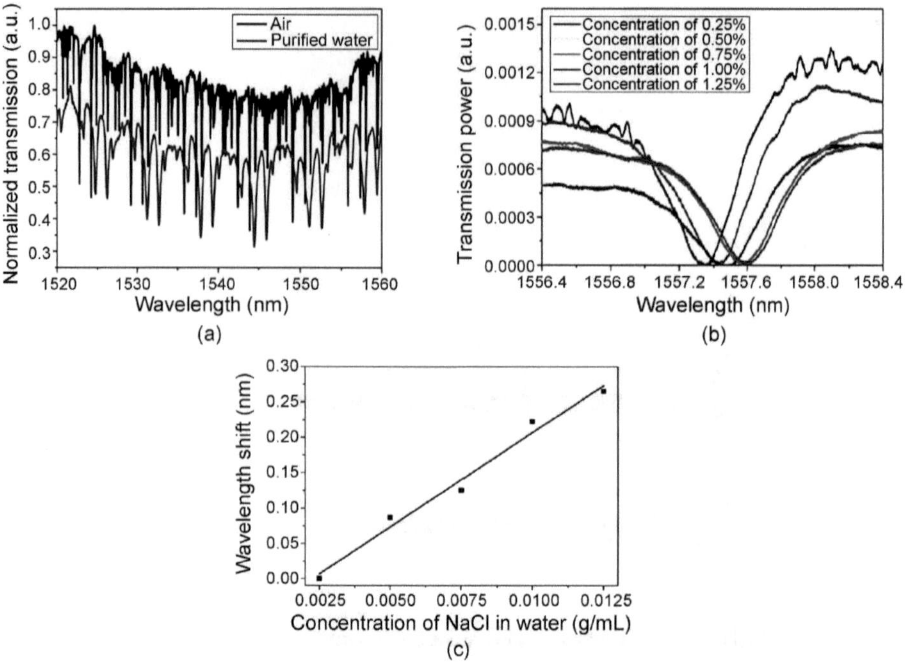

Fig. 3.93: (a) Transmission spectrum of the integrated microtoroid-fiber system measured in air after the CO_2 laser welding (upper curve) and that of the integrated microtoroid-fiber system measured in purified water (lower curve); (b) the close-up view of the spectral dips around 1557 nm wavelength measured at different doping concentrations of salt; (c) the wavelength shift plotted as a function of the doping concentration of salt (squares), which can be well fitted with a linear relationship (curve).

Figure 3.93 shows the results of sensing the refractive index of purified water doped with salt at different concentrations. Specifically, the overall transmission spectra of the microresonator in air and in purified water are compared with the upper and lower curves in Figure 3.93 (a) respectively. It can be seen that compared to the spectra measured in air, the spectral width of each dip becomes larger, and the number of modes in the spectrum measured in water has been significantly reduced, mainly because of the reduced Q-factor and the elimination of the high-order modes, respectively. The measured Q-factor of the microresonator immersed in purified water is only ~3.1×10^3. The severe reduction of the Q-factor in water is mainly caused by two factors. The first limiting factor is that the experiment is carried out at a wavelength of ~1550 nm (i.e. the tunable laser source in our lab), which is highly absorptive in water [291]. The second is that because of the limited range of motion of the nanopositioning stage, the size of the fabricated microtoroid resonator is about 80 μm, which is not large enough to support sufficiently high Q-factors in water as have been investigated before. Actually, in [46], the authors have clearly shown that a fused silica microtoroid with a diameter of ~80 μm exhibits a Q-factor on the level of ~10^3 around 1550 nm in water.

The measured Q-factor of the fabricated resonator ($Q \sim 3.1 \times 10^3$) agrees well with their analysis.

Figure 3.93 (b) shows five spectra recorded in purified water doped with salt at different doping concentrations of 0.25, 0.5, 0.75, 1, and 1.25%. Here, all the concentrations are given in weight percent wt %. One can see that the position of the spectral dip around 1557 nm wavelength shifts significantly with the increasing concentration of the salt, which can be attributed to the change in the refractive index of the water with varying doping concentrations. In addition, Figure 3.93 (c) shows that the position of the spectral dip shifts almost linearly with the increasing doping concentration of salt, which is reasonable because at such a low concentration of salt, the refractive index of the liquid will also vary linearly with the doping concentration.

Based on the data in Figure 3.93 (c), it is evaluated that the sensitivity of the sensor can reach ~220 nm/RIU, yielding a detection limit of 1.2×10^{-4} RIU based on the analysis method provided in [292]. The results are comparable to the best sensors fabricated by the FsLDW so far. However, we could expect large room for improvement of the detection limit of the integrated sensor if we improve the Q-factor of the microresonator with a larger size in liquids and shift the detection light to the visible range.

3.7.5.2 On-chip wavelength-tunable microresonator

In many applications such as microcavity QED or microlasing, precise control of the resonant wavelength of WGM is crucial. The wavelength tuning of WGM has been demonstrated by tactfully tuning the refractive index of the microresonator, which leads to a change in the optical path for the light travelling in the microresonator. Tuning of the refractive index is readily achievable using either thermal-optical or electro-optical effects [293, 294]. In addition, the WGMs are also sensitive to the geometric shape of microresonator, enabling tuning of the resonant wavelength by a slight distortion of the shape of the microresonator using mechanical stretching or squeezing [295, 296]. To facilitate miniaturization and ease of operation, there is a pressing need for the development of a monolithic approach by which the thermal or electric function can be integrated into the microresonator chip.

The fabrication of a microresonator integrated with a microheater on a single fused silica chip by FsLDW is described here. Also demonstrated is on-chip tuning of the resonant wavelength of the microresonator by varying the electric voltage applied to the microheater.

The integrated device is illustrated in Figure 3.94 (a), which contains a microheater and a high-Q microresonator simultaneously fabricated on a fused silica chip using femtosecond laser 3D micromachining. First, the microheater is fabricated by selective metallization on glass surfaces, which is realized using FsLDW followed by electroless plating. The reason for fabricating the microheater prior to the microresonator is to avoid contamination on the microresonator during the electroless plat-

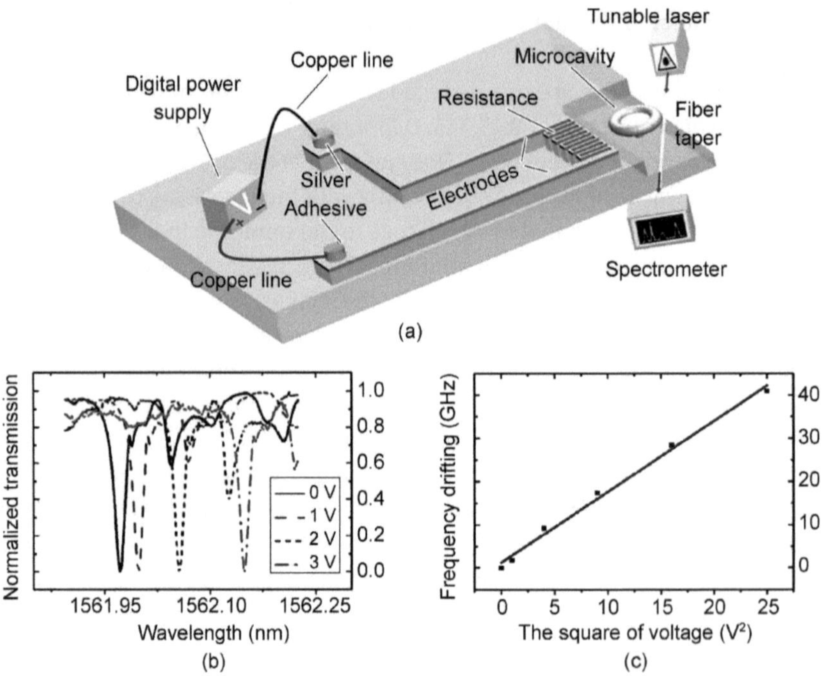

Fig. 3.94: (a) Illustration of layout of the integrated chip; (b) normalized transmission spectra of the microresonator at different electric voltages applied on the microheater; (c) the drift of resonant wavelength as a function of square of the voltage, which can be well fitted with a linear relationship (curve).

ing. Previous investigations have shown that the metal structures are buried in the substrate, which ensures a good adhesion. The total resistance (R) of microheater is measured to be $R = 9.6\,\Omega$. At an electric voltage of 1 V, the power of the microheater is calculated to be ~0.1 W. The microheater is connected to the power supplier by two copper wires that are bonded on the two ends of the microheater using a conductive silver adhesive. This procedure is arranged after the microresonator is fabricated, because the CO_2 laser irradiation employed for surface reflow on the microresonator will cause damage to the silver adhesive.

The transmission spectra of the microresonator in Figure 3.94 are recorded at different electric voltages of 0, 1, 2 and 3 V. The central wavelength of the WGM shifts from 1561.97 nm (the solid line) to 1562.15 nm (the dashed dotted line) due to the temperature change in the microtoroid resonator as the power of the microheater increased from 0 to 0.9 W. Figure 3.94 (c) shows that the drifting of the resonant wavelength is linearly dependent on the V^2 of the microheater, because the joule heating power of the microheater is proportional to V^2. The slope of the fitting line indicates that the tuning rate of the microheater is $1.8\,\text{GHz}/V^2$.

It is noteworthy that the response time of the device is on the order of ~10 s, which significantly surpasses the performance of conventional external heaters. We expect that the integrated microdevice will benefit various applications including tunable laser sources, optical filters, cavity quantum electrodynamics, and so on.

3.7.6 Conclusions

Femtosecond lasers have been used to fabricate microelectrodes and microfluidic and micro-optical components in glass or crystals in a compatible manner. Integration of these basic elements in single substrates by FsLDW offers the possibility of creating many innovative and extraordinary microsystems and devices. Early examples along this direction include the integration of a microlens and microfluidic channels for biosensing, integration of a waveguide-based MZI and microelectrodes for constructing an optical modulator, integration of microheaters and SERS structures on a single chip for Raman sensing, and integration of optical microresonators with a microfiber taper or a microheater for optical controlling/sensing applications.

Femtosecond laser processing is promising due to its ability to realize one-step integration of microfluidics and other microcomponents in glass chips. Glass is an ideal substrate material for many applications mainly because of its high chemical inertness and high optical transparency. The ultimate dream of femtosecond laser integration of biochips is to create a complete 'all-in-one' LOC or optofluidic system in which all the functional components are three-dimensionally integrated in a single exposure step by FsLDW followed by optional batch postprocessing procedures, which are cost-effective for mass production. This is certainly a formidable challenge, but significant progress has been made toward realizing this goal and important technical advances have been demonstrated, as described in this section.

3.8 Summary and outlook

Femtosecond laser microfabrication has become a powerful technique for producing multifunctional microchips. In particular, it allows the incorporation of hybrid functionalities into a monolithic chip to create new exciting devices. Compared with existing techniques, this approach is based on a unique MPA process within an ultrashort duration, which offers the advantages of eliminating thermal diffusion, internal processing, subdiffraction-limited resolution and the capability of multifunctional integration. The flexibility of the direct writing scheme that employs scanning of tightly focused femtosecond laser pulses has enabled the formation of various 3D micro/nanostructures with almost unlimited geometries and configurations, which has shown significant impact on a broad range of applications ranging from optoelectronics, photonics and MEMS to chemical, biological, and medical systems.

In this chapter, we demonstrated a wide variety of hybrid functional integrated microdevices, such as complex 3D micro- and nanofluidic networks, optofluidic chips, opto-electronic modulators, integrated SERS sensors, and so on, in addition to the monofunctional fabrication of microfluidics, micro-optics, and microelectronics. Several issues need to be addressed in the future. First, to meet the requirements of low-loss optics and smooth fluidics applications, improved methods are needed for smoothing the inner surfaces of devices fabricated in porous glass substrates. The existing approaches such as annealing, oxygen/hydrogen flame polishing, and CO_2 laser reflow have enabled the creation of smooth surfaces of optical lenses and microresonators and smooth inner walls of microfluidic channels of relatively large diameters (i.e. diameters of tens of microns), but they become increasingly less effective when the size of microfluidic structure decreases. In addition, although a feature size of $\lambda/20$ has been achieved in porous glass in the transverse direction, it is still challenging to create nanofluidic channels of fully nanometer-scale dimensions in glass substrates. Selective metallization using FsLDW is expected to be applied to a variety of sensing applications such as electrofluidics, SERS sensors, and plasmonics. However, although side wall metallization has been successfully demonstrated [297], the fabrication of 3D freestanding metal structures in closed microfluidic channels has not been achieved. It is expected that these difficulties will be gradually overcome in the future. Finally, glass is an ideal substrate material for such applications mainly because of its high chemical inertness. The ultimate goal of femtosecond laser integration of optofluidics is to create a complete 'all-in-one' LOC or μ-TAS chips in which all the functional components are simultaneously integrated in a single exposure step by FsLDW, followed by optional batch postprocessing procedures, which are cost-effective for massive production. This is certainly a formidable challenge, although significant progress has been made toward realizing this goal and important technical advances have been realized.

In a broad sense, ultrafast laser processing has fundamentally changed materials processing with lasers. It has become a ubiquitous micromachining tool for various materials including metals, semiconductors, ceramics, glass, crystals, polymers, and even soft materials such as biotissues [13]. When being applied for surface structuring, ultrafast laser processing induces novel properties in terms of friction, adhesion, optical absorption and/or reflection, and hydrophobicity [298]. Scanning of tightly focused ultrafast laser pulses in transparent materials enables the formation of 3D micro-/nanostructures with limitless geometries and configurations, which makes it possible to produce functional components and devices for integrated optics and LOC applications [41]. Besides fundamental research, ultrafast lasers have also been employed for several practical and industrial applications, such as photomask repairs, ink nozzle drilling, and medical stent fabrication [299]. These lasers are now also used in the electronics industry for the scribing, patterning, and texturing of glass and semiconductors (e.g. photovoltaic cells and light emitting diode displays) [300]. Due to the limited space, discussion of the above issues, which can be found elsewhere, are not involved in this chapter.

To perfect these technologies, efforts should be focused on the investigation of the underlying mechanisms to achieve better control of the structural changes and/or morphologies in the laser affected zone, refinement of the beam shaping technologies for precise tuning of the geometries of laser written structures and to promote the fabrication efficiency, and to improve the fabrication resolution for precise alignment and coupling between multiple components (e.g. waveguide–waveguide coupling and the intersection of waveguide cores and microfluidic channels). In particular from a technical perspective, early efforts were mainly focused on the spatial or temporal shaping of ultrafast laser pulses, whereas recent interest has shifted to simultaneous spatiotemporal manipulation of the femtosecond laser pulses, which has enabled modification of the peak intensity distribution in the focal area, the elimination of self-focusing, high-efficiency 3D fabrication, and induction of the nonreciprocal writing effect. In addition, the manipulation of the femtosecond laser pulse polarization has enabled the fabrication of polarization-sensitive optical devices and nanofluidic channels in glass [301]. Beam shaping effects have also been exploited in TPP, aimed at either high-speed parallel processing or subdiffraction-limited nanostructuring [302].

As a final note, in the past two decades, scientists and researchers have only devoted laboratory investigations to a particular concentration on the clarification of mechanisms, the development of new fabrication technologies, and extension of the application range. Femtosecond laser processing has long been regarded as an expensive and time-consuming technique; this image has hampered the widespread use of this technique, and efforts should therefore be made to bridge the gap between current fundamental research and future commercial applications of multifunctional microdevices fabricated by FsLDW. Fortunately, significant progress has been achieved in the development of novel ultrafast laser systems. The replacement of Ti:sapphire crystals with fibers or thin disks has not only led to high average powers (i.e. intense ultrafast laser pulses at high repetition rates) that will allow direct laser writing at speeds leveling the order of tens of meters per second, but will also make the ultrafast laser systems more robust and cost-effective [303, 304]. The rich variety of parameters provided by ultrafast laser pulses has allowed femtosecond laser beam shaping to be extensively investigated. For the commercialization of disposable microfluidic sensors, however, further reductions in fabrication cost are required, in addition to improvements to the throughput. In the future, ultrafast laser materials processing will benefit from the rapid development of ultrafast laser systems and ultrafast laser techniques.

Bibliography

[1] M. A. Burns, B. N. Johnson, S. N. Brahmasandra, et al., An integrated nanoliter DNA analysis device, Science **282**(5388) (1998): 484–487.

[2] G. M. Whitesides, The origins and the future of microfluidics, Nature **442** (2006): 368–373.

[3] E. K. Sackmann, A. L. Fulton and D. J. Beebe, The present and future role of microfluidics in biomedical research, Nature **507** (2014): 181–189.

[4] D. Psaltis, S. R. Quake and C. Yang, Developing optofluidic technology through the fusion of microfluidics and optics, Nature **442** (2006): 381–386.

[5] H. Schmidt and A. R. Hawkins, The photonic integration of non-solid media using optofluidics, Nat. Photon **5** (2011): 598–604.

[6] Y. Xia and G. M. Whitesides, Soft lithography, Annu. Rev. Mater. Sci. **28**(1) (1998): 153–184.

[7] J. C. McDonald and G. M. Whitesides, Poly (dimethylsiloxane) as a material for fabricating microfluidic devices. Acc. Chem. Res. **35** (2002): 491–499.

[8] J. W. Choi, S. Kim, R. Trichur, H. J. Cho, et al., A plastic micro injection molding technique using replaceable mold-disks for disposable microfluidic systems and biochips, Proc 5th Int Conf Micro Total Analysis Systems (microTAS)(2001): 411–412.

[9] M. A. Grétillat, F. Paoletti, P. Thiébaud, et al., A new fabrication method for borosilicate glass capillary tubes with lateral inlets and outlets, Sensors Actuators, A **60** (1997): 219–222.

[10] A. Dodge, K. Fluri, E. Verpoorte and N. F. de Rooij, Electrokinetically driven microfluidic chips with surface modified chambers for heterogeneous immunoassays, Anal. Chem. **73**(14) (2001): 3400–3409.

[11] K. Itoh, W. Watanabe, S. Nolte, et al., Ultrafast processes for bulk modification of transparent materials. MRS Bull. **31**(08) (2006): 620–625.

[12] R. R. Gattass and E. Mazur, Femtosecond laser micromachining in transparent materials, Nat. Photon **2**(4) (2008): 219–225.

[13] K. Sugioka and Y. Cheng, Ultrafast lasers: reliable tools for advanced materials processing, Light: Science and Applications **3**(4) (2014):e149.

[14] R. Srinivasan, E. Sutcliffe and B. Braren, Ablation and etching of polymethylmethacrylateby very short (160 fs) ultraviolet (308 nm) laser pulses. Appl. Phys. Lett. **51** (1987): 1285–1287.

[15] S. Küper and M. Stuke, Femtosecond UV excimer laser ablation. Appl. Phys. B **44** (1987): 199–204.

[16] J. V. Rudd, G. Korn, S. Kane, et al., Chirped-pulse amplification of 55-fs pulses at a 1-kHz repetition rate in a Ti-Al2O3 regenerative amplifier. Opt. Lett. **18** (1993): 2044–2046.

[17] B. N. Chichkov, C. Momma, S. Nolte, et al., Short-pulse laser ablation of solid targets. Opt. Commun. **129** (1996): 134–142.

[18] M. F. Yanik, H. Cinar, H. N. Cinar, et al., Neurosurgery: functional regeneration after laser axotomy. Nature **432** (2004): 822.

[19] N. Bärsch, K. Körber, A. Ostendorf and K. H. Tönshoff, Ablation and cutting of planar silicon devices using femtosecond laser pulses. Appl. Phys. A **77** (2003): 237–242.

[20] Y. Nakata, T. Okada and M. Maeda, Fabrication of dot matrix, comb, and nanowire structures using laser ablation by interfered femtosecond laser beams. Appl. Phys. Lett. **81** (2002): 4239–4241.

[21] S. Küper and M. Stuke, Ablation of polytetrafluoroethylene (Teflon) with femtosecond UVexcimer laser pulses. Appl. Phys. Lett. **54** (1989): 4–6.

[22] S. Küper and M. Stuke, Ablation of UV-transparent materials with femtosecond UV excimer-laser pulses. Microelectron. Eng. **9** (1989): 475–480.

[23] J. Krüger and W. Kautek, Femtosecond-pulse visible laser processing of transparent materials. Appl. Surf. Sci. **96–98** (1996): 430–438.

[24] K. M. Davis, K. Miura, N. Sugimoto and K. Hirao, Writing waveguides in glass with a femtosecond laser. Opt. Lett. **21** (1996): 1729–1731.
[25] E. N. Glezer, M. Milosavljevic, L. Huang, et al., Three-dimensional optical storage inside transparent materials. Opt. Lett. **21** (1996): 2023–2025.
[26] W. Watanabe, S. Sowa, T. Tamaki, K. Itoh and J. Nishii, Three-dimensional waveguides fabricated in poly(methyl methacrylate) by a femtosecond laser. Jpn. J.Appl. Phys. **45** (2006):L765–L767.
[27] Y. Hanada, K. Sugioka and K. Midorikawa, UV waveguides light fabricated in fluoropolymer CYTOP by femtosecond laser direct writing. Opt. Express **18** (2010): 446–450.
[28] K. Sugioka and Y. Cheng, Femtosecond laser processing for optofluidic fabrication, LabChip **12** (2012): 3576–3589.
[29] S. Kawata, H. B. Sun, T. Tanaka and K. Takada, Finer features for functional microdevices. Nature **412** (2001): 697–698.
[30] D. F. Tan, Y. Li, F. J. Qi, et al., Reduction in feature size of two-photonpolymerization using SCR500. Appl. Phys. Lett. **90** (2007): 071106.
[31] H. B. Sun, S. Matsuo and H. Misawa, Three-dimensional photonic crystal structures achieved with two-photon-absorption photopolymerization of resin. Appl. Phys. Lett. **74** (1999): 786–788.
[32] S. Maruo, K. Ikuta and H. Korogi, Submicron manipulation tools driven by light in a liquid. Appl. Phys. Lett. **82** (2003): 133–135.
[33] B. B. Xu, Y. L. Zhang, H. Xia, et al., Fabrication and multifunction integration of microfluidic chips by femtosecond laser direct writing. Lab Chip **13** (2013): 1677–1690.
[34] J. Reif, F. Costache, M. Henyk and S. V. Pandelov, Ripples revisited: non-classical morphology at the bottom of femtosecond laser ablation craters in transparent dielectrics. Appl. Surf. Sci. **197–198**, (2002): 891–895.
[35] Q. H. Wu, Y. R. Ma, R. C. Fang, et al., Femtosecond laser-induced periodic surface structure on diamond film. Appl. Phys. Lett. **82** (2003): 1703–1705.
[36] P. Rudolph and W. Kautek, Composition influence of non-oxidic ceramics on self-assembled nanostructures due to fs-laser irradiation. Thin Solid Films **453–454** (2004): 537–541.
[37] J. E. Carey, C. H. Crouch, M. Shen and E. Mazur, Visible and near-infrared responsivity of femtosecond-laser microstructured silicon photodiodes. Opt. Lett. **30** (2005): 1773–1775.
[38] T. Baldacchini, J. E. Carey, M. Zhou and E. Mazur, Superhydrophobic surfaces prepared by microstructuring of silicon using a femtosecond laser. Langmuir **22** (2006): 4917–4919.
[39] Y. Shimotsuma, M. Sakakura, P. G. Kazansky, et al., Ultrafast manipulation of self-assembled form birefringence in glass. Adv. Mat.**22** (2010): 4039–4043.
[40] L. Sansoni, F. Sciarrino, G. Vallone, et al., Polarization entangled state measurement on a chip, Phys. Rev. Lett. **105** (2010): 200503.
[41] R. Osellame, H. J. W. M. Hoekstra, G. Cerullo and M. Pollnau, Femtosecond laser microstructuring: an enabling tool for optofluidic lab-on-chips, Laser Photon. Rev. **5**(3) (2011): 442–463.
[42] S. Kanehira, J. Si, J. Qiu, K. Fujita and K. Hirao, Periodic nanovoid structures via femtosecond laser irradiation. Nano Lett. **5** (2005): 1591–1595.
[43] Y. Shimotsuma, P. G. Kazansky, J. Qiu and K. Hirao, Self-organized nanogratings in glass irradiated by ultrashort light pulses. Phys. Rev. Lett. **91** (2003): 247405.
[44] Y. Liu, M. Shimizu, B. Zhu, et al., Micromodification of element distribution in glass using femtosecond laser irradiation. Opt. Lett. **34** (2009): 136–138.
[45] Y. Cheng, K. Sugioka, M. Masuda, et al., Optical gratings embedded in photosensitive glass by photochemical reaction using a femtosecond laser. Opt. Express **11** (2003): 1809–1816.
[46] W.Yang, P. G. Kazansky and Y. P. Svirko, Non-reciprocal ultrafast laser writing. Nat. Photon. **2** (2008): 99–104.

[47] X. Liu, D. Du and G. Mourou, Laser ablation and micromachining with ultrashort laser pulse, IEEE J. Quantum Electron, **33** (1997): 1706–1716.

[48] M. D. Shirk and P. A. Molian, A review of ultrashort pulsed laser ablation of materials. J. Laser Appl. **10**(1) (1998): 18–28.

[49] A. Plech, P. Leiderer and J. Boneberg, Femtosecond laser near field ablation. Laser Photon. Rev. **3**(5) (2009): 435–451.

[50] K. Sugioka and Y. Cheng, A tutorial on optics for ultrafast laser materials processing: basic microprocessing system to beam shaping and advanced focusing methods. Adv. Opt. Technol. **1**(5) (2012): 353–364.

[51] M. Duocastella and C. B. Arnold. Bessel and annular beams for materials processing. Laser Photon. Rev. **6**(5) (2012): 607–621.

[52] J. Fischer and M. Wegener, Three-dimensional optical laser lithography beyond the diffraction limit. Laser Photon. Rev. **7**(1) (2013): 22–44.

[53] G. Della Valle, R. Osellame and P. Laporta, Micromachining of photonic devices by femtosecond laser pulses. J. Opt. A: Pure and Applied Optics **11** (2009): 013001.

[54] S. Juodkazis, V. Mizeikis and H. Misawa, Three-dimensional microfabrication of materials by femtosecond lasers for photonics applications. J. Appl. Phys. **106**(5) (2009): 051101.

[55] M. Ams, G. D. Marshall, P. Dekker, et al., Investigation of Ultrafast Laser-Photonic Material Interactions: Challenges for Directly Written Glass Photonics. Selected Topics in Quantum Electronics, IEEE **14**(5) (2008): 1370–1381.

[56] M. Ams, G. D. Marshall, P. Dekker, et al., Ultrafast laser written active devices. Laser Photon. Rev. **3**(6) (2009): 535–544.

[57] A. Szameit and S. Nolte, Discrete optics in femtosecond-laser-written photonic structures. J. Phys. B: Atomic, Molecular and Optical Physics **43**(16) (2010): 163001.

[58] J. Thomas, C. Voigtlaender, R. G. Becker, et al., Femtosecond pulse written fiber gratings: a new avenue to integrated fiber technology. Laser Photon. Rev. **6**(6) (2012): 709–723.

[59] F. Chen and J. R. Vazquez de Aldana, Optical waveguides in crystalline dielectric materials produced by femtosecond-laser micromachining. Laser Photon. Rev., 2013.

[60] M. Masuda, K. Sugioka, Y. Cheng, et al., 3-D microstructuring inside photosensitive glass by femtosecond laser excitation. Appl. Phys. A **76**(5) (2003): 857–860.

[61] K. Sugioka, Y. Hanada and K. Midorikawa, Three-dimensional femtosecond laser micromachining of photosensitive glass for biomicrochips. Laser Photon. Rev. **4**(3) (2010): 386–400.

[62] K. Sugioka and Y. Cheng, Integrated microchips for biological analysis fabricated by femtosecond laser direct writing. MRS Bull. **36**(12) (2011): 1020–1027.

[63] R. Buividas, M. Mikutis and S. Juodkazis, Surface and bulk structuring of materials by ripples with long and short laser pulses: recent advances. Progress in Quantum Electronics, 38(3)(2014):119–156.

[64] R. Taylor, C. Hnatovsky and E. Simova, Applications of femtosecond laser induced self-organized planar nanocracks inside fused silica glass. Laser Photon. Rev. **2**(1–2) (2008): 26–46.

[65] E. Stratakis, A. Ranella, M. Farsari, et al., Laser-based micro/nanoengineering for biological applications. Progress in Quantum Electronics **33**(5) (2009): 127–163.

[66] A. Y. Vorobyev and C. Guo, Direct femtosecond laser surface nano/microstructuring and its applications. Laser Photon. Rev. **7**(3) (2013): 385–407.

[67] H. B. Sun and S. Kawata, Two-photon photopolymerization and 3D lithographic microfabrication. 169–273, Berlin Heidelberg, Springer, 2004.

[68] C. N. LaFratta, J. T. Fourkas, T. Baldacchini, et al., Multiphoton fabrication. Angew. Chem. intern. edn. **46**(33) (2007): 6238–6258.

[69] S. Maruo and J. T. Fourkas, Recent progress in multiphoton microfabrication. Laser Photon. Rev. **2**(1–2) (2008): 100–111.
[70] Y. L. Zhang, Q. D. Chen, H. Xia, et al., Designable 3D nanofabrication by femtosecond laser direct writing. Nano Today **5**(5) (2010): 435–448.
[71] M. Malinauskas, M. Farsari, A. Piskarskas, et al., Ultrafast laser nanostructuring of photopolymers: A decade of advances. Physics Reports **533**(1) (2013): 1–31.
[72] Z. Sekkat and S. Kawata, Laser nanofabrication in photoresists and azopolymers. Laser Photon. Rev. **8**(1) (2014): 1–26.
[73] A. Vogel and V. Venugopalan, Mechanisms of pulsed laser ablation of biological tissues. Chem.Rev. **103**(2) (2003): 577–644.
[74] A. Vogel, J. Noack, G. Hüttman, et al., Mechanisms of femtosecond laser nanosurgery of cells and tissues. Appl. Phys. B **81**(8) (2005): 1015–1047.
[75] C. L. Hoy, O. Ferhanoglu, M. Yildirim, et al., Clinical ultrafast laser surgery: recent advances and future directions. Selected Topics in Quantum Electronics, IEEE **20**(2) (2014): 1–14.
[76] H. Misawa and S. Juodkazis (eds.), 3D laser microfabrication: principles and applications, John Wiley and Sons, 2006.
[77] K. Sugioka, M. Meunier and A. Piqué, Laser precision microfabrication, Berlin, Springer, 2010.
[78] K. Sugioka and Y. Cheng (eds.), Ultrafast laser processing: from micro-to nanoscale, Singapore: Pan Standard Publishing Pte. Ltd., 2013.
[79] K. Sugioka and Y. Cheng, Fundamentals of Femtosecond Laser Processing. Femtosecond Laser 3D Micromachining for Microfluidic and Optofluidic Applications, 19–33. London, Springer, 2014.
[80] R. Osellame, G. Cerullo and R. Ramponi, Femtosecond Laser Micromachining: Photonic and Microfluidic Devices in Transparent Materials, 123, Springer, 2012.
[81] T. Brabec and F. Krausz, Intense few-cycle laser fields: Frontiers of nonlinear optics, Reviews of Modern Physics **72**(2) (2000): 545.
[82] M. DiDomenico, Small-Signal Analysis of Internal (Coupling-Type) Modulation of Lasers, J. Appl. Phys. **35**(10) (1964): 2870–2876.
[83] A. Yariv, Internal modulation in multimode laser oscillators, J. Appl. Phys. **36**(2) (1965): 388–391.
[84] D. E. Spence, P. N. Kean and W. Sibbett, 60-fsec pulse generation from a self-mode-locked Ti: sapphire laser, Opt. Lett. **16**(1) (1991): 42–44.
[85] U. Keller, G. W. Thooft, W. H. Knox, et al., Femtosecond pulses from a continuously self-starting passively mode–locked Ti: sapphire laser, Opt. Lett. **16**(13) (1991): 1022–1024.
[86] F. Salin, M. Piché and J. Squier, Mode locking of Ti:Al2O3 lasers and self-focusing: a Gaussian approximation, Opt. Lett. **16**(21) (1991): 1674–1676.
[87] P.Moulton, Ti-doped sapphire: a tunable solid–state laser, Opt.News **8**(6) (1982): 9.
[88] R. L. Fork, C. V. Shank and R. T. Yen, Amplification of 70-fs optical pulses to gigawatt powers, Appl. Phys. Lett. **41**(3) (1982): 223–225.
[89] R. Szipöcs, C. Spielmann, F. Krausz, et al., Chirped multilayer coatings for broadband dispersion control in femtosecond lasers, Opt. Lett. **19**(3) (1994): 201–203.
[90] A. Arai, J. Bovatsek, F. Yoshino, et al., Fiber chirped pulse amplification system for micromachining, Proc. SPIE. 6343 (2006): 63430S
[91] J. Kleinbauer, D. Eckert, S. Weiler and D. H. Sutter, 80 W ultrafast CPA-free disk laser, Proc. SPIE. 6871 (2008): 68711B.
[92] S. V. Marchese, C. R. Baer, A. G. Engqvist, et al., Femtosecond thin disk laser oscillator with pulse energy beyond the 10–microjoule level, Opt. Express **16** (2008): 6397–6407.
[93] TOPTICA Photonics, Ultrafast Fiber Lasers. http://www.toptica.com/

[94] J. C. Diels and W. Rudolph, Ultrashort laser pulse phenomena, Beijing: Academic Press, 2006.
[95] S. Backus, C. G. Durfee, M. M. Murnane, et al., High power ultrafast lasers, Rev. Sci. Instrum. **69**(3) (1998): 1207–1223.
[96] A. M. Weiner, Femtosecond pulse shaping using spatial light modulators, Rev. Sci. Instrum. **71**(5) (2000): 1929–1960.
[97] C. Iaconis and I. A. Walmsley, Spectral phase interferometry for direct electric-field reconstruction of ultrashort optical pulses, Opt. Lett. **23**(10) (1998): 792–794.
[98] D. J. Kane and R. Trebino, Characterization of arbitrary femtosecond pulses using frequency–resolved optical gating, Quantum Electronics, IEEE **29**(2) (1993): 571–579.
[99] R. Trebino, Frequency–Resolved Optical Gating: The Measurement of Ultrashort Laser Pulses: The Measurement of Ultrashort Laser Pulses, Berlin: Springer, 2000.
[100] S. K. Sundaram and E. Mazur, Inducing and probing non–thermal transitions in semiconductors using femtosecond laser pulses, Nat. Mater. **1** (2002): 217–224.
[101] B. C. Stuart, M. D. Feit, S. Herman, et al., Nanosecond-to-femtosecond laser–induced breakdown in dielectrics, Phys. Rev. B **53** (1996): 1749–1761.
[102] L. V. Keldysh, Ionization in the field of a strong electromagnetic wave, Sov. Phys. JETP **20**(5) (1965): 1307–1314.
[103] A. Gleizes, J. J. Gonzalez and P. Freton, Thermal plasma modeling, J. Phys. D: Appl. Phys. **38** (2005): R153–R183.
[104] S. I. Anisimov and B. Rethfeld, Theory of ultrashort laser pulse interaction with a metal, Proc. SPIE. **3093** (1997): 192–203.
[105] P. B. Corkum, F. Brunel, N. K. Sherman, et al., Thermal response of metals to ultrashort pulse laser excitation, Phys. Rev. Lett. **61** (1988): 2886–2889.
[106] M. Fujita and M. Hashida, Applications of femtosecond lasers, Oyo Buturi (Japan) **73** (2004): 178–185.
[107] A. P. Joglekar, H. H. Liu, E. Meyhöfer, G. Mourou and A. J. Hunt, Optics at critical intensity: Applications to nanomorphing, Proc. Natl. Acad. Sci. USA **101** (2004): 5856–5861.
[108] Y. Liao, Y. Shen, L. Qiao, et al., Femtosecond laser nanostructuring in porous glass with sub-50 nm feature sizes, Opt. Lett. **38** (2013): 187–189.
[109] D. Tan, Y. Li, F. Qi, et al., Reduction in feature size of two-photon polymerization using SCR500, Appl. Phys. Lett. **90** (2007): 071106.
[110] J. Xu, Y. Liao, H. D. Zeng, et al., Selective metallization on insulator surfaces with femtosecond laser pulses, Opt. Express **15** (2007): 12743–12748.
[111] Y. Liao, J. Xu, Y. Cheng, et al., Electro-optic integration of embedded electrodes and waveguides in LiNbO3 using a femtosecond laser, Opt. Lett. **33** (2008): 2281–2283.
[112] F. He, H. Sun, M. Huang, et al., Rapid fabrication of optical volume gratings in Foturan glass by femtosecond laser micromachining, Appl. Phys. A **97** (2009): 853–857.
[113] R. Stoian, M. Boyle, A. Thoss, et al., Laser ablation of dielectrics with temporally shaped femtosecond pulses, Appl. Phys. Lett. **80** (2002): 353–355.
[114] S. M. Eaton, H. Zhang, M. L. Ng, et al., Transition from thermal diffusion to heat accumulation in high repetition rate femtosecond laser writing of buried optical waveguides, Opt. Express **16** (2008): 9443–9458.
[115] Y. Wu, C. Y. Wang, W. Jia, X. Ni, M. Hu and L. Chai, Influence of scanning velocity on femtosecond laser direct writing lines on FOTURAN glass, Chin. Opt. Lett. **6** (2008): 51–53.
[116] Y. Shimotsuma, P. G. Kazansky, J. Qiu and K. Hirao, Self–organized nanogratings in glass irradiated by ultrashort light pulses, Phys. Rev. Lett. 91(24)(2003):247405.
[117] P. G. Kazansky, H. Inouye, T. Mitsuyu, et al., Anomalous anisotropic light scattering in Ge–doped silica glass, Phys. Rev. Lett. **82** (1999): 2199–2202.

[118] P. S. Salter and M. J. Booth, Dynamic control of directional asymmetry observed in ultrafast laser direct writing, Appl. Phys. Lett. **101** (2012): 141109(1–4).
[119] H. Zhang, S. M. Eaton and P. R. Herman, Single–step writing of Bragg grating waveguides in fused silica with an externally modulated femtosecond fiber laser, Opt. Lett. **32** (2007): 2559–2561.
[120] S. Matsuo, T. Miyamoto, T. Tomita and S. Hashimoto, Applications of a microlens array and a photomask to the laser microfabrication of a periodic photopolymer rod array, Appl. Opt. **46** (2007): 8264–8267.
[121] T. Kondo, S. Juodkazis, V. Mizeikis, S. Matsuo and H. Misawa, Fabrication of three-dimensional periodic microstructures in photoresist SU-8 by phase-controlled holographic lithography, New. J. Phys. **8** (2006): 250.
[122] L. Li and J. T. Fourkas, Multiphoton polymerization, Mater. Today **10** (2007): 30–37.
[123] M. Straub, M. Afshar, D. Feili, H. Seidel and K. König, Periodic nanostructures on Si (100) surfaces generated by high-repetition rate sub-15 fs pulsed near-infrared laser light, Opt. Lett. **37** (2012): 190–192.
[124] I. A. Walmsley and C. Dorrer, Characterization of ultrashort electromagnetic pulses, Adv. Opt. Photon. **1** (2009): 308–437.
[125] A. Monmayrant, S. Weber and B. Chatel, A newcomer's guide to ultrashort pulse shaping and characterization, J. Phys. B: At. Mol. Opt. Phys. **43** (2010): 103001.
[126] P. G. Kazansky, W. Yang, E. Bricchi, et al., "Quill" writing with ultrashort light pulses in transparent materials, Appl. Phys. Lett. **90** (2007): 151120.
[127] P. Bowlan, U. Fuchs, R. Trebino and U. D. Zeitner, Measuring the spatiotemporal electric field of tightly focused ultrashort pulses with sub-micron spatial resolution, Opt. Express **16** (2008): 13663.
[128] H. F. Li, Z. Zhang, Q. Xu, et al., Characterizing ultrashort optical pulses using second-order nonlinear nanoprobes, Appl. Phys. Lett. **97** (2010): 261108.
[129] M. Mielke, D. Gaudiosi, K. Kim, et al., Ultrafast fiber laser platform for advanced materials processing, J. Laser Micro/Nanoeng. **5** (2010) 53–58.
[130] G. Cerullo, R. Osellame, S. Taccheo, et al., Femtosecond micromachining of symmetric waveguides at 1.5 μm by astigmatic beam focusing, Opt. Lett. **27** (2002): 1938–1940.
[131] Y. Cheng, K. Sugioka, K. Midorikawa, et al., Control of the cross-sectional shape of a hollow microchannel embedded in photostructurable glass by use of a femtosecond laser, Opt. Lett. **28** (2003): 55–57.
[132] M. Ams, G. Marshall, D. Spence and M. Withford, Slit beam shaping method for femtosecond laser direct-write fabrication of symmetric waveguides in bulk glasses, Opt. Express **13** (2005): 5676–5681.
[133] K. Sugioka, Y. Cheng, K. Midorikawa, F. Takase and H. Takai, Femtosecond laser microprocessing with three-dimensionally isotropic spatial resolution using crossed–beam irradiation, Opt. Lett. **31** (2006): 208–210.
[134] M. K. Bhuyan, P. K. Velpula, J. P. Colombier, et al., Single-shot high aspect ratio bulk nanostructuring of fused silica using chirp–controlled ultrafast laser Bessel beams, Appl. Phys. Lett. **104** (2014): 021107.
[135] F. He, H. Xu, Y. Cheng, et al., Fabrication of microfluidic channels with a circular cross section using spatiotemporally focused femtosecond laser pulses, Opt. Lett. **35** (2010): 1106–1108.
[136] D. Kim and P. T. So, High–throughput three-dimensional lithographic microfabrication, Opt. Lett. **35** (2010): 1602–1604.
[137] D. Oron, E. Tal and Y. Silberberg, Scanningless depth-resolved microscopy, Opt. Express **13**(5) (2005): 1468–1476.

[138] G. Zhu, J. van Howe, M. Durst, et al., Simultaneous spatial and temporal focusing of femtosecond pulses, Opt. Express **13**(6) (2005): 2153–2159.
[139] F. He, Y. Cheng, J. Lin, et al., Independent control of aspect ratios in the axial and lateral cross sections of a focal spot for three-dimensional femtosecond laser micromachining, New J. Phys. **13**(8) (2011): 083014.
[140] D. N. Vitek, D. E. Adams, A. Johnson, et al., Temporally focused femtosecond laser pulses for low numerical aperture micromachining through optically transparent materials, Opt. Express **18**(17) (2010): 18086–18094.
[141] D. N. Vitek, E. Block, Y. Bellouard, et al., Spatio-temporally focused femtosecond laser pulses for nonreciprocal writing in optically transparent materials, Opt. Express **18**(24) (2010): 24673–24678.
[142] Y. C. Li, L. C. Cheng, C. Y. Chang, et al., Fast multiphoton microfabrication of freeform polymer microstructures by spatiotemporal focusing and patterned excitation, Opt. Express **20**(17) (2012): 19030–19038.
[143] B. Zeng, W. Chu, H. Gao, et al., Enhancement of peak intensity in a filament core with spatiotemporally focused femtosecond laser pulses, Phys. Rev. A **84**(6) (2011): 063819.
[144] G. Li, J. Ni, H. Xie, et al. Second harmonic generation in centrosymmetric gas with spatiotemporally focused intense femtosecond laser pulses, Opt. Lett. **39**(4) (2014): 961–964.
[145] E. Papagiakoumou, F. Anselmi, A. Bègue, et al., Scanless two-photon excitation of channelrhodopsin-2, Nat. Methods **7**(10) (2010): 848–854.
[146] B. K. Andrasfalvy, B. V. Zemelman, J. Tang, et al., Two-photon single-cell optogenetic control of neuronal activity by sculpted light, Proc. Natl. Acad. Sci. U.S.A. **107**(26) (2010): 11981–11986.
[147] E. Block, M. Greco, D. Vitek, et al., Simultaneous spatial and temporal focusing for tissue ablation, Biomed. Opt. Express **4**(6) (2013): 831–841.
[148] Z. Wang, F. He, J. Ni, et al., Interferometric characterization of pulse front tilt of spatiotemporally focused femtosecond laser pulses, Opt. Express**22**(21) (2014): 26328–26337.
[149] F. He, B. Zeng, W. Chu, et al., Characterization and control of peak intensity distribution at the focus of a spatiotemporally focused femtosecond laser beam, Opt. Express **22**(8) (2014): 9734–9748.
[150] D. Oron and Y. Silberberg, Harmonic generation with temporally focused ultrashort pulses, JOSA. B **22**(12) (2005): 2660–2663.
[151] C. G. Durfee, M. Greco, E. Block et al., Intuitive analysis of space-time focusing with double-ABCD calculation, Opt. Express **20**(13) (2012): 14244–14259.
[152] E. Yew, C. J. Sheppard and P. T. So, Temporally focused wide-field two-photon microscopy: Paraxial to vectorial, Opt. Express **21**(10) (2013): 12951–12963.
[153] B. Sun, P. S. Salter and M. J. Booth, Effects of aberrations in spatiotemporal focusing of ultrashort laser pulses, JOSA A **31**(4) (2014): 765–772.
[154] H. Dana and S. Shoham, Numerical evaluation of temporal focusing characteristics in transparent and scattering media, Opt. Express **19**(6) (2011): 4937–4948.
[155] T. Xi, Z. Zhao and Z. Hao, Filamentation of femtosecond laser pulses with spatial chirp in air, JOSA B **31**(2) (2014): 321–324.
[156] S. Akturk, X. Gu, E. Zeek, et al., Pulse-front tilt caused by spatial and temporal chirp, Opt. Express **12**(19) (2004): 4399–4410.
[157] M. E. Durst, G. Zhu and C. Xu, Simultaneous spatial and temporal focusing for axial scanning, Opt. Express **14**(25) (2006) 12243–12254.
[158] M. E. Durst, G. Zhu and C. Xu, Simultaneous spatial and temporal focusing in nonlinear microscopy, Opt. Commun. **281**(7)(2008): 1796–1805.
[159] A. Brodeur and S. L. Chin, Ultrafast white-light continuum generation and self-focusing in transparent condensed media, J. Opt. Soc. Am. B **16** (1999): 637–650.

[160] M. Beresna, P. G. Kazansky, Y. Svirko, M. Barkauskas and R. Danielius, High average power second harmonic generation in air, Appl. Phys. Lett. **95** (2009): 121502.
[161] M. Mori, E. Takahashi and K. Kondo, Image of second harmonic emission generated from ponderomotively excited plasma density gradient, Phys. Plasmas **9** (2002): 2812.
[162] N. Bloembergen, R. K. Chang, S. S. Jha and C. H. Lee, Optical Second-Harmonic Generation in Reflection from Media with Inversion Symmetry, Phys. Rev. **174** (1968): 813–822.
[163] A. Manz, N. Graber and H. M. Widmer, Miniaturized total chemical analysis systems: A novel concept for chemical sensing, Sens. Actuators, B **1** (1990): 244–248.
[164] G. M. Whitesides, The origins and the future of microfluidics, Nature **442** (2006): 368–373.
[165] Y. Bellouard, A. Champion, B. Lenssen, et al., The femtoprint project, J. Laser Micro/Nanoeng. **7** (2012): 1–10.
[166] F. He, Y. Liao, J. Lin et al., Femtosecond laser fabrication of monolithically integrated microfluidic sensorsin glass, Sensors **14** (2014): 19402–19440.
[167] A. Marcinkevicius, S. Juodkazis, M. Watanabe, et al., Femtosecond laser-assisted three-dimensional microfabrication in silica, Opt. Lett. **26** (2001): 277–279.
[168] S. Matsuo, S. Kiyama, Y. Shichijo, et al., Laser microfabrication and rotation of ship-in-a-bottle optical rotators, Appl. Phys. Lett. **93** (2008): 051107.
[169] M. Masuda, K. Sugioka, Y. Cheng, et al., Direct fabrication of freely movable microplate inside photosensitive glass by femtosecond laser for lab-on-chip application, Appl. Phys. A **78** (2004): 1029–1032.
[170] Y. Li, K. Itoh, W. Watanabe, et al., Three-dimensional hole drilling of silica glass from the rear surface with femtosecond laser pulses, Opt. Lett. **26** (2001): 1912–1914.
[171] K. Ke, E. F. Hasselbrink and A. J. Hunt, Rapidly prototyped three-dimensional nanofluidic channel networks in glass substrates, Anal. Chem. **77** (2005): 5083–5088.
[172] T. H. Elmer, Porous and Reconstructed Glasses, In Engineered Materials Handbook; ASM International: Metals Park, OH, USA, **4** (1992): 427–432.
[173] Y. Liao, Y. Ju, L. Zhang, et al., Three-dimensional microfluidic channel with arbitrary length and configuration fabricated inside glass by femtosecond laser direct writing, Opt. Lett. **35** (2010): 3225–3227.
[174] Y. Liao, J. Song, E. Li, et al., Rapid prototyping of three-dimensional microfluidic mixers in glass by femtosecond laser direct writing, Lab Chip **12** (2012): 746–749.
[175] Y. Liao, Y. Cheng, C. Liu, et al., Direct laser writing of sub-50 nm nanofluidic channels buried in glass for three-dimensional micro-nanofluidic integration, Lab Chip **13** (2013): 1626–1631.
[176] Y. Bellouard, A. Said and P. Bado, Integrating optics and micro-mechanics in a single substrate: a step toward monolithic integration in fused silica. Opt. Express, **13** (2005): 6635–6644.
[177] S. Matsuo, S. Kiyama, Y. Shichijo, et al., Laser microfabrication and rotation of ship-in-a-bottle optical rotators. Appl. Phys. Lett. **93** (2008): 051107.
[178] Y. Ju, Y. Liao, L. Zhang, et al., Fabrication of large-volume microfluidic chamber embedded in glass using three-dimensional femtosecond laser micromachining, Microfluid. Nanofluid. **11** (2011): 111–117.
[179] Y. Bellouard, A. Said, M. Dugan, et al., Fabrication of high-aspect ratio, micro-fluidic channels and tunnels using femtosecond laser pulses and chemical etching, Opt. Express **12** (2004): 2120–2129.
[180] C. Hnatovsky, R. S. Taylor, E. Simova, et al., Polarization-selective etching in femtosec-ond laser-assisted microfluidic channel fabrication in fused silica, Opt. Lett. **30** (2005): 1867–1869
[181] V. R. Bhardwaj, E. Simova, P. P. Rajeev, et al., Optically produced arrays of nano-planes inside fused silica, Phys. Rev. Lett. **96** (2006): 057404(4).

[182] S. Kiyama, S. Matsuo, S. Hashimoto and Y. Morihira, Examination of etching agent and etching mechanism on femotosecond laser microfabrication of channels inside vitreous silica substrates, J. Phys. Chem. C **113** (2009): 11560–11566.
[183] M. Hermans, J. Gottmann and F. Riedel, Selective, Laser-Induced Etching of Fused Silica at High Scan-Speeds Using KOH, J. Laser Micro Nanoeng. **9** (2014): 126–131.
[184] J. Song, X. Wang, X. Hu, et al., Formation mechanism of self-organized voids in dielectrics induced by tightly focused femtosecond laser pulses, Appl. Phys. Lett. **92** (2008): 092904.
[185] D. J. Hwang, T. Y. Choi and C. P. Grigoropoulos, Liquid-assisted femtosecond laser drilling of straight and three-dimensional microchannels in glass, Appl. Phys. A **79** (2004): 605–612.
[186] P. R. Herman, A. Oettl, K. P. Chen and R. S. Marjoribanks, Laser micromachining of transparent fused silica with 1-ps pulses and pulse trains. Proc. SPIE. **3616** (1999): 148–155.
[187] J. R. Anderson, D. T. Chiu, R. J. Jackman, et al., Fabrication of topologically complex three-dimensional microfluidic systems in PDMS by rapid prototyping, Anal. Chem. **72** (2000): 3158–3164.
[188] J. J. Zhu and X. C. Xuan, Particle electrophoresis and dielectrophoresis in curved microchannels, J. Colloid Interface Sci. **340** (2009): 285–290.
[189] A. L. Garcia, L. K. Ista, D. N. Petsev, et al., Electrokinetic molecular separation in nanoscale fluidic channels, Lab Chip **11** (2005): 1271–1276.
[190] A. Abbott, Cell culture: Biology's new dimension, Nature **424** (2003): 870–872.
[191] N. T. Nguyen and Z. G. Wu, Micromixers-A review, J. Micromech. Microeng. 15 (2005): R1, doi:10.1088/0960-1317/15/2/R01.
[192] A. D. Stroock, S. K. Dertinger, A. Ajdari, et al., Chaotic mixer for microchannels, Science **295** (2002): 647–651.
[193] D. Therriault, S. R. White and J. A. Lewis, Chaotic mixing in three-dimensional microvascular networks fabricated by direct-write assembly, Nat. Mater. **2** (2003): 265–271.
[194] T. Yasui, Y. Omoto, K. Osato, et al., Microfluidic baker's transformation device for three-dimensional rapid mixing, Lab Chip **11** (2011): 3356–3360.
[195] S. Wiggins and J. M. Ottino, Foundations of chaotic mixing, Philos. Trans. R. Soc. London Ser. A **362** (2004): 937–970.
[196] P. Carrière, On a three-dimensional implementation of the baker's transformation, Phys. Fluids **19** (2007): 118110.
[197] P. Abgrall and N. T. Nguyen, Nanofluidic devices and their applications, Anal. Chem. **80** (2008): 2326–2341.
[198] M. L. Kovarik and S. C. Jacobson, Nanofluidics in Lab-on-a-Chip Devices, Anal. Chem. **81** (2009): 7133–7140.
[199] T. C. Kuo, D. M. Cannon, M. A. Shannon, P. W. Bohn and J. V. Sweedler, Hybrid threedimensional nanofluidic/microfluidic devices using molecular gates, Sens. Actuators A **102** (2003): 223–233.
[200] S. Jeon, V. Malyarchuk, J. O. White and J. A. Rogers, Optically fabricated three dimensional nanofluidic mixers for microfluidic devices, Nano Lett. **5** (2005): 1351–1356.
[201] S. Lee, R. An and A. J. Hunt, Liquid glass electrodes for nanofluidics, Nat. Nanotechnol. **5** (2010): 412–416.
[202] F. Korte, J. Serbin, J. Koch, et al., Towards nanostructuring with femtosecond laser pulses, Appl. Phys. A **77** (2003): 229–235.
[203] S. Juodkazis, K. Nishimura, S. Tanaka, et al., Laser-induced microexplosion confined in the bulk of a sapphire crystal: evidence of multimegabar pressures, Phys. Rev. Lett. **96** (2006): 166101.
[204] J. O. Tegenfeldt, C. Prinz, H. Cao, et al., Micro-and nanofluidics for DNA analysis, Anal. Biochem. **378** (2004): 1678–1692.

[205] F. He, Y. Cheng, Z. Xu, et al., Direct fabrication of homogeneous microfluidic channels embedded in fused silica using a femtosecond laser, Opt. Lett. **35** (2010): 282–284.
[206] F. He, Y. Cheng, L. Qiao, et al., Two-photon fluorescence excitation with a microlens fabricated on the fused silica chip by femtosecond laser micromachining, Appl. Phys. Lett. **96** (2010): 041108.
[207] F. He, J. Lin and Y. Cheng, Fabrication of hollow optical waveguides in fused silica by three-dimensional femtosecond laser micromachining, Appl. Phys. B **105** (2011): 379–384.
[208] J. Lin, S. Yu, Y. Ma, et al., On-chip three-dimensional high-Q microcavities fabricated by femtosecond laser direct writing, Opt. Express **20** (2012): 10212–10217.
[209] Y. Cheng, K. Sugioka, K. Midorikawa, et al., Three-dimensional micro-optical compo-nents embedded in photosensitive glass by a femtosecond laser, Opt. Lett. **28** (2003): 1144–1146.
[210] Y. Cheng, H. L. Tsai, K. Sugioka, et al., Fabrication of 3D microoptical lenses in photo-sensitive glass using femtosecond laser micromachining, Appl. Phys. A **85** (2006): 11–14.
[211] Z. Wang, K. Sugioka and K. Midorikawa, Three-dimensional integration of microoptical components buried inside photosensitive glass by femtosecond laser direct writing, Appl. Phys. A **89** (2007): 951–955.
[212] L. Qiao, F. He, Y. Cheng, et al., A microfluidic chip integrated with a microoptical lens fabricated by femtosecond laser micromachining, Appl. Phys. A **102** (2011): 179–183.
[213] R. Guo, S. Z. Xiao, X. M. Zhai, et al., Micro lens fabrication by means of femtosecond two photon photopolymerization, Opt. Express **14** (2006): 810–816.
[214] L. Qiao, F. He, C. Wang, et al., Fabrication of a micro-optical lens using femtosecond laser 3D micromachining for two-photon imaging of bio-tissues, Opt. Commun. **284** (2011): 2988–2991.
[215] B. Schenkel, J. Biegert, U. Keller, et al., Generation of 3.8-fs pulses from adaptive compression of a cascaded hollow fiber supercontinuum, Opt. Lett. **28** (2003): 1987.
[216] M. Nisoli, S. D. Silvestri and O. Svelto, Generation of high energy 10 fs pulses by a new pulse compression technique, Appl. Phys. Lett. **68** (1996): 2793.
[217] A. Suda, M. Hatayama, K. Nagasaka and K. Midorikawa, Generation of sub-10-fs, 5-mJ-optical pulses using a hollow fiber with a pressure gradient, Appl. Phys. Lett. **86** (2005): 111116.
[218] A. Paul, R. A. Bartels, R. Tobey, et al., Quasi-phase-matched generation of coherent extreme-ultraviolet light, Nature **421** (2003): 51.
[219] M. Zepf, B. Dromey, M. Landreman, P. Foster and S. M. Hooker, Bright Quasi-Phase-Matched Soft-X-Ray Harmonic Radiation from Argon Ions, Phys. Rev. Lett. **99** (2007): 143901.
[220] E. Esarey, C. B. Schroeder and W. P. Leemans, Physics of laser-driven plasma-based electron accelerators, Rev. Mod. Phys. **81** (2009): 1229.
[221] W. P. Leemans, B. Nagler, A. J. Gonsalves, et al., GeV electron beams from a centimetre-scale accelerator, Nat. Phys. **2** (2006): 696.
[222] S. M. Spillane, T. J. Kippenberg and K. J. Vahala, Ultralow-threshold Raman laser using a spherical dielectric microcavity, Nature **415** (2002): 621–623.
[223] P.Del'Haye, A. Schliesser, O. Arcizet, T. Wilken, R. Holzwarth and T. J. Kippenberg, Optical frequency comb generation from a monolithic microresonator, Nature **450** (2007): 1214–1217.
[224] T. Carmon and K. J. Vahala, Visible continuous emission from a silica microphotonic device by third-harmonic generation, Nat. Phys. **3** (2007): 430–435.
[225] D. W. Vernooy, A. Furusawa, N. Ph. Georgiades, V. S. Ilchenko and H. J. Kimble, Cavity QED with high-Q whispering gallery modes, Phys. Rev. A **57** (1998):R2293–R2296.
[226] T. Aoki, B. Dayan, E. Wilcut, et al., Observation of strong coupling between one atom and a monolithic microresonator, Nature **443** (2006): 671–674.
[227] E. Peter, P. Senellart, D. Martrou, et al., Exciton-photon strong-coupling regime for a single quantum dot embedded in a microcavity, Phys. Rev. Lett. **95** (2005): 067401.

[228] F. Vollmer and S. Arnold, Whispering-gallery-mode biosensing: label-free detection down to single molecules, Nat. Methods **5** (2008): 591–596.
[229] J. Zhu, S. K. Ozdemir, Y.-F. Xiao, et al., On-chip single nanoparticle detection and sizing by mode splitting in an ultrahigh-Q microresonator, Nat. Photon. **4** (2010): 46.
[230] S. L. McCall, A. F. J. Levi, R. E. Slusher, S. J. Pearton and R. A. Logan, Whispering-gallery mode microdisk lasers, Appl. Phys. Lett. **60** (1992): 289–291.
[231] D. K. Armani, T. I. Kippenberg, S. M. Spillane and K. J. Vahala, Ultra-high-Q toroid microcavity on a chip, Nature **421** (2003): 925–928.
[232] C. Gmachl, F. Capasso, E. E. Narimanov, et al., High-power directional emission from microlasers with chaotic resonators, Science **280** (1998): 1556–1564.
[233] A. Serpengüzel, S. Arnold and G. Griffel, Excitation of resonances of microspheres on an optical fiber, Opt. Lett. **20** (1995): 654–656.
[234] J. Lin, Y. Xu, J. Song, et al., Low-threshold whispering-gallery-mode microlasers fabricated in a Nd:glass substrate by three-dimensional femtosecond laser micromachining, Opt. Lett. **38** (2013): 1458–1460.
[235] P. S. Kuo, J. Brevo-Abad and G. S. Solomon, Second-harmonic generation using-quasi-phase-matching in a GaAs whispering-gallery-mode microcavity, Nature Commun. **5** (2014): 3109.
[236] V. S. Ilchenko, A. A. Savchenkov, A. B. Matsko and L. Maleki, Nonlinear optics and crystalline whispering gallery mode cavities, Phys. Rev. Lett. **92** (2004): 043903.
[237] M. Förtsch, J. U. Fürst, C. Wittmann, et al., A versatile source of single photons for quantum information processing, Nature Commun. **4** (2013): 1818.
[238] A. Guarino, G. Poberaj, D. Rezzonico, R. Degl' Innocenti and P. Günter, Electro-optically tunable microring resonators in lithium niobate, Nat. Photon **1** (2007): 407–410.
[239] A. A. Savchenkov, V. S. Ilchenko, A. B. Matsko and L. Maleki, Kilohertz optical resonances in dielectric crystal cavities, Phys. Rev. A **70** (2004): 051804(R).
[240] T.-J. Wang, J.-Y. He, C.-A. Lee and H. Niu, High-quality LiNbO3 microdisk resonators by undercut etching and surface tension reshaping, Opt. Express **20** (2012): 28119–28124.
[241] J. Lin, Y. Xu, Z. Fang, et al., Fabrication of high-Q lithium niobate microresonators using femtosecond laser micromachining, Sci. Rep. **5** (2015): 8072.
[242] J. Lin, Y. Xu, J. Tang, et al., Fabrication of three-dimensional microdisk resonators in calcium fluoride by femtosecond laser micromachining, Appl. Phys. A **116** (2014): 2019.
[243] L. Holland, Vacuum Deposition of Thin Films, Chapman and Hall, London, 1963.
[244] Y. Nakajima, K. Kusuyama, H. Yamaguchi et al., Growth of single-crystal aluminium films on silicon substrates by DC magnetron sputtering, Jpn J. Appl. Phys. **31** (1992): 1860–1867.
[245] A. Jain, K. M. Chi, T. T. Kodas et al., Chemical vapor deposition of copper from hexafluoroacetylacetonato copper(I) – vinyltrimethylsilane deposition rates, mechanism, selectivity, morphology and resistivity as a function of temperature and pressure, J. Electrochem. Soc. **140** (1993): 1434–1439.
[246] J. Zhang, K. Sugioka and K. Midorikawa, Direct fabrication of microgratings in fused quartz by laser-induced plasma-assisted ablation with a KrF excimer laser, Opt. Lett. **23** (1999): 1486–1488.
[247] Y. Hanada, K. Sugioka, Y. Gomi et al., Development of practical system for laser-induced plasma-assisted ablation (LIPAA) for micromachining of glass materials, Appl. Phys. A **79** (2004): 1001–1003.
[248] F. J. Adrian, J. Bohandy, B. F. Kim et al., A study of the mechanism of metal-deposition by the laser-induced forward transfer, J. Vac. Sci. Technol. **5** (1987): 1490–1494.
[249] H. Esrom, J. Zhang, U. Kogelschatz et al., New approach of a laser-induced forward transfer for deposition of patterned thin metal films, Appl. Surf. Sci. **86** (1995): 202–207.

[250] A. P. Joglekar, H. Liu, E. Meyhöfer, G. Mourou and A. J. Hunt, Optics at critical intensity: Applications to nanomorphing, Proc. Natl. Acad. Sci. USA **101** (2004): 5856–5861.
[251] Y. Y. Cao, N. Takeyasu, T. Tanaka, X. M. Duan and S. Kawata, 3D metallic nanostructure fabrication by surfactant-assisted multiphoton-induced reduction, Small **5** (2009): 1144–1148.
[252] J. Xu, Y. Liao, H. D. Zeng, et al., Mechanism study of femtosecond laser induced selective metallization (FLISM) on glass surfaces, Opt. Commun. **281** (2008): 3505–3509.
[253] Y. Liao, J. Xu, H. Y. Sun, et al., Fabrication of microelectrodes deeply embedded in LiNbO3 using a femtosecond laser, Appl. Surf. Sci. **254** (2008): 7018–7021.
[254] Z. Zhou, J. Xu, F. He et al., Surface-enhanced Raman scattering substrate fabricated by femtosecond laser induced co-deposition of silver nanoparticles and fluorescent molecules, Jpn. J. Appl. Phys. **49** (2010): 022703.
[255] Z. Zhou, J. Xu, Y. Liao, et al., Fabrication of an integrated Raman sensor by selective surface metallization using a femtosecond laser oscillator. Opt. Commun. **282** (2009): 1370–1373.
[256] V. M. Dubin, Y. Shacham-Diamand, B. Zhao, P. K. Vasudev and C. H. Ting, Selective and Blanket Electroless Copper Deposition for Ultralarge Scale Integration, J. Electrochem. Soc. **144** (1997): 898–908.
[257] T. Baldacchini, A. C. Pons, J. Pons, et al., Multiphoton laser direct writing of two-dimensional silver structures, Opt. Express **13** (2005): 1275–1280.
[258] T. Tanaka, A. Ishikawa and S. Kawata, Two-photon-induced reduction of metal ions for fabricating three dimensional electrically conductive metallic microstructure, Appl. Phys. Lett. **88** (2006): 081107.
[259] C. B. Schaffer, J. F. García and E. Mazur, Bulk heating of transparent materials using a high-repetitionrate femtosecond laser, Appl. Phys. A **76** (2004): 351–354.
[260] W. Watanabe, S. Onda, T. Tamaki, K. Itoh and J. Nishii, Space-selective laser joining of dissimilar transparent materials using femtosecond laser pulses, Appl. Phys. Lett. **89** (2006): 021106.
[261] G. A. Shafeev, Laser-assisted activation of dielectrics for electroless metal plating, Appl. Phys. A **67** (1998): 303–311.
[262] A. A. Mewe, E. S. Kooij and B. Poelsema, Seeded-growth approach to selective metallization of microcontact-printed patterns, Langmuir **22** (2006): 5584–5587.
[263] O. G. Ramer, Integrated optic electrooptic modulator electrode analysis, IEEE J. Quantum. Electron. **18** (1982): 386–392.
[264] E. L. Wooten, K. M. Kissa, A. Yi-Yan, et al., A review of lithium niobate modulators for fiber-optic communications systems, IEEE J. Sel. Top. Quant. **6** (2000): 69–82.
[265] K. Kordas, J. Bekesi, R. Vajtai, et al., Laser-Assisted Via Hole Metallization in PCB Materials, J. Electron. Mater. **30** (2001): L21-L24.
[266] Z. L. Wang, O. Yaegashi, H. Sakaue, T. Takahagi and S. Shingubara, Bottom-up fill for submicrometer copper via holes of ULSIs by electroless plating, J. Electrochem. Soc. **151** (2004): C781-C785.
[267] M. Fleischmann, P. J. Hendra and A. J. McQuillan, Raman spectra of pyridine adsorbed at a silver electrode, Chem. Phys. Lett. **26** (1974): 163–166.
[268] D. L. Jeanmaire and R.P. van Duyne, Surface Raman spectroelectrochemistry. Part 1.Heterocyclic, aromatic, and aliphatic amines adsorbed on the anodized silver electrode, J. Electroanal. Chem. **84** (1977): 1–20.
[269] M. G. Albrecht and J. A. Creighton, Anomalously intense raman spectra of pyridine at a silver electrode, J. Am. Chem. Soc. **99** (1977): 5215–5217.
[270] K. Kneipp, Y. Wang, H. Kneipp, et al., Single molecule detection using surface-enhanced raman scattering (SERS), Phys. Rev. Lett. **78** (1997): 1667–1670.

[271] S. Nie and S. R. Emory, Probing Single Molecules and Single Nanoparticles by Surface-Enhanced Raman Scattering, Science **275** (1997): 1102–1106.
[272] C. H. Lin, L. Jiang, Y. H. Chai, et al., One-step fabrication of nanostructures by femtosecond laser for surface-enhanced Raman scattering, Opt. Express **17** (2009): 21581–21589.
[273] X. Ma, H. Huo, W. Wang, et al., Surface-enhanced Raman scattering sensor on an optical fiber probe fabricated with a femtosecond laser, Sensors **10** (2010): 11064–11071.
[274] X. Lan, Y. Han, T. Wei, et al., Surface-enhanced Raman-scattering fiber probe fabricated by femtosecond laser, Opt. Lett. **34** (2009): 2285–2287.
[275] C. Monat, P. Domachuk and B. J. Eggleton, Integrated optofluidics: A new river of light, Nature Photon. **1** (2007): 106–114.
[276] A. Crespi, Y. Gu, B. Ngamsom, et al., Three-dimensional Mach–Zehnder interferometer in a microfluidic chip for spatially-resolved label-free detection, Lab Chip **10** (2010): 1167–1173.
[277] N. Bellini, K. C. Vishnubhatla, F. Bragheri, et al., emtosecond laser fabricated monolithic chip for optical trapping and stretching of single cells, Opt. Express **18**(18): 4679–4688.
[278] M. Kim, D. J. Hwang, H. Jeon, K. Hiromatsu, C. P. Grigoropoulos, Single cell detection using a glass-based optofluidic device fabricated by femtosecond laser pulses, Lab Chip **9** (2009): 311–318.
[279] D. Choudhury, W. T. Ramsay, R. Kiss, et al., A 3D mammalian cell separator biochip, Lab Chip **12** (2012): 948–953.
[280] Y. Cheng, K. Sugioka, K. Midorikawa, Microfluidic laser embedded in glass by three-dimensional femtosecond laser microprocessing, Opt. Lett. **29** (2004): 2007–2009.
[281] H. Sun, F. He, Z. Zhou, et al., Fabrication of microfluidic optical waveguides on glass chips with femtosecond laser pulses, Opt. Lett. **32** (2007): 1536–1538.
[282] L. Gui, B. Xu and T. C. Chong, Microstructure in lithium niobate by use of focused femtosecond laser pulses, IEEE Photon. Technol. Lett. **16** (2004): 1337–1339.
[283] J. Burghoff, C. Grebing, S. Nolte and A. Tünnermann, Efficient frequency doubling in femtosecond laser-written waveguides in lithium niobate, Appl. Phys. Lett. **89** (2006): 081108.
[284] J. Thomas, M. Heinrich, J. Burghoff, et al., Femtosecond laser-written quasi-phase-matched waveguides in lithium niobate, Appl. Phys. Lett. **91** (2007): 151108.
[285] L. N. Binh, Lithium niobate optical modulators Devices and applications, J. Cryst. Growth **288** (2006): 180–187.
[286] W. Z. Song, X. M. Zhang, A. Q. Liu, et al., Refractive index measurement of single living cells using on-chip Fabry–Perot cavity, Appl. Phys. Lett. **89** (2006): 203901.
[287] A. M. Armani, R. P. Kulkarni, S. E. Fraser, R. C. Flagan and K. J. Vahala, Label-free, single-molecule detection with optical microcavities, Science **317** (2007): 783–787.
[288] J. M. Ward, P. Féron and S. Nic Chormaic, A taper-fused microspherical laser source, IEEE Photon. Technol. Lett. **20** (2008): 392–394.
[289] F. Vollmer and L. Yang, Review Label-free detection with high-Q microcavities: a review of biosensing mechanisms for integrated devices, Nanophotonics **1** (2012): 267–291.
[290] J. Song, J. Lin, J. Tang, et al., Fabrication of an integrated high-quality-factor (high-Q) optofluidic sensor by femtosecond laser micromachining, Opt. Express **22** (2014): 14792–14802.
[291] G. M. Hale and M. R. Querry, Optical constants of water in the 200-nm to 200-μm wavelength region, Appl. Opt. **12** (1973): 555–563.
[292] I. M. White and X. Fan, On the performance quantification of resonant refractive index sensors, Opt. Express **16** (2008): 1020–1028.
[293] J. M. Ward and S. N. Chormaic, Thermo-optical tuning of whispering gallery modes in Er:Yb co-doped phosphate glass microspheres, Appl. Phys. B **100** (2010): 847–850.
[294] J. P. Rezac and A. T. Rosenberger, Locking a microsphere whispering-gallery mode to a laser, Opt. Express **8** (2001): 605–610.

[295] K. N. Dinyari, R. J. Barbour, D. A. Golter and H. Wang, Mechanical tuning of whispering gallery modes over a 0.5 THz tuning range with MHz resolution in a silica microsphere at cryogenic temperatures, Opt. Express **19** (2011): 17966–17972.

[296] D. Armani, B. Min, A. Martin and K. J. Vahala, Electrical thermo-optic tuning of ultrahigh- microtoroid resonators, Appl. Phys. Lett. **85** (2004): 5439.

[297] J. Xu, D. Wu, Y. Hanada, et al., Electrofluidics fabricated by space-selective metallization in glass microfluidic structures using femtosecond laser direct writing, Lab Chip **13** (2013): 4608–4616.

[298] V. Zorba, E. Stratakis, M. Barberoglou et al., Biomimetic artificial surfaces quantitatively reproduce the water repellency of a lotus leaf, Adv. Mater. **20** (2008): 4049–4054.

[299] H. K. Tönshoff, A. Ostendorf, S. Nolte, F. Korte and T. Bauer, Micromachining using femtosecond lasers, Proc SPIE **4088** (2000): 136–139.

[300] S. Weiler, High-power pico- and femtosecond lasers enable new applications, LaserFocus World **47** (2011): 55–63.

[301] M. Beresna, M. Gecevičius and P. G. Kazansky, Polarization sensitive elements fabricated by femtosecond laser nanostructuring of glass, Opt. Mater. Express **1** (2011): 783–795.

[302] J. Fischer, G. Freymann and M. Wegener, The Materials Challenge in Diffraction-Unlimited Direct-Laser-Writing Optical Lithography, Adv. Mater. **22** (2010): 3578–3582.

[303] A. Tünnermann, T. Schreiber and J. Limpert, Fiber lasers and amplifiers: an ultrafast performance evolution, Appl. Opt. **49** (2010): F71–F78.

[304] J. Gottmann, M. Hermans and J. Ortmann, Microcutting and hollow 3D microstructures in glasses by in-volume selective laser-induced etching (ISLE), J. Laser Micro/Nanoeng. **8** (2013): 15–18.

Derong Li, Xiaohua Lv, Qingming Luo, Shaoqun Zeng

4 Spatial and temporal broadening of a femtosecond laser pulse after angular dispersion

4.1 Introduction

The femtosecond (10^{-15} s) laser has attracted great interest for its ultrashort pulse duration and ultrahigh peak power since its invention in the 1980s [1–8], and is widely used in many fields, such as high energy physics, materials science, chemistry, and biomedical imaging [9–15].

Angular dispersion devices, such as gratings, prisms, or acousto-optical deflectors (AODs), can separate a beam into its component wavelengths, which then propagate in different directions. They can change the shape of the beam spot and the pulse width, and are widely used in many fields such as chirped pulse amplification (CPA), pulse shaping, and material dispersion compensation [16–28]. For example, a grating pair can be used for CPA [16], a prism pair can be used to compensate for the material dispersion inside or outside the laser cavity [17], or a grating pair combined with a pair of lenses and a mask can be used for femtosecond laser pulse reshaping [21]. In particular, the widely used acousto-optical devices, such as AODs [29–31], can be used either as beam scanners or to compensate for material dispersion. Additionally, the characteristics of AODs broaden the pulse of the femtosecond laser considerably, and the beam spot is also significantly distorted, which means that AOD scanning femtosecond lasers are difficult to use over a long period [32, 33]. Therefore, the study of the spatial and temporal broadening of femtosecond laser pulses after angular dispersion, which includes investigation of pulse broadening in the temporal domain and beam spot broadening in the spatial domain, is valuable not only for the development and complementarity of the transmission theory of ultrafast optics, but also for the specific application to femtosecond lasers.

Achieving the limit of the pulse duration is, in the field of ultrafast optics, challenging [1–8]. Therefore, we have focused on the pulse width of a dispersed femtosecond laser, as it is one of the most important parameters in our previous research. Current work in this area is prolific; for example, the pulse width broadening of a femtosecond laser after passing through a single angular dispersion element, a pulse stretcher/compressor, and the pulse width broadening due to residual dispersion in optical systems have been investigated [34–41]. However, there has been far less re-

Derong Li, Xiaohua Lv, Qingming Luo, Shaoqun Zeng (corresponding author), Huazhong University of Science and Technology, Britton Chance Center for Biomedical Photonics, Wuhan 430074, China

https://doi.org/10.1515/9783110304558-004

search on the beam spot shape of a dispersed femtosecond laser beam [42]. This is despite the fact that the spot shape of a femtosecond laser will seriously affect the beam quality, especially when the laser is focused by lenses and, in many applications, such as multiphoton microscopy and micromachining (in which the light sources are femtosecond oscillators or amplifiers), the distortion of the beam spot will reduce the resolution of optical imaging and the accuracy of micromachining [43–54]. For these reasons, the spot shape evolution of a dispersed femtosecond laser beam is an important area in ultrafast optics and worth detailed investigation.

Plane waves [16–26] and Gaussian beams [34–39] are two of the most commonly used models in the study of the propagation characteristics of femtosecond laser pulses. The plane wave is an idealized model as it ignores the beam's spot size and divergence. When the beam spot size is large (on the order of a few centimeters), using the plane wave model combined with the ray tracing method of geometric optics can greatly simplify the analysis of the propagation characteristics of lasers in optical systems. A Gaussian beam is more suitable for describing the characteristics of the laser source as it includes the beam spot size and divergence, and is a more accurate model compared to the plane wave. Using the Gaussian beam model combined with the diffraction integral [34–39] or optics matrix [55–59] method for analysis of the laser beam propagation characteristics, even though the calculation is more complex compared with that using the plane wave, gives results that are generally more accurate. In fact, the Gaussian beam, after some reasonable approximation, can completely degenerate into a plane wave. Therefore, the results calculated using a plane wave model can also be obtained using a Gaussian beam model.

In this chapter, we review the relative progress of investigations into spatial and temporal broadening of a femtosecond laser pulse after angular dispersion. In Section 4.2, we describe the definition of the term 'angular dispersion'. Sections 4.3 and 4.4 are the main content of this chapter; in these we give a detailed analysis of the pulse and beam spot broadening after angular dispersion, reveal the physical mechanisms of each type of broadening, and analyze current research. In Section 4.5, we discuss the applications and some new requirements for angular dispersion devices. It is worth noting that angular dispersion referred to in this chapter only includes first order angular dispersion; the effect of higher order angular dispersion on a femtosecond laser pulse is too complex [60] and not discussed here. In addition, in the analysis of pulse broadening we only consider up to second order dispersion (group delay dispersion, GDD); third and higher order dispersions are also neglected here. When the laser pulse duration is less than 50 fs, the effect of third order dispersion on pulse width is obvious and cannot be ignored [61], thus this chapter is only applicable when the laser pulse is longer than 50 fs.

4.2 Angular dispersion

What is angular dispersion? How does angular dispersion change when a femtosecond laser pulse propagates? Before studying the spatial and temporal profile of dispersed femtosecond laser pulses, we must clearly define angular dispersion, and give an accurate description of its properties. Varju et al. [62, 63] have theoretically and experimentally investigated this problem and their research is summarized in the following section.

4.2.1 Definition

When broad spectrum light (such as a femtosecond laser beam) passes through a wavelength dispersive device (such as a grating or prism), the spectral components are separated perpendicular to the propagation direction of the beam and propagate in different directions; this phenomenon is known as angular dispersion [64, 65]. There are two definitions of angular dispersion in classical optics: the first definition is based on the angle between the propagation directions of different components, and the second definition is based on the angle between the phase fronts of different components [64]. These two definitions have advantages in describing the spatial and temporal characteristics of a femtosecond laser pulse. For example, the first definition is suitable for describing beam spot broadening and spatial chirp [66, 67]; in contrast, the second definition is more suitable for describing group delay dispersion and pulse front tilt [68–73].

Because the phase front is always perpendicular to the propagation direction, the two definitions are identical for plane waves, as shown in Figure 4.1 (a). In addition, for a plane wave, regardless of which definition is adopted, the angular dispersion will always remain the same as the laser pulse propagates. However, for light waves with curved phase fronts, such as Gaussian beams, the two definitions are obviously different, as shown in Figure 4.1 (b). The phase front of a Gaussian beam is a curve, with a radius of curvature that changes as the pulse propagates. The angular disper-

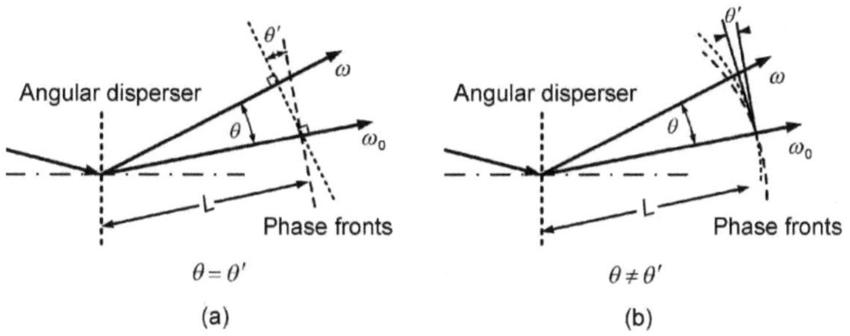

Fig. 4.1: Two definitions of angular dispersion: (a) the plane wave, (b) the Gaussian beam [40].

sion of a Gaussian beam is unchanged under the first definition (as for a plane wave); however, it is continuously changing as the pulse propagates under the second definition (different from that of plane wave) [62]. In Section 4.3, we show that it is exactly this difference in angular dispersion between the plane wave and the Gaussian beam that causes different spatial and temporal characteristics after these two kinds of laser pulses have passed through angular dispersers.

4.2.2 Angular dispersion in Gaussian beams

After the femtosecond laser pulse is dispersed, the angular dispersion parameter obtained using the Gaussian beam model (based on the second definition above) is far different from that of the plane wave model. Varju et al. analyzed the nature of the Gaussian beam, deduced the expression of the relationship between the angular dispersion of the Gaussian beam and the plane wave, and experimentally verified it using a method based on spectrally resolved interferometry [62, 63].

Varju et al.'s work described an expression of the relationship between the angular dispersion of the Gaussian beam and the plane wave [62]

$$\varepsilon' = \varepsilon \frac{d(d + \alpha^2 z) + z_R^2}{(d + \alpha^2 z)^2 + z_R^2}, \tag{4.1}$$

where ε' is the angle between the phase fronts of different spectral components in the Gaussian beam, ε is the angle between the propagation directions of different components (which is the same as that of the plane wave), d is the distance between the beam waist of the Gaussian beam and the disperser, and z is the distance between the disperser and the detector, as shown in Figure 4.2. $z_R = \pi w_0^2/\lambda$ is the Rayleigh length of the Gaussian beam, in which w_0 is the beam waist size, λ is the wavelength in a vac-

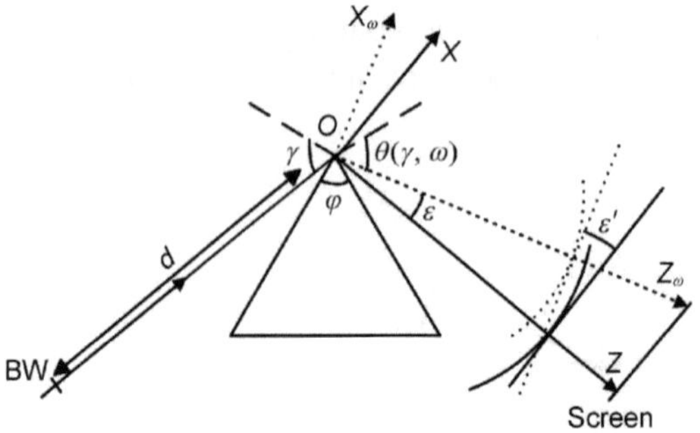

Fig. 4.2: Scheme of a Gaussian beam passing through a prism. The Z axis is defined as the direction of the central frequency ω_0, Z_ω is the direction of the ω component of the spectrum, ε is the angle from Z to Z_ω, ε' is the angle between the phase fronts, BW is the beam waist, and d is the distance between BW and the prism [37].

uum, and α is a parameter that is used to describe the characteristics of the angular dispersion element.

There is a factor $\eta = [d(d + \alpha^2 z) + z_R^2]/[(d + \alpha^2 z)^2 + z_R^2]$ in the expression for the angular dispersion of the Gaussian beam, as shown in equation (4.1). When the propagation distance is far less than the Rayleigh length of the Gaussian beam, i.e. d, $\alpha^2 z \ll z_R$ (which is also called the 'well collimation approximation' [35]), $\eta \approx 1$ and the angular dispersion of the Gaussian beam is approximately equal to that of the plane wave. When the propagation distance is far greater than the Rayleigh length, $\eta \approx 0$ and the angular dispersion of the Gaussian beam tends to 0. When the propagation distance is comparable with the Rayleigh length, $0 < \eta < 1$, which indicates that the angular dispersion of the Gaussian beam is always smaller than that of the plane wave (the absolute value) after the laser pulse is dispersed.

The characteristics of angular dispersion elements can be generally described by the first order derivative of the deflection angle with respect to the angle of incidence γ and the angular frequency [34, 35]

$$\Delta\theta(\gamma, \omega) = \alpha\Delta\gamma + \beta\Delta\omega = \frac{\partial\theta}{\partial\gamma}\Delta\gamma + \frac{\partial\theta}{\partial\omega}\Delta\omega , \tag{4.2}$$

where the parameter α is the angular magnification factor, and β, the first order derivative of the deflection angle with respect to the angular frequency, is the angular dispersion parameter in the first definition.

Varju et al. have experimentally verified the evolution of the angular dispersion of the Gaussian beam using a method based on spectrally resolved interferometry. Their measurements have a resolution of 0.2 µrad/nm. The experimental conditions are as follows [62]: femtosecond laser pulses emitted from a Ti:sapphire oscillator (pulse width 30 fs, repetition rate 76 MHz, spectral bandwidth 54 nm, central wavelength 800 nm) were focused by an achromatic lens ($f = 200$ mm) to produce a Gaussian beam with a beam waist of $w_0 = 120$ µm. The beam was then passed through a fused silica prism (apex angle 45°), changing the parameters d, z, and γ to provide different angular dispersions; a spectrally resolved interferometer was then used to measure the values of angular dispersion. As shown in Figure 4.3 (z is labeled as s in this figure), the experimental results agree well with the theoretical curves calculated from the Gaussian beam model, but differ from those of the plane wave model. It is also shown that the angular dispersion (the absolute value) of the Gaussian beam is always smaller than that of the plane wave under the same conditions (with the same angle of incidence γ and the same propagation distance s); thus the evolution of the angular dispersion of the Gaussian beam is completely verified by the experiments.

Varju et al. found the evolution of the angular dispersion of the Gaussian beam and verified it experimentally [62], but they did not investigate how the angular dispersion of the Gaussian beam broadens the pulse width and beam spot. In the next section we will see that it is exactly the difference in angular dispersion (as shown in equation (4.1)) that leads to the different pulse width evolution of the Gaussian beam and plane wave after passing through the angular dispersion element.

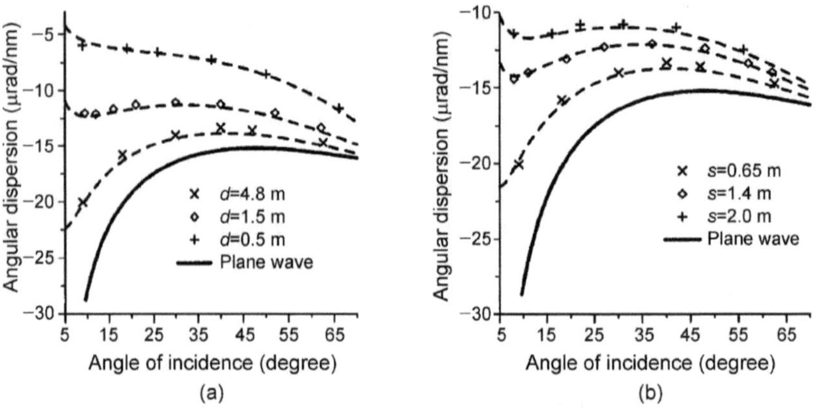

Fig. 4.3: Experimental verification of the evolution of the angular dispersion of the Gaussian beam. *d* is the distance between the beam waist and the prism, *s* is the distance between the prism and the detector, the solid curves are the theoretical values calculated using the plane wave model, the dashed curves are the theoretical values given by the Gaussian beam model, and the symbols represent the experimental results. In (a) $s = 0.6$ m, in (b) $d = 4.5$ m [62].

4.3 Temporal broadening of femtosecond laser pulses after dispersion

The pulse width broadening of dispersed femtosecond laser pulses is essential to the field of ultrafast optics, and the research on this subject is prolific. For example, the arrangement of a grating pair, proposed by Treacy [16], and a prism pair, proposed by Fork [17], were the classical methods for compensating for the material dispersion and then compressing the laser pulse. Martinez [34, 35] established an approach based on the Kirchhoff–Fresnel integral and Fourier transform, in turn based on the Gaussian beam model. He acquired the expression for the pulse width of the dispersed femtosecond laser under the well collimation approximation conditions (i.e. the distance between the beam waist and the angular dispersion element, and the propagation distance after passing through the angular dispersion element are far less than the Rayleigh length of the Gaussian beam). Horvath et al. [36] obtained the pulse shape and pulse width evolution of a Gaussian femtosecond laser pulse after it passed through an idealized grating by numerical simulations; they found that the pulse width tends to a constant and gave a physical explanation of the expression for pulse width when the propagation distance is much longer than the Rayleigh length of the Gaussian beam in the frequency domain. Osvay et al. [40] analyzed the effect of the residual angular dispersion on the pulse width in the CPA system using a plane wave model and experimentally verified their calculations. Zeng et al. [37–39] extended Martinez's work, based on the Gaussian beam model, and obtained the expression for the pulse width of a dispersed femtosecond laser at an arbitrary propagation distance without assuming that the Gaussian beam is well collimated. They experimentally verified their predictions, and also described the evolution of the spectral lateral walk-off and group delay dispersion, which are introduced by angular dispersion at an arbitrary propagation distance, thus revealing the physical mechanism of pulse width broadening by angular dispersion. Furthermore, Zeng et al. extended their analysis to the plane wave and spherical wave models obtained a general expression for the pulse width suitable for the plane wave, Gaussian beam, and spherical wave models, and revealed the relationship and distinction of the physical mechanism of pulse width broadening by angular dispersion between these three models. Table 4.1 summarizes the research on pulse width broadening of a dispersed femtosecond laser. In this section we will review this research, and analyze the relationships of the models.

4.3.1 Pulse broadening after angular dispersion: theoretical analysis

The plane wave model is generally used in rough calculations where the optical ray tracing method is first used to calculate the group delay dispersion introduced by the angular dispersion, and then to estimate the pulse width by the corresponding formula.

The group delay dispersion (GDD) introduced by angular dispersion in the plane wave is [23, 24]

$$\mathrm{GDD}_{\mathrm{PA}} = -k\beta^2 z\,, \tag{4.3}$$

where k is the wave number, β is the angular dispersion parameter from equation (4.2), and z is the propagation distance of the pulse. It is obvious that GDD of the dispersed plane wave is always negative and linearly proportional to the propagation distance. For the plane wave model, the pulse width broadening is completely determined by GDD, which is introduced by angular dispersion [23, 24]

$$\frac{\tau}{\tau_0} = \left[1 + (4\ln 2)^2 \frac{\mathrm{GDD}^2}{\tau_0^4}\right]^{\frac{1}{2}}, \tag{4.4}$$

where τ_0 is the original pulse width defined as the full width at half maximum (FWHM). As mentioned above, although the plane wave model is suitable for the case when the laser beam size is relatively large, more exact calculations need to use the Gaussian beam model. For this purpose, Martinez et al. deduced the expression for the electric field of dispersed Gaussian femtosecond laser pulses at an arbitrary propagation distance using the Kirchhoff–Fresnel integral [35]

$$A(x, z, \omega) = bE(\omega)\exp\left(-\mathrm{i}\frac{kx^2}{2z}\right)\exp\left[\mathrm{i}\frac{kzq(d)}{2q(d+\alpha^2 z)}\left(\frac{x^2}{z^2} + \beta^2\omega^2 + \frac{2x\beta\omega}{z}\right)\right], \tag{4.5}$$

where b is a constant describing the amplitude, $E(\omega)$ is the spectral distribution function of the femtosecond laser pulse, and q is a parameter describing the propagation characteristics of the Gaussian beam in space [74]. Supposing that the angular dispersion only occurs in the XZ plane, the propagation of the Gaussian beam in the

Tab. 4.1: Typical research related to the pulse width broadening of a dispersed femtosecond laser.

Group	Method	Approximations	Expression for pulse width and limited propagation distance	Main achievements
Martinez [34, 35] (1986)	Theoretical analysis based on the Gaussian beam model	Well collimated Gaussian beam	$\frac{\tau}{\tau_0} = \left[(1+u) + 4k^2z^2\beta^4/(1+u)\tau_0^4\right]^{1/2}$ Far less than the Rayleigh length	Obtained the expression for pulse width under well collimation approximation
Horvath et al. [36] (1993)	Numerical simulations based on the Gaussian beam model	The beam waist of the Gaussian beam is located in the entrance of the grating	$\frac{\tau}{\tau_0} = \left[1 + \frac{8\ln 2}{\tau_0^2}\left(\frac{\beta}{\Theta_0}\right)^2\right]^{1/2}$ Far greater than the Rayleigh length	Found the pulse width of the Gaussian beam after angular dispersion tends to a constant
Osvay et al. [40]	Theoretical analysis and experimental verification based on the plane wave model	Plane wave is an idealized model	$\frac{\tau}{\tau_0} = \left[1 + (4\ln 2)^2\frac{\mathrm{GDD}^2}{\tau_0^4}\right]^{1/2}$ Arbitrary propagation distance	Experimentally verified the pulse width broadening by residual angular dispersion in CPA system
Zeng et al. [37–39] (2007–2008)	Theoretical analysis and experimental verification based on the Gaussian beam model	None	$\frac{\tau}{\tau_0} = \left[(1+U) + \frac{V^2}{1+U}\right]^{1/2}$ Arbitrary propagation distance	Obtained the expression for pulse width at arbitrary propagation distance and experimentally verified it

YZ plane is equivalent to that in free space; thus the information from the Y axis is ignored in equation (4.5) to simplify the calculations. Martinez Fourier transformed equation (4.5) to the x–t domain, simplified the expression for the electric field under the well collimation approximation (i.e. d, $\alpha^2 z \ll z_R$), and collected the real parts of the t^2 terms to obtain an expression for the pulse width of a dispersed femtosecond Gaussian laser pulse [34]

$$\frac{\tau}{\tau_0} = \left[(1+u) + \frac{4k^2 z^2 \beta^4}{(1+u)\tau_0^4}\right]^{\frac{1}{2}}, \tag{4.6}$$

where $u = (2\alpha\beta z / w_0 \tau_0)^2$, which Martinez [34] called the spectral lateral walk-off. The different spectral components of the laser pulse are transversely (perpendicular to the propagation direction) spread by angular dispersion. Due to the finite beam size of the Gaussian beam, the observable spectral bandwidth of the laser pulse at each point decreases according to the basic theory of the Fourier transform, and a signal with a smaller bandwidth in the frequency domain must cause a greater duration in the time domain; thus the spectral lateral walk-off will undoubtedly broaden the pulse width of the Gaussian beam. The term $k\beta^2 z$ in equation (4.6) describes the effect of the group delay dispersion: different spectral components propagate in different directions, and this causes relative time delays between different spectral components in the longitudinal axis (i.e. the propagation direction of the laser pulse). Consequently, the entire envelope of the pulse is elongated, and the group delay dispersion will also broaden the pulse width of the Gaussian beam.

Martinez obtained the expression for the pulse width of the dispersed femtosecond Gaussian laser pulse under the well collimation approximation, and pointed out that the pulse width broadening is completely determined by two effects: spectral lateral walk-off and group delay dispersion [34]. He also gave the simulated results describing the pulse width change with propagation distance (with different original values) as shown in Figure 4.4, which shows the pulse width increasing with the propagation distance after angular dispersion. However, due to the use of the well collimation approximation, Martinez's results apply only to the case where the propagation distance of the laser pulse is far less than the Rayleigh length of the Gaussian beam;

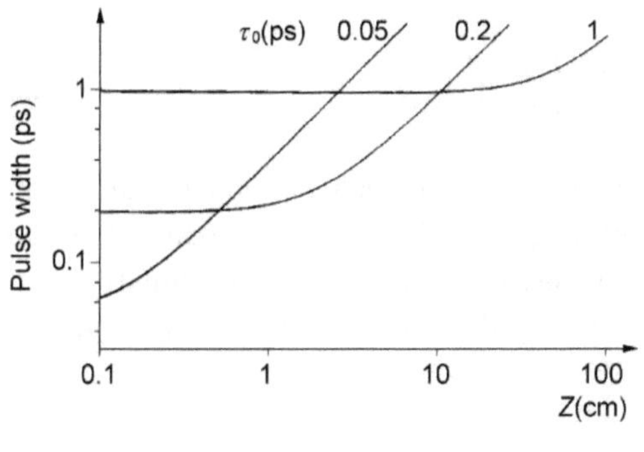

Fig. 4.4: The pulse width as a function of the propagation distance under the well collimation approximation; the three curves correspond to three different original pulse widths [34].

when the propagation distance is beyond this range, equation (4.6) is no longer valid. To investigate the pulse width broadening, as opposed to the theoretically deduced analytical expression, a numerical simulation is an effective method to provide the pulse width values of a femtosecond laser at certain positions, and thus can completely describe the evolution of the width of a dispersed femtosecond laser pulse. Horvath et al. plotted pulse shape at different propagation distances from numerical simulations assuming that the waist of the Gaussian beam is located exactly at the entrance of the grating, and then obtained pulse width as a function of the propagation distance for different angular dispersions (i.e. with different grating constants) (Figure 4.5) [36].

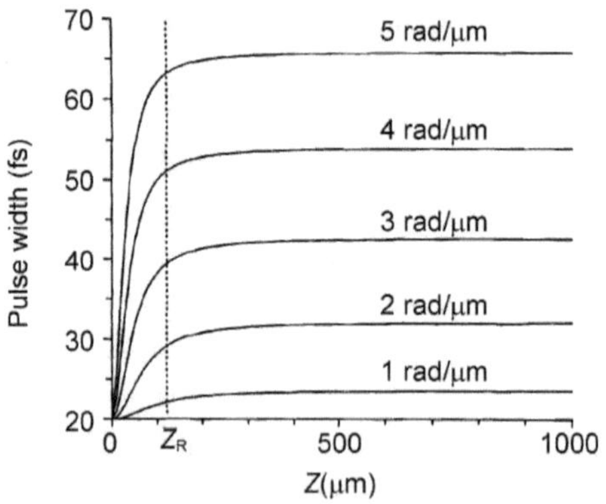

Fig. 4.5: Pulse width as a function of propagation distance; different curves correspond to different angular dispersions [36].

As shown in Figure 4.5, when the propagation distance is far less than the Rayleigh length of the Gaussian beam, the pulse width changes dramatically with propagation distance; this phenomenon agrees well with Martinez's work (as shown in Figure 4.4). However, when the propagation distance is greater than the Rayleigh length, the pulse width changes gradually and finally reaches a constant value; this phenomenon was not predicted by Martinez. Furthermore, the pulse width broadening due to higher angular dispersion is always greater than that due to lower angular dispersion at the same propagation distance.

Horvath et al. gave an expression for the pulse width when the propagation distance is far greater than the Rayleigh length of the Gaussian beam [36]

$$\frac{\tau}{\tau_0} = \left[1 + \frac{8\ln 2}{\tau_0^2}\left(\frac{\beta}{\Theta_0}\right)^2\right]^{\frac{1}{2}}, \tag{4.7}$$

where $\Theta_0 = 2c/w_0\omega_0$ is the intrinsic far field divergence angle of the Gaussian beam. Equation (4.7) indicates that, when the propagation distance is far greater than the Rayleigh length of the Gaussian beam, the pulse width will tend to a constant, thus satisfactorily explaining the phenomenon that the pulse width finally tends to approach a constant as shown in Figure 4.5.

As mentioned above, Martinez's work has a limiting condition, which is the well collimation approximation, and thus the results are suitable only for the case when the propagation distance of the laser pulse is far less than the Rayleigh length of the Gaussian beam. However, the well collimation approximation does not always hold in actual optical systems, and thus the application of this theory is limited [34, 35]. The numerical simulation method of Horvath et al. can determine the pulse width at arbitrary propagation distances, but the expression for the pulse width is only suitable for the case when the propagation distance of the laser pulse is far greater than the Rayleigh length of the Gaussian beam. There is also an approximation used in Horvath et al.'s work where they assume that the waist of the Gaussian beam is exactly located at the entrance of the disperser; this is obviously a special condition and does not hold in general experimental conditions [36]. For the investigation of pulse width broadening, when the propagation distance of the laser pulse is comparable to the Rayleigh length of the Gaussian beam (i.e. the propagation distance is neither far less nor far greater than the Rayleigh length), and the beam waist is located at arbitrary positions, the evolution of the pulse width and its physical mechanism is unclear, and it also lacks an exact formula to describe the pulse width changes. Zeng et al. recently obtained an expression for the pulse width in general conditions without using the above approximations; their work includes Martinez's theory, and the expression obtained can be used to describe the pulse width evolution of the dispersed Gaussian beam at an arbitrary propagation distance, thus completing the theory of the propagation of the laser pulse in ultrafast optics [37, 38].

In fact, the well collimation approximation used in Martinez's work simplifies the calculations to some extent, but it also limits the range over which this theory is applicable [34, 35]. Zeng et al. directly Fourier transformed equation (4.5) to the time domain without using the well collimation approximation, and then obtained the expression for the pulse width of the dispersed Gaussian beam at an arbitrary propagation distance [37]

$$\frac{\tau}{\tau_0} = \left[(1 + U) + \frac{V^2}{1 + U} \right]^{\frac{1}{2}} . \tag{4.8}$$

Equation (4.8) indicates that the pulse width broadening is totally determined by the effects of the spectral lateral walk-off and the group delay dispersion

$$U = \frac{2 \ln 2 z_R^2}{(d + \alpha^2 z)^2 + z_R^2} \left(\frac{2\alpha\beta z}{w_0 \tau_0} \right)^2 , \tag{4.9}$$

where U is the spectral lateral walk-off [37], and

$$V = \frac{(4 \ln 2)}{\tau_0^2} \left[-k\beta^2 z \cdot \frac{(d + \alpha^2 z)d + z_R^2}{(d + \alpha^2 z)^2 + z_R^2} \right] , \tag{4.10}$$

where V is the group delay dispersion [37].

In Martinez's theory [34], the parameter u is only suitable for the case where the propagation distance of the laser pulse is far less than the Rayleigh length of the Gaussian beam, and the parameter U is suitable for an arbitrary propagation distance. We can also see that $U = 2\ln 2z_R^2 u/[(d+\alpha^2 z)^2 + z_R^2]$, where the factor $2\ln 2$ accounts for the difference in the pulse width definition ($1/e^2$ of the magnitude in Martinez's work, and FWHM in Zeng et al.'s work), and the factor $\xi = z_R^2/[(d+\alpha^2 z)^2 + z_R^2]$ can be used to describe the evolution of the spectral lateral walk-off more exactly in general conditions. When the well collimation approximate condition (i.e. d, $\alpha^2 z \ll z_R$) and the same pulse width definition is used, $\xi \approx 1$, and the spectral lateral walk-off simplifies to $U = (2\alpha\beta z/w_0\tau_0)^2$, which is the same term as that of Martinez. Although the spectral lateral walk-off does not appear to be presented in Horvath et al.'s work [36], equation (4.7) actually describes the pulse width broadening by the spectral lateral walk-off in the range of propagation distance far greater than the Rayleigh length of the Gaussian beam. Thus, the parameter U describes a general effect of the spectral lateral walk-off, neither limited to a distance far less than the Rayleigh length nor far outside the Rayleigh length. In Figure 4.6, the factor ξ and the parameter U are plotted as functions of the propagation distance, with $\lambda_0 = 800$ nm, $d = 0$, $w_0 = 0.5$ mm, $z_R = 1$ m, $\tau_0 = 100$ fs, $\alpha = 1$, and $\beta = 0.1$ rad/μm; for comparison, the parameter u in the Rayleigh range in Martinez's work is also plotted [37].

As shown in Figure 4.6, when $\alpha^2 z \ll z_R$, i.e. the propagation distance is far less than the Rayleigh length of the Gaussian beam, the evolution of the spectral lateral walk-off predicted by Martinez and Zeng et al. is the same, and the parameters U and u both increase dramatically with propagation distance z. When the propagation distance approaches, or is larger than, the Rayleigh length, Martinez's theory does not hold, and the spectral lateral walk-off can only be described by the more exact parameter U: the decrease of the factor ξ limits the increasing rate of the parame-

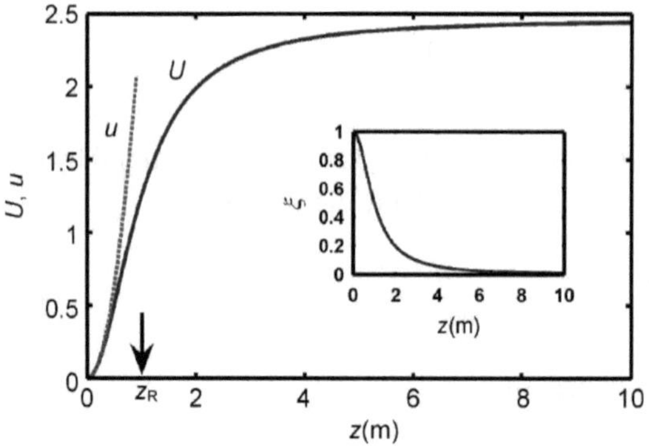

Fig. 4.6: The parameters ξ, U and u plotted as functions of propagation distance, with $0 < \xi < 1$ [37].

ter U, causing it to tend to be stable, until it finally becomes a constant with the value $U = 2\ln 2[2\beta z_R/\alpha w_0\tau_0]^2$ [37].

The group delay dispersion of the dispersed femtosecond laser pulse is described by the parameter V. Although Martinez did not define this parameter [34], there is a corresponding term in equation (4.6) that can be expressed as $v = -2k\beta^2 z/\tau_0^2$, and it is obvious that $V = 2\ln 2v[d(d+\alpha^2 z)+z_R^2]/[(d+\alpha^2 z)^2+z_R^2]$. The factor $2\ln 2$ is also due to the difference in the pulse width definition, and the factor $\eta = [d(d+\alpha^2 z)+z_R^2]/[(d+\alpha^2 z)^2+z_R^2]$ can be used to more exactly describe the evolution of the group delay dispersion of the Gaussian beam after angular dispersion, which is denoted as GDD_{GA}. Compared to the expression for the group delay dispersion given by $-k\beta^2 z$ in Martinez's work, there is a factor η (it is exactly the factor in equation (4.1)) in GDD_{GA}. When the well collimation approximation (i.e. d, $\alpha^2 z \ll z_R$) and the same pulse width definition are used, $\eta \approx 1$, $GDD_{GA} = -k\beta^2 z$, which is the same as Martinez's result (and also the same as in equation (4.3)). The parameter V describes the evolution of the group delay dispersion of the dispersed Gaussian beam at an arbitrary propagation distance, and thus it is suitable for general conditions. The factor η and the parameter V as functions of propagation distance are plotted in Figure 4.7 (the conditions are the same as in Figure 4.6). For comparison, the corresponding parameter v in the Rayleigh range is also plotted here [37].

As shown in Figure 4.7, when $\alpha^2 z \ll z_R$, the evolution of the group delay dispersion predicted by Martinez and Zeng et al. is the same, and the parameters V and v (the absolute values) dramatically increase with propagation distance; beyond this range, Martinez's theory no longer holds, and the group delay dispersion of the dispersed Gaussian beam can only be described by the more exact parameter V. The decrease of the factor η with propagation distance limits the increasing rate of the GDD_{GA}, and when $\alpha^2 z \gg z_R$ the group delay dispersion tends to a constant value of $GDD_{GA} = -k\beta^2 d/\alpha^2$. That is, when the propagation distance of the laser pulse is far greater than the Rayleigh length of the Gaussian beam, the group delay dispersion

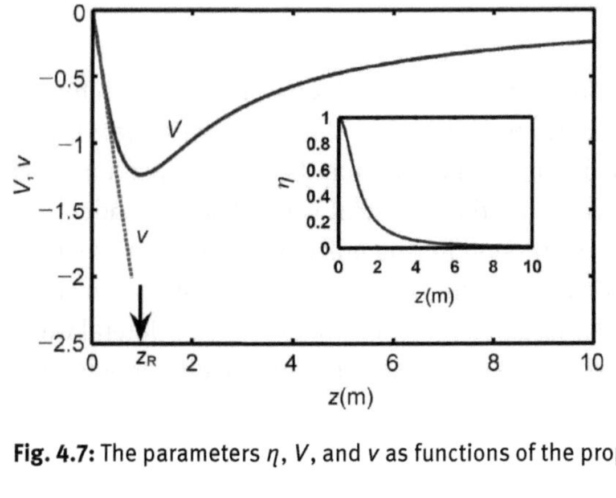

Fig. 4.7: The parameters η, V, and v as functions of the propagation distances with $0 < \eta < 1$ [37].

no longer increases. Considering equation (4.7), Horvath et al. [36] supposed that the beam waist is located at the entrance of the angular dispersion element, i.e. $d = 0$ when $\alpha^2 z \gg z_R$, and the group delay dispersion introduced by the angular dispersion is 0, i.e. the pulse width broadening is entirely determined by the spectral lateral walk-off; this is exactly what is described by equation (4.7).

To validate equation (4.8), we compare the results obtained from equation (4.8) with numerical simulations by Horvath et al. under the same conditions ($\lambda_0 = 800$ nm, $d = 0$, $w_0 = 0.5$ mm, $z_R = 1$ m, $\tau_0 = 100$ fs, $\alpha = 1$, $\beta = 0.1$ rad/μm). The pulse widths obtained are shown as a function of propagation distance in Figure 4.8. As the results from the two methods agree well, equation (4.8) is validated. Additionally, the numerical simulation method used by Horvath et al. must suppose that the waist of the Gaussian beam is located exactly at the entrance of the angular dispersion element [36]; however, this limitation is not necessary, so equation (4.8) can be used for more general conditions.

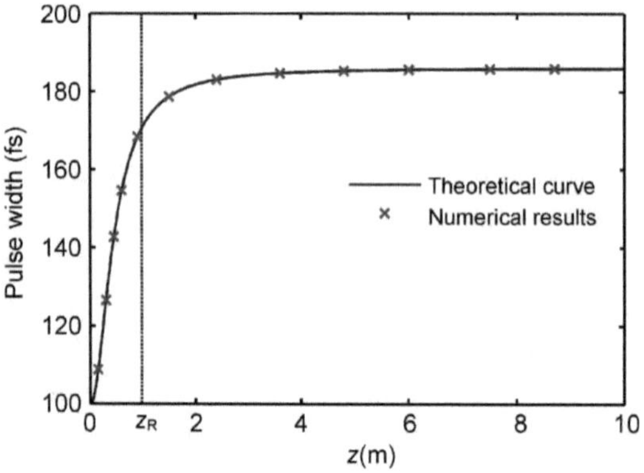

Fig. 4.8: The evolution of the pulse width of the dispersed Gaussian beam, where the symbols *x* are the values from numerical simulations by Horvath et al. [36], while the theoretical curve is plotted using equation (4.8).

Further investigation by Zeng et al. found that the analysis based on the Kirchhoff–Fresnel integral and the Fourier transform established by Martinez is suitable not only for analysis of the pulse width broadening of the Gaussian beam, but also for other optical models, such as the plane wave and the spherical wave. They extended this method and obtained a generalized expression for the pulse width that holds simultaneously for the plane wave, spherical wave and Gaussian beam (as shown in equation (4.8)), and revealed the relationships and distinctions of the spectral lateral walk-off and the group delay dispersion (which are two reasons for pulse width broadening) between the models [39].

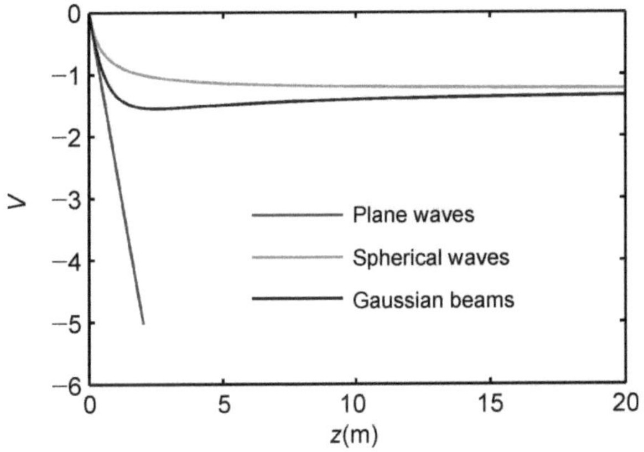

Fig. 4.9: The group delay dispersion of the plane wave, the spherical wave and the Gaussian beam as a function of propagation distance [39].

For the spectral components of the plane wave and the spherical wave distributed in the arbitrary position in free space, the spectral lateral walk-off does not exist in these two models, and the group delay dispersions are respectively expressed as [39]

$$V_P = -\frac{4 \ln 2 \cdot k\beta^2 z}{\tau_0^2}, \tag{4.11}$$

$$V_S = -\frac{4 \ln 2 \cdot \frac{d}{d+\alpha^2 z} k\beta^2 z}{\tau_0^2}. \tag{4.12}$$

The group delay dispersion of the plane wave increases linearly with propagation distance z. The group delay dispersion of the spherical wave increases nonlinearly with z, but finally tends to a constant given by GDD $= -k\beta^2 d/\alpha^2$ (the same as that of the Gaussian beam).

As there is no spectral lateral walk-off in the plane wave and the spherical models, i.e. $U_P = U_S = 0$, the spectral lateral walk-off of the Gaussian beam as a function of propagation distance z is as shown in Figure 4.6, in which the spectral lateral walk-off finally tends to a constant. For $\lambda_0 = 800$ nm, $\tau_0 = 100$ fs, $\alpha = 1$, $\beta = 0.1$ rad/μm, $d = 0.5$ m, and for a Gaussian beam with $w_0 = 0.5$ mm and $z_R = 1$ m, the curves of the group delay dispersion of the plane wave, the spherical wave and the Gaussian beam are shown in Figure 4.9 [39].

As noted above, the group delay dispersion of the plane wave increases linearly with propagation distance; however, the group delay dispersion of the spherical wave and of the Gaussian beam finally tends to a constant of the same value. Based on the generalized expression for pulse width broadening (equation (4.8)), and the expressions for the spectral lateral walk-off and group delay dispersion (equation (4.9), equation (4.12)), the curves for the pulse width evolution of each model are shown in Figure 4.10 [39].

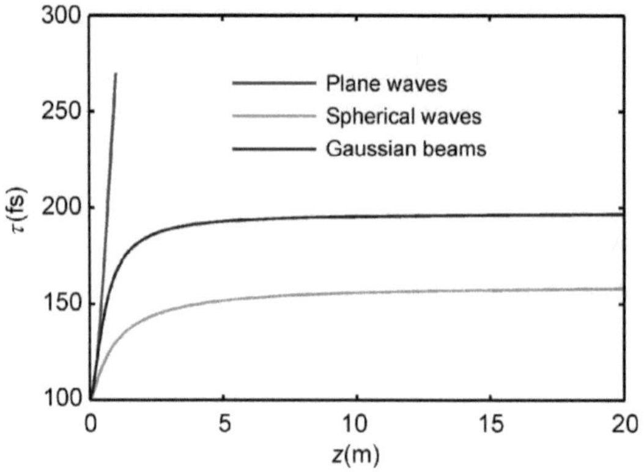

Fig. 4.10: The pulse width of the plane wave, spherical wave, and Gaussian beam as functions of propagation distance [39].

As shown above, for the plane wave, the pulse width always increases with propagation distance z. However, for the spherical wave and the Gaussian beam, the pulse width first increases, then tends to be stable, and finally reaches a constant.

It is worth noting that the pulse width evolution is only described for a femtosecond laser pulse after passing through idealized angular dispersion elements, that is to say the element includes no material dispersion. However, the actual angular dispersion elements generally include material dispersion. For example, when a laser pulse passes through a prism, it must propagate a certain distance in the prism, and thus the prism itself must introduce some material dispersion. The AOD is a device that is not only an angular dispersion element but also produces large material dispersion. These elements can be generally regarded as a combination of the dispersion medium and the angular dispersion element. When investigating the pulse width broadening of a femtosecond laser pulse after it passes through them, the values of the material dispersion must be added to the expression for the group delay dispersion [38]. In the following experiments we will see that the pulse width broadening by an angular dispersion element with material dispersion has obvious differences to that of the idealized angular dispersion element.

4.3.2 Pulse broadening after angular dispersion: experiments and results

The analytical results show that when the spatial modes of a femtosecond laser pulse approach the plane wave (such as a laser beam with a spot size of several centimeters), the Gaussian beam (such as a laser beam with a spot size smaller than 1 millimeter), and the spherical wave (such as a point source formed by a femtosecond laser pulse passing through a pinhole), the corresponding pulse width evolutions after angular dispersion are much different. The experimental verification for the pulse width evo-

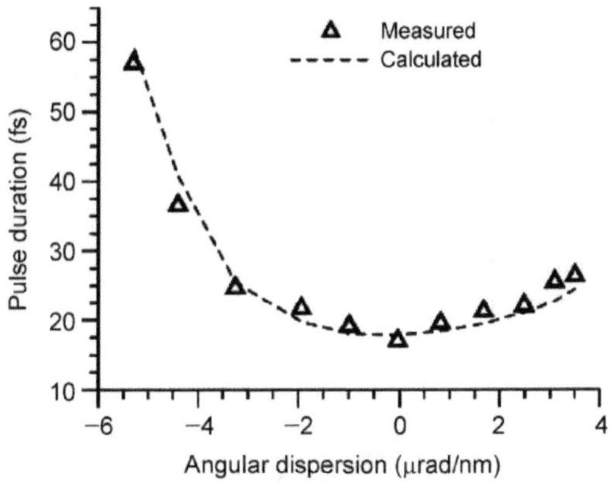

Fig. 4.11: The pulse width of the plane wave as a function of residual angular dispersion, where the dashed line is the theoretical curve and the symbols △ are the experimental results [40].

lution of both the plane wave and the Gaussian beam is comprehensive; however, the pulse width of the spherical wave after passing through the angular dispersion element is too weak to be detected, so the analytical results for the spherical wave have not been verified by experiments.

Osvay et al. [40] measured the pulse width, which is broadened by the residual angular dispersion in the CPA system, under the following conditions: the laser pulse emitted from the Ti:sapphire oscillator is 16 fs, repetition rate 72 MHz, central wavelength 800 nm, the pulse compressor is comprised of two fused silica prisms with apex angles of 68°, and the separation distance of the prisms is 122.2 cm. The effect of changing the angle of incidence of the laser pulse on the second prism is to introduce different residual angular dispersions; the induced pulse width broadening is then measured by a second order interferometer autocorrelator, as shown in Figure 4.11 [40]. The experimental results agree well with the analytical curves based on the plane wave model, thus verifying the evolution of the pulse width of the plane wave after angular dispersion. It is worth noting that, due to the large spot size in the CPA system (generally several centimeters), the plane wave model is sufficient to predict the pulse width of a femtosecond laser after angular dispersion, and the more exact Gaussian beam model is not necessary here.

However, in most optical systems, spot sizes are generally small (less than 1 cm), and the divergence angles of the laser beam cannot be neglected. In these cases, the Gaussian beam model is more suitable for describing the propagation characteristics of the femtosecond laser pulse than the plane wave model. Zeng et al. [37] measured the pulse width of a femtosecond laser pulse after passing through a single prism. The experimental conditions are as follows: the femtosecond laser pulse emitted from the Ti:sapphire oscillator (Spectra-Physics Mai Tai®) is 124 fs, 80 MHz, the central wavelength is 800 nm, the beam waist size is $w_0 = 0.5$ mm, the prism is made of SF10

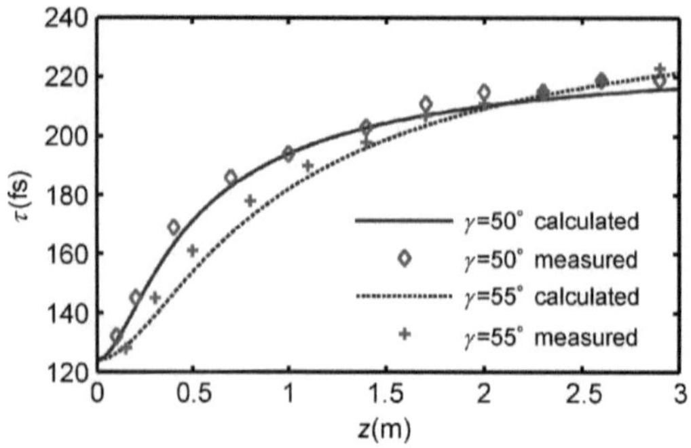

Fig. 4.12: The pulse width evolution of the Gaussian beam after angular dispersion; the solid and dashed lines are the theoretical values, and the symbols are the experimental results [37].

glass with a 60° apex angle, the distance between the beam waist and the prism is $d = 1.6$ m, and the pulse is measured by a two-photon absorption-based autocorrelator (Carpe, APE). The measured pulse widths as a function of propagation distance for angles of incidence $\gamma = 50°, 55°$ are shown in Figure 4.12 [37]. The experimental results agree well with the theoretical curves predicted by the Gaussian beam model, thus verifying the evolution of the pulse width of the Gaussian beam after angular dispersion.

Zeng et al. [38] further verified the pulse width broadening of a femtosecond laser pulse after dispersion by an element with large material dispersion, in which an AOD is chosen as a typical device. The experimental conditions are as follows: the femtosecond laser pulse emitted from the Ti:S oscillator is 124 fs, the repetition rate is 80 MHz, the central wavelength is 800 nm, the beam waist size is $w_0 = 0.5$ mm, the AOD is made of the TeO_2, which is designed to the abnormal Bragg diffraction mode, the distance between the beam waist and the AOD is $d = 1.5$ m, the total material dispersion in the optical system is about 15 000 fs^2, and the pulse width is also measured by a two-photo absorption based autocorrelator (Carpe, APE). The measured values (where the AOD is operated at the central frequency $f = 70$ MHz) are first compared with the theoretical values of the Gaussian beam and the plane wave, as shown in Figure 4.13 [38].

The plane wave model predicted that the pulse width would first decrease to the initial pulse width (124 fs) and then increase, but the measured results agree well with the model for the Gaussian beam in which the pulse width first decreases with increase in propagation distance, with the rate of decrease more gradual than that of the plane wave, then increases a little and finally tends to a constant. The experiments show that, when the spot size is relatively small ($w_0 = 0.5$ mm in the experiments), the prediction of the pulse width by the Gaussian beam model is more accurate than that of the plane wave model. When the AOD operates at different frequencies (which correspond to different angular dispersions), the evolution of the pulse width also changes, as shown in Figure 4.14 [38].

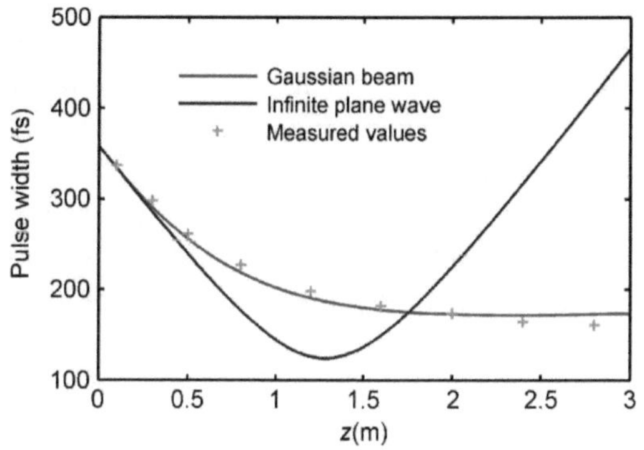

Fig. 4.13: The pulse width evolution when $f = 70$ MHz, showing the theoretical values of the Gaussian beam and the plane wave. The measured pulse width evolution is close to the theoretical curve of the Gaussian beam, but differs from that of the plane wave [38].

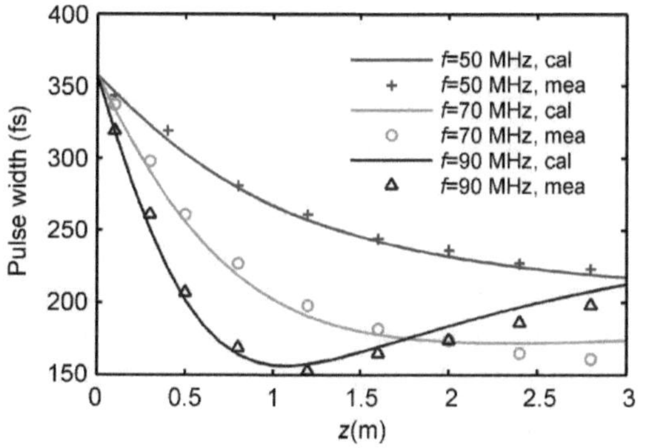

Fig. 4.14: The curves of pulse width for different frequencies of AOD; cal = calculated by the Gaussian beam model; mea = the measured values [38].

When $f = 50$ MHz (the lowest frequency), the pulse width decreases monotonically with the propagation distance; when $f = 70$ MHz (the central frequency) and $f = 90$ MHz (the highest frequency), the pulse width first decreases with propagation distance, then increases and finally tends to a constant. These experiments show that, when the AOD operates at different frequencies, the angular dispersion also changes, thus causing different evolutions of the pulse width. The experimental results agree well with the theoretical values of the Gaussian beam model, but differ from those of the plane wave model, which indicates that the Gaussian beam model is more accurate in analysis of the dispersion characteristics of practical optical systems compared with the plane wave model.

It is interesting to compare Figure 4.14 with Figure 4.5. For the idealized grating in Figure 4.5 (i.e. with no material dispersion), the pulse width is always larger when the angular dispersion is large at the same propagation distance. However, for the AOD (which can be regarded as the combination of a material dispersion medium and an idealized grating), in the range that the propagation distance is far less than the Rayleigh length, the pulse width at the higher frequency (which corresponds to larger angular dispersion) is always smaller than that at the lower frequency. This is due to the larger group delay dispersion provided by the higher frequency, which compensates for the material dispersion of the AOD faster, and thus reduces the pulse width broadening more effectively, in which the group delay dispersion is dominant compared to the spectral lateral walk-off. However, when the propagation distance is far greater than the Rayleigh length of the Gaussian beam, the pulse width at the higher frequency can either be smaller or larger than that of the lower frequency, which is codetermined by the spectral lateral walk-off and the group delay dispersion.

4.3.3 Physical mechanism of pulse broadening after angular dispersion: comparison of the plane wave, spherical wave and Gaussian beam

The theoretical analysis and experiments show that the pulse width evolution of the plane wave, spherical wave and Gaussian beam have obvious differences. In this section, we summarize the differences and their physical mechanisms.

(1) For the three models, the pulse width broadening of a femtosecond laser after passing through the angular dispersion element can be described by the same generalized formula (equation (4.8)), which indicates that the pulse width broadening is entirely determined by the spectral lateral walk-off (which is 0 in the plane wave model and the spherical wave model) and the group delay dispersion [39]. The pulse width of the plane wave always increases with propagation distance; however, the pulse width of the spherical wave and Gaussian beam first increases with propagation distance, then tends to be stable and finally reaches a constant.

(2) For the plane wave and spherical wave, because all the spectral components distribute in the whole space (with infinite beam spot size), after the pulse is dispersed all the spectral components can still be observed in any position; thus, the spectral lateral walk-off is 0, and the pulse width broadening is determined only by the group delay dispersion. However, the Gaussian beam has a finite beam size, thus the spectral components cannot always be observed in any position, and the spectral lateral walk-off is not 0; thus, the pulse width broadening is codetermined by the spectral lateral walk-off and the group delay dispersion.

(3) When the propagation distance is far less than the Rayleigh length of the Gaussian beam ($d, \alpha^2 z \ll z_R$, i.e. the well collimation approximation), the group delay dispersion of the Gaussian beam $V_G = -4\ln 2k\beta^2 z/\tau_0^2$ is the same as that of the plane wave (equation (4.11)). When the propagation distance is far greater than the Rayleigh length, the group delay dispersion of the Gaussian beam $V_G \approx -4\ln 2[d/(d+\alpha^2 z)]k\beta^2 z/\tau_0^2$ is the same as that of the spherical wave (equation (4.12)). This is because that the phase front of the Gaussian beam is close to planar in the near field,

while close to spherical in the far field, and the group delay dispersion is essentially determined by the change of the phase front. The spherical wave and Gaussian beam finally introduce the same group delay dispersion values $V_S = V_G = -4\ln 2k\beta^2 d/\alpha^2\tau_0^2$, which indicate that the group delay dispersion is entirely determined by the parameters of the dispersion element and the position of the light source (for the spherical wave) or the beam waist (for the Gaussian beam). This is because the phase fronts of the various spectral components actually overlap at the surface of the disperser and at infinity, so no GDD is induced in this space. The equivalent point source (or beam waist) is located at a distance d/α^2 in front of the dispersion component. In the reverse propagation direction, the phase fronts of different spectral components do not overlap at this position; thus, the limiting value of the GDD is also determined by this specific distance.

(4) The decrease of the angular dispersion (the second definition of angular dispersion is used here) of the Gaussian beam is the fundamental reason that the pulse width tends to a constant. When the propagation distance increases, the angular dispersion of the Gaussian beam decreases, i.e. the angle between different phase fronts of different spectral components decreases, as shown in equation (4.1). Although the spectral components still spread as the pulse propagates, the rate of increase slows, and the rate of increase of the parameter U also slows. When the angle between different phase fronts is zero, i.e. the angular dispersion is zero, the parameters U and V do not increase further, and the pulse width tends to a constant [37]. It is necessary to point out that the group delay dispersion introduced by the angular dispersion in the Gaussian beam, $\mathrm{GDD_{GA}} = -k\beta^2 z\eta$, has a factor η, which is different to that of the plane wave, $-k\beta^2 z$. η is the factor in equation (4.1) that describes the difference of the angular dispersion between the Gaussian beam and plane wave, further proving that the decrease of the angular dispersion of the Gaussian beam causes the pulse width to tend to a constant after dispersion (this explanation is also suitable for the spherical wave model).

(5) There are two conditions when the Gaussian beam model simplifies to the plane wave model: when the laser beam spot size is infinite; and when the laser beam is well collimated, i.e. d, $\alpha^2 z \ll z_R$. The first condition makes the spectral lateral walk-off of the Gaussian beam disappear, i.e. $U = 0$, as for the plane wave. The second condition makes the group delay dispersion of the Gaussian beam equal to that of the plane wave, i.e. $\mathrm{GDD} = -k\beta^2 z$. Under these two conditions equation (4.8) can be simplified to equation (4.4), which proves that the plane wave model is an approximation of the Gaussian beam model under certain conditions. Furthermore, the theoretical and experimental results show that the results based on the Gaussian beam model are more accurate than those of the plane wave to describe the propagation characteristics of a femtosecond laser pulse after angular dispersion.

(6) It is interesting that when the source of a spherical wave is located exactly at the entrance of the disperser, the spectral lateral walk-off and GDD remain zero regardless of the propagation distance. In this case the pulse is not broadened at all. This is because the phase front propagates uniformly in all directions, and each spectral component retains its spherical shape after dispersion [39].

4.4 Beam spot broadening after angular dispersion

Compared with the comprehensive research on the evolution of the pulse width of a femtosecond laser pulse after dispersion, the beam spot size evolution of a femtosecond laser after dispersion has been almost neglected over a long period. Martinez indicated the beam shape when analyzing the effect of the compressor on a femtosecond laser pulse, but this work is based on a pulse compressor (which comprises a pair of dispersion elements) rather than a single disperser, and is only suitable for the case when the propagation distance is far less than the Rayleigh length of the Gaussian beam [35]. The work of Varju et al. did not reveal how the angular dispersion affects the beam shape [62]. In other applications, most researchers generally calculate the divergence angle of the femtosecond laser beam after dispersion using a geometrical optics method, but do not describe how the beam shape changes. Actually, in many applications, such as multiphoton microscopy and micromachining, dispersers such as prisms, gratings, and AODs are commonly used, and the beam shape is important for the resolution of optical imaging and the accuracy of micromachining [43–54]. Therefore, it is necessary to accurately describe the evolution of the beam shape. Zeng et al. [42] obtained an expression for the spot size of the dispersed beam at an arbitrary distance based on the Gaussian beam model, and found that there are two reasons for beam spot broadening: one is the divergence angle of the Gaussian beam itself, and the other is the spectral lateral walk-off, which is introduced by angular dispersion. The effects of these two factors are different as the beam propagates. Furthermore, the dispersed beam in the far field still propagates in a hyperbolic way, compared with the intrinsic far field divergence of the Gaussian beam. The dispersed beam also amplifies the far field divergence angle, where the magnification factor is determined by the characteristic parameters of the disperser and the laser beam itself. This section will review these studies in detail.

The idealized plane wave model ignores the spot size of the laser beam. In this rough analysis, a laser beam with uniform energy distribution and finite spot size, which is actually a part of an infinite plane wave, is usually assumed, and is called the 'finite plane wave model' [36].

For the finite plane wave model, it is easy to obtain the spot size of the dispersed laser beam

$$w = \frac{w_p}{|\alpha|} + \frac{|\beta|\,\Delta\omega \cdot z}{2}, \tag{4.13}$$

where w_p is the initial radius of the laser beam of the 'finite plane wave', $\Delta\omega$ is the spectral bandwidth, and z is the propagation distance. In equation (4.13), the first term describes spot size changes due to the angle of incidence, and the second term describes spot size changes due to the propagation of the angular dispersion. It is obvious that the spot size of the laser beam increases linearly with propagation distance. Equation (4.13) can be used to estimate the beam spot size, such as in the case where the spot size is relatively large. However, in many other situations, this estimate is not accurate enough, because the laser beam is closer in shape to a Gaussian beam, which is inherently divergent when it propagates in space; this divergence will contribute greatly to the spot size broadening.

Actually, the spot size of the laser beam is included in the above equation (4.5), i.e. the expression for the electric field of the dispersed Gaussian beam obtained by Martinez using the Kirchhoff–Fresnel integral. The analytical expression for the spot size at arbitrary propagation distance can be easily obtained from equation (4.5) by combining all the real parts of the x^2 term [75, 76], in which the well collimation approximation is not used. The spot size is defined as the radius when the intensity falls to $1/e^2$ of the maximum, and is expressed as [42]

$$w = w_0 \left\{ \frac{(1+U)}{\alpha^2} \left[1 + \frac{(d+\alpha^2 z)^2}{z_R^2} \right] \right\}^{\frac{1}{2}} . \tag{4.14}$$

The parameter U describes the spectral lateral walk-off of the dispersed Gaussian beam as shown in equation (4.9). Equation (4.14) describes the spot size evolution of the dispersed femtosecond laser beam at an arbitrary propagation distance. There are two reasons for the beam spot broadening: the intrinsic divergence of the Gaussian beam makes the spot size increase in a hyperbolic way when it propagates in space; and the spectral lateral walk-off introduced by the angular dispersion makes the spot size increase. One can clearly see the physical mechanisms underlying spot size evolution by applying the assumption of no angular dispersion component, i.e. $\alpha = 1$ and $\beta = 0$. In this case, $U = 0$, and equation (4.14) can be simplified to [42]

$$\frac{w}{w_0} = \left[1 + \frac{(d+\alpha^2 z)^2}{z_R^2} \right]^{\frac{1}{2}} . \tag{4.15}$$

This is the hyperbolic manner in which the Gaussian beam travels in free space. When the Gaussian beam is well collimated, i.e. d, $\alpha^2 z \ll z_R$, equation (4.14) can be simplified to [42]

$$\frac{w}{w_0} = \left[\frac{(1+U)}{\alpha^2} \right]^{\frac{1}{2}} . \tag{4.16}$$

In this case the beam spot size evolution is determined only by the spectral lateral walk-off produced by the angular dispersion and is free of the intrinsic Gaussian beam divergence. To see more clearly how these factors affect the beam spot size, its evolution along propagation is simulated by substituting typical parameters into equation (4.14), as shown in Figure 4.15 [42].

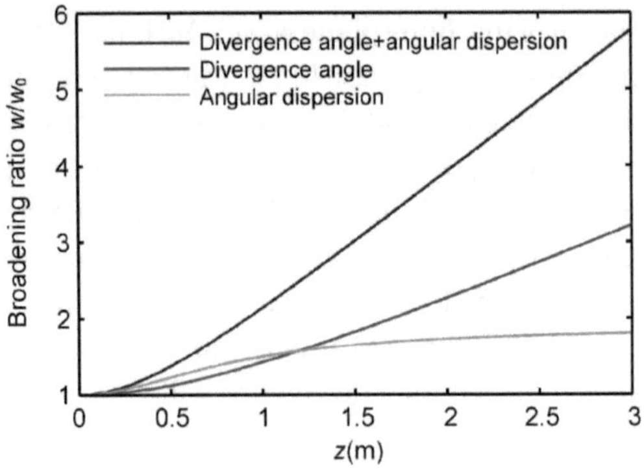

Fig. 4.15: Physical mechanism of the beam spot size evolution of a Gaussian pulse and beam after passing through an angular dispersion element [42].

The three curves in Figure 4.15 represent the beam spot size variation due to the intrinsic Gaussian beam divergence, the spectral lateral walk-off (i.e. the angular dispersion), and the factors working simultaneously. The beam spot broadening is specified by the direct product of the intrinsic beam divergence and the external angular dispersion. It is interesting that when the propagation distance is less than the Rayleigh length, the contribution of the beam size increase from angular dispersion is dominant. This relates to the fact that the Gaussian beam is well collimated inside the Rayleigh length. With the increase in propagation distance, the contribution from intrinsic beam divergence increases, while the contribution from angular dispersion tends to be stable [42].

Zeng et al. measured the spot size of the femtosecond laser beam after dispersion, typically by an AOD. The experimental conditions are as follows [42]: the AOD is a slow shear TeO_2 device with $v = 616$ m/s and the active frequency range is 50–90 MHz. The laser emitted from the Ti:sapphire oscillator (Spectra-Physics Mai Tai®) has an initial pulse width 124 fs, central wavelength 800 nm, spectral bandwidth 10 nm, and waist size $w_0 = 0.5$ mm, corresponding to a Rayleigh length $z_R \sim 1$ m. The optical layout for beam spot size measurement is very simple: a laser pulse with the distance between its beam waist and AOD, $d = 1.5$ m, enters the AOD at a Bragg angle and is diffracted by AO interaction. The first order diffracted beam is then measured at different distances from the AOD using a laser beam analyzer (BeamStar FX50, Ophir) moving on a slip rail. The results when the AOD operates at the central frequency $f_c = 70$ MHz are shown in Figure 4.16; the theoretical results based on the Gaussian beam model and the finite plane wave model are plotted for comparison.

In Figure 4.16 the spot size evolution based on the two models is similar. However, the evolution predicted by the finite plane wave model describes a simple linear increase, as shown in equation (4.13), while the evolution predicted by the Gaussian beam describes a nonlinear increase, as shown in equation (4.14). The rate of increase

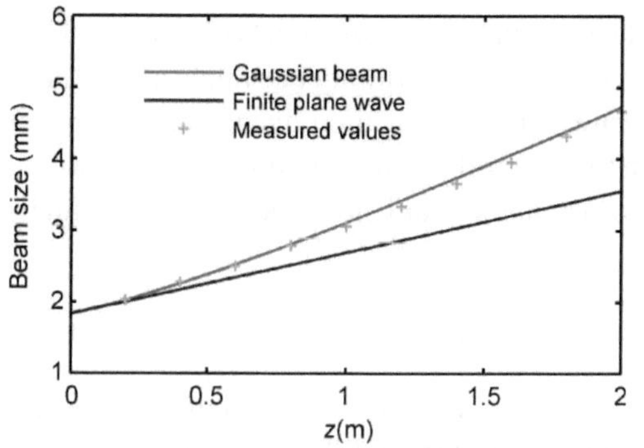

Fig. 4.16: The spot size evolution when the AOD operates at the central frequency. The experimental results agree well with the theoretical results based on the Gaussian beam model [42].

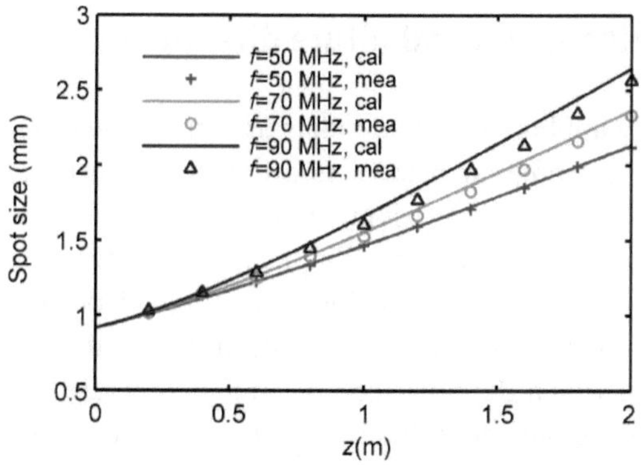

Fig. 4.17: Beam spot size as a function of propagation distance at different AOD frequencies; the curves are calculated from equation (4.14) and the symbols show the experimental results [42].

of the Gaussian beam is greater than that of the finite plane wave due to the intrinsic divergence of the Gaussian beam. The experimental results are closer to the theoretical results of the Gaussian beam, and differ from that of the finite plane wave, thus proving that the Gaussian beam model is more accurate than the finite plane wave model.

A number of values of the frequency of the AOD were used to further verify the spot size evolution at different angular dispersions, and the measured spot size as a function of propagation distance is shown in Figure 4.17 [42].

The experiments are consistent with the theoretical predictions, thus verifying the spot size evolution of the dispersed femtosecond laser beam. With increasing propagation distance the beam spot size increases. For higher frequencies, i.e. greater angular dispersion, the beam spot size is larger than that at lower frequencies at the same distance, and this phenomenon is easy to understand.

Another interesting fact is that in the far field the femtosecond laser beam still has a divergence angle, which corresponds to the intrinsic divergence angle of the Gaussian beam but with a magnification factor. If the propagation distance is far greater than the Rayleigh length ($z \gg z_R$), then the spectral lateral walk-off, U, reaches a constant ($2 \ln 2(2\beta z_R/\alpha w_0 \tau_0)^2$), and the corresponding far field divergence angle can be derived from equation (4.14) [42]

$$\theta' = |\alpha| \sqrt{1 + U} \theta \, , \tag{4.17}$$

where $\theta = \lambda_0 \pi / w_0$ is the intrinsic divergence angle of the Gaussian beam. Equation (4.17) indicates that the dispersed Gaussian beam will still increase in a hyperbolic way in the far field, but the divergence angle is magnified by a factor of $(1 + U)^{1/2} \alpha$.

4.5 Applications of angular dispersion and future directions

Angular dispersion devices are widely applied in laser systems and many measurement devices. In these applications, the angular dispersers are generally used to change the spatial and temporal profile of the laser beam. For example, prisms are used to compensate for dispersion in the cavity of laser oscillators, in which the group delay dispersion introduced by the prism can compensate for the material dispersion inside the cavity, thus avoiding pulse width broadening [17–19]; gratings are used for the compressor in CPA systems, in which the pulse can be stretched before amplification and then compressed after amplification [6, 7]; prisms and gratings are also widely used in spectrometers for separation of the different spectral components transversely, thus making the different spectral components resolvable.

In other applications of femtosecond laser pulses, such as multiphoton microscopy and micromachining [43–54], to satisfy the extreme conditions for multiphoton excitation it is necessary to simultaneously achieve the shortest pulse width in the temporal domain (i.e. the Fourier transform limit) and the smallest focus spot in the spatial domain (i.e. the diffraction limit), thus obtaining the highest density of photons for multiphoton excitation. However, spatial and temporal dispersion in optical systems will dramatically worsen the quality of the laser beam as they broaden the pulse width in time and distort the beam spot in space, making the focused laser beam difficult to use for multiphoton excitation, and the spatial and temporal distortion will also worsen the spatial and temporal resolution of images. In these applications, the angular dispersers are not only used for temporal dispersion compensation (by compressing the pulse width), but also for spatial dispersion compensation (by reshaping the beam spot), and various dispersion compensation methods are essential in this field [43–54]. Especially when the value of the dispersion changes (such as a femtosecond laser beam scanned by an AOD), an angular disperser that can provide a rapid and variable amount of dispersion will be extremely valued, and will enable real time and uniform dispersion compensation and avoid the reduction in resolution when dispersion compensation does not match the optical system.

Although the angular dispersion elements broaden the pulse width and the beam spot, and reduce the resolution of the optical imaging or the accuracy of the micromachining, in some cases angular dispersion elements are used to simultaneously focus the femtosecond laser beam spatially and temporally and consequently improve the resolution [77–81]. By changing the group delay dispersion, the temporal focus will move in the axial direction, thus allowing scanning in the z direction [82, 83]. Similarly, if the angular dispersion elements can provide a fast variable amount of group delay dispersion, the temporal focus will change rapidly in the z direction, and this is very important for developing fast three dimensional multiphoton microscopy to meet the needs of biology and medicine.

Bibliography

[1] J. G. Fujimoto, A. M. Weiner and E. P. Ippen. Generation and measurement of optical pulses as short as 16 fs. Appl. Phys. Lett. **44** (1984): 832–834.

[2] J. M. Halbout and D. Grischkowsky. 12-fs ultrashort optical pulse compression at a high repetition rate. Appl. Phys. Lett. **45** (1984): 1281–1283.

[3] W. H. Knox, R. L. Fork, M. C. Downer, R. H. Stolen and C. V. Shank. Optical pulse compression to 8 fs at a 5-kHz repetition rate. Appl. Phys. Lett. **46** (1985): 1120–1121.

[4] R. L. Fork, C. H. Brito Cruz, P. C. Becker and C. V. Shank. Compression of optical pulses to six femtoseconds by using cubic phase compensation. Opt. Lett. **12** (1987): 483–485.

[5] J. Squier, F. Salin, G. Mourou and D. Harter. 100-fs pulse generation and amplification in Ti:Al2O3. Opt. Lett. **16** (1991): 324–326.

[6] D. E. Spence, P. N. Kean and W. Sibbett. 60-fsec pulse generation from a self-mode-locked Ti:sapphire laser. Opt. Lett. **16** (1991): 42–44.

[7] C. L. Blanc, G. Grillon, J. P. Chambaret, A. Migus and A. Antonetti. Compact and efficient multipass Ti:sapphire system for femtosecond chirped-pulse amplification at the terawatt level. Opt. Lett. **18** (1993): 140–142.

[8] G. Steinmeyer, D. H. Sutter, L. Gallmann, N. Matuschek and U. Keller. Frontiers in Ultrashort Pulse Generation: Pushing the limits in Linear and Nonlinear Optics. Science. **286** (1999): 1507–1512.

[9] A. Scrinzi, M. Y. Ivanov, R. Kienberger and D. M. Villeneuve. Attosecond Physics. J. Phys. B. **39** (2006): R1-R37.

[10] W. Denk, J. H. Strickler and W. W. Webb. Two-photon laser scanning fluorescence microscopy. Science. **248** (1990): 73–76.

[11] K. Svoboda and W. Denk. Photon upmanship: why multiphoton imaging is more than a gimmick. Neuron. **18** (1997): 351–357.

[12] W. R. Zipfel, R. M. Williams and W. W. Webb. Nonlinear magic: multiphoton microscopy in the biosciences. Nat. Biotechnol. **21** (2003): 1369–1377.

[13] K. Svoboda and R. Yasuda. Principles of two-photon excitation microscopy and its applications to neuroscience. Neuron. **50** (2006): 823–839.

[14] F. Helmchen and W. Denk. Deep tissue two-photon microscopy. Nature Methods. **2** (2005): 932–940.

[15] G. McConnell. Improving the penetration depth in multiphoton excitation laser scanning microscopy. J. Biomed. Opt. **11** (2006): 054020.
[16] E. B. Treacy. Optical pulse compression with diffraction gratings. IEEE J. Quantum Electron. QE-**5** (1969): 454–458.
[17] R. L. Fork, O. E. Martinez and J. P. Gordon. Negative dispersion using pairs of prisms. Opt. Lett. **9** (1984): 150–152.
[18] J. P. Gordon and R. L. Fork. Optical resonator with negative dispersion. Opt. Lett. **9** (1984): 153–155.
[19] O. E. Martinez, J. P. Gordon and R. L. Fork. Negative group-velocity dispersion using refraction. J. Opt. Soc. Am. A **1** (1984): 1003–1006.
[20] Z. Bor and B. Racz. Group velocity dispersion in prisms and its application to pulse compression and traveling-wave excitation. Opt. Commun. **54** (1985): 165–170.
[21] A. M. Weiner, J. P. Heritage and E. M. Kirschner. High-resolution femtosecond pulse shaping. J. Opt. Soc. Am. B **5** (1988): 1563–1572.
[22] C. H. Brito Cruz, P. C. Becker, R. L. Fork and C. V. Shank. Phase correction of femtosecond optical pulses using a combination of prisms and gratings. Opt. Lett. **13** (1988): 123–125.
[23] S. Szatmari, G. Kuhnle and P. Simon. Pulse compression and traveling wave excitation scheme using a single dispersive element. Appl. Opt. **29** (1990): 5372–5379.
[24] S. Szatmari, P. Simon and M. Fenerhake. Group velocity dispersion compensated propagation of short pulses in dispersive media. Opt. Lett. **21** (1996): 1156–1158.
[25] M. Nakazawa, T. Nakashima and H. Kubota. Optical pulse compression using a TeO2 acousto-optical light deflector. Opt. Lett. **13** (1988): 120–122.
[26] M. Nakazawa, T. Nakashima, H. Kubota and S. Seikai. Efficient optical pulse compression using a pair of Brewster-angled TeO2 crystal prisms. J. Opt. Soc. Am. B **5** (1988): 215–221.
[27] O. E. Martinez. 3000 times grating compressor with positive group velocity dispersion: application to fiber compensation in 1.3–1.6 um region. IEEE J. Quantum Electron. QE-**23** (1987): 59–64.
[28] S. Akturk, X. Gu, M. Kimmel and R. Trebino. Extremely simple single-prism ultrashort-pulse compressor. Opt. Express **14** (2006): 10101–10108.
[29] R. W. Dixon. Acoustic diffraction of light in anisotropic media. IEEE J. Quantum Electron. QE-**3** (1967): 85–93.
[30] E. I. Gordon. A review of acousto optical deflection and modulation devices. Appl. Opt. **5** (1966): 1629–1639.
[31] J. Xu and R. Stroud, Acousto-Optic Devices: Principles, Design and Applications. John Wiley and Sons, New York, 1992.
[32] W. Denk, D. W. Piston and W. W. Webb. Two-photon molecular excitation in laser-scanning microscopy. In Handbook of Biological Confocal Microscopy, J. B. Pawley ed., Plenum, New York, 1995, pp. 445–458.
[33] V. Iyer, B. E. Losavio and P. Saggau. Compensation of spatial and temporal dispersion for acousto-optic multiphoton laser-scanning microscopy. J. Biomed. Opt. **8** (2003): 460–471.
[34] O. E. Martinez. Pulse distortions in tilted pulse schemes for ultrashort pulses. Opt. Commun. **59** (1986): 229–232.
[35] O. E. Martinez. Grating and prism compressors in the case of finite beam size. J. Opt. Soc. Am. B **3** (1986): 929–934.
[36] Z. L. Horvath, Z. Benko, A. P. Kovacs, H. A. Hazim and Z. Bor. Propagation of femtosecond pulses through lenses, gratings and slits. Opt. Eng. **32** (1993): 2491–2500.
[37] S. Zeng, D. Li, X. Lv, J. Liu and Q. Luo. Pulse broadening of the femtosecond pulses in a Gaussian beam passing an angular disperser. Opt. Lett. **32** (2007): 1180–1182.

[38] D. Li, S. Zeng, X. Lv, et al. Dispersion characteristics of acousto-optic deflector for scanning Gaussian laser beam of femtosecond pulses. Opt. Express **15** (2007): 4726–4734.
[39] D. Li, X. Lv, S. Zeng and Q. Luo. A generalized analysis of femtosecond laser pulse broadening after angular dispersion. Opt. Express **16** (2008): 237–247.
[40] K. Osvay, A. P. Kovacs, Z. Heiner, et al. Angular dispersion and temporal change of femtosecond pulses from misaligned pulse compressors. IEEE J. Sel. Top. Quantum Electron. **10** (2004): 213–220.
[41] K. Osvay, A. P. Kovacs, G. Kurdi, et al. Measurement of non-compensated angular dispersion and the subsequent temporal lengthening of femtosecond pulses in a CPA laser. Opt. Commun. **248** (2005): 201–209.
[42] D. Li, X. Lv, S. Zeng and Q. Luo. Beam spot size evolution of Gaussian femtosecond pulses after angular dispersion. Opt. Lett. **33** (2008): 128–130.
[43] B. K. A. Ngoi, K. Venkatakrishnan, B. Tan, P. Stanley and L. E. N. Lim. Angular dispersion compensation for acousto-optic devices used for ultrashort-pulsed laser micromachining. Opt. Express **9** (2001): 200–206.
[44] J. D. Lechleiter, D. T. Lin and I. Sieneart. Multi-photon laser scanning microscopy using an acoustic optical deflector. Biophys. J. **83** (2002): 2292–2299.
[45] R. D. Roorda, T. M. Hohl, R. Teledo-Crow and G. Miesenbock. Video-rate nonlinear microscopy of neuronal membrane dynamics with genetically encoded probes. J. Neurophysiol. **92** (2004): 609–621.
[46] G. D. Reddy and P. Saggau. Fast three-dimensional laser scanning scheme using acousto-optic deflectors. J. Biomed Opt. **10** (2005): 064038.
[47] V. Iyer, T. M. Hoogland and P. Saggau. Fast functional imaging of single neurons using random-access multiphoton (RAMP) Microscopy. J. Neurophysiol. **95** (2006): 535–545.
[48] P. Saggau. New methods and uses for fast optical scanning. Curr. Opin. Neurobiol. **16** (2006): 543–550.
[49] R. Salome, Y. Kremer, S. Dieudonne, et al. Ultrafast random-access scanning in two-photon microscopy using acousto-optic deflectors. J. Neurosci. Methods **154** (2006): 161–174.
[50] S. Zeng, X. Lv, C. Zhan, et al. Simultaneous compensation for spatial and temporal dispersion of acousto-optical deflectors for two-dimensional scanning with a single prism. Opt. Lett. **31** (2006): 1091–1093.
[51] K. Bi, S. Zeng, S. Xue, et al. Position of the prsim in a dispersion-compensated acousto-optic deflector for multiphoton imaging. Appl. Opt. **45** (2006): 8560–8565.
[52] X. Lv, C. Zhan, S. Zeng, W. R. Chen and Q. Luo. Construction of multiphoton laser scanning microscope based on dual-axis acousto-optic deflector. Rev. Sci. Instrum. **77** (2006): 046101.
[53] S. Zeng, K. Bi, S. Xue, et al. Acousto-optic modulator system for femtosecond laser pulses. Rev. Sci. Instrum. **78** (2007): 015103.
[54] S. Zeng, X. Lv, K. Bi, et al. Analysis of the dispersion compensation of acousto-optic deflectors used for multiphoton imaging. J. Biomed. Opt. **12** (2007): 024015.
[55] O. E. Martinez. Matrix formalism for pulse compressors. IEEE J. Quantum Electron. **24** (1988): 2530–2536.
[56] O. E. Martinez. Matrix formalism for dispersive laser cavities. IEEE J. Quantum Electron. **25** (1989): 296–300.
[57] S. P. Dijaili, A. Dienes and J. S. Smith. ABCD Matrices for dispersive pulse propagation. IEEE J. Quantum Electron. **26** (1990): 1158–1164.
[58] A. G. Kostenbauder. Ray-pulse matrices: A rational treatment for dispersive optical systems. IEEE J. Quantum Electron. **26** (1990): 1148–1157.

[59] M. Nakazawa, H. Kubota, A. Sahara and K. Tamura. Time-domain ABCD matrix formalism for laser mode-locking and optical pulse transmission. IEEE J. Quantum Electron. **34** (1998): 1075–1081.
[60] Y. Nabekawa and K. Midorikawa. High-order pulse front tilt caused by high-order angular dispersion. Opt. Express **11** (2003): 3365–3376.
[61] J. C. Diels and W. Rudolph. Ultrashort Laser Pulse Phenomena, 2nd edn., Academic, San Diego, Calif., 2006.
[62] K. Varju, A. P. Kovacs, K. Osvay and G. Kurdi. Angular dispersion of femtosecond pulses in a Gaussian beam. Opt. Lett. **27** (2002): 2034–2036.
[63] K. Varju, A. P. Kovacs, G. Kurdi and K. Osvay. High-precision measurement of angular dispersion in a CPA laser. Appl. Phys. B **74** (2002): 259–263.
[64] M. Born and E. Wolf. Principles of Optics, 6th edn., Pergamon, Oxford, 1980.
[65] F. A. Jenkins and H. A. White. Fundamentals of Optics, 4th edn., McGraw–Hill, New York, 1976.
[66] X. Gu, S. Akturk and R. Trebino. Spatial chirp in ultrafast optics. Opt. Commun. **242** (2004): 599–604.
[67] S. Akturk, M. Kimmel, P. O'Shea and R. Trebino. Measuring spatial chirp in ultrashort pulses using single-shot Frequency-Resolved Optical Gating. Opt. Express **11** (2003): 68–78.
[68] Z. Bor, B. Racz, G. Szabo, M. Hilbert and H. A. Hazim. Femtosecond pulse front tilt caused by angular dispersion. Opt. Eng. **32** (1993): 2501–2504.
[69] J. Hebling. Derivation of the pulse front tilt caused by angular dispersion. Opt. Quantum Electron. **28** (1996): 1759–1763.
[70] G. Pretzler, A. Kasper and K. J. Witte. Angular chirp and tilted light pulses in CPA lasers. Appl. Phys. B **70** (2000): 1–9.
[71] I. Z. Kozma, G. Almasi and J. Hebling. Geometrical optical modeling of femtosecond setups having angular dispersion. Appl. Phys. B **76** (2003): 257–261.
[72] S. Akturk, X. Gu, E. Zeek and R. Trebino. Pulse-front tilt caused by spatial and temporal chirp. Opt. Express **12** (2004): 4399–4410.
[73] S. Akturk, M. Kimmel, P. O'Shea and R. Trebino. Measuring pulse-front tilt in ultrashort pulses using GRENOUILLE. Opt. Express **11** (2003): 491–501.
[74] A. E. Siegman, Lasers, University Science, Mill Valley, CA, 1986.
[75] S. Akturk, X. Gu, P. Gabolde and R. Trebino. The general theory of first-order spatio-temporal distortions of Gaussian pulses and beams. Opt. Express **13** (2005): 8642–8661.
[76] P. Gabolde, D. Lee, S. Akturk and R. Trebino. Describing first-order spatio-temporal distortions in ultrashort pulses using normalized parameters. Opt. Express **15** (2006): 242–251.
[77] C. J. Zapata-Rodriguez, P. Andres, G. Minguez-Vega, J. Lancis and J. A. Monsoriu. Broadband focused waves with compensated spatial dispersion: transverse versus axial balance. Opt. Lett. **32** (2007): 853–855.
[78] C. J. Zapata-Rodriguez and M. T. Caballero. Isotropic compensation of diffraction-driven angular dispersion. Opt. Lett. **32** (2007): 2472–2474.
[79] E. Tal, D. Oron and Y. Silberberg. Improved depth resolution in video-rate line-scanning multiphoton microscopy using temporal focusing. Opt. Lett. **30** (2005): 1686–1688.
[80] G. Zhu, J. V. Howe, M. Durst, W. Zipfel and C. Xu. Simultaneous spatial and temporal focusing of femtosecond pulses. Opt. Express **13** (2005): 2153–2159.
[81] H. Suchowski, D. Oron and Y. Silberberg. Generation of a dark nonlinear focus by spatio-temporal coherent control. Opt. Commun. **264** (2006): 482–487.
[82] D. Oron, E. Tal and Y. Silberberg. Scanningless depth-resolved microscopy. Opt. Express **13** (2005): 1468–1476.
[83] M. E. Durst, G. Zhu and C. Xu. Simultaneous spatial and temporal focusing for axial scanning. Opt. Express **14** (2006): 12243–12254.

Wei Quan, Xuanyang Lai, Xiaojun Liu, Jing Chen

5 Atomic physics in ultrafast intense laser fields

5.1 Introduction

Since its birth in 1960s, laser has become a very useful tool in advancing our understanding of the fundamental light-matter interaction. In 1990s, the development of ultrashort lasers, represented by all-solid-state Ti:sapphire femtosecond lasers, which are tunable in the visible and near-infrared spectral regions, has shown high impact on spectroscopic and dynamical investigations in different areas, including physics, chemistry, and biology. Nowadays, femtosecond lasers are able to generate bursts of light with durations on the timescale of a few tens of femtoseconds ($1\ \mathrm{fs} = 10^{-15}\ \mathrm{s}$), allowing real time tracing of the nuclear motion in molecules. Additionally, since a femtosecond laser pulse can be produced in such short bursts, the laser field can also be extremely intense, allowing unusual regimes of laser-matter interactions to be studied. For example, the electric field of the intense femtosecond laser pulse can become strong enough to compete with the Coulomb field that binds the electrons to atoms (the Coulombic electric field in a hydrogen atom is about $5.14 \times 10^9\ \mathrm{V/cm}$, corresponding to an equivalent intensity of $3.51 \times 10^{16}\ \mathrm{W/cm^2}$). When irradiated by such an intense laser pulse, the laser-atom interaction enters a completely new strong field region in which a variety of novel atomic phenomena appear, such as multiphoton ionization (MPI), above threshold ionization (ATI), high harmonic generation (HHG), multiple ionization (MI) and molecular Coulombic explosion. The systematic investigations of those atomic processes in intense laser fields continually promote several new branches of science. For example, intense laser driven high harmonic generation has become a routine technique for generating tabletop coherent light sources with pulse durations on the attosecond timescale and wavelengths covering ultraviolet to X-ray region, which significantly advances attosecond science. In this chapter, we will focus on recent progress in strong field atomic physics research, one of the most important frontiers in current atomic, molecular and optical physics.

5.2 Basic concepts in strong field atomic physics

In the context of strong field atomic physics discussed in this chapter, 'strong fields' denote femtosecond laser fields with peak intensities covering $10^{13} - 10^{16}\ \mathrm{W/cm^2}$.

Wei Quan, Xuanyang Lai, Xiaojun Liu, State Key Laboratory of Magnetic Resonance and Atomic and Molecular Physics, Wuhan Institute of Physics and Mathematics, Chinese Academy of Sciences, Wuhan 430071, China
Jing Chen, Institute of Applied Physics and Computational Mathematics, Beijing 100088, China
https://doi.org/10.1515/9783110304558-005

When the laser peak intensity is above 10^{16} W/cm^2, the atom will usually be rapidly ionized during the rising edge of the laser pulse and its outmost electrons will actually not experience the higher peak intensity of the laser pulse. At the other end of the scale, when the laser peak intensity is below 10^{13} W/cm^2, the probability of the higher order processes during the laser-atom interaction discussed in this chapter may become extremely small and the signals (e.g. electrons and ions) will usually be difficult to detect or be accumulated. Therefore, the most studied strong field atomic phenomena, such as MPI, ATI and HHG, usually fall in the intensity range of 10^{13}~10^{16} W/cm^2. In this chapter, we will focus on the physical concepts and processes related to those phenomena.

5.2.1 Multiphoton ionization (MPI)

In 1905, Einstein proposed the photon concept and successfully explained the photoelectric effect. According to his theory, when a piece of metal has light shined on it, photoelectrons may be emitted from its surface, under the condition that the frequency of the light is larger than a threshold. The photoelectron's kinetic energy E_k can be given by

$$E_k = h\upsilon - W\,, \tag{5.1}$$

where h is the Planck constant, υ the light frequency, and W the work function of the metal, corresponding to the atomic ionization threshold. According to equation (5.1), in order to ionize the outmost electron atoms, the photon energy has to be able to overcome the work function. In 1931, shortly after the establishment of quantum mechanics, Goeppert-Mayer predicted that, even if the energy of one photon is smaller than the energy gap between two states, the electron can be excited to a higher state by absorbing more than one photon simultaneously [1]. Now it is well-known that if the upper state is a continumm state, the electron can be ionized via a multiphoton ionization process, as depicted in Figure 5.1. In the framework of the quantum me-

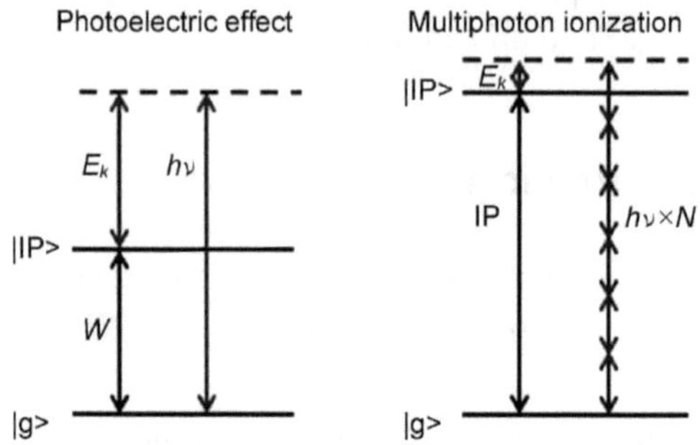

Fig. 5.1: A schematic diagram for the photoelectric effect and multiphoton ionization. $|g\rangle$ denotes the ground state of the atom and IP denotes its ionization potential.

chanical perturbation theory, one can express the multiphoton ionization probability for a nonresonant case as

$$R_N = \sigma_N I^N , \qquad (5.2)$$

where R_N is the total ionization probability, N is the least number of photons to be absorbed to overcome the atomic ionization potential, σ_N is the effective cross section of the N-th order transition and I is the laser intensity.

To observe the multiphoton phenomena, a rather high photon density in the light-atom interaction region is required. This became possible only after the invention of the laser in the 1960s. Shortly after the laser became available, Voronov and Delone observed for the first time the presence of the multiphoton ionization phenomena in the noble gas atomic xenon in 1965 [2]. In their experiments, when the laser electric field was increased to about 1×10^7 V/cm, corresponding to a laser intensity of 1.3×10^{11} W/cm^2, a significant mulitphoton ionization of Xe was observed. For their laser system, one photon energy corresponded to 1.78 eV. The observation of the ionization phenomena meant that at least seven photons had been absorbed by Xe with an ionization potential of 12.13 eV.

Since the effective multiphoton cross section σ_N decreases with the number of absorbed photons, the multiphoton ionization for various atomic targets may be observed at different laser intensities. For example, by employing a laser system with a pulse duration of 50 picoseconds and a wavelength of 1064 nm, the four-photon ionization of Cs atoms, the 11-photon ionization of Xe, and the 22-photon ionization of He can be all studied by setting the laser intensities to about 10^{10}, 10^{13} and 10^{14}–10^{15} W/cm^2 respectively [3].

5.2.2 Above threshold ionization (ATI) and high harmonic generation (HHG)

With further increase in the laser intensity, the interaction of atoms with the intense laser field gives rise to several phenomena that cannot be explained within quantum mechanical perturbation theory. In the following, we will give a brief introduction of two such phenomena, i.e. above threshold ionization (ATI) and high harmonic generation (HHG).

When exposed to a strong laser field, atoms may absorb more photons than necessary for ionization, leading to a series of peaks separated by one photon energy in the photoelectron spectrum. This new phenomenon, termed above threshold ionization (ATI), was first discovered by Agostini et al. [4]. To understand this phenomenon, one has to somehow go back to a general picture of MPI described above.

When the laser intensity is not so high, ATI may still be understood by the lowest order perturbation theory (LOPT). Assuming that the atom absorbs many more photons, e.g. S photons more than the minimum number N, the S-order ATI probability can be expressed in terms of the LOPT,

$$R_{N+S} = \sigma_{N+S} I^{N+S} . \qquad (5.3)$$

Note that the above equation is only approximately held for small S. With the increase of the laser intensity, the ATI photoelectron spectra will exhibit several novel features, e.g. the suppression of the low energy electron peaks, the appearance of a high energy plateau, and a side lobe structure in electron angular distribution, all of which cannot be explained by perturbation theory, indicating that the laser-atom interaction enters a nonperturbative regime. We will discuss these phenomena in more detail in Section 5.5.

High harmonic generation (HHG) represents another typical nonperturbative strong field phenomenon in the context of intense laser-atom interaction. Like ATI, earlier experimental observations of lower order harmonic generation may be understood within quantum mechanical perturbation theory. In 1987, Shore et al. [5] predicted that extremely high order hamonics signals can be emitted when the high energy electron from ATI recombines with the parent ion. Indeed, Ferray et al. [6] and McPherson et al. [7] experimentally observed high order harmonic generation. As HHG is an effective way to achieve tabletop coherent light sources at extreme ultraviolet (XUV) or X-ray wavelengths, it has received much attention since its discovery. In particular, there have been great advances in the past ten years in produing XUV light with attosecond (10^{-18} s) pulse duration based on the HHG technique. It is now becoming possibe to trace the ultrafast electronic motion of matter in its natural timescale. Recent reviews on this research topic can be found in the literature [8–10]. It is worth noting that, physically, both HHG and high order ATI can be well understood within an intuitive electron rescattering scenario, which will be discussed below.

5.2.3 Tunneling ionization and over the barrier ionization

Before an in-depth understanding of the physical picture behind strong field atomic phenomena, for which quantum mechanical perturbation theory completely fails, we will first introduce a basic concept well employed in strong field atomic physics, i.e. tunneling ionization. According to quantum mechanics, a particle can escape through a barrier with a certain probability even when its energy is below the height of the barrier. This phenomenon of tunneling has no counterpart in classical physics. In Figure 5.2 (a), we illustrate the quantum tunneling concept by considering a hydrogen atom in an external static electric field. In this case, the atomic Coulomb potential can be strongly suppressed by the external field and, as a consequence, the initially bound electron may tunnel out through the suppressed Coulomb barrier and become free with some probability.

In 1964, Keldysh made an inspired innovation by applying the tunneling ionization picture, which was usually considered to be valid, to the study of ionization processes in optical fields [11]. For an optical field with a low enough frequency and a high enough intensity, Keldysh assumed that one can take it as a quasistatic field. Therefore, the bound electron in an atom, when subject to an intense laser field, could

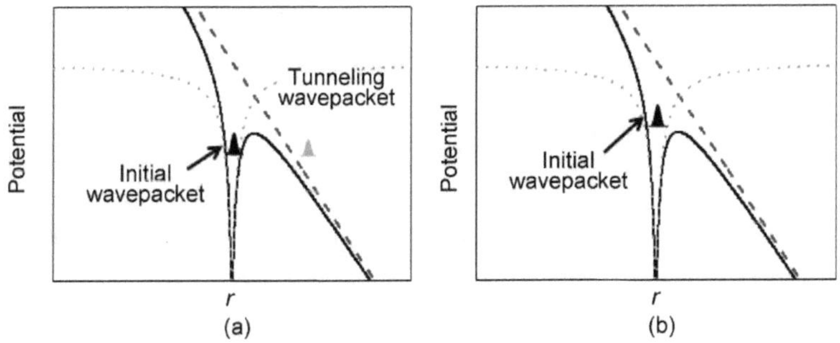

Fig. 5.2: A schematic diagram of tunneling ionization (a) and over the barrier ionization (b) for a hydrogen atom in a static electric field. The dotted line, the dashed line and the solid line denote the atomic Coulomb potential, the static electric field and the distorted Coulomb potential respectively.

also be ionized by tunneling through the barrier created by the combination of the atomic Coulomb potential and the external laser electric field. In this case, the ionization probability only depends on the depth of the barrier and thus on the laser field strength. If treated with a classical approach, considering a laser electric field with an amplitude of E, the barrier width is $l = (E_{\mathrm{IP}}/eE)$, where E_{IP} is the atomic ionization potential and e is the electron charge. The velocity of the electron with energy of E_{IP} is $\sqrt{E_{\mathrm{IP}}/m}$, and the time (t) the electron takes to cross over the barrier will be

$$t = \frac{\sqrt{2mE_{\mathrm{IP}}}}{eE} \tag{5.4}$$

and

$$\omega_t = \frac{eE}{\sqrt{2mE_{\mathrm{IP}}}} \,. \tag{5.5}$$

To qualitatively distinguish the atomic ionization process in intense laser field, a Keldysh parameter (or adiabatic parameter) γ can be defined:

$$\gamma = \frac{\omega}{\omega_t} = \frac{\omega\sqrt{(2mE_{\mathrm{IP}})}}{eE} \,. \tag{5.6}$$

This parameter corresponds to the ratio between the time that the electron goes through the barrier and the oscillating period of the laser field. When $\gamma \ll 1$, i.e. the tunneling time is much less than the laser optical period, the ionization can be considered as a tunneling process as in a static field; on the other hand, when $\gamma \gg 1$, the characteristic timescale for the electron tunneling is larger than the laser optical period. In this case, one has to consider the oscillation of the laser field, where the quasistatic approximation fails and the ionization can still be taken as a multiphoton process.

In 1986, Ammosov, Delone and Krainov further developed this tunneling ionization model based on Keldysh's pioneer work and gave a simple formula (named the ADK formula thereafter) for the calculation of the ionization probability of atoms in an intense laser field [12, 13]. For a linearly polarized laser field E, the ionization probability for an atom initially in an s state can be calculated as

$$W_{\mathrm{ADK}} = \sqrt{\frac{3n^{*3}E}{\pi Z^3}\frac{ED^2}{8\pi Z}} \exp\left(-\frac{2Z^3}{3n^{*3}E}\right) . \tag{5.7}$$

Here $n^* \equiv Z/\sqrt{2E_{\mathrm{IP}}}$ is the effective principle quantum number, Z is the charge of the atomic residue, and $D \equiv (4eZ^3/En^{*4})n^*$. For a circularly polarized laser field, the ionization probability for an s-state is

$$W_{\mathrm{ADK}} = \frac{ED^2}{8\pi Z} \exp\left(-\frac{2Z^3}{3n^{*3}E}\right) . \tag{5.8}$$

These ADK formulae have been commonly adopted in the treatment of tunneling ionization in intense laser fields.

It is noteworthy that, in real experiments, the precondition $\gamma \ll 1$ is usually not well satisfied. Sometimes one may also to judge if the ionization can be taken as a tunneling process:

$$\gamma < 0.5 , \quad \text{and} \quad F < \frac{E_{\mathrm{IP}}^2}{4Z} . \tag{5.9}$$

In reality, for most experiments performed with the commercial Ti:sapphire laser system with infrared wavelength, the condition (5.9) is still difficult to satisfy well, due to the limit of saturation intensity of atomic systems. Nevertheless, as we will show in the following sections, the Keldysh parameter is very instructive in the discussion of the physical mechanism behind strong field ionization of atoms and simple molecules.

Numerous experimental endeavors have been undertaken to test the validity of the tunneling ionization concept. In 1985, Chin et al. [15] investigated the multiphoton ionization phenomenon of Xe atoms using a CO_2 laser. They found that the measured ion yield scales nonlinearly with the laser intensity, and moreover, the ion yield does not change significantly with the laser wavelength. This provided preliminary experimental evidence that the atom may ionize via tunneling in an intense laser field [13, 14]. Later on, Mevel et al. [16] experimetally confirmed the transition from multiphoton ionization to tunneling ionization by recording and comparing the photoelectron spectra of various noble gas atoms at a series of laser intensities.

Their results showed that the photoelectron spectra of Xe exhibit several sharp fine structures, which can be ascribed to the resonance of the laser field with highly excited Rydberg states, indicating that the ionization of Xe falls in a typical multiphoton ionization regime; see Figure 5.3 (a). In contrast, the photoelectron spectrum for He atoms (Figure 5.3 (b)) shows a monotonous drop with electron energy, without any sharp resonance structures. This feature is consistent with the prediction of the tunneling ionization theory.

With the increase in laser intensity, it is predicted that the atomic Coulomb barrier will be further suppressed and the summit of the barrier might be lower than the elec-

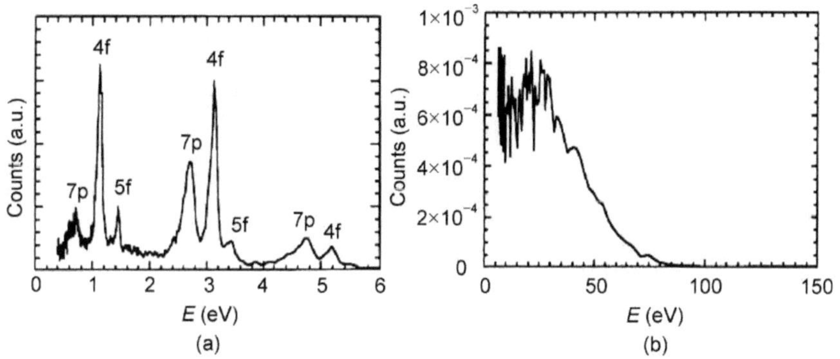

Fig. 5.3: Photoelectron spectra of (a) Xe and (b) He. The corresponding Keldysh parameters (γ) are 1.89 and 0.38 for Xe and He respectively. After [16].

tron energy. In this case, over the barrier ionization (OBI) will occur, as shown in Figure 5.2 (b). Taking the field suppressed Coulomb barrier equal to the electron energy in the ground sate, one gets the critical laser field strength needed for OBI to occur [17]

$$F_{\mathrm{OBI}} = \frac{E_{\mathrm{IP}}^2}{4Z} \,. \tag{5.10}$$

Note that under intensities high enough for OBI to play a role, the ionization probability will be overestimated if tunneling ionization theory is still employed.

5.2.4 Electron rescattering scenario

In 1989, it was found that when a gas target is irradiated by an intense laser pulse, the produced HHG spectra can form a plateau structure with a cutoff energy of $I_{\mathrm{p}} + 3U_{\mathrm{p}}$ [18], where I_{p} is the ionization potential of the gas target, and U_{p} is the pondermotive energy, i.e. the cycle averaged quiver energy of the electron in an oscillating electric field. Later, Schafer et al. [19] found that a similar plateau structure and cutoff also exist in high order ATI spectra. Moreover, in the study of atomic double ionization of noble gas targets in an intense laser field, L'Huillier et al. [20] and Walker et al. [21] found that, for a wide range of laser intensities, the observed doubly charged ion yields cannot be explained by the theoretical calculations based on the single active electron (SAE) approximation, implying that the two electrons are not emitted one by one, but in a nonsequential way. The double ionization experiments indicated that electron correlation should play a signifcant role in the double ionization process. All the above interesting experimental findings have received much attention in the research community and significantly advanced the exploration of the underlying physical mechansim behind strong field atomic ionization.

In 1993, Corkum et al. [22] and Schafer et al. [19] proposed a rescattering scenario (also called a three step model), which can well explain qualitatively the strong field

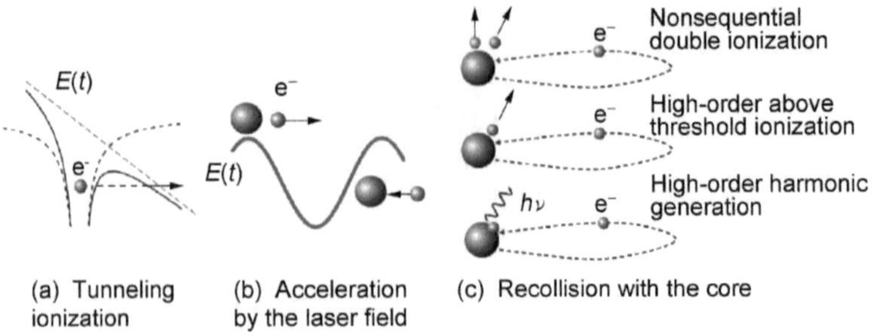

Fig. 5.4: Schematic diagram of rescattering scenario.

atomic phenomena, e.g. HHG, high-order above-threshold ionization (HATI) and non-sequential double ionization (NSDI). According to this scenario (Figure 5.4), in the first step, the electron is tunnel ionized through the lowered Coulomb barrier; in the second step, the freed electron is driven by the laser field, moving far away from the ionic core first and then back to recollide with it; and in the third step, the electron may rescatter elastically upon or recombine with the ionic core, resulting in HATI or HHG. It may also rescatter inelastically with the core, imparting some of its energy to the second still bound electron, causing both of them to become free. This process corresponds to NSDI.

In the past 30 years, there has been increasing experimental evidence that the rescattering scenario can account for most features in intense field-atom interactions, e.g. the plateau structure and cutoff position in HHG and HATI spectra, and the abnormal double charged ion yields in atomic double ionization. It can therefore be considered as a basis for a more comprehensive understanding of a variety of strong field atomic phenomena.

5.3 Relevant experimental techniques

In this section, we will introduce relevant experimental techniques for investigating the ionization dynamics of atoms and molecules subject to intense laser fields, such as multiphoton ionization, above threshold ionization and nonsequential double ionization. In particular, we will discuss ultrafast laser techniques and atomic and molecular spectroscopies.

5.3.1 Ultrafast laser technique

Throughout the history of ultrafast lasers, the pulse duration kept on decreasing, while the laser intensity kept on increasing. In the 1960s, the power of the first ruby laser was on the order of kW and very unstable. The pulse duration was between

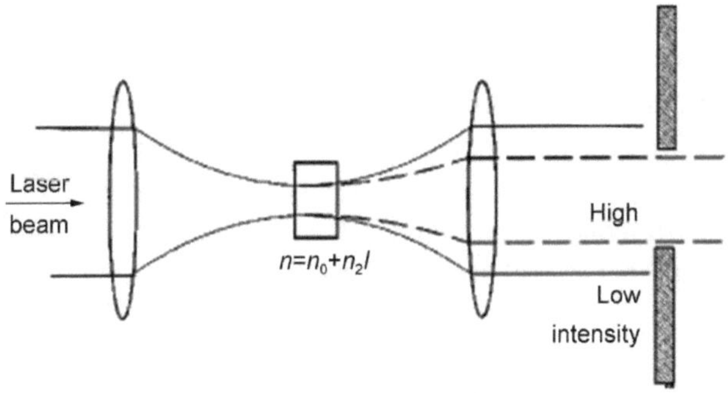

Fig. 5.5: Sketch of the Kerr-lens mode locking mechanism. After [23].

1 ms to 1 s. Shortly after that, thanks to the introduction of the Q-switch technique, the pulse duration was shortened to 10 ns and the power increased from the order of kW to the order of MW. Considering that the laser field with intensity on the order of 10^{13} W/cm^2 or higher is necessary for investigating strong field atomic physics, the laser pulse duration needs to be further compressed and the laser intensity be further increased. Nowadays, the most popular laser system for strong field atomic physics, i.e. the Ti:sapphire laser system, is based on the chirped pulse amplification (CPA) principle. A brief introduction to the working principle of the femtosecond Ti:sapphire laser system will be given below.

Generally, the ultrafast laser system contains two parts: an oscillator and an amplifier. The gain of the oscillator is from a Ti:sapphire crystal pumped by a continuous laser. Ti:sapphire crystals with broad bandwidth (larger than 200 nm) support a wide spectrum of laser outputs, making it possible to generate intense ultrashort laser pulses. Besides the wide spectrum of the laser output, to generate femtosecond laser pulses mode locking of the longitudinal modes is imperative. Currently, all commercial laser systems use the technique of Kerr-lens mode locking (KLM) [23]. The locking mechanism can be understood as follows: when intense laser field is input into a uniform but asymmetric crystal, the third order optical Kerr effect will be generated and the crystal refractive index can be approximately considered as a linear response to the laser intensity,

$$n(I) = n_0 + n_2 I\,. \tag{5.11}$$

Here, n_0 is the linear refractive index of the crystal, n_2 the coefficient of the nonlinear refraction index, and I the laser intensity. When the intensity is larger than a certain threshold intensity [24], the self-focusing effect of the laser field will take place in the crystal, so that the laser beam propagates in a columnar space with a small radius. Since the self-focusing effect is related to the laser intensity, we can modulate the light intensity by limiting the size of the columnar space for the laser propagation, e.g. by limiting the waist radius of the pumping light beam or by adding a hole in the oscillation cavity, as shown in Figure 5.5. When the laser pulse passes the aperture, the laser intensity will be modulated, leading to two side band peaks in the frequency spectrum. The frequency of the two side band peaks will be very close to the frequency of the lower or higher mode, resulting in strong interference, which will effectively compress the pulse duration to the order of femtoseconds.

The power of the oscillator is on the order of nJ, which is too low for most atomic and molecular ionization experiments. Usually, an amplifier can be employed to achieve higher laser intensity. To avoid the nonlinear effect caused by high laser power and the damage to the amplifier crystal, the pulse duration of the output beam from the oscillator is stretched to several hundred picoseconds using a prism or a grating pair. Then the beam is injected into the amplifier, where the beam intensity will be amplified by about six orders of magnitude to several mJ. To realize such a high amplification, the laser beam will go through the Ti:sapphire crystal many times. The pulse will then be compressed back with a prism or with a grating pair to the original duration. This technique, named chirped pulse amplification (CPA), was firstly realized by Mourou et al. in 1985 [25–27] and has promoted the development of ultrafast laser techniques to a great extent.

Generally, there are two kinds of amplifiers, i.e. regenerative amplifiers and multipass amplifiers [28, 29]. For regenerative amplifiers, the gain material is placed into the linear resonator chamber, where the seed pulse can be amplified. The timing sequence of the input and output laser is controlled with two Pockels cells, which can quickly start up and shut down to allow one of the laser pulses with specific polarization direction to pass through based on the Kerr effect. Generally, the seed beam will pass the crystal dozens of times before it passes through the Pockels cells, resulting in serious high order dispersion, which cannot be compensated completely. For multipass amplifiers, because the seed beam passes the crystal fewer times, it is easier to achieve desirable dispersion compensation.

Additionally, because the gain is associated with the laser wavelength and reaches a maximum around 800 nm, the power of the amplified laser field depends on the laser wavelength. After several rounds of amplification, the distribution of the power of the amplified laser becomes narrower in the frequency domain than that of the seed beam, which is usually called the gain narrowing effect and leads to a lower limit (around 20 fs) of the pulse duration.

The output laser pulse usually contains more than 10 optical cycles. For the multicycle laser field, the interaction between the laser field and the atoms or molecules can happen inside any of these cycles. Experimentally, only the averaged results can be measured. To study the ionization dynamics inside one optical cycle, one need few-cycle laser pulses. According to the uncertainty principle, to achieve a few-cycle laser pulse, the spectrum has to be broadened in the frequency domain with a hollow fiber compressor, as is depicted in Figure 5.6. The output laser beam from the amplifier is directed into the noble gas filled vacuum chamber and focused into the entrance of the hollow fiber. During its propagation inside the fiber, the interaction between the noble gas atoms and the laser pulse will generate a self-modulation effect, which will effectively broaden the beam to around 300 nm. Additionally, the laser pulse is positive

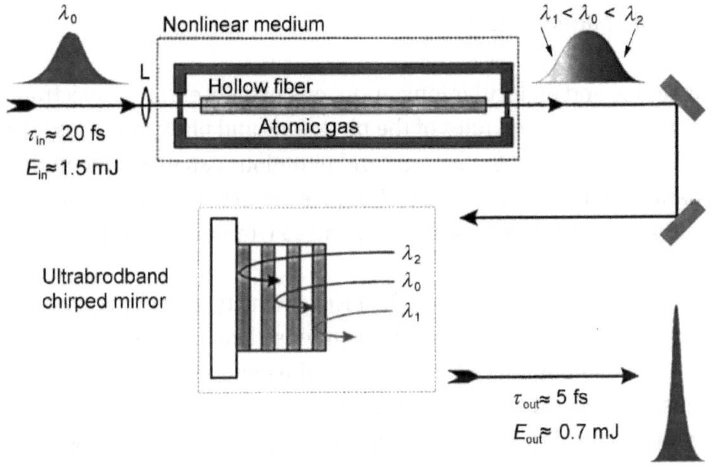

Fig. 5.6: Principle of spectral broadening by self-phase modulation inside gas filled hollow fibers, and of chirp compensation. After [23].

chirped during the spectrum broadening, which must be compensated for to achieve a few-cycle pulse. Hence, the output beam from the fiber is introduced into a set of chirped mirrors, which can add negative chirp to the pulse and compress the pulse duration to less than 5 fs.

Experimentally, the laser beam will be focused before it goes into the target chamber to achieve high intensity. For an ultrashort laser pulse, the high intensity can be achieved with an optical lens. For the popular commercial Ti:sapphire amplifier, the repetition rate is around 1 kHz, the center wavelength is around 800 nm, the single pulse energy is 1 mJ, and the pulse duration is 50 fs, which corresponds to the laser power of 2×10^{10} W. To achieve the laser intensity of 2×10^{14} W/cm^2, the area of the focal spot should be around 10^{-4} cm^2, corresponding to the beam waist of about 60 μm. For an ideal Gaussian beam, the relationship between the radius of the beam waist w_0 and the input beam radius w is given by [30]

$$w_0 = \frac{\lambda f}{\pi w} . \tag{5.12}$$

If the beam radius is 5 mm, it can be calculated that the focal length should be 1200 mm for the wavelength of 800 nm. If a lens with a focal length of 100 mm is employed, the laser intensity can be larger than 10^{16} W/cm^2. Technically, the simple spherical lens cannot achieve an excellent focus spot because of the optical aberration. Usually, an achromatic lens or a parabolic mirror is employed to get a better focusing effect. Along the laser beam, the length of the focus can be defined with the Rayleigh length,

$$Z_R = \frac{\pi w_0^2}{\lambda} . \tag{5.13}$$

The Rayleigh length of a Ti:sapphire laser pulse as mentioned above is about 14 mm.

5.3.2 Photoelectron and photoion spectrometer

To study strong field atomic and molecular ionization, one usually employs spectrometers to measure the momenta and energies of the photoions and photoelectrons. The analysis of the information about photoelectrons and photoions can reveal the underlying physical mechanism. The most popular techniques are time of flight mass spectrometry, photoelectron kinetic energy spectrometry and cold target recoil ion momentum spectroscopy.

By recording the ionization events versus the laser intensity, mass spectroscopy can help us study atomic ionization dynamics in the intense laser field. For example, if the laser intensity is high enough, it was found that the dependence of ionization rate on the laser intensity shows an exponential law instead of a power law, indicating that tunneling ionization is the dominant mechanism. Moreover, as will be discussed in Section 5.6, the nonsequential double ionization and strong electron-electron correlation were found by measuring the dependence of double charged ion yields with respect to the laser intensity, where a 'knee' structure is identified.

Photoelectron kinetic energy spectroscopy is constructed based on the time of flight principle, i.e. the electron kinetic energy is calculated from the time of flight from the interaction point to the detector. Moreover, the angular distribution can be measured if the laser polarization direction can be rotated precisely. To get an accurate energy distribution, it is necessary to shield the residual electric and magnetic fields, which is usually realized with a μm metal shield. Photoelectron kinetic energy spectroscopy is usually used to study above threshold ionization. With this technique, the plateau and the angular distribution of high order above threshold ionization were experimentally observed in the 1990s.

In 2000, cold target recoil ion momentum spectroscopy (COLTRIMS) [31, 32] was developed and greatly promoted the development of atomic and molecular many-body correlation experimental research. In Section 5.6, we will show that the ion momentum distribution and the electron momentum correlation distribution can provide direct experimental evidence that the rescattering is the dominant physical mechanism behind nonsequential double ionization. As a coincident measurement technique, COLTRIMS greatly advanced the understanding of the multiple fragmentation processes in the interaction of strong laser fields with atoms and molecules. To ensure that the photoelectron can be cataloged into the correct reaction channel, it is crucial to reduce the ionization rate to far less than 1. However, this requirement usually results in a very long acquisition time, and thus a laser with high repetition rate and high stability in both wavelength and energy is required.

Generally, there are four steps for the experimental study of strong field atomic physics, including sample preparation, ionization, charged fragment detection, and data analysis, as shown in Figure 5.7. Usually, the steps of sample ionization and charged fragment detection are accomplished inside the vacuum chambers. The ionization step is realized by the interaction of the intense laser field with the atoms and molecules and will not be further discussed here. In the following, we will discuss the other three steps in detail.

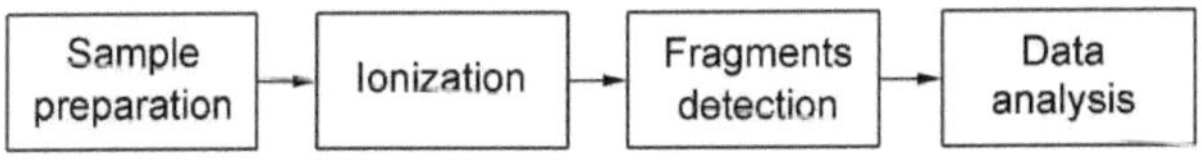

Fig. 5.7: A schematic overview of the experimental procedure.

5.3.2.1 Sample preparation

In the study of atomic and molecular physics, the properties of the single atom and molecule are investigated. Hence, it is necessary to eliminate the interaction between the atom or molecule and its neighbors. To this end, experimentally, the sample is prepared as a dilute gas. If the sample is naturally solid or liquid, it has to be vaporized. For most experiments with ion mass spectroscopy and photoelectron kinetic energy spectroscopy, the sample can be prepared with a leak valve, which is connected to the high pressure cylinder through a pressure regulating assembly. The sample flow introduced into the vacuum chamber can be controlled accurately with the high precision leak valve. The sample introduced in this way can pervade the chamber uniformly, which makes it very convenient to read the sample pressure with a vacuum gauge. Additionally, the interaction point is exclusively determined by the laser focus.

Alkali metals or alkali earth metals must be vaporized with an atomic oven from their natural solid state at room temperature. The atomic oven is composed of two parts: the sample cell and the heating system. To produce high enough atomic vapor pressure, the solid sample inside the sample cell should be heated to some certain temperature. The temperature can be detected with a thermocouple, which can be further applied to control the heating current. There is an aperture in the front of the oven. When the size of the aperture is less than the mean free path of the atoms inside the oven [33], an ideal effusive beam can be emitted with further collimation. Considering the construction of the differential pumping system and the collimation system, the sample preparation process is more complicated compared with the one discussed above. Additionally, the laser focus has to be well adjusted into the beam to ionize the sample, which makes the experimental procedure even more difficult.

In some of experiments, the thermal motion induced by ion momentum distribution is even wider than that produced by the laser field. Therefore, to study the photoion momentum distribution for atoms and molecules subject to intense laser fields, the target should be introduced with a supersonic beam, by which the temperature of the beam is cooled so low that the influence of thermal motion can be ignored. The principle can be comprehended as the high pressure sample noble gas atoms passing through the supersonic nozzle by adiabatic expansion, where the thermal energy is transferred into translational kinetic energy, resulting in cooling of the beam. After collimation with a skimmer and a series of pinholes, the well collimated supersonic beam is achieved. Compared with the two former sample preparation methods, the supersonic beam leads to a narrower thermal distribution.

5.3.2.2 Charged fragment detection

The photofragments from the interaction between laser field and atoms or molecules can be acquired with a detection system. Here, a brief introduction on this topic will be given.

There are several detecting schemes, such as electric sector, magnetic sector, quadrupole, time of flight mass spectrometer, etc. Here, we will focus on the time of flight scheme, which is fast, compact, not sensitive to the dimension error, and able to acquire multiple kinds of particles simultaneously. The only drawback of the scheme is that the resolution is not as good as the others. Recently, with the rapid development of the relevant techniques, the resolution of spectrometers based on the time of flight principle has been improved significantly. Figure 5.8 shows the geometry of a typical time of flight (TOF) mass spectrometer [34]. During the experiment, the atoms and molecules are ionized by the laser field in the field free region. After a well controlled delay, a pulse electric field is applied to extract the ions into the acceleration region. The accelerated ions are detected by the detector after another field free flight. The time spent in flight by the ion is recorded with high resolution. Analysis reveals that this time is closely related to the charge-to-mass ratio, which means that the charge-to-mass ratio can be determined with the time of flight. The mass resolution can be improved with the delayed pulse electric field and with combination of acceleration and field free regions.

If the number of charged particles is large, it is possible to acquire the signal with a Faraday cup. For weak signals, it is usually measured by electron multipliers, Channeltron electron multipliers (CEMs) and multichannel plates (MCPs). The latter will be discussed in detail. With MCP detectors, the weak signals can be amplified millions of times. Equipped with right anode, it is even possible to realize imaging. Figure 5.9

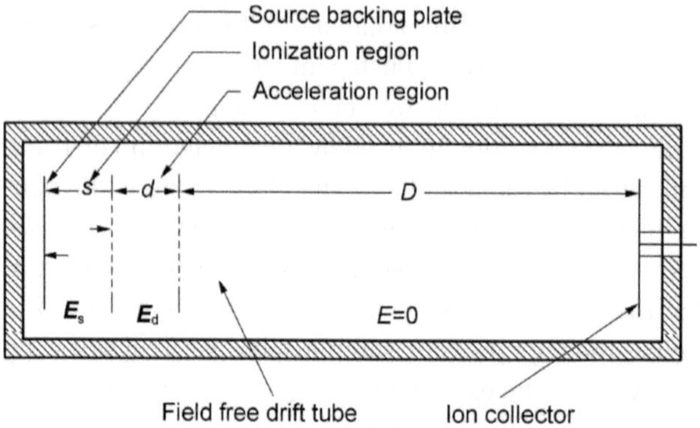

Fig. 5.8: Geometry of a typical TOF mass spectrometer. After [34].

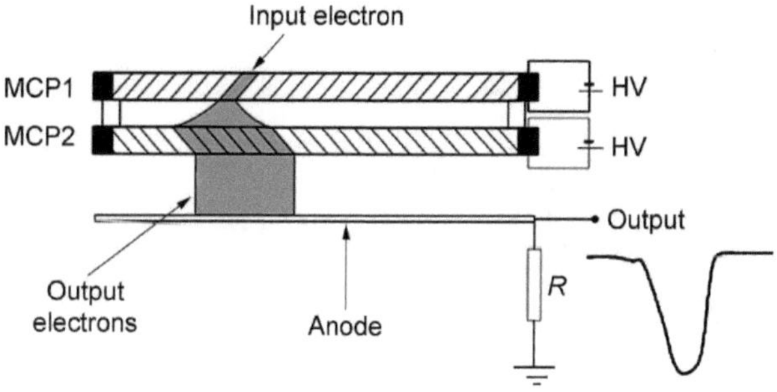

Fig. 5.9: A schematic diagram of a MCP detector. After [35].

shows the principle of the MCP detector [33, 35], where the incident electron enters one of the channels of the MCP plates. With the acceleration of high electric field, the electron hits the wall of the channels and emits secondary electrons, which can induce even more electrons. These electrons will be amplified further in the second MCP plate and the amplified electrons transformed into a significant signal pulse, which will be eventually collected with a detector anode and outputted by a coupling circuit. In this way, one single electron or ion can induce a significant signal pulse.

Figure 5.10 shows the principle of a typical photoelectron kinetic energy spectrometer [36]. The laser beam is focused into the interaction region. The photoelectrons fly into the field free region and are detected by the MCP detector through a pinhole. With precise measurement of the time of flight and distance, we can calculate the kinetic energy. If the polarization direction of the laser beam can be altered with high resolution, it is possible to acquire the angle resolved kinetic energy distribution. The pinhole between the interaction region and the field free region can be employed to select ionization events inside the laser focus, and thus, the distribution of the laser intensity can be scanned with careful tuning of the lens position.

Recently, a detection scheme termed velocity mapping imaging (VMI) was developed [37], with which the angle resolved kinetic energy distributions can be measured with high energy resolution in a single event. Figure 5.11 gives a schematic diagram of the electron lens for VMI spectroscopy.

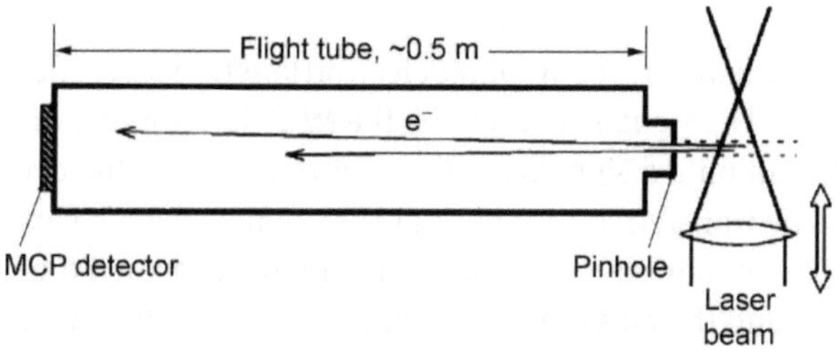

Fig. 5.10: Photoelectron kinetic energy spectrometer. After [36].

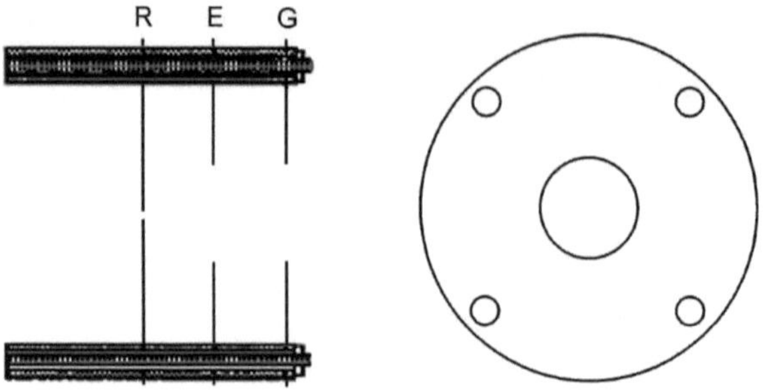

Fig. 5.11: A schematic diagram of the electron lens for VMI spectroscopy. After [37].

The electron lens is composed of three annual electric field plates with identical outer diameter. The inner diameter of plate R is small enough to calibrate the supersonic beam. The inner diameters of plate E and G are identical and larger than that of R. The ionization happens inside the region between R and E. The inhomogeneous electric field generated by E, G and R can focus all the electrons with identical vector velocities into one point in the detector. Generally, to reap a desirable focus effect, the voltage for E is supposed to be 0.7 times that for R. In the meantime, G is grounded. One obvious difference between VMI and kinetic energy spectroscopy is that the position sensitive detector is only necessary for VMI. The sign of the electric field can be chosen to be negative or positive, depending on the charge polarity of the particle. For electron lenses, the voltage of R and E should be negative.

COLTRIMS is more complicated than any of the spectrometers discussed above. Its principle is sketched in Figure 5.12. The supersonic beam crosses the laser beam in the interaction area, where the atoms or molecules are ionized. The photoelectrons and photoions produced during ionization are accelerated towards their own detectors. A pair of Helmholtz coils is employed to confine the transverse movement of the photoelectrons to achieve the detection with 4π solid angle. Generally, MCP detectors with delay-line anodes for position encoding are equipped to measure the impact times and positions of photoelectrons and photoions, which can be used to retrieve their three-dimensional momenta.

5.3.2.3 Data analysis

The data analysis procedure for mass spectroscopy, photoelectron kinetic energy spectroscopy and COLTRIMS is similar. The signals detected by the MCP detectors are amplified by the preamplifiers. The amplified signals are then introduced into the constant fraction discriminators to be transferred into normal NIM signals, which carry the precise time information of the photofragments. According to the measured time, the momenta, the kinetic energy and the mass-to-charge ratio can be calculated. For example, in a photoelectron kinetic energy spectrometer, if the electron has traveled for the length of L with the time of t, its energy can be calculated with the equation

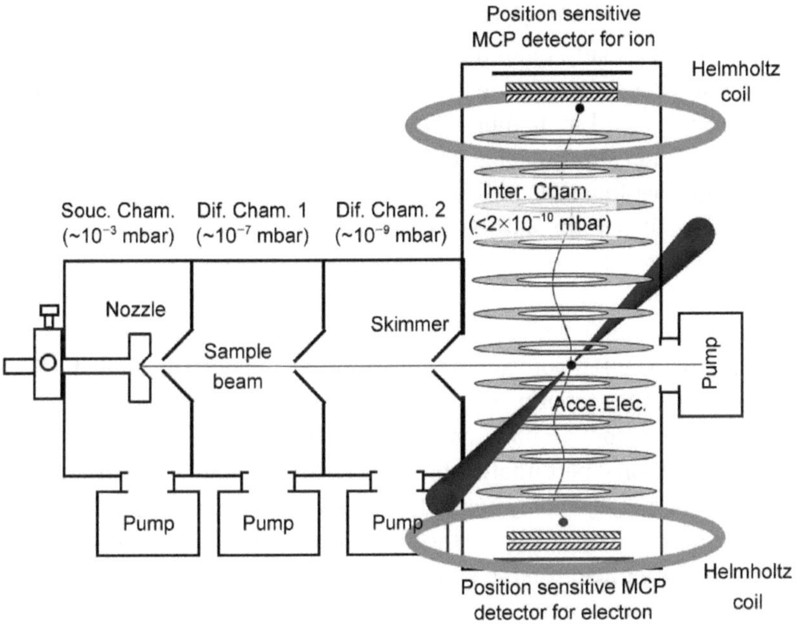

Fig. 5.12: A schematic diagram of COLTRIMS.

$v = L/t$. For signals from the MCP detector with a delay-line anode for position imaging, the spatial distributions can be retrieved, which can be further used to obtain the transverse momenta of the photofragments. All the time and position information can be stored and processed one by one in list mode by a personal computer.

5.4 Theoretical approaches

The above threshold ionization phenomenon found by Agostini et al. in 1979 [4] has aroused great interest in the study of atomic physics in strong laser fields. Following this discovery, people performed extensive and in-depth theoretical investigations on the interaction between intense laser fields and atoms and achieved great success in the description and interpretation of various phenomena in the strong field of atomic physics, which greatly deepened our understanding of behavior of materials in strong laser fields. The interaction of atoms with strong laser fields represents an interaction between electromagnetic fields and matter, or more specifically, the interaction between electromagnetic fields and electrons. For electromagnetic interaction, we already have a very mature and precise theory, namely quantum electrodynamics (QED), so the ionization process should in principle be studied by the theory of quantum electrodynamics. However, the laser field is usually treated as a classical field due to the following reasons. First of all, the laser field is a highly coherent light source. When the laser intensity gets up to 10^{13} W/cm^2, the estimated number of photons per unit phase space, n, is about 10^{18}. Usually, quantum fluctuation is inversely proportional to the number n, so it is accurate enough to treat the laser field as a classical field. Secondly, the existing experiments and simulations show that when the laser intensity is lower than 10^{22} W/cm^2, it is not strong enough to produce electron-positron pairs. Thus, in the range of current laser intensity, the electron can be treated as a classical particle. In fact, there are a small amount of works in which the light fields are quantized and the ATI is studied with the scattering theory (for more details, see [38–44]). Note that their results are almost the same as those obtained from the quantum S-matrix theory. In addition, it is not very convenient to study the interaction between the laser pulse and electron with QED theory and it is relatively complicated to study the time dependent system with the scattering theory. So here we mainly introduce theoretical models and numerical methods within the framework of quantum mechanics.

Furthermore, traditional perturbation theory is no longer valid in the strong laser field and thus it is difficult to employ methods under the framework of quantum mechanics, which brings great challenges to the theoretical study of strong field physics. Considering that the laser field used in experiments is usually in the infrared region and even in the mid-infrared region, the corresponding photon energy is about 1 eV (e.g. for the commonly used Ti:sapphire laser, the wavelength $\lambda = 800$ nm, $\hbar\omega = 1.55$ eV). In general, the ionization potential of atoms and molecules is about a dozen electron volts or more and the kinetic energy of the electron in the laser field, i.e. the ponderomotive potential, is much larger than the photon energy, and thus tens or several hundred photons are involved in strong field atomic physics processes. According to the correspondence principle of quantum and classical perspectives, these extreme multiphoton processes can be treated with classical methods. Thus, a classical model is being developed to study the interaction of atoms with strong laser fields and great progress has been made in this field. In particular, the classical model could provide a very clear picture of the interaction of atoms and molecules with intense laser fields.

Additionally, a semiclassical model has been developed to study the interaction between atoms and strong long wavelength laser fields. In this model, the electron is ionized by tunneling through a barrier, which, to a certain extent, retains quantum properties of the interaction processes. After tunneling ionization, the motion of the electron in the atomic potential and the laser field could be described with classical Newtonian equation of motion. This semiclassical model greatly simplifies the calculation and provides a clear physical picture of strong laser field ionization dynamics and thus has been widely used in investigations of strong field atomic physics.

Currently, the main theoretical methods in the study of the interactions of atoms and molecules with strong laser fields include the time dependent Schrödinger equation (TDSE), quantum mechanics S-matrix theory, and fully classical and semiclassical models. These methods will be introduced in the following sections. Furthermore, in Section 5.6, we will discuss the important progress these theoretical models have made in explaining strong field physical experiments.

5.4.1 Numerical solution of time dependent Schrödinger equation

According to the discussion above, the most direct and accurate method to study the interaction of atoms with intense laser fields is the numerical solution of time dependent Dirac equation. For the current laser fields, we usually concerned with neutral atoms or low charged ions. Because the effect of the magnetic field of the laser fields on the electrons is almost negligible in comparison with the electric field effect, the nonrelativistic Schrödinger equation is accurate enough to describe the processes without considering the relativistic effect and magnetic field effect.

Depending on the form of the electromagnetic interaction, i.e. the gauge of the electromagnetic interaction, the Schrödinger equation can be written in the following two forms:

$$i\hbar\frac{\partial}{\partial t}\Psi(r,t) = -\frac{\hbar^2}{2m}\nabla^2\Psi(r,t) - \frac{i\hbar e}{m}\mathrm{A}\cdot\nabla\Psi(r,t) + \frac{e^2}{2m}A^2\Psi(r,t) + V(r)\Psi(r,t) \quad (5.14)$$

$$i\hbar\frac{\partial}{\partial t}\Psi(r,t) = -\frac{\hbar^2}{2m}\nabla^2\Psi(r,t) - eE(t)\cdot r\Psi(r,t) + V(r)\Psi(r,t)\ , \quad (5.15)$$

which correspond to the two different forms of interaction

$$V_{\mathrm{I}}^{A} = -\frac{i\hbar e}{m}A\cdot\hat{p} + \frac{e^2}{2m}A^2 \quad (5.16)$$

$$V_{\mathrm{I}}^{L} = -eE\cdot r\,. \quad (5.17)$$

Here V_{I}^{A} and V_{I}^{L} are written in the velocity gauge and the length gauge respectively. Because the electromagnetic interaction is gauge invariant, the calculations with formulas (5.16) and (5.17) are equivalent in principle and the wave functions of the two solutions differ only by a phase factor $\exp(\frac{ie}{\hbar}\mathrm{A}\cdot\mathrm{r})$, which doesn't change the physical results. However, due to the limit of calculation accuracy, the results obtained from the two gauges will not be exactly the same. Moreover, due to the different forms of interaction, the steps of time and space are different in the calculations. Here, we introduce a numerical solution of the formula (5.15) in the length gauge form with the differential method.

First, we discuss the amount of calculations usually required in the numerical methods. Previous works concerning the interaction between light and atoms usually study the transition between bound states, so the range of the wave functions in the space is limited. But in strong field atomic physics, people mainly study the ionization of the bound electrons. For the ionized electron, the range of the wavefunction is unlimited. Thus, people need to evaluate how large the spatial range should be in the calculations to achieve sufficient accuracy. After electrons are ionized from bound states to continuous states by laser fields, the ionized electrons will oscillate in the alternating laser field, where the oscillation amplitude

$$\alpha_0 = \frac{E_0}{\omega^2}\,. \quad (5.18)$$

Here E_0 is the laser electric field amplitude and ω is the laser frequency. Previous experience shows that the spatial range in the calculation should be at least two to three times α_0 to obtain a reliable ionization rate. For the laser field with wavelength 800 nm and intensity 1×10^{15} W/cm^2, α_0 is about 52 a.u. (atomic units), and thus the spatial range in the calculation is about 100~150 a.u. To further calculate the photoelectron spectra and angular distributions, the spatial range in the calculation should be greatly increased to about 10 times α_0. In addition, because of the limited spatial range in the calculation, the electron wavefunction may reflect at the boundaries, which seriously affects the calculation accuracy. Usually, an absorption potential or mask function is added in the region close to the boundaries to remove the particles that reach the boundaries.

Here we will take the interaction of hydrogen atoms with linearly polarized laser field as an example to introduce the numerical solution of TDSE [45]. This system is azimuthally symmetric with respect to the polarization axis and the magnetic quantum number is a good quantum number, and thus the TDSE calculations are performed in a two dimensional space. For this case, the TDSE becomes (in atomic units)

$$i\hbar\frac{\partial}{\partial t}\Psi(r,t) = \left\{-\frac{1}{2}\nabla^2 - \frac{1}{r} - zE_0 f(t)\sin(\omega t)\right\}\Psi(r,t)\ , \tag{5.19}$$

where the wavefunction ψ is expanded in the spherical coordination,

$$\Psi(\mathrm{r},\theta,\phi,\mathrm{t}) = \sum_{l=0}\Phi_l(r,t)\,Y_l^0(\theta,\phi)\ . \tag{5.20}$$

Next, Φ is dispersed as

$$\Phi_l(r,t) \longrightarrow \Phi_l(r_j,t) \longrightarrow \Phi_l^j(t)\ , \tag{5.21}$$

where

$$r_j = (j-0.5)\Delta r\ . \tag{5.22}$$

Here, Δr is the grid size in the r-direction. If we let

$$g_l^j = r_j\Phi_l^j\ , \tag{5.23}$$

a final differential equation is thus obtained after some algebraic procedure:

$$\begin{aligned} i\frac{\partial}{\partial t}g_l^j &= -\frac{1}{2(\Delta r)^2}\left\{c_j g_l^{j+1} + c_{j-1}g_l^{j-1} - 2d_j g_l^j\right\} + \left(\frac{l(l+1)}{2r_j^2} - \frac{1}{r_j}\right)g_l^j \\ &\quad + E_0 f(t)\, r_j \sin(\omega t)\left(a_l g_{l+1}^j + a_{l-1}g_{l-1}^j\right) \\ &= (H_0 g)_l^j + (H_I g)_l^j\ , \end{aligned} \tag{5.24}$$

where

$$c_j = \frac{j^2}{j^2-\frac{1}{4}}\ ,\quad d_j = \frac{j^2-j+\frac{1}{2}}{j^2-j+\frac{1}{4}}\ ,\quad a_l = \frac{l+1}{\sqrt{(2l+1)(2l+3)}}\ . \tag{5.25}$$

In the differential equation (5.24), the g_1^j and $g_1^{j\pm1}$ are coupled into l by H_0, i.e. the Hamilton of hydrogen, while g_1^j and $g_{l\pm1}^j$ are coupled into j by H_I, i.e. the interaction Hamilton between atom and external field. The time propagation of the equation can be solved by the Peaceman–Rachford method:

$$g_1^j(t+\Delta t) = [I + iH_0\tau]^{-1}[I + iH_I\tau]^{-1}[I - iH_I\tau][I - iH_0\tau]g_1^j(t)\,, \tag{5.26}$$

where I is unit matrix and $\tau = \Delta t/2$. The interaction H_I is considered in the middle of each time step. Because all matrices in (5.26) are the triangular matrices, these equations could be solved by the shooting method.

In order to solve the TDSE, we should obtain the initial wavefunction of the bound electron. Before the interaction of atom with laser pulse, i.e. the envelope of the laser pulse $f(t) = 0$, the system is in the ground state of the field free hydrogen atom. Because TDSE is transformed from the differential equation into the difference equation, the ground state is dependent on the range of space and the grid step. When the grid step tends to infinitely small and the spatial range tends to infinitely large, the solved ground state will become a real ground state of the hydrogen atom. The ground state of the difference equation can be obtained by solving the field free Schrödinger equation with imaginary time propagation algorithms.

First, we have a test ground wavefunction Ψ, which can be expanded with the different eigenstates of a hydrogen atom (the eigenvalues are denoted by E_i)

$$\Psi(t) = \sum_i c_i e^{-iE_it}\Psi_i\,. \tag{5.27}$$

After the imaginary time transform $t \longrightarrow -i\tau$,

$$\Psi(t) = \sum_i c_i e^{-E_i\tau}\Psi_i\,. \tag{5.28}$$

Because the energy of the ground state is the lowest in all bound states, the amplitudes of the excited states will decay quickly with the propagation of τ until finally only the ground state is left. Therefore, a pure ground state can be obtained without the influence of the excited states.

In the following, we will take the ionization of hydrogen atoms in the linearly polarized laser field as an example to introduce the basic numerical method. The electric field of the laser field is E_0 with envelope $f(t)$ and frequency ω. The time step in the calculation depends on ω and E_0. Usually, the time step is about 1/200 to 1/1000 of the optical cycle. The envelope $f(t)$ increases linearly at the beginning for some optical cycles and then remains constant for another 10–20 optical cycles. Generally speaking, some transition states (e.g. the excited states) would be generated during the increase of the pulse and then decay quickly when $f(t)$ remains constant during the first few cycles. After that, the ground state of hydrogen will decay exponentially. In this case, we can define the stable ionization rate. During the propagation of equation (5.28), a mask function, in the form of $\cos^{1/8}$, is used in the region close to the boundaries to remove

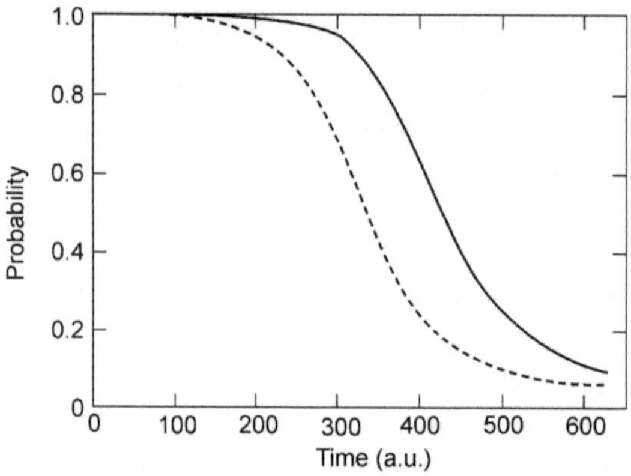

Fig. 5.13: The absolute value of the wavefunction (solid line) and the occupation number of the ground state as a function of the time (in atomic units). The photon energy is 0.2 Hartree and the maximal laser intensity is 1.75×10^{14} W/cm^2. After [46].

the wav packet close to the boundaries in order to avoid the reflection of wave packets from the boundaries. Typical calculated results are shown in Figure 5.13.

5.4.2 Fully classical trajectory method

In the classical method, the motion of electrons in the ionic potential and external field is solved with the Newtonian equation of motion [47]. The simplest case is the motion of an electron for a hydrogen atom in a laser field, where the Newtonian equation of motion is given by

$$\frac{d^2 r}{dt^2} = E(t) - \nabla \frac{1}{r} \,. \tag{5.29}$$

The solution of the Newtonian equation of motion is also related to the initial condition of the electron. To simulate the hydrogen in the strong laser field, the initial condition of the electron is described in a microcanonical distribution, and the electron trajectories from the different conditions are solved with the classical trajectory Monte Carlo (CTMC) method. Finally, we study the ionization of atoms in strong laser fields by analyzing the final energy and momentum of each trajectory. The distribution of the initial conditions is given by the following method. For each electron with given energy, the values of five parameters are selected at random in the following regions:

$$\begin{gathered} -\pi \le \phi \le \pi \,, \quad -1 \le \cos\theta \le 1 \,, \quad -\pi \le \eta \le \pi \,, \\ 0 \le \varepsilon^2 \le 1 \,, \quad 0 \le \theta_n \le 2\pi \,, \end{gathered} \tag{5.30}$$

where ε is the orbital eccentricity, θ_n is an orbital parameter that is proportional to the time, and ϕ, θ and η are the Euler angles. The random distribution of θ_n corresponds to the uniform distribution of the phase in the periodic trajectory of the electron. In addition, it is more convenient to use u rather than θ_n, since it can be obtained with Kepler equation

$$\theta_n = u - \varepsilon \sin u\,. \tag{5.31}$$

In this case, the initial condition of the electron trajectory could be obtained with the following three steps: (1) put the trajectory with ε in an arbitrary direction, e.g. in the y-z-plane with the main axis along z; (2) let the angle of the electron trajectory be u; and (3) rotate the orbital with the Euler angles ϕ, θ and η. Therefore, the initial coordinates and momenta of the electron are given by the following equations:

$$\vec{C}^0 = A\vec{C}^0_0 \quad \text{and} \quad \vec{P}^0 = A\vec{P}^0_0 \tag{5.32}$$

$$\vec{C}^0_0 = \begin{bmatrix} 0 \\ a(1-\varepsilon^2)^{\frac{1}{2}} \sin u \\ a(\cos u - \varepsilon) \end{bmatrix}, \quad a = \frac{Z}{2U}, \tag{5.33}$$

where Z is the nuclear charge and U is the binding potential of the atom, and

$$\vec{P}^0_0 = \begin{bmatrix} 0 \\ b\,(1-\varepsilon^2)^{\frac{1}{2}} \cos u/(1-\varepsilon\cos u) \\ -b \sin u/(1-\varepsilon \cos u) \end{bmatrix}, \quad b = (2mU)^{\frac{1}{2}}, \tag{5.34}$$

where m is the effective mass of electrons in the atom,

$$A = \begin{bmatrix} -\sin\phi\sin\eta + \cos\phi\cos\theta\cos\eta & -\sin\phi\cos\eta - \cos\phi\cos\theta\sin\eta & \cos\phi\sin\theta \\ \cos\phi\sin\eta + \sin\phi\cos\theta\cos\eta & \cos\phi\cos\eta - \sin\phi\cos\theta\sin\eta & \sin\phi\sin\theta \\ -\sin\theta\cos\eta & \sin\theta\sin\eta & \cos\theta \end{bmatrix} \tag{5.35}$$

For a two-electron system, the Newtonian equation of motion is given by

$$\frac{d^2\boldsymbol{r}_i}{dt^2} = \boldsymbol{E}(t) - \nabla\left(V^i_{\text{ne}} + V_{\text{ee}}\right), \tag{5.36}$$

where the subscripts $i = 1$ and 2 denote the first and second electrons respectively. The ionic potential and the Coulomb potential between two electrons are

$$V^i_{\text{ne}} = -\frac{Z_{\text{eff}}}{|r_i|}, \quad V_{\text{ee}} = -\frac{1}{|r_1 - r_2|}, \tag{5.37}$$

where Z_{eff} is the effective charge of the ion and r_i denotes the position of the ith electron. Before solving equation (5.29), we also need to obtain the initial conditions of the two electrons, which can be also described with microcanonical distribution (for more details, see [49]). It is worth pointing out that in contrast to quantum theory, even in the case where there is no external field, the two electrons can interact with each other in the classical model, resulting in the exchange of energy. One of the electrons might be ionized automatically, leading to the unphysical process. To solve this problem, one usually replaces the Coulomb potential V_{ne}^{i} with a softcore potential, i.e.

$$V_{\mathrm{ne}}^{i} = -\frac{Z_{\mathrm{eff}}}{\sqrt{r_i^2 + a^2}} . \tag{5.38}$$

If the softening parameter a satifies

$$a > -\frac{2}{E_{\mathrm{g}}} , \tag{5.39}$$

where E_{g} is the energy of the ground state of the two electrons, then autoionization will not occur.

Obviously, it is much easier to solve the Newton equation (5.29) than the Schrödinger equation. Recently, lots of progress has been made in the study of the ionization of atoms and molecules in strong laser fields with the fully classical model [50–58], which greatly helps us to understand the ionization of atoms and molecules in strong laser fields, especially the dynamics of two electron ionization.

5.4.3 Semiclassical approach

The semiclassical method is suitable for systems with high laser intensity and long wavelength, i.e. the Keldysh parameter in equation (5.6) is much smaller than one. For this case, the tunneling time of the electron in the presence of the laser field is smaller than one optical cycle, and thus the electric field of the laser field can be considered as a static field and the ionization rate can be obtained from the tunneling ionization rate in the static field. The quasistatic Ammosov–Delone–Krainov (ADK) tunneling rate is given by equation (5.7). As we can see from this equation, the tunneling rate increases exponentially with the laser intensity, i.e. the stronger the laser electric field, the larger the ionization rate. Thus, the ionization is most significant at the peak of laser electric field. In terms of semiclassical theory, the bound electron could be ionized by tunneling through a barrier formed by Coulomb potential and laser field at any time, after which it evolves along a specific trajectory according to classical mechanics. The weight of each orbit is the tunneling rate. Thus, the weight becomes largest when the electric field is maximal. The tunneling exit after ionization is about $r \sim E_{\mathrm{IP}}/|e\varepsilon|$. In strong laser fields, the distance between the exit and core is some atomic units. Thus, the initial condition of the electron in the calculation could be taken as

$$x(t = t_0) = 0 , \quad \dot{x}(t = t_0) = 0 . \tag{5.40}$$

After that, the electron evolves along the classical trajectory in the presence of the laser field. If the electron runs directly to the detector without further interaction with the core, the electron is called a direct electron. On the other hand, if the electron returns and recollides with the core, we call this kind of electron a rescattering electron. For the photoelectron ionized via tunneling in a strong laser field, the oscillation is much larger than the atom radius, and thus the influence of the Coulomb potential on the photoelectron can be neglected during the oscillation of the electron in the laser field. For the laser field with a vector potential $A(t)$, the momentum of the electron ionized at the time t_0 is given by

$$mv(t) = e(A(t_0) - A(t)) \equiv p - eA(t)\,, \tag{5.41}$$

where $e = -|e|$ is the electron charge. According to equation (5.41), the electron momentum consists of two parts: the drift momentum, $p \equiv eA(t_0)$, and the oscillation momentum, $-eA(t)$, which oscillates with the vector potential of the laser field. After one cycle average, the kinetic energy becomes

$$E = \frac{m}{2}\left\langle v(t)^2\right\rangle_T = \frac{p^2}{2m} + \frac{e^2}{2m}\left\langle A(t)^2\right\rangle_T \equiv E_{\text{drift}} + U_{\text{p}}\,, \tag{5.42}$$

where U_{p} is the ponderomotive energy and its value is

$$U_{\text{p}} = \frac{e^2}{2m}\left\langle A(t)^2\right\rangle_T\,. \tag{5.43}$$

Thus, the ponderomotive energy is the average energy of the oscillating electron in one optical cycle, which can be considered as a potential in the focal area.

If the duration of the laser pulse is long enough, the intensity of the laser field doesn't change very much until the electron leaves the focus of the laser, and then the final momentum of the electron consists of the drift momentum and the momentum obtained from acceleration of the ponderomotive potential. If the laser pulse is too short, the pulse is over before the electron can travel any appreciable distance, and the final momentum is only the drift momentum $p = eA(t_0)$. For a femtosecond laser pulse (e.g. the duration is less than 100 fs), most of electrons won't travel any appreciable distance, and thus the final momentum of the electron is determined by the drift momentum. For the linear polarization, when the ionization rate is maximal at the peak of the electric field, the vector potential of the laser field is vanishing. Therefore, the ionization amplitude of the electron with small momentum is most pronounced in the photoelectron spectrum. With the increase of the energy of the photoelectron, the vector potential also increases. Correspondingly, the electric field decreases and the ionization rate becomes small, which is consistent with experimental observations.

After the integration of equation (5.41), we can obtain

$$x(t) = \frac{e}{m}\left((t - t_0)\,A(t_0) - \int_{t_0}^{t} d\tau A(\tau)\right). \tag{5.44}$$

For linear polarization, the electron returns to the core at time $t_1 (t_1 > t_0)$ if $x(t_1) = 0$, $y(t)z(t) \equiv 0$. Therefore, the return time t_1 can be solved as a function of the tunneling time t_0. The energy of the electron at the return time is given by

$$E_{\text{ret}} = \frac{e^2}{2m}\left[A(t_1) - \text{A}(t_0)\right]^2. \tag{5.45}$$

The maximal return energy can be calculated as $E_{\text{ret,max}} = 3.17U_p$ with the ionization time and the return time $\omega t_0 = 108°$ and $\omega t_1 = 342°$ respectively.

For HATI, the electron will recollide with the core after returning to it, obtaining more energy from the laser field. For example, when the electron returns to the core at time t_1, the electron momentum is $mv(t_1 - t_0) = e(A(t_0) - A(t_1))$. After back-rescattering, the momentum becomes $e(-A(t_0) + A(t_1))$. The final momentum of the electron at the detector is $e[2A(t_1) - A(t_0)]$ after being further accelerated by the laser field. According to equation (5.41), the kinetic energy becomes

$$E_{\text{bs}} = \frac{e^2}{2m}[2A(t_1) - A(t_0)]^2 . \tag{5.46}$$

According to equation (5.46), the maximal energy is $E_{\text{bs, max}} = 10.007\ U_p$, for $\omega t_0 = 105°$ and $\omega t_0 = 352°$. Thus, to obtain the maximal return energy or the maximal back-rescattering energy, the phase of the laser field at the tunneling ionization time should be close to the peak of the electric field and the return time is kept close to the cross of the electric field. It can be well understood that when the electron returns to the core at the time close to zero and then back-rescatters, the direction of the momentum changes. Next, the electron is further accelerated by the laser field to obtain the maximal kinetic energy, as shown in Figure 5.14.

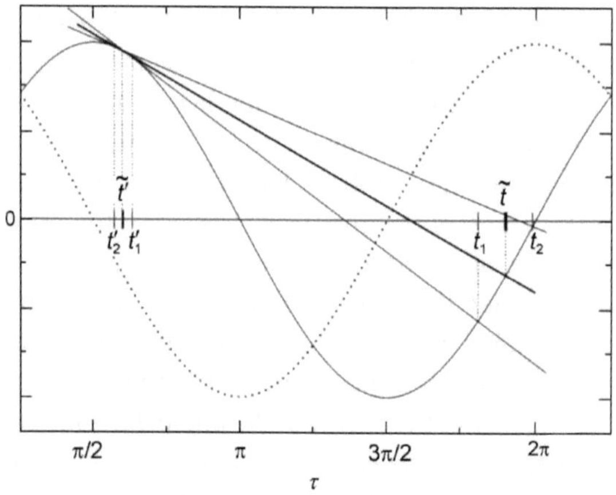

Fig. 5.14: The evolution of the electron return time t as a function of tunneling time t'. After [59].

Figure 5.14 shows how to solve the return time of electrons in a laser field of sine form following a pictorial method. For the sine laser field, the relation of the return time t_1 and the tunneling time t_0 is given by [59]

$$\sin \omega t_1 - \sin \omega t_0 = \omega(t_1 - t_0)\cos \omega t_0 . \tag{5.47}$$

Equation (5.47) shows that the slope of the line along the tunneling time and return time is the same as the tangent direction at the tunneling time. As shown in the pictorial method in Figure 5.14, a tangent from the electric field at the tunneling time makes the intersection of the electric field with the line correspond to the return time. Based on this picture, it is easy to see that, in comparison with the electron orbit with the maximal return energy, the electron ionized at the early time t_2' returns to the core at the late time t_2 and thus has a longer travel time. In contrast, the electron ionized at a late time t_1' returns to the core at an earlier time t_1 and thus has a shorter travel time.

In the model above, there is an important assumption that the influence of the Coulomb potential on the motion of the electron is completely neglected. When the Coulomb potential is considered, the trajectory cannot be written in the analytic form. Thus, people have developed a Coulomb-correct semiclassical model. Here, we take double ionization of an atom in elliptical polarization as an example. For the linear polarization, we can let ellipticity $\varepsilon = 0$, while for the single ionization, all of the quantities related to the second electron are neglected. In this model, it is assumed that the outermost bound electron is ionized by tunneling through the barrier formed by the Coulomb potential and the electric field of the laser field. Later, the interaction between the ionized electron and the second electron is fully determined by the Newtonian equation of motion (5.36). Here, the elliptical polarized laser field is denoted by $E(t) = (E_x(t), 0, E_z(t))$, where $E_x(t) = a(t)E_{0x}\sin\omega t$, $E_z(t) = a(t)E_{0z}\cos\omega t$ with the envelope function $f(t)$ as

$$f(x) = \begin{cases} 1\,, & t \le 10T \\ \cos^2\frac{(t-10T)\pi}{6T}\,, & 10T \le t \le 13T \\ 0\,, & t \ge 13T\,, \end{cases} \tag{5.48}$$

where T is the optical period and ω is the frequency. The ellipticity of the laser field is defined as $\varepsilon \equiv E_{0x}/E_{0z} < 1$. The ionization potentials of the first and second bound electrons are $E_{\mathrm{IP}1}$ and $E_{\mathrm{IP}2}$ respectively. The ionic potential and the Coulomb potential between the two electrons are given by equation (5.37).

In order to solve (5.36), the initial conditions for the two electrons, i.e. the initial positions and velocities, have to be set up. The initial position of the first electron can be obtained from the tunnel ionization theory. The Schrödinger equation for an electron in a uniform field E can be reformulated as [60–62]

$$\frac{d^2\Psi}{d\eta^2} + \left(\frac{E_{IP1}}{2} + \frac{1}{2\eta} + \frac{1}{4\eta^2} + \frac{E\eta}{4}\right)\Psi = 0 \tag{5.49}$$

in parabolic coordinates. Physically, equation (5.49) describes the electron tunneling through a one dimensional potential

$$U(\eta) = -\frac{1}{4\eta} - \frac{1}{8\eta^2} - \frac{E_z\eta}{8} \tag{5.50}$$

with energy $K = -E_{\mathrm{IP}}/4$. Thus, the outer turning point η_0 of the potential $U(\eta)$, known as the 'tunnel exit', can be determined by $K = U(\eta)$. In our calculation for an elliptically polarized field, the initial condition of the first electron is determined as follows. First, the electric field of the laser is confined in the x-z plane. At each tunneling moment t_0, the z axis is rotated to be parallel to the instantaneous electric field. The initial condition is obtained in the rotated coordinates (the same as in the linearly polarized field) and then projected to the original coordinates. For the first electron, the initial position in the rotated coordinate is $x'_{10} = y'_{10} = 0$, $z'_{10} = -\frac{1}{2}\eta_0$ and hence the coordinates in the original coordinate are $x_{10} = -\frac{\eta_0}{2}\sin\{\arctan[\varepsilon\tan(\omega t_0)]\}$, $y_{10} = 0$ and $z_{10} = -\frac{\eta_0}{2}\cos\{\arctan[\varepsilon\tan(\omega t_0)]\}$, where t_0 represents the ionization time. The initial longitudinal velocity at the ionization time t_0 is assumed to be zero, while a nonzero initial velocity v_{per} perpendicular to the laser polarization direction is given in the calculation. The corresponding initial velocities are $v'_{1x0} = v_{per}\cos\theta$, $v'_{1y0} = v_{per}\sin\theta$, $v'_{1z0} = 0$, where θ is the angle between v_{per} and x axis in the rotated coordinates. Projection into original coordinates gives the initial velocity of the first electron: $v_{1x0} = v_{per}\cos\theta\cos\{\arctan[\varepsilon\tan(\omega t_0)]\}$, $v_{1y0} = v_{per}\sin\theta$, $v_{1z0} = -v_{per}\cos\theta\sin\{\arctan[\varepsilon\tan(\omega t_0)]\}$. The weight of each trajectory is evaluated by $w(t_0, v_{per}) = w(0)\overline{w(1)}$, where w_0 is the ADK rate given by equation (5.7), and $\overline{w(1)}$ can be expressed by

$$\overline{w(1)} = \frac{v_{per}(2E_{IP1})^{\frac{1}{2}}}{E\pi}\exp\left[-\frac{v_{per}^2(2E_{IP1})^{\frac{1}{2}}}{E}\right]. \tag{5.51}$$

For the second electron (bound electron), its initial conditions are determined by a microcanonical distribution as discussed in Section 5.4.2.

In the model calculation, more than 10^6 initial points are randomly distributed in the parameter space $-\frac{\pi}{2} < \omega t_0 < \frac{\pi}{2}$, $v_{per} > 0$, $0 < \theta < \frac{\pi}{2}$ for the first electron, and in the microcanonical distribution for the second electron. This two electron system is evolved until the end of the laser field according to the classical Newtonian equation (equation (5.36)) and double ionization events are identified by an energy criterion.

5.4.4 Quantum mechanical S-matrix method

A TDSE of the interaction of atoms and molecules with a strong laser field can be solved not only by the numerical method, but also by the S-matrix theory. The numerical solution of the TDSE is accurate, but it needs a large amount of computation and it is not easy to extract a clear physical picture. However, the analytic S-matrix theory needs less computation and gives a very clear physical picture, hence it attracts lots of attention. The S-matrix theory was developed along with the development of strong field atomic and molecular physics. Because of the Coulomb potential, it can't obtain the analytic solution, so some approximations have to be made. Considering the nonlinear and nonperturbative characteristics of the strong field atomic process, people developed the nonperturbation S-matrix theory to study the interaction of electrons with laser fields, i.e. the strong field approximation (SFA) model.

According to quantum theory, the transition amplitude from the initial state $|\psi_0\rangle$ of an atom to the final state with drift momentum p at the time t is given by

$$M_{\mathrm{p}} = \lim_{t\to\infty, t'\to\infty} \left\langle \psi_{\mathrm{p}}(t)|U(t,t')|\psi_0(t')\right\rangle . \tag{5.52}$$

Here $U(t,t')$ is the time evolution operator of the Hamiltonian

$$H(t) = -\frac{1}{2m}\nabla^2 - er\cdot\varepsilon(t) + V(r) , \tag{5.53}$$

which includes the atomic binding potential $V(r)$ and the interaction term $-er\cdot\varepsilon(t)$. Furthermore, we introduce the Hamiltonians for an atom without the field and for a free electron in the laser field,

$$H_{\mathrm{a}} \equiv H_{\mathrm{atom}} = -\frac{1}{2m}\nabla^2 + V(r) \tag{5.54}$$

$$H_{\mathrm{f}}(t) \equiv H_{\mathrm{field}}(t) = -\frac{1}{2m}\nabla^2 - er\cdot\varepsilon(t) . \tag{5.55}$$

The corresponding time evolution operators are denoted by U_{a} and U_{f}. In equation (5.52), $|\psi_0\rangle$ is the ground state of the atomic Hamiltonian H_{a}, and $|\psi_{\mathrm{p}}\rangle$ is a scattering state with asymptotic momentum p of the Hamiltonian H_{f}, which is known as the Volkov state [63]. In the length gauge, the analytic form of the Volkov state is

$$\psi_{\mathrm{p}}^{V_v}(t) = |p - eA(t)\rangle e^{-\mathrm{i}S_{\mathrm{p}}(t)} , \tag{5.56}$$

where $|p - eA(t)\rangle$ is a plane wave state and its representation in coordinate space is

$$\left\langle r|p - eA(t)\right\rangle = (2\pi)^{-\frac{3}{2}} e^{\mathrm{i}\cdot(p-eA(t))\cdot r} \tag{5.57}$$

$$S_{\mathrm{p}}(t) = \frac{1}{2m}\int_{t_0}^{t} d\tau[p - eA(t)]^2 . \tag{5.58}$$

Note that the choice of the lower limit of the integral in equation (5.58) is not important. It will only introduce a phase that does not contribute to any observable quantity.

The time evolution operator $U(t,t')$ satisfies the integral equation (i.e. Dyson equation) and it is convenient to generate perturbation expansion. Here, the time evolution operator can be expanded with respect to either the interaction term $H_{\mathrm{I}}(t) = -er\cdot\varepsilon(t)$ or the binding potential $V(r)$,

$$U\left(t,t'\right) = U_{\mathrm{a}}\left(t,t'\right) - \mathrm{i}\int_{t'}^{t} d\tau U(t,\tau) H_{\mathrm{I}}(\tau) U_{\mathrm{a}}\left(\tau,t'\right) \tag{5.59}$$

$$U\left(t,t'\right) = U_{\mathrm{f}}\left(t,t'\right) - \mathrm{i}\int_{t'}^{t} d\tau U_{\mathrm{f}}(t,\tau) V U\left(\tau,t'\right) . \tag{5.60}$$

In equation (5.59), the terms U and U_a in the integral on the right-hand side can be exchanged. Similarly, the terms U and U_f in the integral on the right-hand side of (5.60) can also be exchanged. Inserting equation (5.59) into equation (5.52), we can obtain

$$M_p = -i \lim_{t\to\infty} \int_{-\infty}^{t} d\tau \left\langle \psi_p(t) \left| U(t,\tau) H_I(\tau) \right| \psi_0(\tau) \right\rangle , \tag{5.61}$$

which is still exact. An important simplification we now introduce is the SFA, i.e. $|\psi_p\rangle$ and U are replaced by $|\psi_p^{V_V}\rangle$ and U_f respectively. As a result, equation (5.61) can be simplified as

$$M_p = -i \int_{-\infty}^{\infty} dt_0 \left\langle \psi_p^{V_V}(t_0) | H_I(t_0) | \psi_0(t_0) \right\rangle . \tag{5.62}$$

The physical meaning of this transformation is that, after the electron has been promoted into the continuum at time t_0 due to the interaction $H_t(t_0) = -er \cdot \varepsilon(t_0)$, it no longer feels the atomic potential. This process leads to the direct electron, and equation (5.62) is called Keldysh–Faisal–Reiss (KFR) amplitudes. With further mathematical transformation, the amplitude (5.62) can then be rewritten as

$$M_p = -i \int_{-\infty}^{\infty} dt_0 \left\langle \psi_p^{V_V}(t_0) | V(r) | \psi_0(t_0) \right\rangle . \tag{5.63}$$

For sufficiently high intensity (small Keldysh parameter), the saddle point method (method of steepest descent) can be used to solve the integral in equation (5.63). Since $\psi_p^{V_V}(t) = |p - eA(t)\rangle e^{-iS_p(t)}$ and $\psi_0(t) = |\psi_0\rangle e^{iE_{IP}t}$, this amounts to determining the solutions of

$$\frac{d}{dt}\left[E_{IP}t + S_p(t)\right] = E_{IP} + \frac{1}{2}[p - eA(t)]^2 = 0 . \tag{5.64}$$

Assuming the solution of equation (5.64) is t_s, the amplitude (5.63) can then be written as

$$M_p \propto \sum_n \delta\left(\frac{p^2}{2m} + E_{IP} + U_p - n\omega\right) \times \sum_s \left(\frac{2\pi i}{S_p''(t_s)}\right)^{\frac{1}{2}} e^{i[E_{IP}t_s + S_p(t_s)]} \left\langle p - eA(t_s) | V | \psi_0 \right\rangle , \tag{5.65}$$

where S_p'' denotes the second derivative of the action (5.58) with respect to time. In equation (5.58), the sum is only for the interval $0 < \mathrm{Re}\, t_s < T$, where T is the period of the laser field. The δ function in equation (5.65) comes from the sum of infinite periods of the laser pulses. It can be understood as the interference of the contributions from different periods and photoelectron spectra that are nonvanishing only for the photoelectron with some specific discrete energies

$$E_p \equiv \frac{p^2}{2m} = n\omega - U_p - E_{IP} . \tag{5.66}$$

In order to consider the rescattering electron, we should study the interaction of the free electron with the ionic potential. Inserting equation (5.60) into equation (5.61) and employing SFA again, i.e. with $|\psi_{\rm p}\rangle$and$U$ replaced by $|\psi_{\rm p}^{V_{\rm V}}\rangle$ and $U_{\rm f}$ respectively, we can obtain the following equation after some algebra:

$$M_{\rm p} = -{\rm i} \int_{-\infty}^{\infty} dt_1 \int_{-\infty}^{t_1} dt_0 \left\langle \psi_{\rm p}^{V_{\rm V}}(t_1) | V U_{\rm f}(t_1, t_0) V | \psi_0(t_0) \right\rangle . \tag{5.67}$$

The physical content of the amplitude (29) corresponds to a recollision scenario: the electron is promoted into the continuum at some time t_0; it then propagates in the continuum subject to the laser field until at the later time t_1 it returns in the range of the binding potential, where the electron is scattered into its final Volkov state. The time evolution operator is expanded in terms of the Volkov states

$$U_{\rm f}(t_1, t_0) = \int d^3k |\psi_{\rm k}^{V_{\rm V}}(t_1)\rangle \left\langle \psi_{\rm k}^{V_{\rm V}} t_0 \right| . \tag{5.68}$$

Inserting equation (5.68) into equation (5.67), we obtain

$$M_{\rm p} \sim \int_{-\infty}^{\infty} dt_1 \int_{-\infty}^{t_1} dt_0 \int d^3k e^{{\rm i}S_{\rm p}(t_1,t_0,k)} m_{\rm p}(t_1, t_0, k) , \tag{5.69}$$

where $m_{\rm p}(t_1, t_0, k) = \langle p - eA(t_1)|V|k - eA(t_1)\rangle\langle k - eA(t_0)|V|\psi_0\rangle$. For ATI, the action is

$$S_{\rm p}(t_1, t_0, k) = -\frac{1}{2m}\int_{t_1}^{\infty} d\tau [p - eA(\tau)]^2 - \frac{1}{2m}\int_{t_0}^{t_1} d\tau [k - eA(\tau)]^2 + \int_{-\infty}^{t_0} d\tau E_{\rm IP} . \tag{5.70}$$

The saddle point method can also be used to solve equation (5.69) and the saddle points can be obtained from the following equations:

$$\begin{cases} [k - eA(t_0)]^2 = -2mE_{\rm IP} \\ (t_1 - t_0)k = -\int_{t_1}^{t_0} d\tau\, eA(\tau) \\ [k - eA(t_0)]^2 = [p - eA(t_0)]^2 . \end{cases} \tag{5.71}$$

The first condition enforces energy conservation at the time of tunneling. The second condition ensures that the electron returns to its parent ion, and the third one expresses that, on this occasion, the electron rescatters elastically into its final state. Assuming the saddle point equations have the forms (t_{1s}, t_{0s}, k_s), $(s = 1, 2, \dots)$, the matrix element can be written as

$$M_{\rm p} \sim \sum_{\rm s} \left(\frac{(2\pi{\rm i}\hbar)^5}{\det\left(\frac{\partial^2 S_{\rm p}}{\partial q_j^{(s)} \partial q_k^{(s)}}\right)_{j,k=1,\dots,5}} \right)^{\frac{1}{2}} e^{{\rm i}S_{\rm p}(t_{1s}, t_{0s}, k_s)} m_{\rm p}(t_{1s}, t_{0s}, k_s) . \tag{5.72}$$

For periodic fields, there is more than one solution for a given final kinetic energy. Usually, there is a pair of orbits in one period. These solutions correspond to the rescattering trajectories of the tunneling ionized electron that oscillates in the laser field and then recollides with the ionic core. The final transition amplitude is the coherent sum of these orbits, which are also named 'quantum orbits'. Usually, the orbit in the first optical period is called the short orbit, while the orbits with longer travel time are named the long orbits. Their physical effect will be discussed later. The final transition amplitude produces a δ function as in equation (5.65). To obtain the ATI spectrum, the process is split into two parts. First, one needs to solve formula (5.71) for the saddle points and then obtain the transition amplitude by inserting the saddle points into equation (5.72). Note that only the transition matrix element of a single final state with momentum p is considered here, rather than the entire wavefunction of the final state. Since $E_{\mathrm{Ip}} > 0$, the saddle points are complex. For $E_{\mathrm{IP}} = 0$, the initial velocity of the electron orbit at the time of tunneling vanishes according to equation (5.71). If $E_{\mathrm{IP}} \neq 0$, but $\gamma^2 = \frac{E_{\mathrm{IP}}}{2U_{\mathrm{p}}}$ is much smaller than one, the imaginary part of the solutions of equation (5.71) is also far less than the real part, and thus the solutions are close to the results of the classical model. The solutions of the stable quantum orbits in the saddle point approximation have the form

$$mx(t) = \begin{cases} (t - t_{0s})\, k_s - \int_{t_{0s}}^{t} d\tau\, eA(\tau)(\mathrm{Re}\, t_{0s} \leq t \leq \mathrm{Re}\, t_{1s}) \\ (t - t_{1s})\, p - \int_{t_{1s}}^{t} d\tau\, eA(\tau)(t \geq \mathrm{Re}\, t_{1s})\,. \end{cases} \tag{5.73}$$

The electron orbit is a function of real time t, while the conditions $x(t_1) = 0$ and $x(t_0) = 0$ are for complex times t_0 and t_1. Thus, the orbit from equation (5.73) as a function of real time does not start from the origin, but from a certain position away from the origin, which is called the tunnel exit. In contrast to the ionization time, the return time is a good approximation and the electron returns to the origin at this time.

Here, we will discuss the conditions of validity of the SFA. After the expansion of the transition matrix element into the second term, it has the form

$$\begin{aligned} M_{\vec{p}} = -\mathrm{i} \lim_{t \longrightarrow \infty} \Bigg\{ & \int_{-\infty}^{t} dt' \left\langle \chi_{\mathrm{p}}(t) \right| U^{(V)}(t, t')\, H_{\mathrm{I}}(t') \left| \phi_0(t') \right\rangle \\ & -\mathrm{i} \int_{-\infty}^{t} dt' \int_{-\infty}^{t'} dt'' \left\langle \chi_{\mathrm{p}}(t) \right| U^{(V)}(t, t')\, V U^{(V)}(t', t'')\, H_{\mathrm{I}}(t'') \left| \phi_0(t'') \right\rangle \Bigg\} \\ = -\mathrm{i} & \int_{-\infty}^{\infty} dt_0 \left\langle \chi_{\mathrm{p}}^{V}(t_0) \right| H_{\mathrm{I}}(t_0) \left| \phi_0(t_0) \right\rangle \\ & + \mathrm{i}^2 \int_{-\infty}^{\infty} dt_0 \int_{-\infty}^{t_0} dt_1 \int dp' \left\langle \chi_{\mathrm{p}}^{V}(t_0) \right| V \left| \chi_{p'}^{V}(t_0) \right\rangle \left\langle \chi_{p'}^{V}(t_1) \right| H_{\mathrm{I}}(t_1) \left| \phi_0(t_1) \right\rangle\,. \end{aligned} \tag{5.74}$$

Clearly, the second term of the transition amplitude on the right side of equation (5.74) comes from the contribution of the ionic potential. The condition of validity of the SFA is that the contribution of the ionic potential is negligible in comparison with the direct ionization amplitude, i.e. the second term is much smaller than the first term. Now we estimate the ratio of the two terms. Note that the expression, $\langle\chi_{p'}^V(t_1)|H_{\mathrm{I}}(t_1)|\varphi_0(t_1)\rangle$ in the second term is the same as the integral in the first term. So, the integral of this part is generally considered to be the same as the first term, and we just need to estimate

$$\left\langle\chi_{\mathrm{p}}^V(t_0)|V|\chi_{p'}^V(t_0)\right\rangle . \tag{5.75}$$

Assuming that the effective region of the ionic potential is R and $|\chi_{p'}^V(t_0)\rangle$ is a plane wave with the normalized volume λ^3, where λ is the laser wavelength, one thus obtains

$$\left\langle\chi_{\mathrm{p}}^V(t_0)|V|\chi_{p'}^V(t_0)\right\rangle \approx \frac{R^3}{\lambda^3} . \tag{5.76}$$

Obviously, the validity of SFA depends on the interaction range of the ion potential. If the effective ionic potential is 1 a.u. and the wavelength of the laser field is 800 nm, the value of equation (5.76) is about 10^{-12}. Under this condition, the SFA is valid. But, if the ionic potential is a long range potential, e.g. the Coulomb potential, the SFA is strictly not applicable. We will discuss it further later.

In the Section 5.4.1, concerning the numerical solution of TDSE, we discussed the selection of gauge of the interaction. Although the interaction of electric field is gauge invariant, the results from the length gauge and the velocity gauge are not exactly the same. In SFA theory, the differences between the two gauges are more significant due to approximations adopted in the theory. Thus, how to select the gauge in the calculations becomes a key issue. Due to the page limit, we will not further discuss it here. For more details, see references [11, 65, 66].

5.5 Key experimental facts

In 1979, the discovery of above threshold ionization (ATI) indicated that the interaction between atoms and the laser field enters into a new nonperturbative regime. During the past thirty years, the atomic physics in intense laser fields has been well studied due in part to the great progress achieved in ultrafast intense laser technology: the pulse duration was shortened from the order of nanoseconds and picoseconds to the order of femtoseconds, the repetition rate was raised from the order of hertz to the order the kilohertz, and the laser intensity from the order of 10^{13} to much stronger than 10^{16} W/cm^2. In the meantime, spectroscopic techniques were also developed and optimized from the ability of single particle detection to coincident and differential measurement. These improvements in experimental techniques promoted the development of intense field atomic physics research, where new experimental phenomena and physical effects are frequently discovered. Here, we will discuss important experimental progress and focus on two typical ionization phenomena: above threshold ionization (ATI) and nonsequential double ionization (NSDI).

5.5.1 Above threshold ionization

Here the most important progress on the study of atomic ATI will be discussed, including the discovery of ATI, peak suppression for low order ATI, low energy structure (LES), Freeman resonance, high order ATI plateau and resonance like enhancement (RLE).

As discussed previously, ATI is the phenomenon by which an atom, subjected to an intense laser field, may absorb many more photons than the minimum necessary for ionization. The extra energy is converted to kinetic energy, resulting in evenly spaced peaks in the photoelectron kinetic energy spectrum (PES). A typical ATI spectrum is shown in Figure 5.15 (a), where the peak with highest yield corresponds to the lowest order ATI. With the increase of the kinetic energy, the peak amplitude drops quickly. ATI was first discovered by Agostini in 1979 [4] when measuring the photoelectron kinetic energy spectra of Xe. After that, ATI became a hot topic in the strong field physics community.

5.5.1.1 Peak suppression for low order ATI

The discovery of ATI has attracted great attention in strong field physics [72–74]. With the rapid development of electronic techniques, high resolution timing devices were equipped widely in laboratories around the world, which improved the resolution of photoelectron kinetic energy spectroscopy to a great extent [73]. Compared with the experimental scheme employed by Agostini in 1979, PES based on the time of flight principle is more convenient and efficient. In the meantime, the rapid development of laser technique has made the laser system with higher intensity commercially available. Under the circumstance mentioned above, people investigated the evolution of relative strengths of the different ATI orders with respect to the laser intensity. The earlier experimental results demonstrate that the yields of higher order ATI can be even higher than those of lower order ATI. In 1983, the experimental work of Kruit et al. [75] revealed that the amplitude of the low order ATI peaks degrade, or even disappear, with rising laser intensity, which is termed peak suppression for low order ATI. This result is in contrast to the prediction of simple perturbative theory. In 1986, further experimental exploration on this issue was performed by Yergeau et al. [67], as shown in Figure 5.15 (b).

The physical origin of peak suppression is the influence of the laser field with spatial Gaussian distribution and the ponderomotive acceleration to the photoelectrons [3, 67, 76]. As discussed above, a free photoelectron subject to the intense laser field will oscillate along the field and feel the ponderomotive potential of the laser

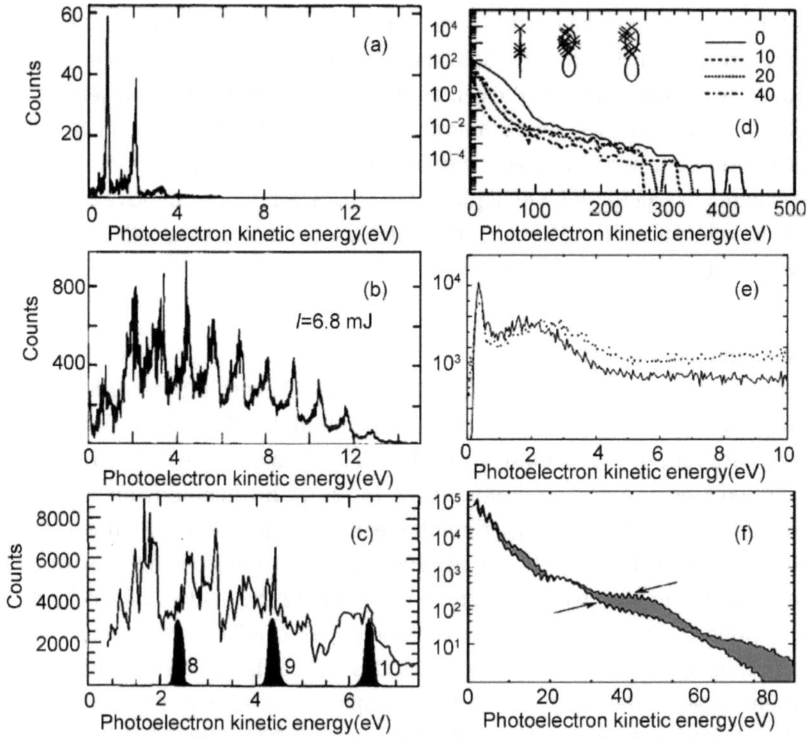

Fig. 5.15: Important ATI experimental results: (a) typical ATI spectrum for Xe, after [67]; (b) low order ATI peak suppression, after [67]; (c) Freeman resonance in ATI spectra, after [68]; (d) angle resolved ATI spectra and photoelectron angular distribution, after [69]; (e) low energy structure in ATI spectra, after [70]; and (f) resonance-like enhancement structure in high order ATI spectra, after [71].

field, as described with equation (5.43). With simple formula transformation, we will get

$$U_{\mathrm{p}} = \left\langle \frac{e^2 E^2}{4m\omega^2} \right\rangle , \tag{5.77}$$

where $\langle\,\rangle$ means one optical cycle average, and E is the laser electric field. Within the laser field, only if the energy absorbed by the electron is larger than $I_{\mathrm{p}} + U_{\mathrm{p}}$, can the photoelectron be ionized and detected by the spectrometer [3, 77], where I_{p} is the ionization potential of the system. The kinetic energy of the photoelectron can be calculated with equation (5.66). However, the typical time required for a free photoelectron to escape the laser focus is around 10 ps. For the laser electric field with pulse duration considerably longer than the time, the photoelectron can travel out of the focus and 'slide' down the whole potential before the laser field is turned off. During this process, the photoelectron will be accelerated and acquires a kinetic energy of U_{p}. Therefore, the energy lost during the ionization will be compensated precisely during the flight out of the focus. In the experiment performed by Yergeau et al. [67], the laser pulse duration was long enough and the positions of the peaks in ATI spectra were not related to the laser intensity. Considering the Gaussian spatial distribution of the laser field, the high intensity region is close to the center of the focus and possesses a very small volume, which, in turn, means fewer atoms are involved. High laser intensity will give rise to large ponderomotive energy, which raises the effective ionization potential and reduces the yields. As calculated with equation (5.66), larger ponderomotive potential will contribute to lower order ATI peak. Therefore, the yields of lower order ATI peak can be suppressed, as shown in Figure 5.15 (b).

5.5.1.2 Freeman resonance structure

At the end of the 1980s, it was possible to study atomic ATI with picosecond or even femtosecond laser fields. In 1987, Freeman et al. [78] found that if the pulse duration is less than 1 picosecond, the individual ATI peaks break up into narrow fine structures, which are termed Freeman resonance structures. In 1991, a further study on this issue was performed [68]. The classical experimental results are depicted in Figure 5.15 (c). The measured photoelectron kinetic energy spectra for laser electric fields with two kinds of pulse duration are given in Figure 5.15 (c), where the peaks labeled by 8, 9 and 10 correspond to the measurements with 100 ps laser pulse and the curve with fine structures above them is measured with a 300 fs laser pulse. In the case of 100 ps, there are evenly spaced narrow peaks in the ATI spectra for Ar, while in the case of 300 fs, each ATI peak is shifted and breaks up into narrow fine structures.

As with the discussion on peak suppression, we will compare the time that it takes for the electron to fly away from the focus and the pulse duration to understand the difference caused by the pulse duration. For a 100 ps pulse, the photoelectron has enough time to be accelerated before the laser field turned off. So, the shift of the photo electron kinetic energy by the ponderomotive shift of the ionization potential can be compensated by the acceleration of the ponderomotive potential. The linewidth of the ATI peaks is not related to the laser intensity. For 300 fs, the laser electric field vanishes before the electron escapes and the influence of the ponderomotive shift of the ionization potential cannot be compensated. Furthermore, the involvement of Rydberg states resonance makes the situation even more complicated, which results in complicated structures in the ATI spectra.

The fine subpeak structures in Figure 5.15 (c) are termed Freeman resonance structures, which can be attributed to the resonance between Rydberg states and the ground state. The physical mechanism is illustrated in Figure 5.16. When the laser electron field is turned off, the energy levels are shown with short horizontal lines in Figure 5.16. With increasing principle quantum number, the energies of the levels become higher and higher, while the spacing between two adjacent levels becomes smaller and smaller. After the laser electric field is turned on, the AC Stark effect will shift the levels with respect to the laser intensity. For Rydberg states, the AC Stark shift is approximately equal to the ponderomotive energy, as shown with oblique lines in

Figure 5.16 for the Rydberg states around $n = n_0$. The energy level will be shifted upward if the laser intensity is increased. When the laser intensity is I_{n_0}, the shifted level possesses the identical energy to that of the N-photon dressed ground state. For the laser intensity of I_{n_0+1} and I_{n_0-1}, the other two Rydberg states of $n = n_0 + 1$ and $n = n_0 - 1$ are in resonance with the dressed ground states. During a real experiment, because there is a temporal and spatial distribution of the laser electric field, if the laser intensity is higher than I_{n_0-1} there will be resonances involved in multiple levels, including the Rydberg states of $n = n_0$, $n = n_0 + 1$ and $n = n_0 - 1$, which can lead to a series of fine peaks in the ATI spectra.

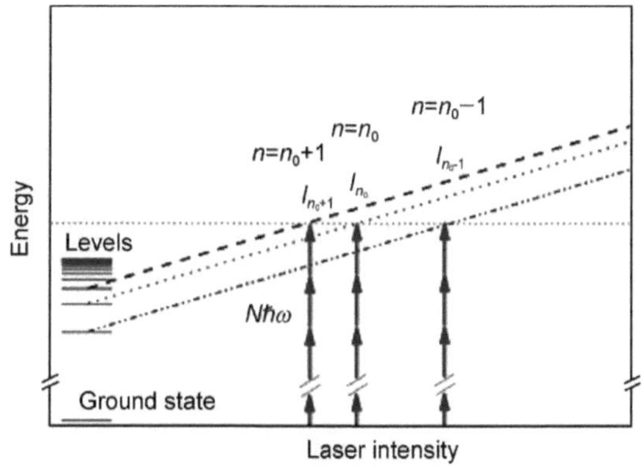

Fig. 5.16: A schematic diagram for the mechanism of Freeman resonance.

5.5.1.3 High order ATI plateau

In the late 1990s, the femtosecond laser system with repetition rate of kHz became more and more popular, which provided an efficient tool to study the high energy ATI photoelectron. Compared with low energy electrons, the yields of high energy electrons will be two to three orders of magnitude lower, and a laser system with high repetition rate becomes a prerequisite for such studies. In 1994, Paulus et al. [79] measured the photoelectron kinetic energy spectra extending to very high energy. Surprisingly, they found a plateau structure for all the noble gas atoms, in which the electron yields vary slowly with respect to kinetic energy until a high energy cutoff. This result is in sharp contrast to the low energy electron, which drops rapidly with respect to kinetic energy. Two years later, Walker et al. [69] systematically studied ATI for He. Their experimental results are shown in Figure 5.15 (d). With increasing photoelectron kinetic energy, the yields of electrons in the kinetic energy interval of 0–100 eV drop rapidly. If the energy is raised further, a plateau emerges, which extends to a high energy of around $10U_\text{p}$. Figure 5.15 (d) also shows the evolution of photoelectron kinetic energy distribution with respect to the angle between the laser polarization and the direction of the time of flight of the spectrometer. If the angle is increased, the plateau becomes shorter and the energy cutoff becomes lower. The insets of Figure 5.15 (d) give the angular distribution at three distinct kinetic energies, which become narrower for low energy electrons and much wider for the electrons in the plateau region. In fact, the earlier experimental work of Yang et al. [80] revealed that the photoelectron kinetic energy distributions for Xe possess a maximum at the angle of 45° for electron energy of about $9U_P$. These results look very strange, but they can all be well reproduced with the rescattering model. In [81], the calculation of the angular distribution of photoelectrons based on the semiclassical procedure is performed and the calculation results are in agreement with the experimental observation [80].

5.5.1.4 Low energy structure

Recently, the femtosecond optical parametric amplifier, which can generate wavelength tunable mid-infrared femtosecond laser pulses, became available. This extended the experimental study of strong field atomic physics from near-infrared, corresponding to the wavelength of Ti:sapphire laser systems (around 800 nm), to the mid-infrared. From previous sections, we know that there are two ways to get into a deep tunneling ionization regime: increasing the laser intensity and decreasing the laser frequency (i.e. increasing the laser wavelength). Considering the saturation intensity, there is a restriction to the laser intensity that can be applied to study the atomic ionization. In fact, it is impossible to get into the tunneling ionization regime with the laser field at 800 nm for the atoms with low ionization potentials. Nevertheless, in the mid-infrared wavelength regime, it is possible for most of the atoms to get into the tunneling ionization regime at moderate intensity as the Keldysh parameter is more sensitive to the laser wavelength. Therefore, it is operative to study atomic ionization in a deep tunneling regime with a mid-infrared laser field.

In 2008, Colosimo et al. [82] experimentally investigated the photoelectron kinetic energy spectra (PES) and high order harmonic generation (HHG) of noble gas atoms at wavelengths from 800 to 3600 nm. They found that the cutoff of both the PES and HHG is proportional to the square of the wavelength, which confirms the validity of the rescattering model for the covered wavelengths.

In 2009, Blaga et al. [83] and Quan et al. [70] independently found a very interesting hump structure in a low energy region of the PES for several kinds of noble gas atoms and simple molecules. The low energy ATI spectra for Xe at 2000 nm are depicted in Figure 5.15 (e) for the laser intensity of 8.0×10^{13} W/cm^2 (dotted line) and 6.4×10^{13} W/cm^2 (solid line), where the hump structure is clearly shown. This structure is mainly comprised of two parts. One part is a sharp peak below 1 eV and barely dependent on the laser intensity. The other part is relatively broad and appears at high kinetic energy. Both the width and the position of this part depend on the laser intensity and wavelength. This undocumented structure is different from the well accepted tunneling ionization picture and the quantum mechanical calculation based on SFA. Interestingly, this structure can be well reproduced with the numerical solution of the TDSE. This structure is termed the low energy structure (LES) and has attracted great attention. It is worth noting that Blaga et al. [83] observed only the higher part of the structure, while the Quan et al. [70] revealed two parts of the structure. In 2012, Wu et al. [84] explored further the low energy structure with COLTRIMS and found both the high energy and low energy parts of LES, which will be discussed in detail in Section 5.6 from the theoretical viewpoint.

5.5.1.5 The resonance-like enhancement of high order ATI

Until now, most experimental findings of atomic ionization in the intense laser field could be well understood with the rescattering model. However, due to the complexity of the intense field-atom interactions, there are still some open questions in this field. One of these is the resonance-like enhancement of high order ATI.

Shortly after the discovery of the plateau of high order ATI by Paulus et al. [79], Hertlein et al. [85] and Hansch et al. [86] independently found that part of the PES, i.e. at the energy interval of 6–8 times ponderomotive energy, will be enhanced significantly with the increase of the laser intensity, which is termed resonance-like enhancement (RLE). Figure 5.15 (f) depicts the PES for Ar at the laser intensity of 1.1×10^{14} W/cm^2 and 1.4×10^{14} W/cm^2, where the yields in the energy interval of 30–56 eV rise rapidly with increasing intensity. The kinetic energy of this structure does not change with respect to the laser intensity, which is very similar to Freeman resonance. One may then ask: If the physical origin behind RLE is the same as that of Freeman resonance? Stimulated by this question, some groups performed investigations theoretically and experimentally. Detailed discussion on this issue will be given in Section 5.6.

5.5.2 Nonsequential double ionization

Atoms subject to intense laser fields can not only eject a single electron after absorbing multiple photons, which gives rise to ATI as described above, but can also emit two or even more electrons simultaneously. More interestingly, many experiments demonstrate that the electrons are not ejected sequentially one by one under some circumstances, i.e. the two or more valence electrons can be ionized nonsequentially, which is termed nonsequential double (multiple) ionization. NSDI has attracted a lot of attention since its discovery because it involves electron-electron correlation. Here we will discuss the important experimental findings concerning atomic NSDI.

5.5.2.1 Ion yields

In 1983, Huillier et al. [20] measured the yields of the doubly charged Xe with respect to the laser intensity and found that Xe^{2+} is produced nonsequentially. Their experimental results revealed that the correlation of the electrons is crucial for the ionization process. After that, a lot of efforts were made to explore double ionization for He and

Xe. It was found that the measured doubly charged ion yields can be several orders of magnitude higher than the calculation based on single active electron approximation [21, 87, 93] and a 'knee' structure appears in some specific intensity interval. Figure 5.17 (a) depicts the yields of single and doubly charged ions with respect to the laser intensity for He. As we can see, the 'knee' structure emerges for the intensity lower than 3×10^{15} W/cm^2, which can be taken as direct experimental evidence for nonsequential double ionization.

To elaborate the physical mechanism behind the 'knee' structure, Dietrich et al. [94] experimentally studied the double charged ion yields with respect to the laser ellipticity. It was revealed that the yields of NSDI drops quickly with the increasing laser ellipticity. They also found that the high order harmonic generation amplitude possesses similar ellipticity dependence. Their experimental results demonstrate that it is possible that NSDI and high order harmonic generation share similar underlying physics. This is the first experimental evidence that the physical mechanism of NSDI is based on the rescattering model.

Although the yields of NSDI drops rapidly with increasing ellipticity for most noble gas atoms, the 'knee' structure has been identified for more complicated molecules, including NO [95] and the alkali atom Mg [96] subject to circularly polarized laser fields, which is apparently against the intuitive rescattering model and resulted in a new round of debate. Interestingly, the following classical calculation [54, 97] and semiclassical calculation [98] demonstrate that these abnormal phenomena can be understood with the rescattering model [54, 97].

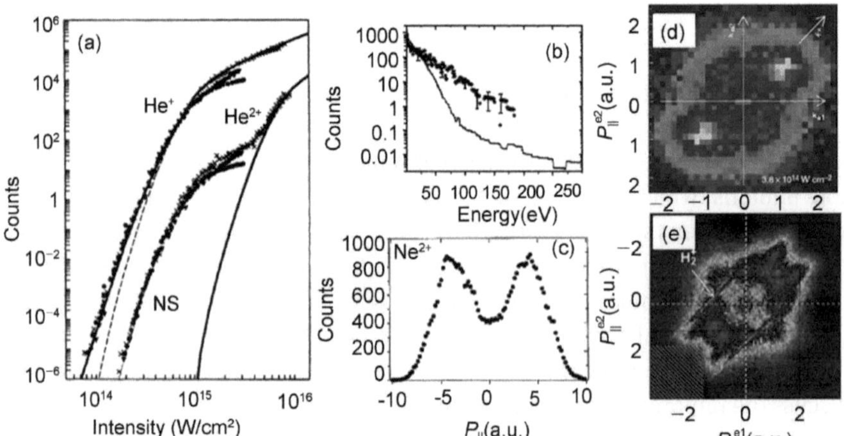

Fig. 5.17: Important NSDI experimental results: (a) single (double) ionization rates as a function of laser intensity for He, after [87]; (b) kinetic energy spectra for photoelectrons coincident to single and double charged He (dotted line for He^{2+}, solid line for He^{+}), after [88]; (c) momentum distribution of Ne, after [89, 90]; (d) momentum correlation between the two emitted electrons coincident to Ar^{2+} ions from double ionization, after [91]; and (e) momentum correlation between the two emitted electrons coincident to He^{2+} from double ionization, after [92].

5.5.2.2 Photoelectron kinetic energy distribution

To understand the physical mechanism behind NSDI deeply, it is necessary to perform systemically measurements, including photoelectron kinetic energy spectra or momentum spectra, to obtain more information. Recently, the photoelectron kinetic energy spectra of NSDI have been acquired for He [88], Ar [99, 100] and Xe [99, 101]. During the experiments, the laser intensity was chosen to be within the interval of NSDI. It was revealed that the averaged electron kinetic energy is much larger than that of single ionization. In Figure 5.17 (b), the kinetic energy spectra for single and double ionization of He are compared. The experimental observation of the electron yields from NSDI drops with increasing kinetic energy and is coincident to the theoretical prediction, which further supports rescattering as the underlying physical mechanism.

5.5.2.3 Ion momentum distribution

The development of COLTRIMS paved the way for the clarification of the physical mechanism of NSDI. This spectroscopy method was employed by Weber et al. [90] and Moshammer et al. [89] to measure the recoil ion momentum distribution for He and Ne. The momentum distribution of the doubly charged Ne in the polarization direction, i.e. the longitudinal direction, is shown in Figure 5.17 (c), where a double peak structure can be identified. In the direction perpendicular to the laser polarization, i.e. the transverse direction, there is a single peak structure in the momentum distribution of the double charged ion. These results can be well reproduced with the rescattering model. According to this model, the photoelectron will come back to the core when the laser electric field is around zero and impact the core, giving rise to double ionization. During this process, both electrons can achieve significant momentum from the laser electric field, which leads to the double peak structure in the polarization direction. This double peak structure deviates from the predictions of both shake off [21] and collective tunneling mechanisms [102], which were also proposed to explain the 'knee' structure.

The systematic investigation of doubly charged ion momentum for He, Ne and Ar was performed by Jesus et al. [103] and obvious dependences on the species were revealed. The double peak structure is only visible for Ne, but not for He and Ar. One possible explanation relies on the mechanism of recollision excitation with subsequent field ionization (RESI). After that a photoelectron is accelerated by the laser electric field and goes back to the core, the impact between the electron and the core can further lead to double ionization or excite the ionic core to excited states depending on its kinetic energy. The former process is called electron impact ionization (EII), which can give rise to the double peak structure; and the latter process is termed RESI, where the excited core can be further ionized at the next laser field maximum, resulting in a single peak in the doubly charged ion momentum distribution. The ratio between the contribution of EII and RESI is distinct for different noble gas atoms, which leads to the difference on the doubly charged ion momentum distribution. Rudenko et al. [104] measured the multicharged ion (up to quadruple charged ion) momentum distribution for Ne and Ar. The transition from double peak structure to a single peak one was observed for Ne^{2+}, which indicates the transition from nonsequential double ionization to sequential double ionization. As with the doubly charged ion, the dependence on species persists for multicharged ions, and the double peak structure is prominent for Ne, but not for Ar. In fact, there are significant yields around zero for Ar^{3+} and the double peak structure is barely visible for Ar^{4+}, which means that the dominant mechanism for multiple ionization for Ar is RESI.

5.5.2.4 Photoelectron momentum correlation distribution

To fully understand the physical mechanism behind NSDI, it is instrumental to acquire the photoelectron momentum correlation distribution. The most direct procedure to obtain the momentum information of the two electrons is to measure the two photoelectrons in a single event. However, due to restrictions imposed by the dead time of the detector, usually only one electron and one doubly charged ion are measured for most of the experiments. With this procedure, the momentum of the second electron can be calculated with the momentum conservation equation, where the tiny momentum contribution from the photons is ignored. The first electron momentum correlation distribution is measured by Weber et al. [91], as shown in Figure 5.17 (d). When the laser intensity is lower than 3.8×10^{14} W/cm^2, NSDI can be taken as the dominant mechanism. In this process, the two ejected electrons fly along the same direction and give rise to the distributions in the first and third quadrants. If the laser intensity is increased, the above feature vanishes because the sequential double ionization becomes dominant. These experimental results indicate that the rescattering mechanism is the physical origin of NSDI. The electron momentum correlation distributions for He were recently obtained by two groups [92, 105], as shown in Figure 5.17 (e). A notably V-shaped structure, or finger-like pattern, can be identified in these high resolution measurements. This structure can be attributed to the Coulomb interaction of the final electrons and the EII dynamics.

So far, the photoelectrons from NSDI appear in the first and third quadrants of photoelectron momentum correlation distribution with momenta far from zero. These results can be well understood with the rescattering model [106]. Recently, Liu et al. [107] performed the double ionization of Ar and Ne subject to a laser field with low intensity. In this case, as the maximum kinetic energy (i.e. $3.17U_p$) that the electron acquired from the laser field is lower than the ionization potential of the second electron, RESI becomes the dominant ionization channel. Their results show that, in contrast to those of higher intensity, the photoelectrons appear in the second and fourth quadrants, which means that the two electrons prefer to emit in opposite directions. This is not the case for Ne [108] and the physical origin of this result is still under debate. The semiclassical calculation attributes this result to multiple recollisions with core [108]. Recently, Hao et al. [109] clarified that the interference of different ionization channels can lead to the opposite flying directions of the two ejected electrons with quantum S-matrix theory. More theoretical and experimental research on this issue is expected in the near future.

5.6 Comparison between experiment and theory

For the past decades, theoretical and experimental studies contributed equally to our understanding of the dynamics of atoms and molecules subject to intense laser fields. The new experimental findings always stimulated the development and improvement of theory while the theoretical simulations provided clear explanation and prediction for experiments. In this section, we will make a thorough comparison between experiment and theory in order to get deep physical insights into the experimental facts discussed above.

5.6.1 Low energy peaks in ATI spectra

At the time of discovery of ATI by Agostini, Reiss et al. [111] performed the theoretical investigation of the ionization rates, photoelectron energy distributions and angular distributions for atoms subject to intense laser fields with long wavelengths within the SFA in the velocity gauge. In Figure 5.18, the measured (a) [110] and calculated (b) [111] PES for Xe subject to a circularly polarized laser field at a wavelength of 1064 nm and intensity of 1.9 (in the unit of U_p/ω) are depicted. As one can see, the agreement between the theoretical and expeirmental results is excellent. Schafer et al. [112] calculated the electron energy distribution for H subject to a linearly polarized laser field at the wavelength of 532 nm and intensity of 2×10^{13} W/cm^2, as shown in Figure 5.18 (c). In the electron energy distribution, the evenly spaced ATI peaks are well fromed and their amplitudes drop gradually with the electron energy, in agreement with the experimental measurements.

In fact, the peak suppression discussed above should be elaborated for the cases of linearly and circuarly polarized laser fields [65]. In Figure 5.19 (a), the evolution of the yields of the various ATI channels (with different photons absorbed) with respect to the laser intensity in a circularly polarized laser field is shown. The contribution from each channel first increases with the intensity and then decreases when the saturation intensity is reached. It is noteworthy that before the saturation of channel n, the contribution from channel $n + 1$ is already larger than that from channel n, giving rise to peak suppresion for the low energy part in the spectra. This is shown in Figure 5.19 (b), in which the contribution of channels $n = 9$ and $n = 10$ is apparently larger than that of $n = 8$. The situation is different in a linearly polarized laser field, as shown in Figure 5.19 (c). In this case, the contribution of each channel increases with

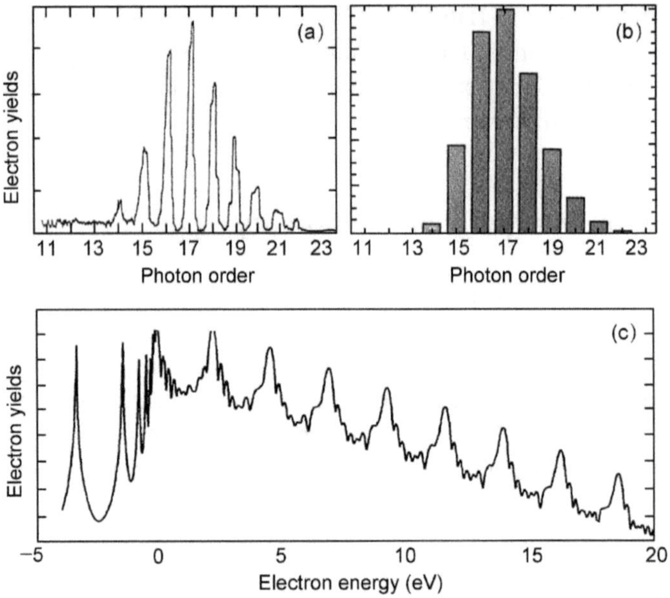

Fig. 5.18: Measured and calculated photoelectron kinetic energy spectra for Xe subject to a circularly polarized laser field: (a) experimental measurement, after [110]; (b) calculation based on S-Matrix theory, after [111]; and (c) electron energy distribution for Xe subject to a linearly polarized laser field calculated by numerically solving the TDSE, after [112].

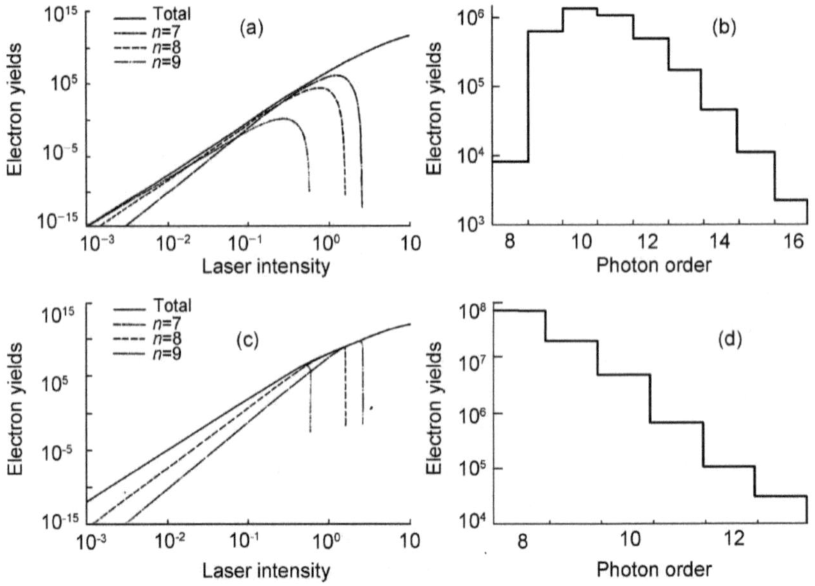

Fig. 5.19: (a) and (c): electron yields from various orders of ATI ($n = 7$, 8 and 9) as a function of laser intensity for circularly and linearly polarized laser fields respectively. (b) and (d): electron yields for different orders of ATI at the laser intensity of 1 (in the unit of U_p/ω) and wavelength of 1064nm for circularly and linearly polarized laser fields respectively. After [65].

respect to the laser intensity until it suddenly drops to zero without any sign of saturation when the final energy of the electron approches zero. The contribution from the the lower order channel is always larger than that from higher channels, giving rise to a monotonous decrease of the electron yields with respect to the number of photons absorbed for the case of a linearly polarized laser field, as shown in Figure 5.19 (d). This result is in line with the calculation by numerically solving the TDSE, as shown in Figure 5.18 (c). Note that no focus averaging or ponderomotive acceleration have been considered in the calculation. These two effects may cause peak suppresion in the low energy part of the photoelectron spectrum for atoms subject to a linearly polarized laser field.

5.6.2 High energy plateau in ATI spectra

As discussed in Section 5.5, there is a plateau in high order ATI spectra, which can be explained within the rescattering model. In fact, the semiclassical model can reproduce the experimental measurements with high precision. For example, in 1996, Walker et al. [69] systematically studied the evolution of the PES and angular distribution with respect to the laser intensity. The Keldysh parameter was kept far lower than 1 to ensure that ionization took place in the tunneling ionization regime. Their experimental results are shown in Figure 5.20 (a). One year later, Sheehy et al. [113] performed the investigation on this issue theoretically and experimentally. They reproduced the measured angle resolved PES with a semiclassical calculation considering an accurate potential function of the ionic core. The PES in Figure 5.20 is composed of two parts. One is in the kinetic energy interval of 0–2 U_p, where the electron yields drop rapidly with the electron energy. In the con-

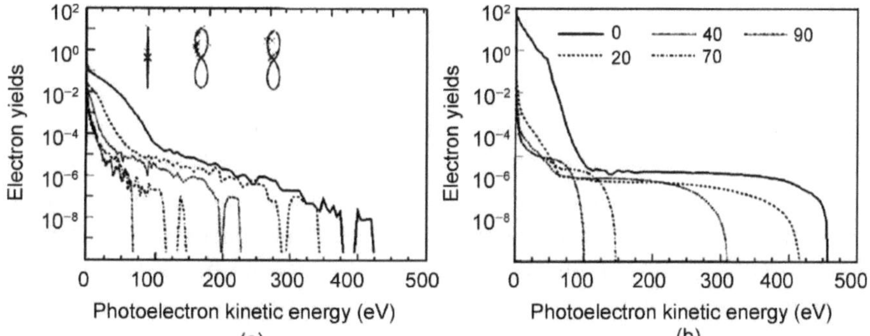

Fig. 5.20: Angle resolved photoelectron kinetic energy spectra. (a) The measured PES for He at the wavelength of 780 nm and intensity of 4.5×10^{14} W/cm^2 for five emission angles. (b) Semiclassical calculation results with parameters identical to those of (a). The angular distributions for the corresponding electrons are shown in the inset of (a). After [113].

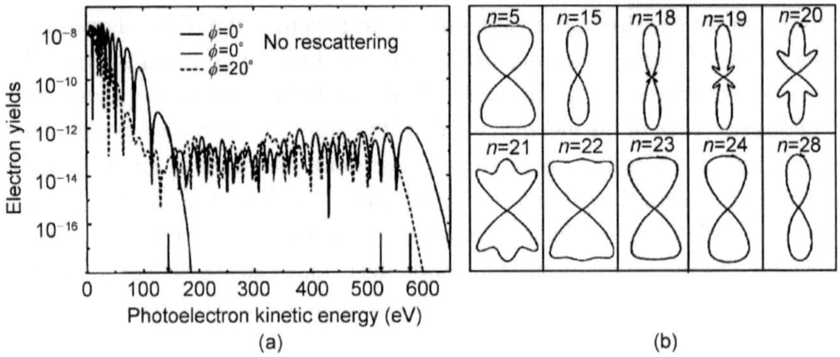

Fig. 5.21: The calculated PES and angular distributions with the SFA. (a) PES for the emission angles of 0° and 20° for He. For the calculation of the PES with 0°, the results with or without contributions from rescattering electrons are compared. After [114]. (b) The angular distributions of different ATI orders for Xe, after [115].

text of semiclassical theory, this part of PES comes from the direct electron, which is ionized without further interaction with the core. In this case, simple calculation shows that the electron yield decreases rapidly with electron kinetic energy. The other part is in the energy interval of 8–10 U_p, which can be taken as the result of the elastic scattering of the photoelectron by the ionic core. The photoelectron yields varied gently with respect to the electron energy in this energy interval. According to the calculation in Section 5.4.3, the photoelectron tunneling ionized at $\omega t_0 = 108°$ will come back to the core at $\omega t_1 = 342°$ and it will collide with the core with the maximum energy of $E_{\text{ret,max}} = 3.17\ U_p$. If this electron is scattered backward, it may be accelerated further to reach a kinetic energy of up to 10 U_p. The rescattering model can explain the relationship between the cutoff energy and the emission angle of the electron shown in Figure 5.20.

Except for the semiclassical model, the SFA was also employed to study the high order ATI spectra. In 1997, Lohr et al. [114] reproduced the PES for He by employing the SFA considering a zero range potential. The important features of calculated PES, such as the beginning energy, the range and the height of the plateau, match the experimental results very well. The PES for He subject to the laser field at the photon energy of 1.58 eV and the laser intensity of 10^{15} W/cm^2 are depicted in Figure 5.21 (a), where two cases for emission angles of 0 and 20 degrees are considered. For the calculation of the PES in 0 degrees, the results with or without contributions from rescattering electrons are compared. The calculation results with equation (5.62), where the rescattering of the electron by the core is not included, are consistent with semiclassical calculation. Both theories predict a rapid decrease in the electron yields with the electron kinetic energy in the interval of 0–2 U_p. The calculation results with equation (5.67), where the rescattering of the electron by the core is included, show a very flat

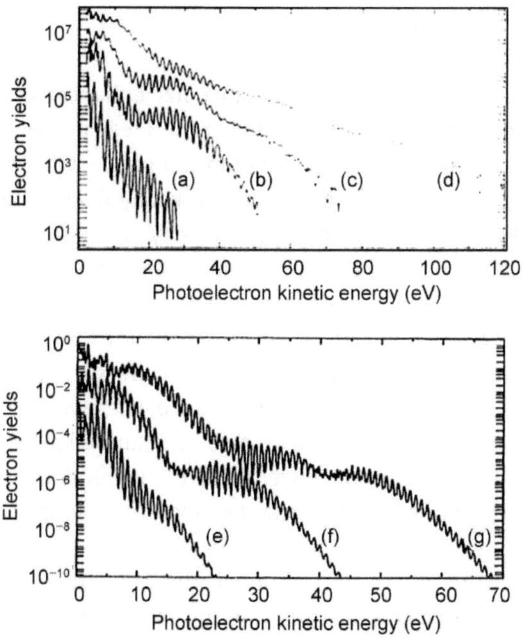

Fig. 5.22: The evolution of photoelectron kinetic energy spectra for Ar with respect to the laser intensity. The upper panel is for the experimental measurements with laser pulse duration of 40 fs and wavelength of 630 nm. The laser intensities are: (a) 6.4×10^{13} W/cm^2; (b) 1.2×10^{14} W/cm^2; (c) 2.4×10^{14} W/cm^2; and (d) 4.4×10^{14} W/cm^2. The lower panel is the calculation results with numerical solution of the time dependent Schrodinger equation. In the calculation, the photon energy is adopted as 2eV, and the pulse duration is 20 optical cycles. The laser intensities are: (e) 5.6×10^{13} W/cm^2; (f) 1.3×10^{14} W/cm^2; and (g) 2.2×10^{14} W/cm^2. After [79].

plateau and the energy of its cutoff depends on the emission angle. The cutoff energy will become lower for a larger emission angle. In 1994, Becker et al. [115] performed the calculations for the angular distributions of different ATI orders with a theory derived from the SFA. Their results are shown in Figure 5.21 (b), which are in accordance with the experimental measurements [115]. As discussed above, the SFA can reproduce the main features of the experimental results and suggests an identical physical picture to the semiclassical model.

The physical picture of the rescattering model was confirmed by numerically solving the TDSE with a SAE. In 1994, Paulus et al. [79] found that the plateau of high order ATI spectra extends to higher energy with increasing laser intensity, as shown in the upper panel of Figure 5.22. The experimental results were well reproduced by numerically solving the one dimensional TDSE, as shown in the lower panel of Figure 5.22. This investigation verified that the plateau of high order ATI spectra can be understood in the framework of the SAE, which is consistent with the rescattering model.

5.6.3 Resonance-like enhancement in ATI spectra

Generally, there are two kinds of theories to explain the resonance-like enhancement (RLE) in high order ATI. One explanation is that it is similar to the Freeman resonance [78], where the RLE structures originate from multiphoton resonance with intensity dependent excited bound states [116–123]. The excited electron may absorb one more photon to be ionized, which possesses a nonzero initial velocity and is rescattered by the core multiple times. In this process, the electron may achieve an energy of 6 U_p–8 U_p. in the laser field. From the viewpoint of quantum mechanics, the interference of the scattered electron wavefunction leads to the RLE structures. Several groups have carried out the calculations with the procedure of numerically solving the TDSE and found that the excited states can be identified in resonance whenever RLE structures appear. In the upper panel of Figure 5.23, the calculated PES can well reproduce the measurements in a large energy scale. In the lower panel of Figure 5.23, the details within each ATI order can also be replicated. This result support the resonance mechanism of RLE structures.

The other theory is under the framework of the SFA. Although the RLE structures cannot be replicated with the simplest semiclassical rescattering model [81], they

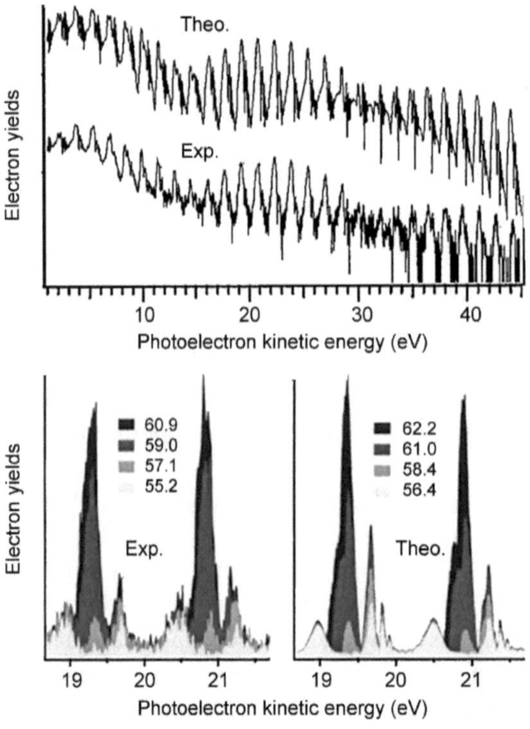

Fig. 5.23: Comparison of the measured and calculated PES for Ar. After [116].

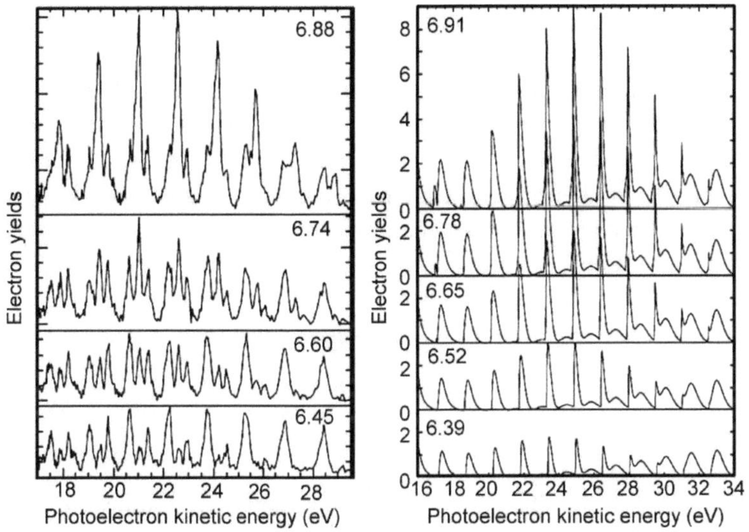

Fig. 5.24: Comparison of measured (left) and calculated (right) photoelectron kinetic energy spectra for Ar. After [85] and [124].

emerge if the constructive interference of many electron orbits is included in the calculation [71, 124–130]. As shown in the right panel of Figure 5.24, the RLE structures in the calculated PES are in line with the experimental measurements shown in the left panel of Figure 5.24. The energy difference between the effective ionization potential and the ground state is precisely equal to the energy of n times photon energy for the specific laser intensity of I_n. The condition for the constructive interference is that the laser intensity is equal to I_n. If the laser intensity $I > I_n$ and the absorption of n photons is not sufficient for the atom to be ionized, the ionization channel of n photons is thus closed. When the laser intensity is slightly larger than I_n, very low energy electrons will emerge and be rescattered by the Coulomb potential of the core many times alongside the oscillation of the laser field. The interference of these orbits leads to RLE structure in the PES. It is noteworthy that the constructive interference is crucial in both theories. However, there are no excited states involved in the second theory, which is the most important difference between the two theories.

Although a lot of experimental endeavors have been made, it is still not possible to exclude any of the theoretical explanations discussed above. Paulus et al. [71] and Milošević et al. [130] investigated the RLE structures in high order ATI for noble gas atoms subject to a multicycle laser field and explained their experimental measurements with SFA theory. In 2003, Grasbon et al. [131] found that the RLE structures disappear if a few-cycle laser field rather than a multicycle laser field was applied. These experimental results are consistent with the theory where the RLE is closely related to the interference of multireturn orbitals. In 2010, Cornaggia [132] further experimentally investigated the PES for several simple molecules, such as H_2, N_2, O_2,

CO_2, N_2O, etc. The results demonstrated that the RLE structures only exist in PES for H_2. In 2013, Quan et al. [133] systematically studied the high order ATI spectra with respect to laser intensity and found the RLE structures in the PES for N_2 but not for O_2. With SFA theory, Quan et al. reproduced the experimental results. The theoretical analysis indicated that the observed difference between N_2 and O_2 can be attributed to the different symmetry of their ground state wavefunctions. It is noteworthy that the difference may not exist within the resonance mechanism. Therefore, this latest result supports the multiple orbitals interference mechanism in the framework of SFA theory. At present, the physics behind RLE structures is still under debate and further experimental and theoretical explorations on this issue are required.

5.6.4 Low energy structure in ATI spectra

The low energy structure in the PES is one of the most important findings for ATI. With an intense laser pulse with a rather long wavelength (e.g. in the mid-infrared regime), atomic ionization falls well into the tunneling ionization regime. Quan et al. [70] reproduced the experimental measurements with a semiclassical model considering the ionic Coulomb potential, as shown in Figure 5.25. Panel (a) shows the calculation results based on the semiclassical model. The five curves from top to bottom are calculated with different parameters: (1) without Coulomb potential, laser wavelength of 2000 nm; (2) with Coulomb potential, laser wavelength of 2000 nm; (3) with Coulomb potential, laser wavelength of 1500 nm; (4) with Coulomb potential, laser wavelength of 1250 nm; and (5) with Coulomb potential, laser wavelength of 800 nm.

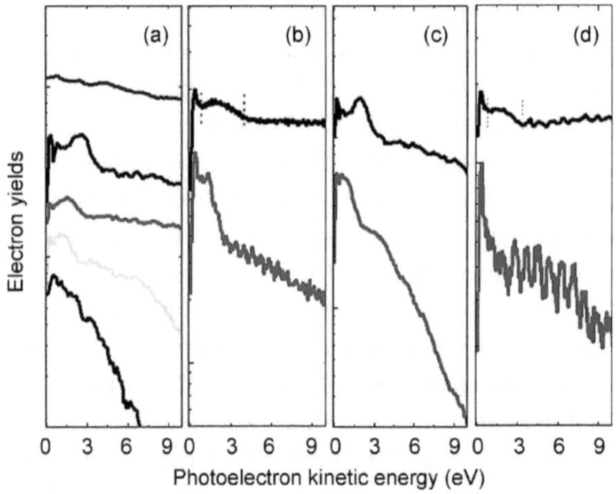

Fig. 5.25: Calculated (a, c) and measured (b, d) low energy ATI spectra for Xe. See text for details. After [70].

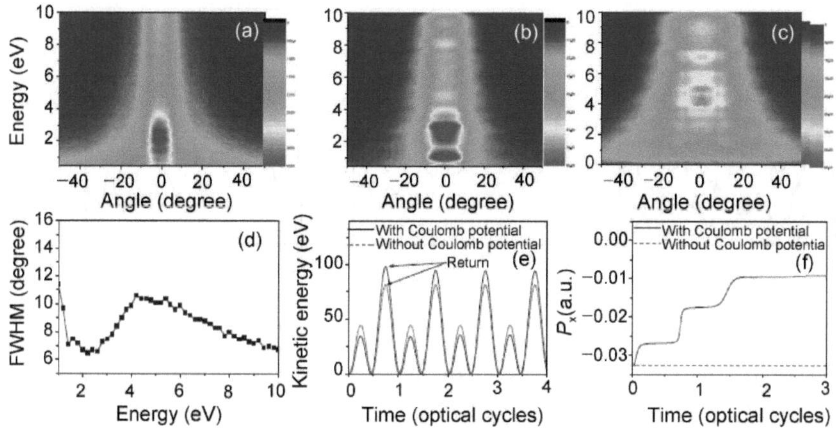

Fig. 5.26: (a) Measured angle resolved PES for Xe. (b), (c) Calculated angle resolved PES for Xe with and without considering the Coulomb field respectively. (d) Full width at half maximum of measured angle resolved PES. (e) The evolution of tunnel ionized electron energy with respect to time. (f) The evolution of the transverse momentum of a photoelectron with respect to time. The laser intensity is 8.0×10^{13} W/cm^2 and the wavelength is 2000 nm. After [134].

The laser intensity is 8.0×10^{13} W/cm^2. Panel (b) shows the experimental measurements. The laser wavelength is 2000 nm and laser intensities are 6.4 (upper curve) and 3.2 (lower curve) $\times$ 10^{13} W/cm^2. Panel (c) shows the semiclassical calculation results with parameters identical to those of (b). Panel (d) shows the experimental measurements at a wavelength of 1500 nm and laser intensities of 10.0 (upper curve) and 4.0 (lower curve) $\times$ 10^{13} W/cm^2. As shown, the measured PES (b) can be well reproduced by the calculation (c). Note that the photon energy is very small at a mid-infrared laser wavelength compared with that at 800 nm, so the spacing between the ATI peaks is relatively smaller. The laser focusing effect further makes the ATI peak barely visible and leads to smooth PES curves, which is in good agreement with experimental results, as shown in Figures 5.25 (b) and (d).

For the high energy part of the low energy structure (HLES), we can understand its physical origin from Figure 5.26 [134]. In Figure 5.26 (a), the measured angle resolved PES are shown, where an appreciable peak can be identified in the energy interval of 0.5–4 eV, corresponding to the dip of the curve in Figure 5.26 (d). These results are additional evidence for the underlying mechanism of the HLES. The semiclassical calulations with Coulomb potential included can reproduce the experimental measurements, as shown in Figure 5.26 (b), but the calculation without Coulomb potential failed to reproduce the HLES. The kinetic energy (determined dominantly by the longitudinal momentum) and the transverse momentum of the electron trajectory with final emission angle of 1.3° and of 3.5 eV is depicted with respect to time in Figures 5.26 (e) and (f). For comparison, the trajectories with identical parameters are depicted where the only difference is the inclusion or exclusion of the Coulomb potential in the cal-

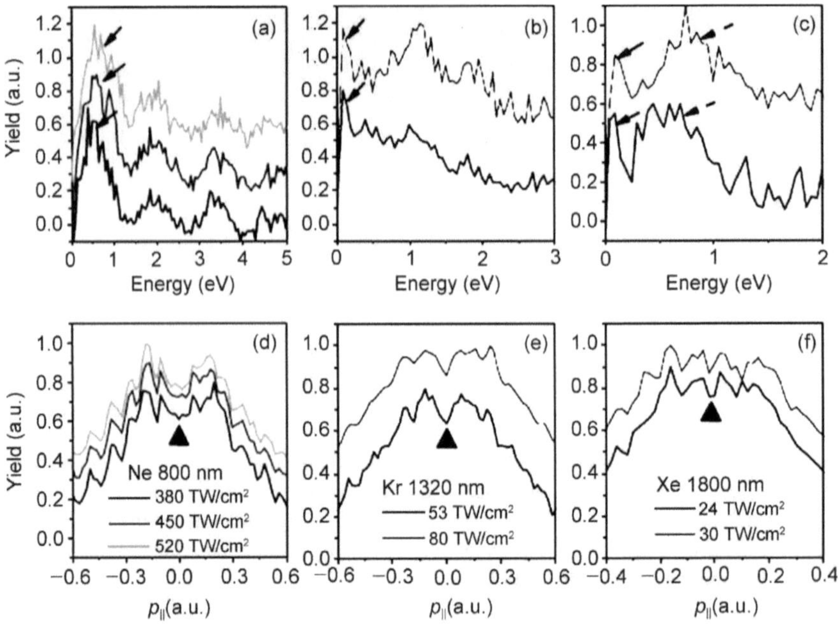

Fig. 5.27: The very low energy structure and the corresponding longitudinal momentum distributions for noble gas atoms Ne (a, d), Kr (b, e) and Xe (c, f). The solid arrows indicate the very low energy structure and the dashed arrows the HLES. After [84].

culation. With the comparison, it is easy to discover that the Coulomb potential can affect both the longitudinal and transverse momenta of the electron. For the longitudinal momentum, the Coulomb potential shifts the distribution. For example, as shown in panel (e), the electron kinetic energy is decreased at the instant of 0.25 times optical periods and increased at 0.75 times optical periods. Note that the Coulomb potential can shorten the time span from the tunneling ionization to the return, which modifies the electron dynamics significantly. As for the transverse momentum, the Coulomb potential can reduce the transverse momentum significantly, which is closely related to the distance between the electron and the core. As shown in Figure 5.26 (f), the transverse momentum degrades significantly at three moments, when the electron moves nearest to the core. The Coulomb potential lowers the emission angle from 6.5 degrees, which is obtained in calculation without Coulomb potential, to 1.3 degrees. This the emission angle leads to a narrower distribution in Figure 5.26 (a) than in Figure 5.26 (b). Furthermore, although the final energy of the electron is 3.5 eV in Figure 5.26 (e), the energy of the electron when it returns to the core can be up to 90 eV (~3 U_p), which means that even for the photoelectron with high return energy, the Coulomb potential can affect its dynamics significantly. The detailed theoretical analysis can be found in [134, 135].

The PES and longitudinal momentum distributions measured with COLTRIMS for noble gas atoms subject to laser fields with different wavelengths and intensities are shown in Figure 5.27. For the photoelectron kinetic energy spectra at 800 nm, the ATI peaks can be identified clearly, which fade out when the wavelength is increased to 1800 nm. In the meantime, HLES becomes more and more significant. Note that the very low energy structures (VLES) below 1 eV are indicated in all the spectra. The energy of VLES is around 0.5 eV at 800 nm and in the interval of $0.1 - 0.2$eV when the wavelength is increased to 1320 and 1800 nm. Interestingly, the double hump structures appear in all the longitudinal momentum distributions. There is a minimum at zero momentum and a pronounced hump on each side that consists of a series of peaks. The peak of the hump corresponds to the VLES in PES.

These experimental results can be reproduced by numerically solving the TDSE, as shown in Figure 5.28. If the short range potential is employed, instead of the long range Coulomb potential, then all the LES disappear, including HLES and VLES, which means that VLES is also the result of interactions between the photoelectron and the long range Coulomb potential. In 2002, Chen et al. [135] calculated the longitudinal momentum distributions for He and the results are qualitatively in agreement with Figure 5.27. According to their analysis, the double hump structure can be attributed to multireturn low energy electrons. Kästner et al. [136, 137] and Wu et al. [84] proposed a 'bunching' mechanism for the VLES. According to this theory, the tunneling photoelectron ionized within a certain interval of the laser phase can go back to the core with zero momentum. Moreover, the momentum that the electron acquired from the laser field may be totally compensated by the Coulomb field, which leads to a peak structure (VLES) in the momentum and energy distributions, as shown in Figure 5.29. In 2014, theoretical work with a softened potential was performed by Lin et al. [138], which demonstrated that HLES is due to the larger range of the Coulomb potential, and VLES is due to the nearer Coulomb potential. Note that these electrons possess very low kinetic energy and locate farther than 10 a.u. from the core during the interaction. In contrast, to produce high order ATI with backscattering electrons, the distances have to be around several a.u. Detailed theoretical analysis can be found in [84, 136, 137].

As discussed above, the semiclassical calculation can reproduce the experimental data and also the results by numerically solving the TDSE. This can be understood because the LES appears in the PES for atoms ionized in a deeply tunneling ionization regime, where the wavelength is long enough and the photon energy is much less than the ponderomotive energy U_{p}. According to the discussion in Section 5.4, the semiclassical model is supposed to work well. In contrast, there are problems in the treatment of LES with the SFA because in this approximation, the ionic potential is completely ignored. According to the semiclassical analysis, the ionic Coulomb potential is crucial for LES. Unfortunately, there is no convergence for the S-matrix calculation with long range Coulomb potential, as described in equation (5.67). Therefore, it is expected that the SFA fails to reproduce LES [83]. Yan et al. [139] improved the quantum orbit proce-

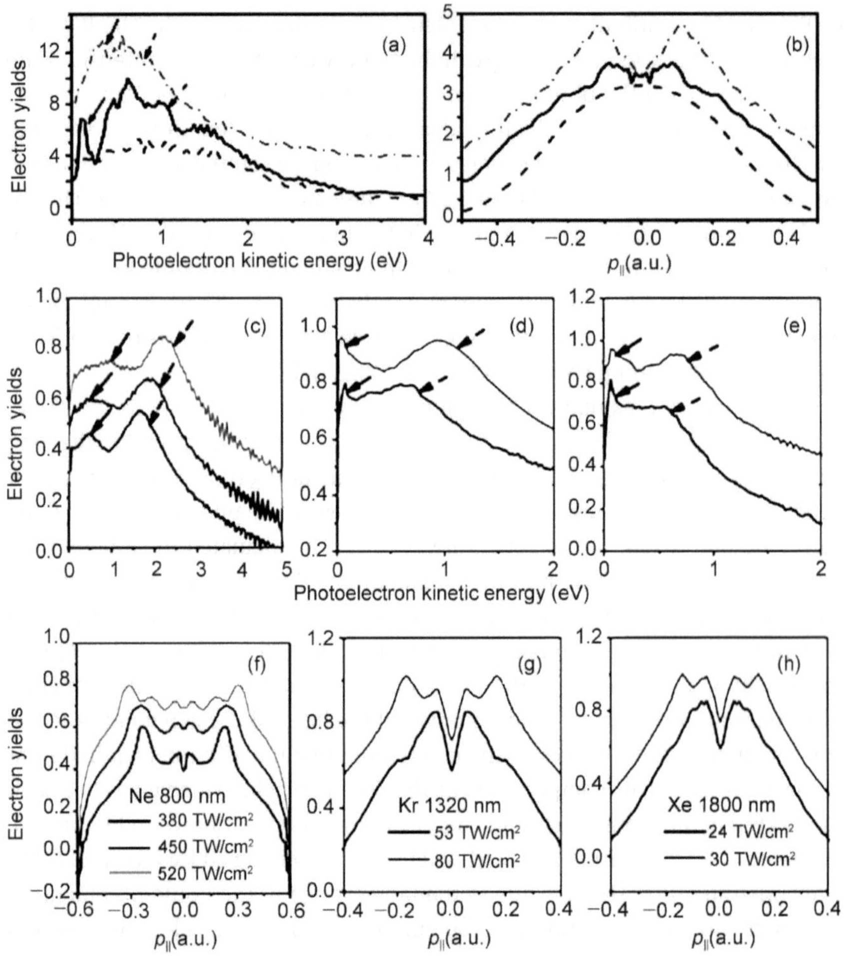

Fig. 5.28: (a), (b) The calculated PES and longitudinal momentum distributions by numerically solving the TDSE. The solid curve in the middle is the calculation result with the model potential for Xe. The upper dash-dot line is the calculation result with the hydrogenic potential model. The lower dash line is the calculation result with short range potential. (c), (d), (e), (f), (g), and (h) are PES and longitudinal momentum distributions calculated with the semiclassical model. (c) and (f) Ne at 800 nm; (d) and (g) Kr at 1320 nm; (e) and (h) Xe at 1800 nm. After [84].

dure based on the SFA and calculated 10^8 quantum orbits in the combination field of the Coulomb field and laser field. They eventually reproduced the experimental measurements with equation (5.72) employed to calculate the transition matrix elements. Guo et al. [140] managed to avoid the divergence problem by including ground state depletion in their calculation procedure and achieved the HLES with equation (5.67). They further studied the scaling of LES with respect to the Keldysh parameter and the ponderomotive energy to elucidate the significance of long range Coulomb potential in LES. It is noteworthy that all the calculations mentioned above can only reproduce

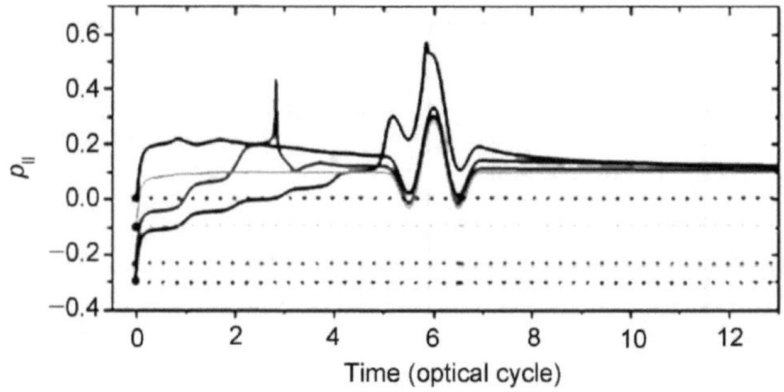

Fig. 5.29: Evolution of the longitudinal momenta of photoelectrons with respect to time for four different trajectories calculated with a semiclassical model. The solid lines are for the calculation with Coulomb potential considered, while the dotted lines are for the calculation without Coulomb potential.

the HLES. To reproduce VLES, it is necessary to include higher orders of the S-matrix, which poses a great challenge because of the divergence problem.

5.6.5 Doubly charged ion yields from NSDI

The theoretical method for NSDI is much more difficult than for ATI mentioned above. In the study of ATI process, a single active electron approximation can be well employed. But for NSDI process, it involves electron-electron interaction and is a typical three body Coulomb system strongly perturbed by a strong laser field. Due to the increase of the spatial dimensions, the requirement of calculation capacity becomes much larger than that of a single electron. Furthermore, since the exact form of the interaction between the electron and ion is not clear, it is difficult to study the NSDI process theoretically [141, 142]. Currently, the widely used approaches can be separated into two categories [141]. One is called the 'single step' theory, including numerical solution of the TDSE as well as the completely classical methods. Note that both methods are first principle theory. The other category is the 'multistep' theory, including the S-matrix theory and semiclassical theory. These methods can all reproduce (at least qualitatively) the experimental observations well, but they also have some shortcomings. Numerical calculation of the TDES is time consuming, and in practice, a variety of approximations have to be employed. In addition, TDSE simulation cannot give an intuitively physical picture. The calculation of the completely classical method is also time consuming and usually its result cannot be quantitatively well compared with the experimental observation. The shortcoming of the multistep model is that the physical processes are studied with the lack of a solid physical principle. Here, we take ion yield of NSDI, ion momentum distribution along laser polarization direction, and correlated electron momentum distribution as examples to compare these kinds of theoretical calculations with experimental results.

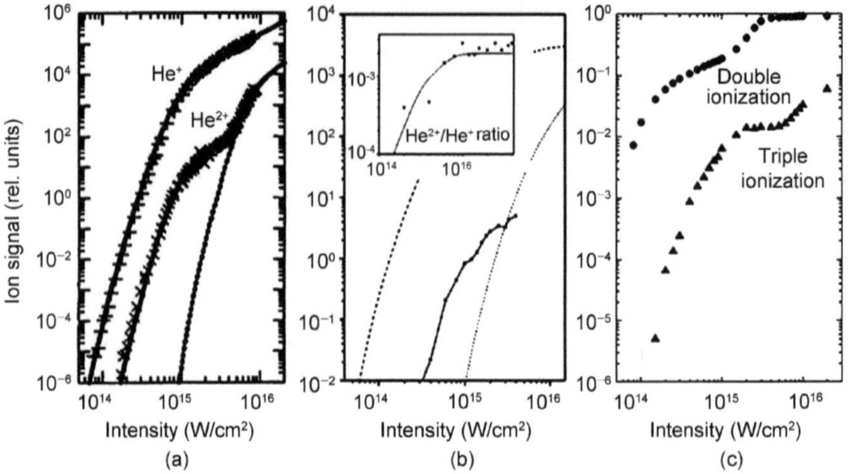

Fig. 5.30: Ionization yields as a function of laser intensity. (a) Ion yields for single and double ionization of He for the laser fields with $\lambda = 780$ nm, and pulse duration of 160fs. After [143]. (b) ADK simulation of the ion yields for single and double ionization of He. Inset shows the ratio of yields of He^{2+} and He^{+}. After [61]. (c) Yields of the single and double ionization of a model atom as a function of laser intensity with a completely classical method. After [53].

Figure 5.30 shows the experimental observations and theoretical calculations of the yields of single ionization and double ionization of He atoms in linearly polarized laser fields as a function of laser intensity. In Figure 5.30 (a), the symbols + and x denote the measured yields of the singly and doubly charged He ions respectively. The solid line is for the SFA calculation and the dashed line represents the yield of the He^{2+} ions by a sequential ionization process based on SAE approximation. The measured ion yield of He^{2+} is many orders of magnitude higher than the prediction based on SAE, indicating that the production of He^{2+} involves a strong electron-electron correlation.

The calculation of the S-matrix theory can very well reproduce the experimental results in a wide range of laser intensities. The theory includes not only the free electron from above threshold ionization, but also the second electron impact-ionized by the first electron when it recollides with the ionic core in the presence of the laser field, as well as the energy exchange between the two electrons [143]. In general, the laser pulse used in the S-matrix theory is a monochromatic laser field. For the laser pulse with duration 160 fs in experiments, the infinitely long pulse is a good approximation.

In Figure 5.30 (b) the dotted lines and dashes correspond to the tunneling ionization rates of He and He^{+} with the ADK model, while the filled circles represent the results calculated with the semiclassical model. The inset shows the ratio of yields of He^{2+} and He^{+} as a function of laser intensity, where the filled circles are the results including the rescattering processes and the solid line is a fit to experimental data [61]. The semiclassical method qualitatively reproduces the experimental data. For the ratio of yields between He^{2+} and He^{+}, the semiclassical method is especially consistent with the data. In the calculation, it is assumed that the outermost electron is ionized by tunneling ionization, and the motion of the tunnel-ionized electron is completely determined by the classical Newtonian equation in the presence of the laser field and ionic Coulomb potential. The ionized electron might return to the core and collide with the second electron, leading to the simultaneous ionization of the two electrons. The interaction of the two electrons is the Coulomb interaction.

Figure 5.30 (c) exhibits the ion yields for the single and double ionization of a model atom as a function of laser intensity calculated with a completely classical method. The filled circles represent the yield of doubly charged ions, and the filled triangles denote the yield of the doubly charged ions [53]. Even though the quantum effect is not considered, the yield of doubly charged ions from the classical method as a function of laser intensity is qualitatively consistent with the data. Moreover, the completely classical method also predicts that the yield of triple ionization exhibits the so-called 'knee' signature of NSDI. Compared with the S-matrix theory, the advantages of the semiclassical and completely classical methods are that a clear physical mechanism can be obtained by analyzing the classical trajectory, and the form of the laser field can be easily changed. Even so, the shortcomings of the two methods are also obvious, i.e. quantum effects cannot be considered.

5.6.6 Ion momentum distribution from NSDI

Another important experimental measurement in NSDI is the differential momentum distribution of doubly charged ions. For some specific laser intensities, NSDI will exhibit a double hump structure in the ion momentum distribution along the laser polarization direction and the maximal yield appears at the positions with nonvanishing momenta. Figure 5.31 (a) shows the measured momentum distribution of doubly charged ions of He along the laser polarization direction at intensity of 6.6×10^{14} W/cm^2, wavelength of 800 nm and pulse duration of 220 fs. The ionization yields are maximal at the momentum around ± 2 a.u.

Figure 5.31 (b) shows the momentum distribution of doubly charged ions of He calculated with the semiclassical method. The laser parameters are the same as that in (a). Semiclassical calculations qualitatively reproduce the experimental observations. Particularly after smoothing of the results to reduce the statistical error, the curve shows a double hump structure as in the experimental data. A closer inspection shows that the positions of the peaks in the calculated momentum distribution of He^{2+} are larger than that of the data, while the yield near zero momentum is lower than the experimental observation. Figure 5.31 (c) shows the S-matrix simulation of the momentum distribution. The amplitude of the maximal momentum, the positions of the peak in the distribution, and the width of the momentum are in good agreement with the data. The only difference is that near zero momentum, the calculated yields

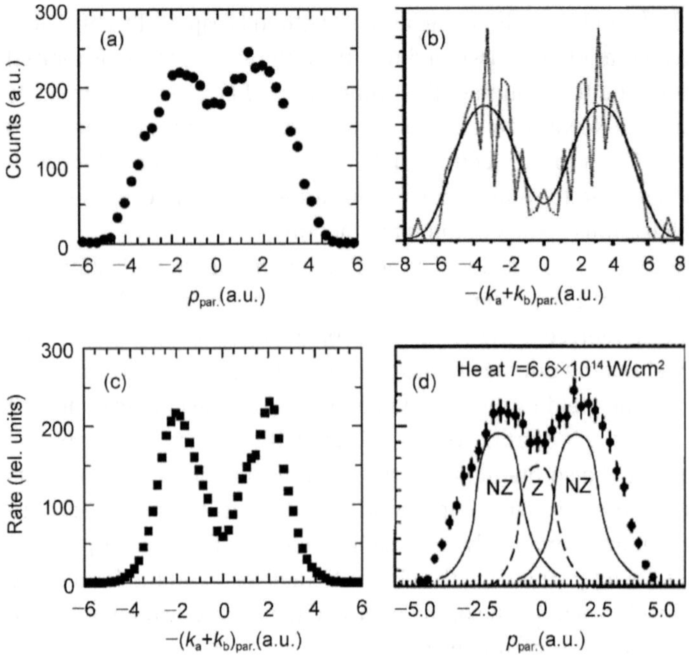

Fig. 5.31: Measured and calculated momentum distributions of doubly charged ions in the direction of the laser polarization. (a) Experimental data of He^{2+}. After [90]. (b) Semiclassical simulations. The laser parameters are the same as in (a). After [144]. (c) Simulation with S-matrix theory. After [145]. (d) Simulation with completely classical method and its comparison with the data. After [52].

are significantly lower than the experimental results. The reason may be that the calculation does not consider the impact-excitation double ionization channel [145]. Figure 5.31 (d) shows the results with the completely classical method, as well as the data. The advantages of classical calculations are that they can be used to back-analyze the various electron trajectories. The trajectories can be categorized into two classes: one may produce a single peak (indicated by Z) at zero momentum; the other will produce a double hump structure (represented by NZ) in positive and negative symmetry at nonvanishing momenta. The experimental observation of the double hump structures can be well understood by the combination of these two kinds of trajectories, as shown in Figure 5.31 (d).

5.6.7 Correlated electron momentum distribution from NSDI

In 2007, Staudte et al. [92] and Rudenko et al. [105] found that the correlated electron momentum distributions from NSDI of He exhibit a finger-like pattern. The theoretical analysis [92, 146] shows that the finger-like structure is related to the electron-electron repulsion in the double ionization process. Figure 5.32 (a) shows the measurement

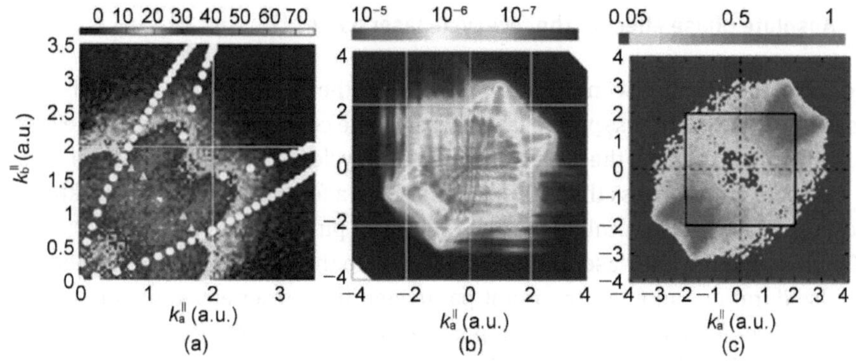

Fig. 5.32: Measured and calculated correlated electron momentum distributions from double ionization of He. (a) Experimental data. After [92]. (b) TDSE simulations. After [92]. (c) Semiclassical calculation results. After [146]. The parameters used in (b) and (c) are the same as those in (a). $k_a^{\parallel}$ and $k_b^{\parallel}$ represent the momenta of the two electrons along the laser polarization direction.

by Staudte et al. with a linearly polarized laser field, where $I = 4.5 \times 10^{14}\,\mathrm{W/cm^2}$, $\lambda = 800$ nm and pulse duration 40 fs. The white line represents the $2\sqrt{U_p}$ cutoff momentum predicted by the classical model. As shown in Figure 5.32 (a), the momentum distribution extends beyond the classical limit. Figure 5.32 (b) shows the TDSE simulation, which qualitatively agrees with the data. In order to reduce the amount of calculation, the motion of two electrons is confined to the laser polarization direction. This simplified approach may give rise to a deviation from the experimental results. Figure 5.32 (c) is the semiclassical calculation, where the black curve denotes the $2\sqrt{U_p}$ cutoff momentum. The semiclassical calculation with the Coulomb potential qualitatively reproduces the finger-like structure. However, if the Coulomb potential is replaced by a short range potential, the finger-like structure disappears. This fully shows that the long range Coulomb interaction between the two electrons is the physical origin of the finger-like structure [146].

5.7 Current research frontiers

In the previous sections, we introduced two typical physical phenomena: above threshold ionization and nonsequential double ionization, which are generated by the interaction of atoms with multicycle laser pulses at near-infrared wavelengths with the intensity in the range of 10^{13}–$10^{14}\,\mathrm{W/cm^2}$. Recently, with the rapid development of ultrafast laser technology, the study of strong field atomic physics has further extended in depth and scope. Some new research fields are gradually being developed. In the following, we will briefly introduce some recent progress in strong field atomic physics.

5.7.1 Absolute phase effect of the few-cycle laser pulse

At the end of the twentieth century, laser pulses with the durations less than 10 fs were achieved [147, 148]. For Ti:sapphire laser systems, the central wavelength of the laser pulse is about 800 nm and the period of one optical cycle is 2.7 fs. Thus, the laser pulse with the pulse duration less than 10 fs contains only a few optical cycles. This kind of femtosecond laser pulse is called the few-cycle laser pulse.

Compared to the long laser pulses, in addition to the laser wavelength, intensity, and pulse duration, another important parameter, the carrier-envelope phase (also called the absolute phase), plays an important role in the interaction between atoms and the laser field. A few-cycle laser field can be usually written in the following form:

$$E(t) = E_0(t)\sin(\omega t + \varphi)\,, \tag{5.78}$$

where $E_0(t)$ denotes the pulse envelope, ω is the laser oscillating frequency (or carrier frequency), t is the time, and φ is the carrier-envelope phase, i.e. the absolute phase. When the absolute phase is varied, the electric field of the few-cycle laser will be changed significantly. Figure 5.33 shows the change of the electric field of the few-cycle laser field with the absolute phase φ. When $\varphi = 0$, the maximum electric field of the laser field is along the positive direction, while when $\varphi = \pi$, the maximum electric field of the laser field is along the negative direction. In both cases, the components of the electric fields along the positive and negative directions are asymmetric. Only when $\varphi = \pi/2$, are the positive and negative components of the electric field of the laser pulse symmetric. Because the interaction of atoms with the strong laser field, e.g. above threshold ionization and nonsequential double ionization processes, is closely associated with the electric field form of the laser field, can the strong field atomic process be controlled by changing the electric field form through the absolute phase.

Using the quasistatic model, Dietrich et al. [152] first studied the influence of the absolute phase of the circularly polarized few-cycle laser pulse on the photoionization process. Their simulation results showed that the photoelectron angular distri-

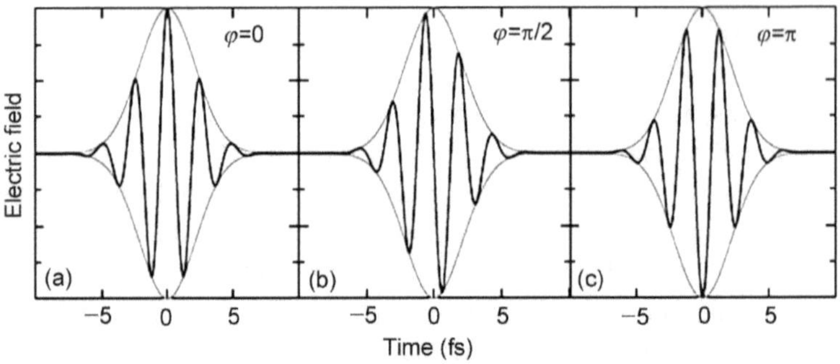

Fig. 5.33: Electric field of the few-cycle laser field as a function of time with the different absolute phases. (a) $\varphi = 0$, (b) $\varphi = \pi/2$, (c) $\varphi = \pi$. After [149].

bution is strongly dependent on the absolute phase. Thus, they proposed measuring the absolute phase of the laser pulse in experiment using the peak position of the photoelectron angular distribution. In 2001, Paulus et al. [149] first studied the absolute phase effect on the photoionization process during the interaction of atoms with the few-cycle laser pulse. In the experiment, the noble gas atoms were ionized by the circularly polarized laser pulse and the photoelectrons emitted along two opposite directions were detected by two detectors. The absolute phase of laser pulses can be revealed after the correlation analysis of the number of photoelectrons detected by the two detectors. The experimental results show that the absolute phase effect is obvious for a few-cycle laser pulse with duration of 6 fs, but the absolute phase effect cannot be seen once the pulse duration increases to 8 fs. Moreover, this effect becomes weak after considering the focus volume effect, and the longer the laser pulse, the much weaker this effect. Milošević et al. [153] analyzed the experimental results with SFA theory and proposed a method to measure the pulse duration of the few-cycle laser pulse without phase stabilization.

Based on the previous works, in 2003 Paulus et al. [150] proposed an experimental method to measure the absolute phase of few-cycle laser pulses. They used a pair of glass wedges to optimize the dispersion and adjust the CEP of the few-cycle laser pulse, and additionally systematically measured and compared the CEP dependence of the ATI spectra emitted along the opposition directions. In Figure 5.34, for the linearly polarized laser field with pulse duration of 5 fs, central wavelength of 760 nm, and intensity around 10^{14} W/cm^2, the experimental results show that the photoelectron spectra along the opposition directions change with absolute phase of the few-cycle laser pulse. In particular, for high energy electrons, the ratio of the electron yields emitted in the different directions approaches one order of magnitude. The dramatic changes of the high energy electron spectra with the absolute phase can be well understood with the rescattering model. First, due to the exponential dependence of the tunneling ionization probability on the laser electric field, for the few-cycle laser pulse the photoionization is most likely to occur in one or two optical cycles near pulse peak. Second, in such an optical cycle, only electrons ionized by tunneling in a very narrow time may return to the parent ion. After the elastic collisions with ion, the electron is accelerated and gains more energy from the laser electric field, leading to the generation of the high energy electron. These two constraints suggest that the high energy photoelectron strongly depends on the absolute phase of the few-cycle laser pulse. Combined with semiclassical calculations and experimental results, they achieve a precise measurement of the absolute phase of the few-cycle laser pulse through the ratio of high energy electrons emitted along two different directions.

In 2005, Lindner et al. [154] experimentally showed the 'double slit' interference of the ionized electron wave packets in the time domain with the phase stabilized few-cycle laser pulses. In the experiment, they carefully selected the absolute phase of the few-cycle laser pulses such that the detected electrons are from different time 'windows' in the laser pulse. The interference of the electron wave packets from different

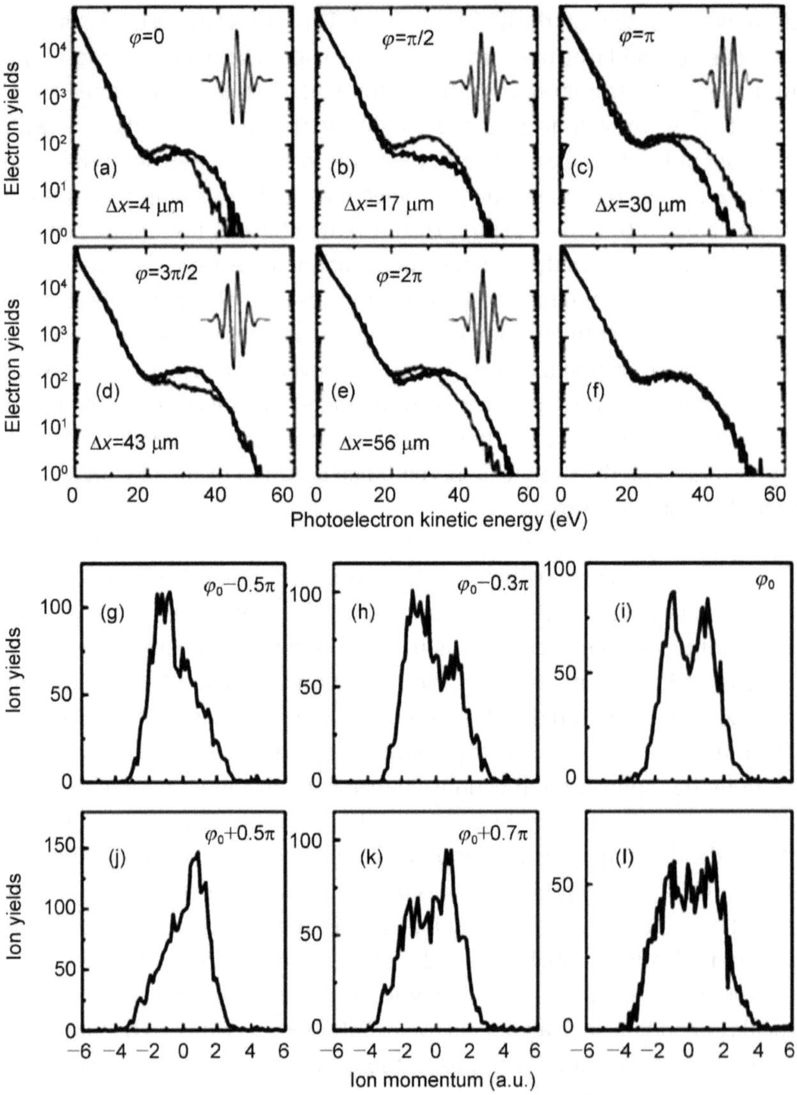

Fig. 5.34: (a), (b), (c), (d) and (e) PES of Xe along the positive and negative polarization directions of the laser pulse with different absolute phases. (f) PES along the positive and negative polarization directions of the laser pulse without phase stabilization. After [150]. (g), (h), (i), (j), (k) Momentum distribution of Ar^{2+} along the laser polarization direction with different absolute phases. (l) Momentum distribution of Ar^{2+} along the laser polarization direction without phase stabilization. After [151].

channels results in significant interference patterns in the above ionization threshold photoelectron spectra. Their experiments also showed that the control of the coherent ultrafast electron behavior in the suboptical-cycle timescale can be achieved by modulating the absolute phase.

Like above threshold ionization, the atomic NSDI process also strongly depends on the absolute phase of few-cycle laser pulses. In 2004, Liu et al. [151] studied the momentum distribution of Ar^{2+} using ion-electron momentum correlation spectra. They found that for the laser pulse with a pulse duration of 5 fs, a central wavelength of 760 nm, and intensity around 3.5×10^{14} W/cm^2, the ion momentum distribution along the laser polarization shows a significant asymmetry, and this asymmetry can be modulated with the absolute phase, as shown in Figure 5.34. In fact, before this experimental work, Liu et al. [155] had developed a semiclassical model based on the electron rescattering to study the absolute phase dependence of the photoelectron momentum distribution in nonsequential double ionization. They found that with the variation of the absolute phase, the electron momentum distribution gradually moves from the first quadrant to the third quadrant. Because the measured momentum of the double charged ion is equal to the sum of two electron momenta, the experimental results agree well with theoretical predictions.

The atomic photoionization process is sensitive to the laser pulse duration and the absolute phase, which can be used for the real time diagnosis of the few-cycle femtosecond laser pulse. For example, in 2009, based on Paulus et al.'s works, Wittmann et al. [156] further developed an absolute phase measurement method with the high order above threshold ionization, and realized a high precision measurement of the phase, i.e. less than 10 mrad. In the experiment, they set up a 'stereo-ATI', similar to Paulus et al.'s works, to obtain the absolute phase value by measuring and comparing the photoelectron spectra along two opposite directions of the laser polarization. Since the interaction region allows a high density of the gas target, enough high-energy electrons can be obtained for a single laser shot, and thus an accurate measurement of the absolute phase for a single laser pulse can be achieved. Sayler et al. also used the stereo-ATI to achieve a real time accurate measurement of laser pulse duration for the few-cycle laser pulse [157].

Although there are still some technical challenges to measure absolute phase for the current stereo-ATI technology, for example the influence of the pulse duration and the intensity fluctuation on the measurement results cannot be fully eliminated, the stereo-ATI technology for measurement of the absolute phase has greatly promoted the study of the electron dynamic behavior in the subcycle time scale [158–160]. The latest developed stereo-ATI technology based on the solid state sample [161] is free of vacuum systems, making it more practical and convenient.

5.7.2 Atomic ionization by superintense laser field

In previous sections, we mainly discussed the interaction of atoms with intense laser of intensity around 10^{13}–10^{16} W/cm^2. With further increase of the laser intensity, many highly charged ions will emerge. Additionally, the relativistic effect and magnetic field effect will also make the interaction of atoms with strong laser fields more complicated. Recently, attention has been paid to atomic ionization by the superintense laser field. For example, in 2005 Palaniyappan et al. [162] experimentally measured the ion yields of the noble gas atoms as a function of laser intensity in the range 10^{14}–10^{18} W/cm^2.

As shown in Figure 5.35, by comparing the experimental results with the ADK simulations, the ions of Ne^{2+}, Ne^{3+}, Ne^{4+} and Ne^{5+} are mainly generated from the nonsequential double ionization in a certain intensity range, which can be seen from the 'knee' structure in the figure. The experimental results were simulated with a three dimensional semiclassical model under relativistic conditions. Theoretical analysis shows that when the light intensity is higher than 10^{17} W/cm^2, the Lorentz force of the magnetic field of the laser field begins to affect the behavior of electrons, which leads to the suppression of the rescattering. Almost at the same time, Giubbini et al. [163, 164] measured the ion yields of noble gas atoms in the laser intensity range

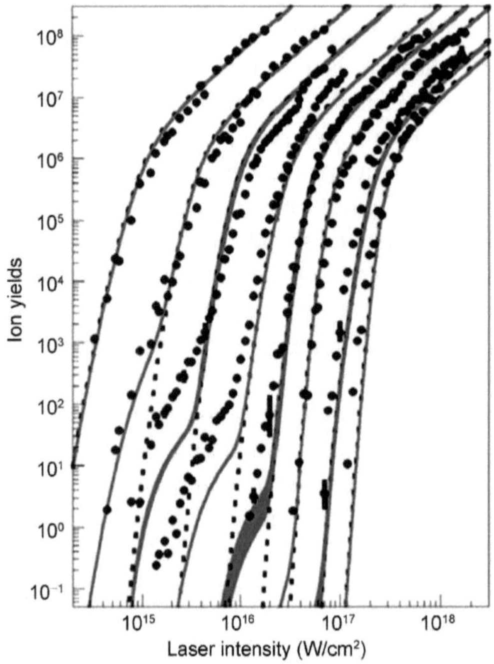

Fig. 5.35: Ion yields (from Ne^{+} to Ne^{8+} from left to right) as a function of the laser intensity. Filled circles: experimental data; dotted line: ADK theoretical model calculations; solid line: a three dimensional rescattering model calculation including the relativistic effects. After [162].

10^{16}–10^{18} W/cm^2 and observed highly charged ions (up to 15^+). In their theoretical analysis, the relativistic effects are also taken into account. They found that when laser light intensity is close to the relativistic region, the widely used rescattering mechanism for the nonsequential double ionization is largely suppressed. The possible reasons are that the electron impact ionization cross-section decreases with increasing charge state, or the diffusion effect of the return electron wave packet increases and the magnetic component of the laser field influences the dynamics of the electron.

5.7.3 Molecular ionization in intense laser fields

Compared with the simple atom, the molecule system has a complex electronic structure and multiple atomic nuclei. The behavior of molecules in strong laser fields is significantly different from that of atoms. How the complex structure of a molecule affects its dynamic behavior in a strong laser field is a hot and challenging issue in current research.

The influence of molecular structure on the ionization is first found in the measurement of the total ionization probability of molecules. In the mid-1990s, it was found that in the near-infrared femtosecond laser field, the ionization probability of some molecules is much smaller than that of their companion atoms (the atom with similar ionization potential as the molecule), i.e. it exhibits strong ionization suppression [165]. Figure 5.36 shows the ionization probability of the diatomic molecules (N_2 and O_2) and the noble gas atoms (Ar and Xe) measured by Guo et al. [166] as a function of laser intensity. Note that the molecule N_2 and the atom Ar chosen in the experiment have similar ionization potential (N_2: 15.58 eV, Ar: 15.76 eV), while the molecule O_2 has similar ionization potential to the atom Xe (O_2: 12.8 eV, Xe: 12.76 eV). As one can see from the figure, for N_2 and Ar, the changes in the ionization probability with laser intensity are almost the same, but surprisingly, compared with atom Xe, the ionization probability of the molecule O_2 is significantly lower, showing a strong ionization suppression. This molecular ionization suppression phenomenon has attracted wide attention in the strong field physics community and many theoretical models have been developed in the past ten years. For example, in the molecular-ADK (MO-ADK) model [167] the appearance of the molecular ionization suppression is attributed to different asymptotic behaviors of the ground state wavefunctions. Another explanation attributes the ionization suppression to the multielectron screening effect in molecules [168]. Quantum mechanical S-matrix theory however predicts that this phenomenon is related to the multicenter structure of molecule, i.e. the interference between photoelectron wave packets emitted from the different atomic centers leads to ionization suppression [169]. Figure 5.36 shows the quantum mechanical S-matrix calculation, which well reproduces the experimental results. Furthermore, in 2012 Lin et al. [170] experimentally and theoretically studied molecular ionization suppression. By measuring and comparing the ionization yields of the molecules (O_2 and N_2) and

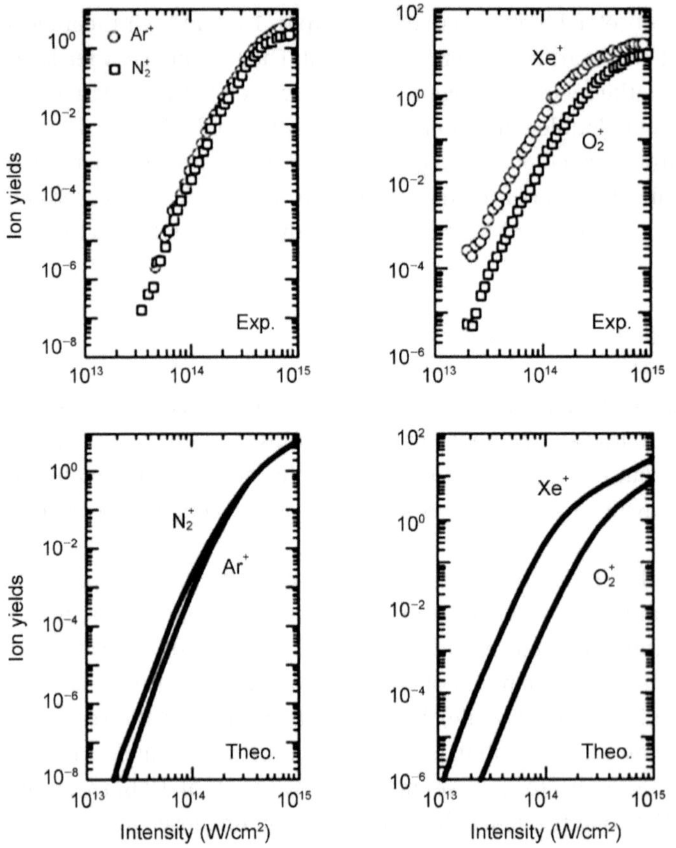

Fig. 5.36: Single ionization probabilities of N_2, Ar and O_2, Xe as a function of the laser intensity for a laser pulse with a pulse duration of 30 fs and a central wavelength of 800 nm. Measured results are shown in the upper panel. After [166]. S-matrix calculation is in the lower panel. After [169].

their companion atoms (Xe and Ar) in intense near-infrared and mid-infrared laser fields, they found that the molecular ionization suppression with respect to its companion atom strongly depends on the laser intensity and wavelength. No suppression was seen in the molecule N_2 in comparison with its companion atom Ar. The study strongly supports the emergence of molecular ionization suppression from molecular two-center interference, while both the molecular ADK formulation and multielectron screening model failed to predict the observed wavelength dependence of ionization suppression.

The structure of diatomic molecules also affects the nonsequential double ionization process. In Figure 5.37, Eremina et al. [171] measured the correlated electron momentum distribution from nonsequential double ionization of molecules N_2 and O_2. Experimental results show that, as with the noble gas atoms, the electron momentum correlation distribution of the double ionization of the N_2 molecule is mainly in the

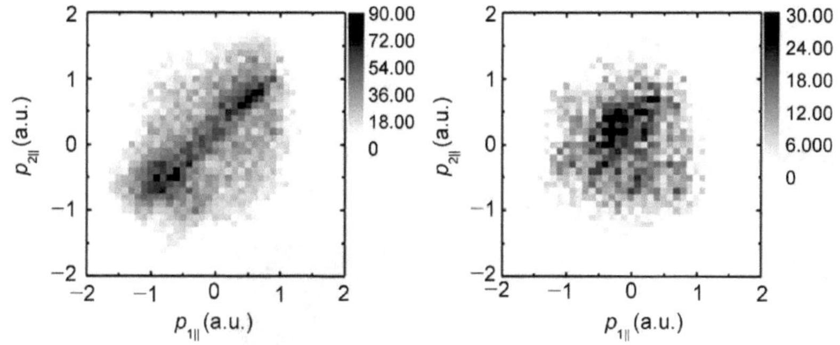

Fig. 5.37: Experimental correlated electron momentum distribution from nonsequential double ionization of molecular N_2 at the laser intensity of 1.5×10^{14} W/cm^2 (left) and O_2 at the laser intensity 1.7×10^{14} W/cm^2(right). The laser wavelength is 800 nm and the pulse duration is 35 fs. After [171].

first and third quadrants, while for molecular O_2, the electron momentum distribution is in the near-zero momentum region. In their work, they implemented the molecular structure and the double atomic center interference effects into the rescattering model. The simulation results very well reproduced the experimental results, suggesting that the molecular structure significantly affects the correlated electron emission.

According to the tunneling ionization theory, the ionization probability depends exponentially on the ionization potential of the system. As for a complex molecular system, the energy of the highest occupied molecular orbital (HOMO) may be not too much larger than that of the other occupied molecular orbitals (e.g. HOMO-1, HOMO-2, etc.). Therefore, the contribution from different molecular orbitals needs to be considered in the ionization processes of the molecule. Especially for some specific molecular orientations, the contribution from HOMO-1 or even HOMO-2 molecular orbitals to strong field ionization may become significant. In 2008, McFarland et al. [172] measured the high order harmonic generation of molecular N_2 with different orientations and observed the contribution from the HOMO-1 molecular orbital to the tunneling ionization. In 2009, Akagi et al. [173] experimentally studied the ionization behavior contributions from different molecular orbitals of HCl. For the HCl molecule, the energy of the highest occupied molecular orbital is −12.747 eV, while the energy of the second highest occupied molecular orbital is −16.265 eV, relatively close to the former. In experiments, they measured the fragments from the HOMO-1 channel to determine the contribution from HOMO-1. Figure 5.38 shows that after the ionization from the HOMO state, the generated ion is in the ground state ($X^2\Pi$) without dissociation. In contrast, the ionization of molecular orbital HOMO-1 state will generate the excited state ($A^2\Sigma^+$) ion, which will be dissociated by coupling to the $2^2\Sigma^+$ state. In experiments, the contribution from the HOMO-1 state to the ionization process can be observed by detecting the atomic ion signal from dissociation.

During the past decade, it became possible to use the intense field driven tunnel-ionized electron beam to probe molecular structures and ultrafast dynamics pro-

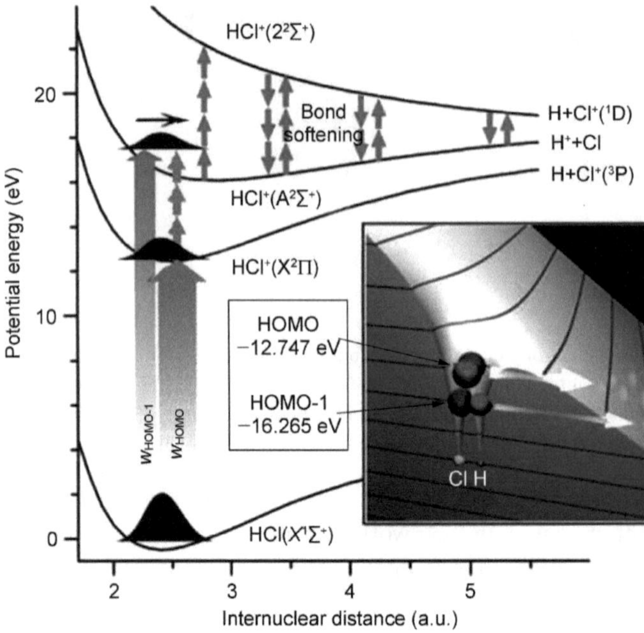

Fig. 5.38: Ionization and dissociation of molecular HCl subject to an intense laser field. After [173].

cesses, as the ionized electron beam has ultrahigh temporal and spatial resolution. In 2002, Niikura et al. [174] first proposed a 'molecular clock' scheme by analyzing the electron wave packet features from hydrogen molecules in a strong laser field. By comparing the proton kinetic energy spectrum from the dissociation of molecular hydrogen in the linear polarization and elliptical polarization fields, they obtained information on the vibrating wave packet of H_2^+ induced by recollision of the tunnel-ionized electrons. In combination with theoretical calculations, they revealed that the ionized electron wave packet has high current density and short pulse (~1 fs) features. One year later, using the correlation properties of the rescattering electron wave packet and the parent ion from the strong field ionization of molecular hydrogen, and by measuring the kinetic energy distribution of protons from molecular dissociation with different laser wavelengths, they observed the vibration of the nuclear wave packet of D_2^+ from the multiphoton ionization of D_2 with a time and spatial resolution as high as 200 as and 0.05 Å respectively [175].

As previously mentioned, because of the multiatomic center nature of molecules, the intense field ionized electron in molecules may come from different atoms. As in Young's double slit interference of light, the ionized electron wave packets from the different atomic centers may interfere with each other. The interference pattern is closely associated with the molecular structure, e.g. nuclear distance. Based on this picture, in 1996 Zuo et al. [176] proposed an effective method to probe the structure and dynamics of molecules by laser induced electron diffraction. Later, this method was further developed and improved [177]. In 2008, Meckel et al. [178] experimentally studied the molecular structure and the dynamics using laser induced electron diffraction. They measured the three dimensional electron momentum distributions of the diatomic molecules N_2 and O_2 spatially aligned with intense laser pulse, and furthermore obtained information on the ground state of the electron orbit and the position of the parent ion. After that, Okunishi et al. [179] and Kang et al. [177] experimentally measured the angular distribution of the high order ATI of randomly oriented molecules and further revealed the relationship between the electron angular distribution and molecular structure in its ground state in terms of quantum mechanical S-matrix theory. In 2012, Blaga et al. [180] studied the ultrafast change of the bond length of the molecular ions N_2^+ and O_2^+ with time using the laser induced electron diffraction scheme. According to the rescattering picture, the return time of the tunnel-ionized electron wave packet induced by the laser field is related to the laser wavelength. By measuring and comparing the diffraction patterns from the scattering of the tunneling-ionized electron with the parent ion at different wavelengths, information on the time evolution of the molecular bonds can be obtained. In their work, they measured the electron momentum spectra at three different mid-infrared wavelengths and the corresponding electron scattering cross-section was extracted from the electron momentum spectra. Furthermore, by comparing the experimental results with theoretical calculations, a best-fit value of bond length of the molecular ion at various wavelengths was extracted from the experimental data. Their study showed that with the laser induced electron diffraction method, one may simultaneously detect the dynamics of electrons and ions with an unprecedented time resolution of a few femtosecond and spatial resolution of subangstroms.

5.7.4 Strong field atomic physics in XUV and X-ray regimes

In the previous sections, we mainly introduced the progress of strong field atomic physics in the long wavelength range (near-infrared and mid-infrared band). For long laser wavelength, the electron motion is relatively fast in comparison with the oscillation of the electric field of the laser pulse. Thus, it is reasonable to consider the laser field as a quasistatic field for which the semiclassical or even classical model can be used to understand the interaction of atoms with strong laser fields and the dynamic processes involved. Recently, with the development of free electron laser technology [183–185], laser pulses with wavelengths in the range from extreme ultraviolet (XUV) to X-rays, pulse durations of a few femtoseconds, and peak laser intensities of 10^{18} W/cm^2 or even higher, can be achieved, which provides a powerful tool to study strong field atomic physics at short wavelengths.

In 2009, Richter et al. [181] systematically studied the ionization behavior of different noble gas atoms with extreme ultraviolet pulses generated by the free electron laser in Hamburg (FLASH) with wavelength of 13.7 nm (corresponding to photon en-

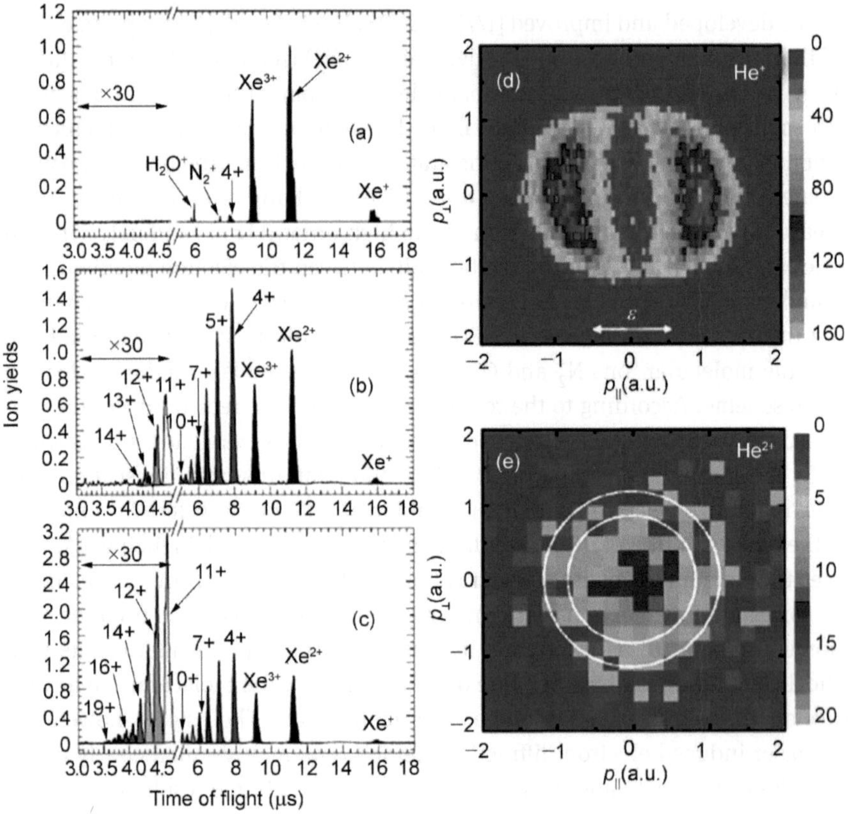

Fig. 5.39: (a), (b) and (c): Time of flight mass spectrum of Xe produced by the free electron laser. The photon energy is 90.5 eV, the laser pulse duration is about 10 fs, and the laser intensity is (a) 2.5×10^{12} W/cm^2, (b) 1.7×10^{15} W/cm^2, and (c) 2.0×10^{15} W/cm^2. After [181]. (e) and (f): Recoil ion momentum distribution of single atomic ionization (e) and double ionization (f) of He atoms in the laser pulse with a photon energy of 44 eV, a pulse duration of 25 fs, and intensity of 1×10^{14} W/cm^2. After [182].

ergy 90.5 eV) and laser intensity up to 2×10^{15} W/cm^2. Their experiments showed that the ionization processes of the noble gas atoms in the XUV band shows a strong dependence on the species. For example, for the atoms Ne, Ar, and Kr, highly charged ions (e.g. charge number of 7) can be observed with the highest laser intensity, while for atomic Xe, charge numbers can be as high as 19, just as shown in the left panel of Figure 5.39. The highly nonlinear ionization behavior of Xe atoms cannot be understood with the standard models developed in the optical wavelength, e.g. the tunneling ionization theory based on the quasistatic field approximation. In the short wavelength laser field, the ionization process is significantly affected by many factors, e.g. inner-shell electron structure of the atom, electron correlation and resonance. The observed highly nonlinear ionization behavior of atomic Xe may be related to the giant resonance effect of its 4d state.

Similar to strong field ionization in the optical band, the measurement of the ion yields cannot provide much information about the physical mechanism of ionization. Recently, the differential ion and electron momentum distributions have also been measured. With recoil ion momentum spectroscopy, Rudenko et al. [182] studied the simplest two-photon two-electron ionization process of the two-electron system, i.e. the He atom, using the laser pulse generated by the free electron laser (FLASH) with photon energy of 44 eV and laser intensity of up to 10^{14} W/cm^2. They experimentally measured the momentum distributions of He^+ and He^{2+}. They found that the two momentum distributions are significantly different; see the right panel in Figure 5.39. Theoretical calculations show that the distribution of double charged ions originates from the nonsequential ionization process. The ion distribution concentrated at near-zero momentum indicates that two electrons tend to show back-to-back emission (opposite momentum direction), a significant anticorrelation feature. This work provides a benchmark to develop the many-body quantum theory for few-photon and few-electron interaction systems and is helpful to further understand electron correlation in the XUV regime.

With the further decrease of the wavelength of the free electron laser pulse to the X-ray band, the the ionization of the inner shell electrons will become dominant, which will generate numerous electron relaxation processes, e.g. Auger and the cascaded processes, resulting in the generation of highly charged ions and more complex interactions between atoms and the strong laser field. Young et al. [186] studied the interaction of the noble gas atoms with an X-ray laser pulse generated by an X-ray free-electron laser (Linac Coherent Light Source, LCLS) (light intensity 10^{18} W/cm^2, wavelength in the range 1.5~0.6 nm, photon flow about 10^{15} photons/Å^2). Considering that the binding energy of the 1s electron of an Ne atom is 870 eV, Young et al. chose three typical photon energy values: 800 eV, lower than the ionization threshold; 1050 eV, higher than the ionization threshold; and 2000 eV, higher than the ionization threshold for all charge states of neon. They experimentally measured the time of flight mass spectra of the ionization of neon. By comparing the data and the theoretical simulations, the laser pulse parameters, such as the photon flow and the pulse duration, can be determined. Their results showed that even though the laser intensity gets up to 10^{18} W/cm^2, the interaction of atoms with the X-ray laser pulse mainly takes place through a series of single photon absorption process. Recently, Fukuzawa et al. [187] extended the study of the interaction of radiation with X-ray laser pulses to the hard X-ray laser pulse generated by the X-ray free electron laser (SPring-8 Angstrom Compact free electron Laser, SACLA). They studied the multiphoton and multielectron ionization of the noble gas atom Xe in the hard X-ray range with photon energy up to 5.5 keV. Their results showed that the Xe atom can absorb five or even more hard X-ray photons and the number of the ionized electron can get up to 26. Due to the strong electron-electron interaction inside atoms, each photon can cause five electron ionizations on average.

5.8 Summary and outlook

In this chapter, recent progress in strong field atomic physics has been reviewed, and the importance of atomic structure and electron-electron correlation has been demonstrated in the ionization dynamics of atoms subject to intense laser fields. Most recent advances include wide employment of few-cycle laser pulses in this research area, enabling the exploration of the intense field-atom interaction, e.g. the NSDI process [159], with an unprecedented time resolution of attoseconds. Meanwhile, by employing elliptically polarized light, one adds an additional dimension to the interaction, providing more details on the strong field atomic processes [188–192] and facilitating the understanding of electronic dynamics on attosecond timescales [193–196].

Nevertheless, the research progress reviewed in this chapter is mainly constrained to the near-infrared wavelength region and to simple atoms and molecules. The recent development of ultrafast optical parameter amplifiers (OPAs) in the mid-infrared wavelength region enables the exploration of the wavelength scaling of the semiclassical aspects in intense laser-atom interactions, while the rapid development of free electron laser technology, which can produce intense laser pulses in extreme ultraviolet and even X-ray regions, will reveal novel atomic physics phenomena. In this case, the inner-shell electrons will be first ionized, followed by an Auger process. A full understanding of this ionization process involves the interplay of the inner-shell electronic structure, electron-electron correlations and the resonance.

Moreover, in the past few years, considerable progress has been made towards the understanding of how complex systems, such as molecules, clusters and even solids, interact with intense laser fields. Many novel effects, such as multiatomic center interference, molecular alignment and many-body effects, start to play a role in interaction processes. However, it is still an open question whether the well accepted theoretical models, such as strong field approximation (SFA) and the rescattering model, which were developed mainly in studies on the simple atomic and molecular systems, can still apply to the complex systems. In particular, the understanding of electron-electron correlation in this context is still largely an open question. All in all, one expects that the study of atomic and molecular dynamics in strong laser fields will continue to be one of the most rich and challenging topics within atomic, molecular and optical physics in the years to come.

Bibliography

[1] M. Goeppert-Mayer. Über Elementarakte mit zwei Quantensprüngen. Ann. Phys., **9** (1931): 273–295.

[2] G. S. Voronov and N. B. Delone. Ionization of xenon atom by electric field of ruby laser emission. JETP Lett., **1** (1965): 66–68.

[3] G. Mainfray and C. Manus. Multiphoton ionization of atoms. Rep. Prog. Phys., **54** (1991): 1333–1372.

[4] P. Agostini, F. Fabre, G. Mainfray, G. Petite and N. K. Rahman. Free-free transitions following 6-photon ionization of xenon atoms. Phys. Rev. Lett., **42** (1979): 1127–1130.

[5] B. W. Shore and P. L. Knight. Enhancement of high optical harmonics by excess-photon ionization. J. Phys. B, **20** (1987): 413–423.

[6] M. Ferray, A. Lhuillier, X. F. Li, et al., Multiple-harmonic conversion of 1064-nm radiation in rare-gases. J. Phys. B, **21** (1988):L31-L35.

[7] A. McPherson, G. Gibson, H. Jara, et al., Studies of multiphoton production of vacuum ultraviolet-radiation in the rare-gases. J. Opt. Soc. Am. B, **4** (1987): 595–601.

[8] P. Agostini and L. F. DiMauro. The physics of attosecond light pulses. rep. prog. phys., **67** (2004): 813.

[9] P. B. Corkum and F. Krausz. Attosecond science. Nat. Phys., **3** (2007): 381–387.

[10] F. Krausz and M. Ivanov. Attosecond physics. Rev. Mod. Phys., **81** (2009): 163–234.

[11] L. V. Keldysh. Ionization in field of a strong electromagnetic wave. Sov. Phys. -JETP, **20** (1965): 1307–1314.

[12] M. V. Ammosov, N. B. Delone and V. P. Krainov. Tunnel ionization of complex atoms and atomic ions in a varying electromagnetic-field. Zh. Eksp. Teor. Fiz., **91** (1986): 2008–2013.

[13] N. B. Delone and V. P. Krainov. Tunneling and barrier-supression ionization of atoms and ions in a laser radiation field. Usp. Fiz. Nauk., **168** (1998): 531–549.

[14] V. P. Krainov, W. Xiong and S. L. Chin. An introductory overview of tunnel ionization of atoms by intense lasers. Las. Phys., **2** (1992): 467–475.

[15] S. L. Chin, F. Yergeau and P. Lavigne. Tunnel ionization of Xe in an ultra-intense CO_2-laser field (10^{14} w cm^{-2}) with multiple charge creation. J. Phys. B, **18** (1985): L213-L215.

[16] E. Mevel, P. Breger, R. Trainham, et al., Atoms in strong optical-fields – evolution from multiphoton to tunnel ionization. Phys. Rev. Lett., **70** (1993): 406–409.

[17] X. M. Tong and C. D. Lin. Empirical formula for static field ionization rates of atoms and molecules by lasers in the barrier-suppression regime. J. Phys. B, **38** (2005): 2593–2600.

[18] X. F. Li, A. Lhuillier, M. Ferray, L. A. Lompre and G. Mainfray. Multiple-harmonic generation in rare-gases at high laser intensity. Phys. Rev. A, **39** (1989): 5751–5761.

[19] K. J. Schafer, B. Yang, L. F. Dimauro and K. C. Kulander. Above threshold ionization beyond the high harmonic cutoff. Phys. Rev. Lett., **70** (1993): 1599–1602.

[20] A. L'Huillier, L. A. Lompre, G. Mainfray and C. Manus. Multiply charged ions induced by multiphoton absorption in rare-gases at 0.53 μm. Phys. Rev. A, **27** (1983): 2503–2512.

[21] FittinghoffDN, P. R. Bolton, B. Chang and K. C. Kulander. Observation of nonsequential double ionization of helium with optical tunneling. Phys. Rev. Lett., **69** (1992): 2642–2645.

[22] P. B. Corkum. Plasma perspective on strong-field multiphoton ionization. Phys. Rev. Lett., **71** (1993): 1994–1997.

[23] T. Brabec and F. Krausz. Intense few-cycle laser fields: Frontiers of nonlinear optics. Rev. Mod. Phys., **72** (2000): 545–591.

[24] V. E. Zakharov and A. B. Shabat. Exact theory of 2-dimensional self-focusing and one-dimensional self-modulation of waves in nonlinear media, sov. Phys.-JETP, **34** (1972): 62–69.

[25] A. Dubietis, G. Jonusauskas and A. Piskarskas. Powerful femtosecond pulse generation by chirped and stretched pulse parametric amplification in bbo crystal. Opt. Commun., **88** (1992): 437–440.

[26] D. Strickland and G. Mourou. Compression of amplified chirped optical pulses, opt. Commun., **56** (1985): 219–221.

[27] P. Maine, D. Strickland, P. Bado, M. Pessot and G. Mourou. Generation of ultrahigh peak power pulses by chirped pulse amplification, IEEE J. Quantum Electron., **24** (1988): 398–403.

[28] J. Kawanaka, K. Yamakawa, H. Nishioka and K. Ueda. 30-MJ, diode-pumped, chirped-pulse Yb:Ylf regenerative amplifier, Opt. Lett., **28** (2003): 2121–2123.

[29] J. F. Xia, J. Song and D. Strickland. Development of a dual-wavelength Ti:sapphire multi-pass amplifier and its application to intense mid-infrared generation, Opt. Commun, **206** (2002): 149–157.
[30] A. E. Siegman. Lasers, Oxford, Oxford University Press, 1986, 669–671.
[31] R. Dorner, V. Mergel, O. Jagutzki, et al., Cold target recoil ion momentum spectroscopy: A 'momentum microscope' to view atomic collision dynamics, Phys. Rep., **330** (2000): 95–192.
[32] J. Ullrich, R. Moshammer, A. Dorn, et al., Recoil-Ion and electron momentum spectroscopy: Reaction-microscopes, rep. prog. phys., **66** (2003): 1463–1545.
[33] R. J. Elliott. Laser spectroscopy of barium and strontium in high magnetic fields. London University of London, 1996.
[34] W. C. Wiley and I. H. McLaren. Time-of-flight mass spectrometer with improved resolution. Rev. Sci. Instrum., **26** (1955): 1150–1157.
[35] J. L. Wiza. Microchannel plate detectors. J. Nucl. Instrum. Meth., **162** (1979): 587–601.
[36] P. Hansch and L. D. VanWoerkom. High-precision intensity-selective observation of multiphoton ionization: a new method of photoelectron spectroscopy. Opt. Lett., **21** (1996): 1286–1288.
[37] A. Eppink and D. H. Parker. Velocity map imaging of ions and electrons using electrostatic lenses: application in photoelectron and photofragment ion imaging of molecular oxygen. Rev. Sci. Instrum., **68** (1997): 3477–3484.
[38] D. S. Guo and T. Aberg. Quantum electrodynamical approach to multiphoton ionization in the high-intensity H-field. J. Phys., **21** (1988): 4577–4591.
[39] D. S. Guo, T. Åberg and B. Crasemann. Scattering theory of multiphoton ionization in strong fields. Phys. Rev. A, **40** (1989): 4997–5005.
[40] L. Gao, X. Li, P. Fu, R. R. Freeman and D. S. Guo. Nonperturbative quantum electrodynamics theory of high-order harmonic generation. Phys. Rev. A, **61** (2000): 063407.
[41] J. Gao, D. S. Guo and Y. S. Wu. Resonant above-threshold ionization at quantized laser intensities. Phys. Rev. A, **61** (2000): 043406.
[42] J. Chen, S. G. Chen and J. Liu. Quantum electrodynamic approach to multiphoton ionization in strong fields. J. Phys. B, **36** (2003): 1559–1572.
[43] B. Wang, L. Gao, X. Li, D. S. Guo and P. Fu. Frequency-domain theory of high-order above-threshold ionization based on nonperturbative quantum electrodynamics. Phys. Rev. A, **75** (2007): 063419.
[44] B. Wang, Y. Guo, J. Chen, Z. C. Yan and P. Fu. Frequency-domain theory of nonsequential double ionization in intense laser fields based on nonperturbative qed. Phys. Rev. A, **85** (2012): 023402.
[45] K. C. Kulander, K. J. Schafer and J. L. Krause. Time-dependent studies of multiphoton processes, atoms in intense laser fields, San Diego, Academic, 1992, 247–298.
[46] K. C. Kulander. Multiphoton ionization of hydrogen: A time-dependent theory. Phys. Rev. A, **35** (1987): 445–447.
[47] C. R. Feeler and R. E. Olson. Single ionization of Ne by intense laser fields. J. Phys. B, **33** (2000): 1997–2003.
[48] J. S. Cohen. Comment on the classical-trajectory monte carlo method for ion-atom collisions. Phys. Rev. A, **26** (1982): 3008–3010.
[49] L. Meng, C. O. Reinhold and R. E. Olson. Electron removal from molecular hydrogen by fully stripped ions at intermediate energies. Phys. Rev. A, **40** (1989): 3637–3645.
[50] R. Panfili and J. H. Eberly. Comparing classical and quantum dynamics of strong-field double ionization. Opt. Express, **8** (2001): 431–435.
[51] R. Panfili, S. L. Haan and J. H. Eberly. Slow-down collisions and nonsequential double ionization in classical simulations. Phys. Rev. Lett., **89** (2002): 113001.

[52] P. J. Ho, R. Panfili, S. L. Haan and J. H. Eberly. Nonsequential double ionization as a completely classical photoelectric effect. Phys. Rev. Lett., **94** (2005): 093002.
[53] P. J. Ho and J. H. Eberly. In-plane theory of nonsequential triple ionization. Phys. Rev. Lett., **97** (2006): 083001.
[54] X. Wang and J. H. Eberly. Elliptical polarization and probability of double ionization. Phys. Rev. Lett., **105** (2010): 083001.
[55] X. Wang, J. Tian and J. H. Eberly. Angular correlation in strong-field double ionization under circular polarization. Phys. Rev. Lett., **110** (2013): 073001.
[56] Y. Zhou, Q. Liao and P. Lu. Mechanism for high-energy electrons in nonsequential double ionization below the recollision-excitation threshold. Phys. Rev. A, **80** (2009): 023412.
[57] Y. Zhou, Q. Liao and P. Lu. Asymmetric electron energy sharing in strong-field double ionization of helium. Phys. Rev. A, **82** (2010): 053402.
[58] Y. Zhou, C. Huang, Q. Liao and P. Lu. Classical simulations including electron correlations for sequential double ionization. Phys. Rev. Lett., **109** (2012): 053004.
[59] W. Becker, F. Grasbon, R. Kopold, et al., Above-threshold ionization: From classical features to quantum effects. Adv. At. Mol. Opt. Phys., **48** (2002): 35–98.
[60] B. B. Hu, J. Liu and S. G. Chen. Plateau in above-threshold-ionization spectra and chaotic behavior in rescattering processes. Phys. Lett. A, **236** (1997): 533–542.
[61] L. B. Fu, J. Liu, J. Chen and S. G. Chen. Classical collisional trajectories as the source of strong-field double ionization of helium in the knee regime. Phys. Rev. A, **63** (2001): 043416.
[62] J. Chen, J. Liu and S. G. Chen. Rescattering effect on phase-dependent ionization of atoms in two-color intense fields. Phys. Rev. A, **61** (2000): 033402.
[63] D. M. Wolkow. Über eine Klasse von Lösungen der Diracschen Gleichung. Z. Phys., **94** (1935): 250–260.
[64] M. Lewenstein, K. C. Kulander, K. J. Schafer and P. H. Bucksbaum. Rings in above-threshold ionization: A quasiclassical analysis. Phys. Rev. A, **51** (1995): 1495–1507.
[65] H. R. Reiss. Effect of an intense electromagnetic field on a weakly bound system. Phys. Rev. A, **22** (1980): 1786–1813.
[66] J. H. Bauer. Qualitatively different theoretical predictions for strong-field photoionization rates. Phys. Rev. A, **84** (2011): 025403.
[67] F. Yergeau, G. Petite and P. Agostini. Above-threshold ionization without space-charge. J.Phys. B, **19** (1986):L663-L669.
[68] R. R. Freeman and P. H. Bucksbaum. Investigations of above-threshold ionization using subpicosecond laser-pulses. J. Phys. B, **24** (1991): 325–347.
[69] B. Walker, B. Sheehy, K. C. Kulander and L. F. DiMauro. Elastic rescattering in the strong field tunneling limit. Phys. Rev. Lett., **77** (1996): 5031–5034.
[70] W. Quan, Z. Lin, M. Wu, et al., Classical aspects in above-threshold ionization with a midinfrared strong laser field. Phys. Rev. Lett., **103** (2009): 093001.
[71] G. G. Paulus, F. Grasbon, H. Walther, R. Kopold and W. Becker. Channel-closing-induced resonances in the above-threshold ionization plateau. Phys. Rev. A, **64** (2001): 021401(R).
[72] P. Agostini, M. Clement, F. Fabre and G. Petite. Multi-photon ionization involving multi-photon continuum-continuum transitions. J. Phys. B, **14** (1981):L491-L495.
[73] P. Kruit, J. Kimman and M. J. Vanderwiel. Absorption of additional photons in the multi-photon ionization continuum of xenon at 1064, 532 and 440 nm. J. Phy. B, **14** (1981): L597–L602.
[74] F. Fabre, P. Agostini, G. Petite and M. Clement. Angular-distribution in above-threshold ionization of Xe at 0.53 μm. J. Phys. B, **14** (1981): L677-L681.
[75] P. Kruit, J. Kimman, H. G. Muller and M. J. Vanderwiel. Electron-spectra from multiphoton ionization of xenon at 1064, 532, and 355 nm. Phys. Rev. A, **28** (1983): 248–255.

[76] M. Protopapas, C. H. Keitel and P. L. Knight. Atomic physics with super-high intensity lasers. Rep. Prog. Phys., **60** (1997): 389–486.
[77] J. Kupersztych. Inverse half-bremsstrahlung in multiphoton ionization of atoms in intense light-beams. Europhys. Lett., **4** (1987): 23–28.
[78] R. R. Freeman, P. H. Bucksbaum, H. Milchberg, et al., Above-threshold ionization with subpicosecond laser-pulses. Phys. Rev. Lett., **59** (1987): 1092–1095.
[79] G. G. Paulus, W. Nicklich, H. L. Xu, P. Lambropoulos and H. Walther. Plateau in above-threshold ionization spectra. Phys. Rev. Lett., **72** (1994): 2851–2854.
[80] B. R. Yang, K. J. Schafer, B. Walker, et al., Intensity-dependent scattering rings in high-order above-threshold ionization. Phys. Rev. Lett., **71** (1993): 3770–3773.
[81] G. G. Paulus, W. Becker, W. Nicklich and H. Walther. Rescattering effects in above-threshold ionization – a classical-model. J. Phys. B, **27** (1994): L703–L708.
[82] P. Colosimo, G. Doumy, C. I. Blaga, et al., Scaling strong-field interactions towards the classical limit. Nat. Phys., **4** (2008): 386–389.
[83] C. I. Blaga, F. Catoire, P. Colosimo, et al., strong-field photoionization revisited. Nat. Phys., **5** (2009): 335–338.
[84] C. Y. Wu, Y. D. Yang, Y. Q. Liu, et al., Characteristic spectrum of very low-energy photoelectron from above-threshold ionization in the tunneling regime. Phys. Rev. Lett., **109** (2012): 043001.
[85] M. P. Hertlein, P. H. Bucksbaum and H. G. Muller. Evidence for resonant effects in high-order ati spectra. J. Phys. B, **30** (1997): L197–L205.
[86] P. Hansch, M. A. Walker and Van L. D. Woerkom. Resonant hot-electron production in above-threshold ionization. Phys. Rev. A, **55** (1997): R2535–R2538.
[87] B. Walker, B. Sheehy, L. F. DiMauro, et al., Precision measurement of strong field double ionization of helium. Phys. Rev. Lett., **73** (1994): 1227–1230.
[88] R. Lafon, J. L. Chaloupka, B. Sheehy, et al., Electron energy spectra from intense laser double ionization of helium. Phys. Rev. Lett., **86** (2001): 2762–2765.
[89] R. Moshammer, B. Feuerstein, W. Schmitt, et al., Momentum distributions of Ne^{N+} ions created by an intense ultrashort laser pulse. Phys. Rev. Lett., **84** (2000): 447–450.
[90] T. Weber, M. Weckenbrock, A. Staudte, et al., Recoil-ion momentum distributions for single and double ionization of helium in strong laser fields. Phys. Rev. Lett., **84** (2000): 443–446.
[91] T. Weber, H. Giessen, M. Weckenbrock, et al., Correlated electron emission in multiphoton double ionization. Nature, **405** (2000): 658–661.
[92] A. Staudte, C. Ruiz, M. Schöffler, et al., Binary and recoil collisions in strong field double ionization of helium. Phys. Rev. Lett., **99** (2007): 263002.
[93] S. Larochelle, A. Talebpour and S. L. Chin. Non-sequential multiple ionization of rare gas atoms in a Ti:sapphire laser field. J. Phys. B, **31** (1998): 1201–1214.
[94] P. Dietrich, N. H. Burnett, M. Ivanov and P. B. Corkum. High-harmonic generation and correlated two-electron multiphoton ionization with elliptically polarized light. Phys. Rev. A, **50** (1994): R3585–R3588.
[95] C. Guo and G. N. Gibson. Ellipticity effects on single and double ionization of diatomic molecules in strong laser fields. Phys. Rev. A, **63** (2001): 040701.
[96] G. D. Gillen, M. A. Walker and Van L. D. Woerkom. Enhanced double ionization with circularly polarized light. Phys. Rev. A, **64** (2001): 043413.
[97] F. Mauger, C. Chandre and T. Uzer. Recollisions and correlated double ionization with circularly polarized light. Phys. Rev. Lett., **105** (2010): 083002.
[98] L. B. Fu, G. G. Xin, D. F. Ye and J. Liu. Recollision dynamics and phase diagram for nonsequential double ionization with circularly polarized laser fields. Phys. Rev. Lett., **108** (2012): 103601.

[99] J. L. Chaloupka, J. Rudati, R. Lafon, et al., Observation of a transition in the dynamics of strong-field double ionization. Phys. Rev. Lett., **90** (2003): 033002.
[100] E. R. Peterson and P. H. Bucksbaum. Above-threshold double-ionization spectroscopy of argon. Phys. Rev. A, **64** (2001): 053405.
[101] B. Witzel, N. A. Papadogiannis and D. Charalambidis. Charge-state resolved above threshold ionization. Phys. Rev. Lett., **85** (2000): 2268–2271.
[102] U. Eichmann, M. Dörr, H. Maeda, W. Becker and W. Sandner. Collective multielectron tunneling ionization in strong fields. Phys. Rev. Lett., **84** (2000): 3550–3553.
[103] de V. L. B. Jesus, B. Feuerstein, K. Zrost, et al., Atomic structure dependence of nonsequential double ionization of He, Ne and Ar in strong laser pulses. J. Phys. B, **37** (2004): L161–L167.
[104] A. Rudenko, K. Zrost, B. Feuerstein, et al., Correlated multielectron dynamics in ultrafast laser pulse interactions with atoms. Phys. Rev. Lett., **93** (2004): 253001.
[105] A. Rudenko, V. L. B. de Jesus, T. Ergler, et al., Correlated two-electron momentum spectra for strong-field nonsequential double ionization of He at 800 nm. Phys. Rev. Lett., **99** (2007): 263003.
[106] A. Becker, R. Dorner and R. Moshammer. Multiple fragmentation of atoms in femtosecond laser pulses. J. Phys. B, **38** (2005): S753–S772.
[107] Y. Liu, S. Tschuch, A. Rudenko, et al., Strong-field double ionization of Ar below the recollision threshold. Phys. Rev. Lett., **101** (2008): 053001.
[108] Y. Q. Liu, D. F. Ye, J. Liu, et al., Multiphoton double ionization of Ar and Ne close to threshold. Phys. Rev. Lett., **104** (2010): 173002.
[109] X. Hao, J. Chen, W. Li, et al., Quantum effects in double ionization of argon below the threshold intensity. Phys. Rev. Lett., **112** (2014): 073002.
[110] P. H. Bucksbaum, M. Bashkansky, R. R. Freeman, T. J. McIlrath and L. F. DiMauro. Suppression of multiphoton ionization with circularly polarized coherent light. Phys. Rev. Lett., **56** (1986): 2590–2593.
[111] H. R. Reiss. Electron spectrum in intense-field photoionization. J. Opt. Soc. Am. B, **4** (1987): 726–730.
[112] K. J. Schafer and K. C. Kulander. Energy analysis of time-dependent wave functions: Application to above-threshold ionization. Phys. Rev. A, **42** (1990): 5794–5797.
[113] B. Sheehy, R. Lafon, M. Widmer, et al., Single- and multiple-electron dynamics in the strong-field tunneling limit. Phys. Rev. A, **58** (1998): 3942–3952.
[114] A. Lohr, M. Kleber, R. Kopold and W. Becker. Above-threshold ionization in the tunneling regime. Phys. Rev. A, **55** (1997):R4003-R4006.
[115] W. Becker, A. Lohr and M. Kleber. Effects of rescattering on above-threshold ionization. J. Phys. B, **27** (1994): L325–L332.
[116] M. J. Nandor, M. A. Walker, L. D. Van Woerkom and H. G. Muller. Detailed comparison of above-threshold-ionization spectra from accurate numerical integrations and high-resolution measurements. Phys. Rev. A, **60** (1999):R1771-R1774.
[117] H. G. Muller and F. C. Kooiman. Bunching and focusing of tunneling wave packets in enhancement of high-order above-threshold ionization. Phys. Rev. Lett., **81** (1998): 1207–1210.
[118] H. G. Muller. Tunneling excitation to resonant states in helium as main source of superponderomotive photoelectrons in the tunneling regime. Phys. Rev. Lett., **83** (1999): 3158–3161.
[119] H. G. Muller. Numerical simulation of high-order above-threshold-ionization enhancement in argon. Phys. Rev. A, **60** (1999): 1341–1350.
[120] E. Cormier, D. Garzella, P. Breger, et al., Above-threshold ionization contrast and channel closure in argon. J. Phys. B, **34** (2001): L9–L17.
[121] J. Wassaf, V. Véniard, R. Taïeb and A. Maquet. Strong field atomic ionization: Origin of high-energy structures in photoelectron spectra. Phys. Rev. Lett., **90** (2003): 013003.

[122] R. M. Potvliege and Vučić S. High-order above-threshold ionization of argon: Plateau resonances and the floquet quasienergy spectrum, phys. Rev. A, **74** (2006): 023412.
[123] R. M. Potvliege and S. Vucic. Freeman resonances in high-order above-threshold ionization, J. Phys. B, **42** (2009): 055603.
[124] R. Kopold, W. Becker, M. Kleber and G. G. Paulus. Channel-closing effects in high-order above-threshold ionization and high-order harmonic generation. J. Phys. B, **35** (2002): 217–232.
[125] R. Kopold and W. Becker. Interference in high-order above-threshold ionization. J. Phys. B, **32** (1999): L419–L424.
[126] S. V. Popruzhenko, P. A. Korneev, S. P. Goreslavski and W. Becker. Laser-induced recollision phenomena: Interference resonances at channel closings. Phys. Rev. Lett., **89** (2002): 023001.
[127] E. Hasovic, M. Busuladzic, A. Gazibegovic-Busuladzic, D. B. Milosevic and W. Becker. Simulation of above-threshold ionization experiments using the strong-field approximation. Las. Phys., **17** (2007): 376–389.
[128] D. B. Milošević, Hasović E, Busuladžić M, Gazibegović-Busuladžić A and W. Becker. Intensity-dependent enhancements in high-order above-threshold ionization. Phys. Rev. A, **76** (2007): 053410.
[129] D. B. Milosevic, E. Hasovic, S. Odzak, et al., Wavelength dependence of channel-closing enhancements in high-order above-threshold ionization and harmonic generation. J. Mod. Opt., **55** (2008): 2653–2663.
[130] D. B. Milosevic, W. Becker, M. Okunishi, et al., Strong-field electron spectra of rare-gas atoms in the rescattering regime: enhanced spectral regions and a simulation of the experiment. J. Phys. B, **43** (2010): 015401.
[131] F. Grasbon, G. G. Paulus, H. Walther, et al., Above-threshold ionization at the few-cycle limit. Phys. Rev. Lett., **91** (2003): 173003.
[132] C. Cornaggia. Enhancements of rescattered electron yields in above-threshold ionization of molecules. Phys. Rev. A, **82** (2010): 053410.
[133] W. Quan, X. Lai, Y. Chen, et al., Resonancelike enhancement in high-order above-threshold ionization of molecules. Phys. Rev. A, **88** (2013): 021401.
[134] W. Chu, M. Y. Wu, B. Zeng, et al., Unexpected breakdown of the simple man's model for strong-field photoionization in the high-energy recollision region. Phys. Rev. A, **85** (2012): 021403.
[135] J. Chen and C. H. Nam. Ion momentum distributions for he single and double ionization in strong laser fields. Phys. Rev. A, **66** (2002): 053415.
[136] A. Kästner, U. Saalmann and J. M. Rost. Energy bunching in soft recollisions revealed with long-wavelength few-cycle pulses. J. Phys. B, **45** (2012): 074011.
[137] A. Kästner, U. Saalmann and J. M. Rost. Electron-energy bunching in laser-driven soft recollisions. Phys. Rev. Lett., **108** (2012): 033201.
[138] Z. Lin, M. Wu, W. Quan, et al., Coulomb effect on low-energy structure in above-threshold ionization spectra induced by mid-infrared laser pulses. Chin. Phys. B, **23** (2014): 023201.
[139] T. M. Yan, S. V. Popruzhenko, M. J. VrakkingJ and D. Bauer. Low-energy structures in strong field ionization revealed by quantum orbits. Phys. Rev. Lett., **105** (2010): 253002.
[140] L. Guo, S. S. Han, X. Liu, et al., Scaling of the low-energy structure in above-threshold ionization in the tunneling regime: Theory and experiment. Phys. Rev. Lett., **110** (2013): 013001.
[141] W. Becker, X. Liu, P. J. Ho and J. H. Eberly. Theories of photoelectron correlation in laser-driven multiple atomic ionization. Rev. Mod. Phys., **84** (2012): 1011–1043.
[142] C. Figueira de Morisson Faria and X. Liu. Electron-electron correlation in strong laser fields. J. Mod. Opt., **58** (2011): 1076–1131.

[143] A. Becker and F. H. M. Faisal. Interplay of electron correlation and intense field dynamics in the double ionization of helium. Phys. Rev. A, **59** (1999): R1742–R1745.
[144] J. Chen, J. Liu, L. B. Fu and W. M. Zheng. Interpretation of momentum distribution of recoil ions from laser-induced nonsequential double ionization by semiclassical rescattering model. Phys. Rev. A, **63** (2000): 011404.
[145] A. Becker and F. H. M. Faisal. Interpretation of momentum distribution of recoil ions from laser induced nonsequential double ionization. Phys. Rev. Lett., **84** (2000): 3546–3549.
[146] D. F. Ye, X. Liu and J. Liu. Classical trajectory diagnosis of a fingerlike pattern in the correlated electron momentum distribution in strong field double ionization of helium. Phys. Rev. Lett., **101** (2008): 233003.
[147] J. P. Zhou, G. Taft, C. P. Huang, et al., Pulse evolution in a broad-bandwidth Ti-sapphire laser. Opt. Lett., **19** (1994): 1149–1151.
[148] M. Nisoli, S. DeSilvestri, O. Svelto, et al., Compression of high-energy laser pulses below 5 fs. Opt. Lett., **22** (1997): 522–524.
[149] G. G. Paulus, F. Grasbon, H. Walther, et al., Absolute-phase phenomena in photoionization with few-cycle laser pulses. Nature, **414** (2001): 182–184.
[150] G. G. Paulus, F. Lindner, H. Walther, et al., Measurement of the phase of few-cycle laser pulses. Phys. Rev. Lett., **91** (2003): 253004.
[151] X. Liu, H. Rottke, E. Eremina, et al., Nonsequential double ionization at the single-optical-cycle limit. Phys. Rev. Lett., **93** (2004): 263001.
[152] P. Dietrich, F. Krausz and P. B. Corkum. Determining the absolute carrier phase of a few-cycle laser pulse. Opt. Lett., **25** (2000): 16–18.
[153] D. B. Milosevic, G. G. Paulus and W. Becker. Phase-dependent effects of a few-cycle laser pulse. Phys. Rev. Lett., **89** (2002): 153001.
[154] F. Lindner, M. G. Schätzel, H. Walther, et al., Attosecond double-slit experiment. Phys. Rev. Lett., **95** (2005): 040401.
[155] X. Liu and C. Figueira de Morisson Faria. Nonsequential double ionization with few-cycle laser pulses. Phys. Rev. Lett., **92** (2004): 133006.
[156] T. Wittmann, B. Horvath, W. Helml, et al., Single-shot carrier-envelope phase measurement of few-cycle laser pulses. Nat. Phys., **5** (2009): 357–362.
[157] A. M. Sayler, T. Rathje, W. Muller, et al., Real-time pulse length measurement of few-cycle laser pulses using above-threshold ionization. Opt. Express, **19** (2011): 4464–4471.
[158] M. Kubel, K. J. Betsch, N. G. Johnson, et al., Carrier-envelope-phase tagging in measurements with long acquisition times. New J. Phys., **14** (2012): 093027.
[159] B. Bergues, M. Kubel, N. G. Johnson, et al., Attosecond tracing of correlated electron-emission in non-sequential double ionization. Nat. Commun., **3** (2012): 813.
[160] N. Camus, B. Fischer, M. Kremer, et al., Attosecond correlated dynamics of two electrons passing through a transition state. Phys. Rev. Lett., **108** (2012): 073003.
[161] T. Paasch-Colberg, A. Schiffrin, N. Karpowicz, et al., Solid-state light-phase detector. Nat. Photon., **8** (2014): 214–218.
[162] S. Palaniyappan, A. DiChiara, E. Chowdhury, et al., Ultrastrong field ionization of Ne^{N+} ($N \leq 8$): Rescattering and the role of the magnetic field. Phys. Rev. Lett., **94** (2005): 243003.
[163] E. Gubbini, U. Eichmann, M. Kalashnikov and W. Sandner. Core relaxation in atomic ultrastrong laser field ionization. Phys. Rev. Lett., **94** (2005): 053602.
[164] E. Gubbini, U. Eichmann, M. Kalashnikov and W. Sandner. Strong laser field ionization of Kr: first-order relativistic effects defeat rescattering. J. Phys. B, **38** (2005): L87–L93.
[165] A. Talebpour, C. Y. Chien and S. L. Chin. The effects of dissociative recombination in multiphoton ionization of O_2. J. Phys. B, **29** (1996):L677-L680.

[166] C. Guo, M. Li, J. P. Nibarger and G. N. Gibson. Single and double ionization of diatomic molecules in strong laser fields. Phys. Rev. A, **58** (1998):R4271-R4274.
[167] X. M. Tong, Z. X. Zhao and C. D. Lin. Theory of molecular tunneling ionization, phys. Rev. A, **66** (2002): 033402.
[168] C. L. Guo. Multielectron effects on single-electron strong field ionization, Phys. Rev. Lett., **85** (2000): 2276–2279.
[169] J. Muth-Bohm, A. Becker and F. H. M. Faisal. Suppressed molecular ionization for a class of diatomics in intense femtosecond laser fields. Phys. Rev. Lett., **85** (2000): 2280–2283.
[170] Z. Lin, X. Jia, C. Wang, et al., Ionization suppression of diatomic molecules in an intense mid-infrared laser field. Phys. Rev. Lett., **108** (2012): 223001.
[171] E. Eremina, X. Liu, H. Rottke, et al., Influence of molecular structure on double ionization of N_2 and O_2 by high intensity ultrashort laser pulses, Phys. Rev. Lett., **92** (2004): 173001.
[172] B. K. McFarland, J. P. Farrell, P. H. Bucksbaum and M. Guhr. High harmonic generation from multiple orbitals in N_2. Science, **322** (2008): 1232–1235.
[173] H. Akagi, T. Otobe, A. Staudte, et al., Laser tunnel ionization from multiple orbitals in HCl. Science, **325** (2009): 1364–1367.
[174] H. Niikura, F. Legare, R. Hasbani, et al., Sub-laser-cycle electron pulses for probing molecular dynamics, Nature, **417** (2002): 917–922.
[175] H. Niikura, F. Legare, R. Hasbani, et al., Probing molecular dynamics with attosecond resolution using correlated wave packet pairs. J. Nature, **421**(6925) (2003): 826–829.
[176] T. Zuo, A. D. Bandrauk and P. B. Corkum. Laser-induced electron diffraction: A new tool for probing ultrafast molecular dynamics. Chem. Phys. Lett., **259** (1996): 313–320.
[177] M. Lein. Molecular imaging using recolliding electrons. J. Phys. B, **40** (2007):R135-R173.
[178] M. Meckel, D. Comtois, D. Zeidler, et al., Laser-induced electron tunneling and diffraction. Science, **320** (2008): 1478–1482.
[179] M. Okunishi, R. Itaya, K. Shimada, et al., Two-source double-slit interference in angle-resolved high-energy above-threshold ionization spectra of diatoms. Phys. Rev. Lett., **103** (2009): 043001.
[180] C. I. Blaga, J. L. Xu, A. D. DiChiara, et al., Imaging ultrafast molecular dynamics with laser-induced electron diffraction. Nature, **483** (2012): 194–197.
[181] M. Richter, M. Y. Amusia, S. V. Bobashev, et al., Extreme ultraviolet laser excites atomic giant resonance. Phys. Rev. Lett., **102** (2009): 163002.
[182] A. Rudenko, L. Foucar, M. Kurka, et al., Recoil-ion momentum distributions for two-photon double ionization of He and Ne by 44 eV free-electron laser radiation. Phys. Rev. Lett., **101** (2008): 073003.
[183] W. Ackermann, G. Asova, V. Ayvazyan, et al. Operation of a free-electron laser from the extreme ultraviolet to the water window. Nat. Photon., **1** (2007): 336–342.
[184] T. Shintake, H. Tanaka, T. Hara, et al. A compact free-electron laser for generating coherent radiation in the extreme ultraviolet region. Nat. Photon., **2** (2008): 555–559.
[185] P. Emma, R. Akre, J. Arthur, et al. First lasing and operation of an angstrom-wavelength free-electron laser. Nat. Photon., **4** (2010): 641–647.
[186] L. Young, E. P. Kanter, B. Krässig, et al. Femtosecond electronic response of atoms to ultra-intense X-rays. Nature, **466** (2010): 56–61.
[187] H. Fukuzawa, S. K. Son, K. Motomura, et al. Deep inner-shell multiphoton ionization by intense X-ray free-electron laser pulses. Phys. Rev. Lett., **110** (2013): 173005.
[188] X. Lai, C. Wang, Y. Chen, et al., Elliptical polarization favors long quantum orbits in high-order above-threshold ionization of noble gases. Phys. Rev. Lett., **110** (2013): 043002.
[189] M. Wu, Y. Wang, X. Liu, et al., Coulomb-potential effects in nonsequential double ionization under elliptical polarization. Phys. Rev. A, **87** (2013): 013431.

[190] M. Li, Y. Liu, H. Liu, et al., Subcycle dynamics of Coulomb asymmetry in strong elliptical laser fields. Phys. Rev. Lett., **111** (2013): 023006.
[191] D. Shafir, H. Soifer, C. Vozzi, et al., Trajectory-resolved Coulomb focusing in tunnel ionization of atoms with intense, elliptically polarized laser pulses. Phys. Rev. Lett., **111** (2013): 023005.
[192] P. Salieres, B. Carre, L. Le Deroff, et al., Feynman's path-integral approach for intense-laser-atom interactions. Science, **292** (2001): 902–905.
[193] P. Eckle, A. N. Pfeiffer, C. Cirelli, et al., Attosecond ionization and tunneling delay time measurements in helium. Science, **322** (2008): 1525–1529.
[194] P. Eckle, M. Smolarski, P. Schlup, et al., Attosecond angular streaking. Nat. Phys., **4** (2008): 565–570.
[195] A. N. Pfeiffer, C. Cirelli, M. Smolarski, et al., Timing the release in sequential double ionization. Nat. Phys., **7** (2011): 428–433.
[196] A. N. Pfeiffer, C. Cirelli, M. Smolarski, et al., Attoclock reveals natural coordinates of the laser-induced tunnelling current flow in atoms. Nat. Phys., **8** (2012): 76–80.

Shufeng Wang, Kang Meng, Qihuang Gong

6 Ultrafast spectroscopic techniques and ultrafast photophysics in polymers

6.1 Introduction to ultrafast spectroscopy

6.1.1 Development of femtosecond ultrafast lasers

The world's first laser was born in the 1960s [1]. On 15 May 1960, Theodore H. Maiman successfully made the first laser with a ruby, whose working wavelength is 0.6943 μm. It opened a new door for people to explore nature. As a new type of light source, which was coherent, it deeply influenced the whole scientific world, including physics, chemistry, biology, materials, and information science. It also prompted new interdisciplinary research. Optics itself developed new study fields, such as nonlinear optics and quantum optics.

Ultrafast pulsed laser sources have very short pulse duration, so that light energy concentrates in an ultrashort timescale, generating ultrahigh peak power. Shortening the laser pulse width is an important aim of laser technology. At first, short laser pulses were limited by Q-switch, with an output power of ~10^6 W and pulse width of 10^{-9}–10^{-10} s [2, 3]. During the mid-1960s, mode locking technology was invented. The mode locking ruby laser and neodymium glass laser created ultrashort laser pulses of picosecond duration. Then femtosecond dye lasers, which had broadband spectra, generated short light pulse of ~100 femtosecond (10^{-15} s, fs) [4]. The further multiplying of pulse peak power was realized by chirped pulse amplification (CPA) appearing in the same period [5, 6]. This technique disperses the laser spectra in order to stretch the pulse and reduce peak power safely amplify the pulse energy. Finally, the pulse is compressed back to a femtosecond one. This reduces the nonlinear effect and avoids the damage due to a reduced intense electric field inside the gain medium. The technique increased peak laser pulse power by five to six orders of magnitude, or even higher, while keeping the pulse width. This technology is still a major way to obtain high pulse energy femtosecond laser sources. In 1990s, the invention of all-solid-state femtosecond lasers, such as Ti:sapphire lasers, greatly increased their accessibility and commercialization, making the wide application of ultrafast lasers possible. This led to the fast developments in optical science and applications. Ti:sapphire femtosecond lasers using Kerr-lens mode locking technology (KLM) greatly simplified the structure of laser systems, making the miniaturization of femtosecond laser light sources

Shufeng Wang, Kang Meng, Qihuang Gong (corresponding author), Peking University, School of Physics, Beijing 100871, China

https://doi.org/10.1515/9783110304558-006

possible, improving their stability at the same time [7]. Using dispersion control and compensation methods, such as chirped mirror, we can make full use of the ultrawide fluorescent spectrum of Ti:sapphire to achieve ultrashort femtosecond pulses of a few cycles (4 ~ 5 fs) [8].

6.1.2 Ultrafast spectroscopic techniques

Spectroscopy is an important method in optical science, which uses the light spectrum to study various substances and their interactions. The spectrum represents electromagnetic radiation arranged according to wavelength, frequency, photon energy, wave number, and so on, with relative intensity variation. Spectroscopy provides information on electronic structures, such as energy levels of atoms and molecules, densities of states, molecular geometries, the nature of chemical bonds, and reaction kinetics. It is a method for both qualitative and quantitative analysis.

The development of ultrashort pulsed lasers is the basis of femtosecond ultrafast spectroscopy. This technique can probe initial reactions or processes in the fields of physics, chemistry, and biology that steady state or Q switching laser technologies can't reach. For example, the vibration, bond breaking and formation process of chemical bonds occur in the femto- to picosecond scale, requiring systems capable of femtosecond time resolution.

6.1.2.1 Fluorescence upconversion

Fluorescence upconversion is a spectroscopy technology using sum frequency in nonlinear optical crystals (NLO crystals, such as BBO (barium borate), LBO (lithium borte), and $LiIO_3$) to convert and detect fluorescence dynamics. The experimental setup is shown in Figure 6.1. The 800 nm femtosecond laser pulses produced by a femtosec-

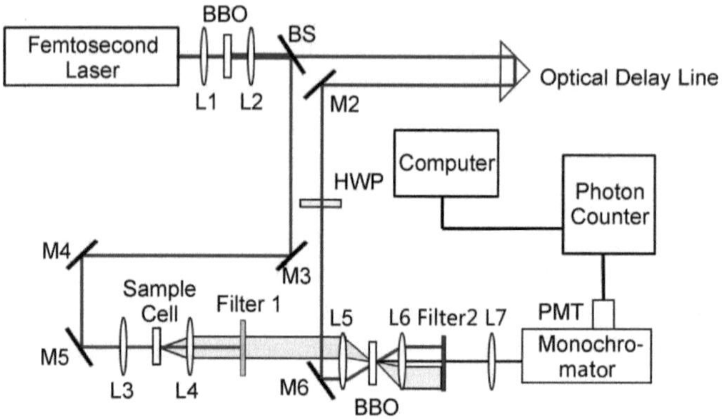

Fig. 6.1: Fluorescence upconversion setup.

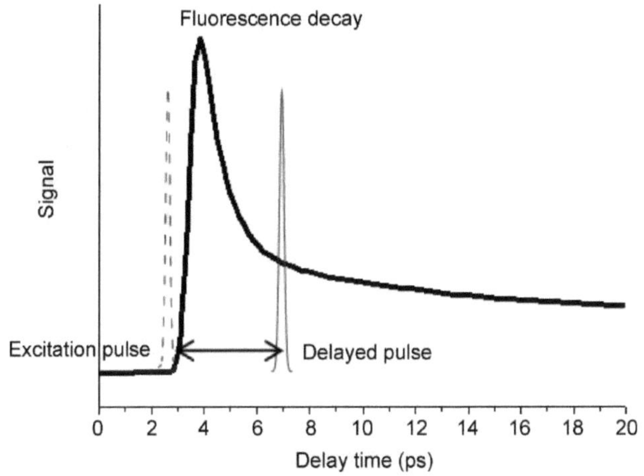

Fig. 6.2: The principle of fluorescence upconversion.

ond oscillator are partially frequency doubled through a BBO crystal to get a 400 nm light source for pumping the sample and generating fluorescence. The fluorescence is then focused on a second NLO crystal. The remaining 800 nm light pulses are also focused onto this second NLO crystal after passing an optical delay line, acting as gate pulses. When the fluorescence and gate pulses are spatially overlapped on this crystal, their sum frequency is produced. The energy of generated photons is the sum energy of fundamental frequency photons and fluorescence photons, extended into the ultraviolet range. The fluorescence participating in the sum frequency is the part that temporally overlaps with the gate pulse. The amplitude of sum frequency signal is proportional to the fluorescent strength when it overlaps with the gate pulse. By scanning the optical delay line, we can change the relative time delay between the fundamental frequency pulse and fluorescence, so as to obtain the relative intensity of fluorescence relaxation at various time delays. By tuning the angle of the second BBO crystal, we can achieve sum frequency phase matching of the fundamental and different fluorescence wavelength. The principle for time resolved detection is shown in Figure 6.2.

The sum frequency process on nonlinear optical crystals requires phase matching of fluorescence and gate pulse photons (energy and momentum conservation) [9]

$$\omega_S = \omega_F + \omega_G \,, \tag{6.1}$$

$$\boldsymbol{k}_S = \boldsymbol{k}_F + \boldsymbol{k}_G \,, \tag{6.2}$$

where k is the wave vector, ω is angular frequency, and S, F, and G represent sum frequency, fluorescence, and the fundamental frequency light respectively. In collinear or approximate collinear cases, the conservation of momentum (phase matching) condition can be written as:

$$\frac{n_g}{\lambda_G} + \frac{n_L}{\lambda_L} = \frac{n_S}{\lambda_S}\,. \tag{6.3}$$

Because the sum frequency occurs only at the gate pulse timescale, the time resolution of this technology can reach the order of a femtosecond pulse's width, which greatly exceeds the resolution of photoelectric detectors. In addition, better signal-to-noise ratio (SNR) can be achieved as the ultraviolet signal has nearly zero background noise. It should be noted that the fluorescence pulse and gate pulse overlapping in space will result in a weakening in time resolution. Thicker BBO crystals will also reduce the time resolution of fluorescence upconversion.

6.1.2.2 Fluorescent optical parameter amplification

The fluorescence upconversion technology is suitable for detecting the fluorescence dynamic process because of its high signal-to-noise ratio. However, as it is limited by the NLO crystal's phase matching, we can only obtain fluorescence decay at a single wavelength for one scan. Additionally, this method is not sensitive to weak fluorescence, which limits the detection for samples of low luminous intensity. Therefore, a different kind of ultrafast fluorescence detection method, time resolved fluorescent optical parametric (FOPA) amplification technology, can be applied [10, 11].

The FOPA technique uses the optical parametric amplification process to directly achieve wideband spectral dynamics of weak fluorescence signals. When weak fluorescence enters the nonlinear crystal at the same time as another higher energy photon under phase matching conditions, the weak fluorescence will be amplified. Strong pump light converts its photon into a signal photon at the same frequency as the fluorescent photon and an idle light photon. This conversion process increases the number of fluorescence photons, and thus amplifies fluorescence. The efficiency of the amplification can reach ~10^6. Experiments verify that amplified fluorescence and the incident weak fluorescence are linearly related, which correctly represents fluorescence kinetics. In addition, by choosing a specific cutting angle of the nonlinear optical crystal, we can obtain homogenous phase matching conditions in a broad spectral range, i.e. all the fluorescence in the range can be homogenously amplified synchronously. The time resolution of this method is also decided by the pulse width of the pump light. The setup is shown in Figure 6.3.

To have effective FOPA, high pump pulse energy is needed, such as that obtained from a chirped pulse amplification (CPA) system. The femtosecond pulses are frequency doubled through a BBO crystal, and then used as the pump for the optical parametric process. When frequency doubled light pulses passing through the BBO crystal are above an intensity threshold, circular cone radiation will be created (parametric superfluorescence produced by pulsed laser, shown as circular cone radiation). It has been found that when the crystal rotates along its vertical axis, the radius of the colored cone radiation changes. The blue edge changes faster than the red edge. Therefore, when the crystal rotates, multiple wavelengths, from green to red, overlap

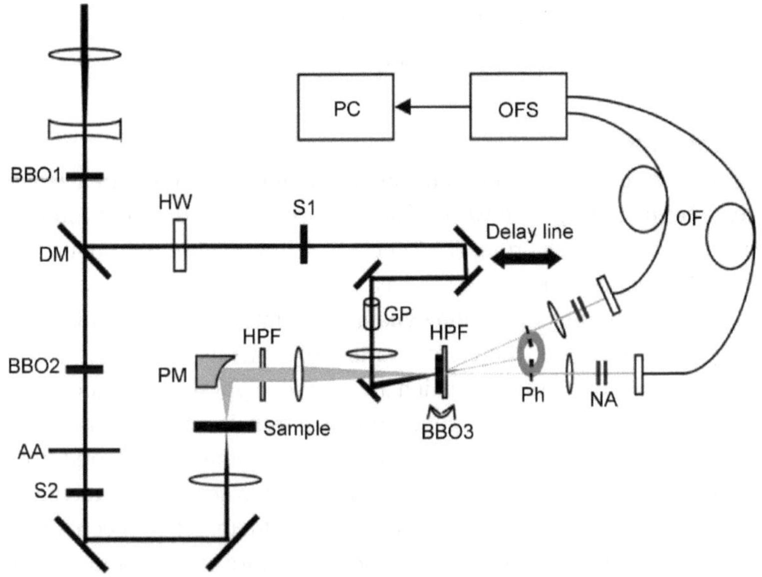

Fig. 6.3: The setup of time resolved optical parametric amplification technology.

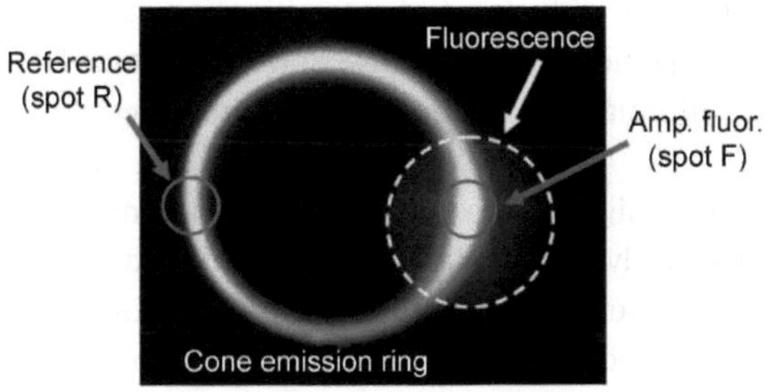

Fig. 6.4: The broadband FOPA based on cone emission [13].

in space at a certain angle, as shown in Figure 6.4. The crystal angle in which multiple wavelengths overlap is the working angle for broadband FOPA. When weak fluorescence is injected along this angle as a seed light, bright amplified fluorescence can be observed [12]. After amplification, fluorescence is recorded by spectrometer. By changing the delay line, we can directly read time resolved emission spectra (TRES).

Since this technology has ultrahigh amplification (~10^6), it is sensitive to a small number of fluorescent photons. Spectra obtained by a single gate pulse can display discrete photons. To obtain a continuous and smooth transient fluorescence spectrum, integration and averaging are required. The high amplification also requires stability of the laser light source during integration. Experiments have proved that the fluctuation of conical emission (out of the amplified area) is proportional to that of pump energy and amplified fluorescence. So, by referring to the fluctuation of conical radiation, we can effectively correct the FOPA signal. This greatly reduces the integration time, and improves the signal-to-noise ratio.

6.1.2.3 Transient absorption

The two ultrafast time resolved spectroscopic methods mentioned above are used to detect the fluorescence of a sample. However, fluorescence is emitted from the lowest excited state according to the Kasha rule. The intensity of fluorescence is proportional to the S_1 population. Femtosecond time resolved fluorescence spectrum detection thus studies the evolution of the population dynamics of emissive excited states and the related environment. However, there are a variety of nonemissive excited states and other products in molecular systems, which are related to physical, chemical and biological processes. After excitation, the population will transfer and redistribute to other states, producing various products. Although some processes transferred from emissive to nonradiative products can be detected by the fluorescence, their dynamics cannot be fully observed. This means that time resolved fluorescence spectra have their limitations. As an example, using fluorescence spectra, we can observe organic photovoltaic devices for exciton cooling, diffusion, exciton annihilation, charge separation and other physical process related with excitons, which are emissive. However, charge separation states, polarons, and their diffusion and recombination cannot be observed [14, 15]. In Cytochrome C, the ligand separation of +3 and +2 iron ions [16], and in some proteins, light induced conformation variations do not directly correlate with fluorescence emission [17].

The nonradiative states can transit to higher states by absorbing photons. Transient absorption then can be applied to study excited state population dynamics. Radiative states can also be detected in transient absorption in the form of stimulated emission. Therefore, this method can be used to detect ultrafast dynamics processes more comprehensively. The transient absorption method is a powerful way to study the ultrafast dynamical processes of complex systems.

The setup of the transient absorption technique is shown in Figure 6.5. A strong pump beam excites the sample at $t = 0$, and populates the electronic states of the sample. Another probe light passes through an optical delay line (single wavelength

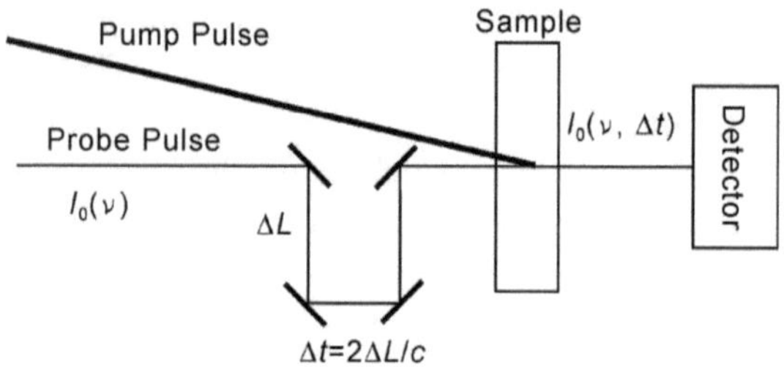

Fig. 6.5: The light path diagram of a transient absorption spectrum.

or supercontinuum, with a much lower energy than that of pump light) and overlaps with the pump light at same spot on the sample, at the time delay Δt. The probe pulse has different transmittance according to the time delay. The recorded intensity of this probe light is then compared with that when pump light is absent, to achieve the variation in transmittance, ΔT. The intensity of the probe light before going through the sample is $I_0(\nu)$, and the sample thickness is l. According to the principle of the Beer–Lambert law, light intensity after going through the sample is [18]

$$I(\nu, 0) = I_0(\nu) \times 10^{-\varepsilon_\nu N(0) l} . \tag{6.4}$$

After time Δt,

$$I(\nu, \Delta t) = I_0(\nu) \times 10^{-\varepsilon_\nu N(\Delta t) l} , \tag{6.5}$$

where ε_ν is the molar absorption coefficient of samples at frequency ν. $N(0)$ is the population at t_0 and $N(\Delta t)$ is the population when absorbing the probe photon. The absorption value OD at $t = 0$ and $t = \Delta t$ can be expressed as

$$\mathrm{OD}(n, 0) = \log \frac{1}{T(n, 0)} , \tag{6.6}$$

$$\mathrm{OD}(n, \Delta t) = \log \frac{1}{T(n, \Delta t)} , \tag{6.7}$$

where T is transmittance. According to equations (6.6) and (6.7), the variation of OD is

$$\Delta\mathrm{OD}(n, \Delta t) = \log \frac{T(n, 0)}{T(n, \Delta t)} = \log \frac{I(n, 0)}{I(n, \Delta t)} . \tag{6.8}$$

Together with (6.4) and (6.5) we can get

$$\Delta\mathrm{OD}(n, \Delta t) = \varepsilon_\nu \, [N(\Delta t) - N(0)] \, l . \tag{6.9}$$

For a thin film whose absorption coefficient and thickness are fixed, the change of the absorption is proportional to the population of a certain energy level after delay Δt. The change of the population shows the evolution of dynamic processes over time.

Although we can obtain the dynamic processes of electronic states in complex systems by transient spectroscopy, determining the products and distinguishing the overlap of their absorption in spectra is more difficult. In addition, the products may have small population numbers or absorption cross-sections, requiring high signal-to-noise ratio in detection. Optical chopper, lock-in amplifier, boxcar, reference light, and improved light source stability can be applied to effectively reduce the noise of the probe signal. The supercontinuum probe is generated by a nonlinear process in transparent media, while the tunable single wavelength probe is generated through multiple steps (e.g. OPA). The stability is influenced by light source fluctuation and the environment. By splitting the probe light into two beams, one acts as a reference beam going through a sample area where there is no pump light. This reference beam helps eliminate pulse-to-pulse instability. Monitoring the changes of the pump intensity is also an effective way to improve the signal-to-noise ratio.

6.1.2.4 Streak camera

The femtosecond ultrafast spectroscopy techniques above can give us time resolved spectra to the femtosecond pulse width limit. However, they demand very precise adjustment of the system. They are complicated and have low efficiency in data collection. The longest timescale is up to few nanoseconds, limited by optical delay line. For some processes that occur between femtoseconds and nanoseconds, such as macromolecular structure evolution, we can use slower but a more efficient acquisition method: the streak camera. The streak camera has many advantages, including simple equipment, picosecond time resolution, and easy adjustment. The whole time resolved fluorescence spectrum can be obtained in a single scan. Because it is not an all-optical technique, the limit of time resolution is picoseconds [19, 20]. However, through different configurations, the instrument can detect a wide range of dynamics occurring over ranges from picoseconds to milliseconds.

Picosecond time resolution can be achieved in the streak camera by high speed spatial electric field scanning. It can also be combined with an external spectrometer for temporal spectral resolution. The working principle is shown in Figure 6.6. Incident light goes through the spectrometer, and separates in space at the exit. It then goes through the lens and enters the slit of streak camera. Different spectral components distribute along the slit, so that each wavelength can be independently detected. The light then is imaged onto the cathode of the streak tube to generate photoelectrons. These photoelectrons fly through the streak tube while being accelerated by a perpendicular field linearly varying with time. Then the photons (electrons) at different time delays experience different perpendicular accelerating fields, corresponding to different deflection angles. When the electrons arrive at a two dimensional microchannel plate, they are separated in space according to the initial photon energy and time delay. Through multiplication, the electrons bombard the screen and generate the image, with one dimension for spectrum direction and the other for time. Therefore, the temporal delay of the fluorescence is converted into the spatial distribution of the brightness. So, with a streak camera, we can get a complete fluorescence spectral-temporal two dimensional map at once.

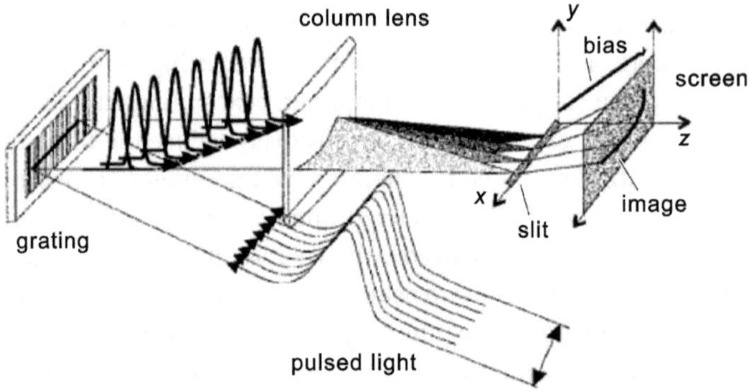

Fig. 6.6: The principle of the streak camera.

6.1.2.5 Instrument response function, IRF

Instruments respond to an ultrashort pulse with an output of a certain width and shape. For the femtosecond spectroscopy techniques mentioned above, their temporal resolution limit is the pulse width. Due to geometric configuration or dispersion, this response is likely to be broadened. For a streak camera, the femtosecond optical pulse will be broadened to picoseconds or longer. This kind of response can be referred to as instrument response function (IRF). For temporally continuous signals, the decays are observed as convolutions of the signal $I(t)$ and the IRF. We must deconvolute the signal to get the ultrafast dynamic process. A signal described by a multiexponential model can be written as

$$I(t) = \sum_{i=1}^{n} A_i \exp\left(\frac{-t}{\tau_i}\right) , \tag{6.10}$$

in which A_i is the amplitude of the i-th exponential decay, and τ_i is the life of the i-th exponential decay. The data $F(t)$ obtained by the experiment is

$$F(t) = \int_{-\infty}^{+\infty} \mathrm{IRF}\left(t'\right) \sum_{i=1}^{n} A_i \exp\left(-\frac{t-t'}{\tau_i}\right) dt' . \tag{6.11}$$

This convolution for other relaxation models, such as stretched exponential or a power law, has a similar form.

To access true ultrafast dynamics processes, we must measure the IRF precisely. In ultrafast systems, we need to ensure that the optical path for IRF measurement is consistent with that of signal measurement. In the streak camera, we must ensure the consistency of the measurement conditions, e.g. that the spectral slit width is consistent with the temporal axis slit width. Ultrafast IRF can be measured by the self-correlation experimental method, while in a streak camera system the scattering of the pump pulse at the sample can be taken as the source for measuring IRF. By deconvolution, we can finally get the ultrafast dynamics process in various models.

6.2 Ultrafast photophysics in polymers

6.2.1 Introduction on optoelectronic materials and their physics

Photophysical processes are induced by light excitation of atoms and molecules, including excited state internal conversion, molecular twisting, proton transferring, intersystem crossing, etc. These processes depend on the states being excited and their temporal evolution. Therefore, the core issues in molecular photophysics are electronic level evolution via structure and the correlation of electronic states, including ground states and excited states.

The electronic levels of molecules relate to the configuration of atoms in the environment, so they vary with modification, structure, environment, and many other

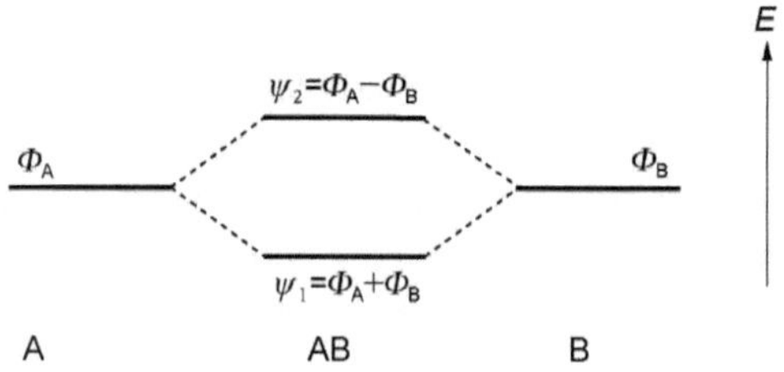

Fig. 6.7: The bonding orbit and antibonding orbit in diatomic molecules.

factors. The linear combination of atomic orbitals-molecular orbital (LCAO-MO) theory can be applied to explain the formation of molecular orbits (MOs), which is a linear combination of atomic orbits (AOs). Each of the valence electrons of the molecule makes a contribution. These electrons can move in the conjugated molecular region and do not belong to the individual atoms any more. The sharing of atomic orbits forms MOs and new energy levels for molecules. Demands must be met for proximity energy, symmetrical matching, equality with number of AOs, Pauli exclusion principle and so on. MO wavefunction ψ can be used to describe the corresponding molecular motion, for example two identical atomic orbits Φ_A and Φ_B generate two molecular orbits by interaction, which are shown in Figure 6.7. One MO energy is lower than the AO, forming the bonding orbit. The other is higher, forming the antibonding orbit.

Primary MOs include the bonding π and σ orbits, the n orbits of lone pair electrons, and the antibonding π^* and σ^*orbits. The π orbits are generated by the parallel overlay of atomic p orbits. Parallel p orbits of multiple adjacent atoms in a nodal plane of the molecular plane can eventually form a continuous π orbit, the extended π bond, which is the main influence on the energy levels of conjugated molecules and conjugated polymers. n orbits are nonbonding orbits in lone pair electrons containing heteroatoms. For example, such one pair p electrons exist on the nonbonding orbits in molecules containing oxygen, sulfur and nitrogen atoms. The π orbits and n orbits are involved in visible spectrum absorption. The molecular skeleton is composed of σ orbits. Two s, an s and a p, or two p orbits overlapping can form such σ orbits. It is a main form of bonding between atoms, forming stable compounds. The σ bonding orbit results in a large gap between energy levels, corresponding to the ultraviolet absorption transition. For the compounds without conjugate electronic structure, the absorption occurs in the ultraviolet region, which is a behavior of σ bonding absorption. As for the compounds with conjugated electronic structure, the absorption shifts to the visible region, and varies with the conjugated structure. For example, planarization of conjugated molecules may enhance conjugation, resulting in a red-shifted spectrum. Conversely, the twisting of the molecular chains has the opposite effect. Aggregation of polymer chains can extend the conjugation from molecular chains to intermolecular bonds, thereby enhancing conjunction and spectral shifting. With this change, we can tell the degree of aggregation of the molecular chains.

The molecules of organic polymers can transit from a ground state to an excited state by light excitation of certain wavelengths. The change of electron distribution contributes to the changes of the forces between atoms, all of which might result in variations in molecular structure. New physical and chemical changes will emerge alongside the excited state. All the responses determine the material functions and the realization of applications. In optical functional materials, the generation of excited states is the beginning of photoelectric conversion. After this, different phenomena will occur, including light induced electron transfer, photoisomerization, resonance energy transfer, proton transfer, intersystem crossing, chemical bond breaking, restructuring and so on. It is complicated to determine the temporal relationships for these processes. In term of dynamics of these processes, there are great differences in the order of magnitude over which they take place, as the excited state dynamics act on timescales from femtoseconds (fs) up to milliseconds (ms). The whole procedure, from photon absorption to the accomplishment of charge transport, contains a series of sequential or parallel processes, which will be described in detail in the following section.

Throughout photoelectric conversion, the first step is to obtain energy by excitation. The conventional view is that all the significant photochemical and photophysical processes begin at the activation of the lowest excited singlet state (S_1) or the lowest triplet excited state (T_1). The ground state (S_0) in molecules will be excited to a singlet excited state, for example S_1 and S_2, by absorbing the corresponding energy of a photon. Molecules in higher excited states will relax to the lowest singlet excited state by internal conversion, due to the overlapping of the vibrational levels and electronic energy levels.

Deactivation is caused by electrons leaving the S_1 excited state. This process can be exposed to many different chemical and physical processes, including radiative transition, nonradiative transition, energy transfer, electron transfer, intersystem crossing, proton transfer, and chemical reactions. The first four processes are particularly important in photoelectric materials.

(1) Radiative transition

Radiative transition is the process by which molecules in the excited state jump back to the ground state by emitting a photon, including fluorescence and phosphorescence. Fluorescence is the light generated by transition between electronic states of the same multiplicity, e.g. by the common $S_1 \rightarrow S_0$ transition, which is in the timescale of nanoseconds (10^{-10}–10^{-8} s) corresponding to spontaneous emission lifetime. P3HT, the commonly used polythiophene material in polymer photovoltaic devices, has a fluorescence lifetime of ~600 ps [21], which can be detected by means of time resolved fluorescence spectra or photon induced absorption spectra. Phosphorescence is induced by the transition between a triplet state and a singlet state, typically the $T_1 \rightarrow S_0$ transition. The phosphorescence transition is forbidden according to the transition selection rules. However, it can occur because of the coupling of spin and orbit. The decay of phosphorescence is much slower than that of fluorescence with a lifetime of microseconds to seconds or even longer. In excited state dynamics, the revolution of molecules will lead to variation of the population in the S_1 state, or molecular structure and energy levels, which are faster than fluorescence decay, and cannot be detected by means of ultrafast fluorescence spectra. Therefore, pump probe and other measurements are instead needed to detect processes after transition from the S_1 state.

(2) Nonradiative transition

Nonradiative transitions refer to the transitions from the excited state to the ground state without emitting a photon. They occur at the isoenergetic points of the different electronic potential surfaces. The position of the points relates to vibrational energy levels. The higher vibration energy levels of the lower electronic state coincide with the lower vibrational energy levels of the higher electronic state, which makes the electron in the higher excited state transfer to the ground state, and then cool down to a lower vibrational energy level by relaxation. No photons are emitted as the system energy remains constant. Besides transition to the ground state, the transition from higher excited state to lower excited state is also nonradiative, for example $S_{\mathrm{n}} \to S_1$. This kind of process occurs at very fast timescales of picoseconds. The S_1 state is the main resource for photophysical processes and photochemistry. Intersystem crossing (ISC) is also nonradiative. It is the transition between a triplet state and a singlet state with electron spin inversion, for example, $S_1 \to T_1$.

(3) Energy transfer

Energy transfer is the process by which the energy of one molecule gained by excitation transfers to another molecule. It is possible that one molecule transits to the ground state, and the neighbor one transits from the ground state to excited state by accepting energy from the first molecule. It is also possible that the neighbor molecule transits from its excited state to higher state, e.g. singlet exciton annihilation in a polymer system. For macromolecular materials, the molecular scale is much greater than the interchain distance and the conjugated chromophore scale. In such systems, an exciton represents a localized excited state over chain segments, and is the unit for energy transfer.

Förster resonance energy transfer (FRET) [22] is the most discussed energy transfer method in conjugated polymer systems. The transfer rate is inversely proportional to the sixth power of the distance r, which is between the donor and acceptor

$$k_T = \frac{1}{\tau_{\mathrm{D}}^0}\left(\frac{R_0}{r}\right)^6 , \tag{6.12}$$

in which τ_D^0 is the excited state lifetime of the donor without energy transfer, and R_0 is the Förster characteristic radius

$$R_0 = 0.2108\left[\kappa^2 \Phi_D^0 n^{-4}\int_0^\infty I_D(\lambda)\varepsilon_A(\lambda)\lambda^4 d\lambda\right]^{\frac{1}{6}}, \tag{6.13}$$

in which κ^2 is the orientation factor, Φ_D^0 is the quantum yield of donor without energy transfer, n is the average refractive index, $I_D(\lambda)$ is the fluorescence spectroscopy of the donor, and $\varepsilon_A(\lambda)$ is molar absorption coefficient of acceptor.

FRET is the process by which the energy donor dipole transits, followed by acceptor excitation. Another kind of energy transfer without emissive dipole transition is called Dexter-type energy transfer. In this model, the excited electron of the donor molecular transfers to the nonground state of acceptor molecules directly, and a ground state S_0 electron of the acceptor molecule conversely transfers to the S_0 of the donor molecule. The result is equivalent to the situation in which the donor molecule stays in the ground state, and the acceptor molecule in the excited state. This kind of energy transfer requires that the donor and acceptor are sufficiently close (< 1 nm), thus exhibiting electron orbit overlay. For this reason, Dexter-type energy transfer is also called short range energy transfer. The rate of this energy transfer shows exponential decay as the distance between donor and acceptor increases, and the rate of decay is faster than FRET:

$$k_{\text{Dexter}} = KJ\exp\left(\frac{-2R_{\text{DA}}}{L}\right), \tag{6.14}$$

in which J is overlay integration of the absorption and emission spectrum, K is the experiment related factor, R_{DA} is distance between donor and acceptor, and L is the radius of van der Waals forces.

In polymer materials, the question of which kind of transfer is dominant remains controversial. According to the phenomenon, energy transfer can be divided into two different types, downhill energy transfers and isoenergetic energy transfers, by the initial injected energy and relationship of energy levels. Hot exciton energy transfer occurs between excitons at high excited state energy and the adjacent chromophores. As hot excitons have the energy to transcend the barrier, they transfer very fast, in several ps or even in the sub-ps timescale [23–26]; the isoenergetic energy occurs after hot excitons have lost excess vibrational energy. This energy transfers to nearby chromophores with similar energy levels. The rate is relatively slow, in the 10–100 ps range, since the driving force for overcoming the barrier is from thermal fluctuations, etc. [27, 28].

Exciton annihilation is another kind of energy transfer process under high excitation density, both for singlet and triplet excitons. When the excitons are close, singlet-singlet annihilation and singlet-triplet annihilation occurs. In these processes, the energy from an excited exciton transfers to another, with the former transiting back to the ground state and the latter transiting to a higher excited state.

(4) Electron transfer

Electron transfer refers to the process by which a donor molecule in the excited state offers an electron to a molecule in the ground state (oxidative electron transfer), or as an electron acceptor receives an electron from a molecule in ground state (reductive electron transfer), thereby generating an ion pair. In polymer photovoltaic devices, after light absorption, excitons transport out excited state electrons, and eventually a current output forms, completing the electron transfer process. In the polymer or fullerene (charge acceptor), the rate of electron transfer is rather fast, generally less than 100 fs, which can reach 100% efficiency; the reverse electron transfer is much slower.

6.2.2 Fundamental photophysics in polymers

6.2.2.1 Generation of excitons

When a polymer absorbs a photon, it will produce an exciton, which is an excited state with an electron and a hole coinciding in space. The existence of excitons indicates that the Coulomb attraction force is very strong. This is different from semiconductor materials. Because of the diversity of polymers and their preparation methods and conditions, the exciton binding energy directly relates to the specific material. To ensure the separation of positive and negative charges, either additional energy must be injected or and electric field must be applied. Therefore, the exciton binding energy is used to describe the energy difference between the excited state and free charge, which has been verified in many experiments [29, 30]. However, if free charge is directly generated by light excitation in semiconductors, no intermediate state forms.

Polymer chains have a certain periodicity, so they have some similar characteristics as inorganic semiconductors, and they are called one dimensional semiconductor materials. Optical excitations can be taken as interband transitions. The excitation process of many polymers, which is of Frenkel type, is different from semiconductor materials. (Frenkel exciton refers to materials with a small dielectric constant.) The electron and hole have a strong Coulomb interaction, so that they form a small exciton, similar to the scale of a repeat unit. Exciton binding energy is usually 0.1–1 eV. The corresponding Wannier–Mott exciton in semiconductors is present in a large dielectric constant material. The dielectric shielding effect reduces the electron-hole Coulomb interaction, which makes the exciton radius larger than the periodic structure, so its bound energy can be very small, about 0.01 eV, which makes the exciton unstable. Therefore, the choice of whether to employ the semiconductor band model or the molecular exciton model depends on the exciton binding energy. In the molecular exciton model, the electron and hole have a strong interaction, so charge separation requires additional energy to occur. Of particular note in organic semiconductors, the polymer MEH-PPV has a 0.7(±0.15) eV binding energy [31], and P3HT has a ~0.3 eV binding energy [32]. Smaller (0.06 eV) and larger (1 eV) binding energy polymers also exist [33].

The lower the degree of polymerization is the absorption bluer. The inhomogeneous broadening of absorption spectra of conjugated polymers is related to the spectral unit (chromophore) scale, and the distribution of length determines the distribution of the corresponding energy levels. Generally speaking, for long molecular chains, torsion and bending constitute defects in the molecular chain, breaking the conjugation of molecules, thus forming the boundary of chromophores. The quantification calculations show that only the chemical defects that make the conjugated carbon skeleton saturated (such as the double bond being replaced by a single bond saturation, etc.) can break the conjugation and then form the boundaries of chromophores. This single bond can be produced by chemical synthesis. The structural defects, for example the bending of the molecule in the plane, the molecular plane bending or twisting, and the cis and trans isomers, can change the degree of delocalization, but do not clearly define the boundary of chromophores, resulting in termination of conjugation. The absorption may even shift to the red when the plane bends [34, 35].

The above models and analysis are based on a static ground state configuration. By researching the change of the interatomic forces after excitation, it was discovered that the rearrangement of atoms in the excited state will cause excitation localized to the chromophore. Due to the long molecular chain owning the lowest energy, excitons tend to be localized to these long conjugated structures. These stabilized excited state electron distributions also relate to the structure defects. Excited conjugated structures will be divided into regions along the spectrum. The bent structure of the molecular plane may become an excited state trap due to low energy, making the exciton localized. In cis isomers the defects are more complicated as both traps and nontraps are likely to occur. Obviously, the excited state localization time associates with the atomic rearrangement time, which is on the sub-picosecond scale. For example, for a polythiophene, PTOPT, pump probe anisotropy relaxation studies reveal a process of about 37 fs [36]. This process includes the initial rapid cooling and localization, similar to the timescale of molecular vibration of C-C bonds.

Because of the complexity of the polymer morphology, excitons generated by excitation may not be the only result. If higher energy photons are injected for excitation, for example in P3HT films, polaron pairs may be generated within 100 fs by direct conversion of hot excitons [37]. Polymers that contain both a charge donor and acceptor have a wide absorption band due to the donor-acceptor dipole. Therefore, this benefits the absorption of sunlight in organic photovoltaic devices. The location of positive and negative charges along this polymer chain only partially overlaps. Therefore, the form of the excitation is more likely to be polaron pairs [38].

6.2.2.2 Energy transfer and exciton migration

The energy transfer occurs following the formation of the photogenerated excited state. First, the excitons are generated, which cover a few chain segments. Next, exciton and energy propagation leads to the spatial position change. For an isolated molecular chain, exciton migration is intrachain; if the polymer chains aggregate, transport includes interchain migration. Transfer can occur during the entire lifetime until it reaches the bound sites, which are low energy points in the propagation path, such as dimers and other structures. It can also separate on the way and form electron-hole pairs, which deactivate the original excited states. Such deactivations may be due to impurities. If the deactivation sites do not emit photons, i.e. excitons transfer to the ground state via nonradiative transition, there will be a loss of energy; if these sites are emissive, we can achieve the required emission of light of a certain wavelength by designing this particular impurity, which is important for organic light-emitting diodes (OLEDs) and other light-emitting devices. For example, the white OLED spectrum needs to be similar to the sun's radiation, but the polymer emission spectrum of a single material cannot cover the entire visible region because of the limited gap width value. Therefore, doping with other chromophores in blue-emitting OLED devices, such as green and red fluorophores, is necessary. When the device is excited, the energy transfers from the polymer host to the doped chromophores, resulting in multicolor light emission in one device, which appears white. Therefore, the energy transfer is of great significance for the application of the device. The key design points deal with the relationship of energy levels between host polymer and doping chromophores, requiring adjustments to the material's microstructure configuration, thus affecting the efficiency of energy transfer and other characteristics of the emission spectrum. In the OLED device, electrons and holes recombine in special sites. The principle of organic photovoltaic devices is opposite to OLED devices, whose principle is that excitons generated by light excitation are converted into free charges. Therefore, excitons need to be effectively transmitted to the specific sites to conduct charge separation. In a photovoltaic device, the design of sites and the configuration of polymers are both important and indispensable steps. The transport of charge is also affected by polymer configuration or charge receptor, and the overall design can lead to efficient charge separation.

The conjugation of polymers affects exciton transporting properties. In the transferring process, the positive and negative charges are bound together by the Coulomb interaction, forming excitons and transfer energy in the form of dipole resonance. In Förster energy transfer, the donor and acceptor chromophores transfer energy via dipole resonance in space. The emission spectra of donors are usually redshifted (Stokes shifted) compared to absorption, and overlap with the absorption spectrum of the acceptor. The degree of overlap, the distance between the donor and acceptor, and the orientation determine the rate of energy transfer. The rate is proportional to the sixth power of the distance, and strongly correlates with the orientation (orientation factor changes from 0–4). Thus, energy transfer is closely related to the configuration of the molecular chains, which can be used for observation and investigation of the molecular chain configuration. The Förster model, which treats the dipole as a point dipole, requires that the scale of the chromophore is much smaller than the distance between chromophores. However, in the polymer, the chromophore includes several segments, making it difficult to meet this requirement. Transition density approximation and linear dipole approximation can overcome this deficiency. When the transition density between sites is small, such as in triplet exciton transitions, Dexter-type energy transfer can occur. This refers to the charge exchange process caused by orbital overlap.

The spectral overlap usually discussed in energy transfer is based on steady state emission and absorption spectra. Actually, in excited states, spectral dynamics is very complicated, since ultrafast (< 100 fs) processes involve intramolecular vibrational relaxation and exciton self-trapping; excited state restructuring and thermally induced configuration rearrangements occur on picosecond timescales. These nonthermal equilibrium processes are much shorter than the fluorescence lifetime. Therefore, emission in ultrafast processes is a very small portion of the steady state fluorescence spectra. However, its role on energy transmission is critical. The contribution of these nonthermal equilibrium process to the proportion of energy transfer depends on their own rate of energy and electron transfer processes.

The rate of initial energy transfer depends on the absorbed photon energy. Because of the diversity of polymer configurations, chromophores of different conjugation lengths exist. A smaller chromophore has a higher energy level than a larger chromophore, so its absorption and emission shift to blue. When excitation is at the blue side of the absorption band, these smaller chromophores are more likely to be excited than the larger ones. Their radiation has a large overlap with the absorption of the longer polymer chain. If these long chains are close, and have similar orientation, then energy transfer will occur very rapidly. The energy will continuously transmit to larger chromophores until it finds the lowest energy site [39]. Usually, the lowest energy sites are called traps. The rapid transferring of energy to the large chromophores is what we called hot exciton transfer.

If the excitation is at the red side of the absorption band, these larger chromophores are more likely to be excited than the smaller ones. In this case the energy of the excitons is similar to excitons located in the trap. This situation can be seen as a state in which the hot exciton transfer process is isoenergetic. The energy of adjacent sites is not less than the excited site, so the emission and absorption spectra overlap is small. There is only a small probability of energy transfer to the surrounding chromophores. However, due to the role of thermal fluctuations and spectral homogeneity broadening, the spectral overlap increases for donor and acceptor, enhancing energy transfer. This situation requires that the excitation point and surrounding sites have similar energy, so this transfer is called isoenergetic energy transfer. This transport process requires a material with good uniformity. In uniform or ordered materials energy can transmit for a long distance; in disordered materials, excitons are more easily trapped.

These two energy transfer process can be well separated from their dynamics. The hot exciton energy transfer process is usually completed within a few picoseconds after excitation, and some can even be shorter than a picosecond. Isoenergetic energy transfer processes continue for hundreds of picoseconds. The time and frequency of these two processes jumping between sites are also quite different. By transient absorption for anisotropic spectra and spectral moving, the two energy transfers can be observed in polythiophene polymer materials, such as PDOPT. Within the initial 1 ps, anisotropy decreases rapidly and the spectra shift to red, which is ascribed to hot exciton energy transfer. For the following tens of picoseconds or even longer timescales, the anisotropy continues to decrease with isoenergetic energy transfer. This process is associated with the excitation wavelength: if the excitation photon energy is higher, the hot exciton energy transfer process has a high energy starting point, and thus a longer relaxation time; if excitation photon energy is lower, the initial excitation is already close to the end of the hot exciton energy transfer, and thus this process will end much faster.

Energy transfer will make particles in the excited state move to the low energy sites, and the spectrum can be observed shifting over time in the dynamic spectra. The evolution of polarization anisotropy, including transient absorption and fluorescence anisotropy, reflects the changes of dipole orientation in the energy transfer process. The hot exciton energy transfer process can be observed by evolution of population, spectral shift, polarization anisotropy dynamics and so on; the isoenergetic energy transfer process manifests during relatively slower anisotropy evolution, and spectral shift behavior is not significant. However, the evolution of the energy transfer and anisotropy depends on different aspects of the specific polymer configuration. For example, the hot exciton transfer process in a uniform, isotropic material can be very efficient, so that the anisotropy decays rapidly to zero. The isoenergetic transfer process still continues after that, even though no anisotropy can be observed any more. In this case, the anisotropy dynamics cannot be seen along the energy transfer. Thus, in highly disordered systems, such as a polymer film, there are multiple channels for energy transfer, and the anisotropy will diminish after the initial several transitions, while the energy transfer process continues. On the other hand, in ordered polymer systems, anisotropy reaches a nonzero constant value, even after energy transfer has been completed. Polymer aggregation means the polymer locally orders as crystals. If ordered regions exist as lower energy sites, then efficient energy transfer is not possible, so that anisotropy is maintained. Therefore, by means of fluorescence spectral shift, anisotropy studies, and control of excitation conditions, we are able to study anisotropic or isotropic materials efficiently.

Through comparison between intrachain and interchain energy transfer processes, it is generally considered that intrachain transfer is inefficient and slow while the interchain transfer is highly efficient and occurs at high speed. In the Förster energy transfer model, there are chain segments connected in line within chromophores. Therefore, the coupling is weak. Using transition density and a wire dipole, a stronger interaction and faster rate of transition can be obtained. However, a large number of experimental and theoretical studies have shown that the interchain and intrachain energy transfer processes have similarly high rates [40, 41]. The high rate of the interchain process is due to the multiple transporting channels, which are in relation to the specific chain arrangement. For example, polymers containing small side chains can be closely clustered, so that energy transfers fast. However, if the polymer contains large side groups, then molecular chains cannot effectively contact, so energy transfer is limited [42].

The exciton transport distance of polymers can be observed by fluorescence decay with an added quenching mechanism. As an effective charge acceptor, PCBM and many other fullerene derivatives may be used as quenchers to observe exciton transport. Excitons in P3HT can be transported for a distance of 2.6–2.7 nm [43–45]. There are some disadvantages of employing fullerenes, e.g. fullerenes tend to diffuse into the polymer, blurring the interface, and resonance energy transfer may occur. Samuel et al., using TiO_2 as a charge acceptor (quencher) to overcome the above difficulties, obtained a diffusion distance of ~8.5 nm for P3HT [46].

6.2.2.3 Exciton annihilation

In polymers, two different types of energy transfer processes can be found in the excited state. As described above, excitons transfer from the excited site to the nonexcited site. The hot exciton transfer process results in continuous redshift of the emission spectrum, and lower energy sites become occupied, making the transfer slower and slower. Another transfer process is excitons giving their energy to already excited sites. This occurs at high concentrations of excitons, and is called singlet-singlet excitons annihilation (*SS*). The two excitons collide with only one live exciton while the other exciton is quenched to the ground state. This causes the number of particles in the excited state to be lowered. If the singlet excitons can evolve into triplet excitons, then singlet-triplet (*ST*) and triplet-triplet (*TT*) exciton annihilation can also occur in the material.

Singlet exciton annihilation is a common phenomenon in molecular crystals, photosynthesis systems, *J*-aggregates, conjugated polymers and so on [47, 48]. This process takes into account overlap of emission spectra and the excited state absorption spectrum, which manifests as a redshifted wide absorption band with respect to the ground state absorption. After annihilation, the high energy excitons return to thermal equilibrium of the excited state by intramolecular vibrational relaxation via the thermal conversion process. This quenching process requires a relatively high excitation density; at a lower density, the excitation annihilation phenomenon will be significantly weakened. However, if the excitons in the polymer can be transmitted for a long distance, then even in low density situations exciton annihilation can occur.

The exciton-exciton annihilation process will lead to quenching of one exciton, while the other exciton in is preserved in a high excited state, interrupting the original spectrum moving process. For anisotropy studies, excitons initially have the high dipole orientation. Due to the energy transfer associated with the chromophore orientation, these excitons can be strongly quenched, resulting in a very fast anisotropy relaxation. Thus, rapid decay of anisotropy associated with the light intensity can be observed.

This complex process is nonlinear, reflecting exciton transfer speed and concentration. It can be used for the study of dynamics of excitons, i.e. energy transfer processes. Under a certain pump intensity, annihilation is coupled with diffusion processes. In the initial excited state energy transfer, the exciton annihilation is mixed with the spectral shift and anisotropy relaxation. The velocity of the exciton in the hot exciton energy transfer process varies with its position in energy level, which makes the collision occur, so that exciton annihilation rate is energy dependent. Anisotropic changes affect the mutual exciton orientation over time, and therefore can affect the annihilation rate. This makes observation of the initial stages of the energy transfer phenomena quite complex. When the system reaches thermal equilibrium and uniform orientation, the spectral shift and anisotropy no longer reflect the excited state energy transfer process. In this case, the exciton annihilation can still influence the evolution of an excited state population, so that it can be used to study this process. The exciton annihilation process is more likely to occur in the film than in the solution, because the excitons can move in three dimensional space in the film but only in one dimension in solution, which greatly reduces the probability of encountering other excitons. To obtain a similar annihilation rate, a much higher pump intensity is needed in solutions than in films.

At a higher intensity of excitation, singlet and triplet states can quench singlet states, and this is called singlet-singlet annihilation (*SS*) or singlet-triplet annihilation (*ST*). Which one is dominant depends on whether the excitation light source is a continuous or pulsed light, the efficiency of intersystem crossing, and the annihilation rate. In the pulse excitation condition, if the pulse width is less than the lifetime of the singlet excited state, the instantaneous excitation energy is high and the average intensity is low, meaning singlet annihilation is the main process. *SS* annihilation behaves as a nonexponential decay of the excited state. High average power is required for *ST* annihilation, and due to the triplet state cannot be directly generated by light excitation, but is instead generated by accumulated intersystem crossing from singlets, or by conversion from singlet annihilation. *ST* annihilation leads to a single exponential decay of excited states, strongly correlated with the intensity of excitation. In *ST* annihilation, the transfer of the triplet state excitation is much lower than the singlet one, but also more efficient. However, due to the high mobility of singlet excitons, the overlap of the absorption spectrum of triplets and the emission spectrum of singlets makes the distant Förster energy transfer process happen, which quenches the singlet excitons efficiently [49].

With a femtosecond or picosecond laser source, singlet exciton annihilation can be used to study time domain exciton dynamics. The method requires laser pulses of shorter duration than the average lifetime of singlet excitons. Singlet exciton annihilation occurs when the light intensity reaches a certain level, called the intensity threshold of singlet exciton annihilation. When the excitation density in MEH-PPV reaches 3 μJ/cm²/pulse (7×10^{12} photons/cm²/pulse), the singlet exciton quenching process appears prominently in exciton dynamics. As described above, the quenching efficiency depends not only on the incident light intensity, but also on the specific configuration of the material. Different material preparation methods vary the aggregation of molecular chains, resulting in different singlet exciton transfer processes, and thus affecting the efficiency of singlet exciton annihilation. This means that in the literature, experimental results are also different, even for the same polymer. For example, for MEH-PPV films prepared with either chlorobenzene or tetrahydrofuran as solvents, the exciton annihilation rate between the two is an order of magnitude different [50].

For a relatively high efficiency intersystem crossing polymer, under certain conditions of pulsed excitation, singlet-triplet exciton annihilation can also be observed. To distinguish this from singlet-singlet annihilation and singlet-triplet annihilation, we need to change the pulse energy and repetition rate. By maintaining the pulse energy and changing the pulse frequency, the singlet-triplet annihilation process will be affected, but not singlet annihilation. Then, by fixing pulse frequency and changing pulse energy, singlet annihilation can be studied. For example, in polythiophene POMeOPT, due to the sulfur atom, the intersystem crossing efficiency reaches 50% [51]. The exciton lifetime of this system is 270 ps. For 6 μJ/cm²/pulse (2×10^{12} photons/cm²/pulse), the average light intensity is less than 0.5 W/cm², and either *SS* or *ST* annihilation is obvious. In 2.5 kHz frequency, the *SS* annihilation process appears as nonexponential decay, which gradually becomes faster as the pulse energy increases.

Exciton-charge collision can also induce annihilation. At high pump intensity, when charges are effectively accumulated, effective resonance energy transfer can occur between excitons and charge states, i.e. annihilation happens. The strength of exciton-charge annihilation relates to the product of concentrations of charge and exciton, and a quadratic relationship to the incident light intensity [52]. For high average intensity light, charge states can be accumulated and be observed. Charge pairs can directly result in the formation of excited state quenching [53, 54]. Exciton-charge annihilation can be described by Förster energy transfer.

6.2.2.4 Generation of charges

The presence of electrons and holes in conjugated polymers will cause atomic arrangement (chemical bonds) to change, with the formation of polarons. In photoelectric conversion materials, polarons are the form of carriers. Polymer molecules with polarons differ from the nonexcited polymer molecules in the spatial configuration. The existence of polarons in molecular chains can be judged by this difference. Similar to the exciton, the polaron is also in the form of a given domain, which occupies the scale of several repeat units. For example, in MEH-PPV, the polaron occupies approximately 2–3 repeat units [55–57]. The polaron band energy is inversely proportional to the length of the oligomer, which is a similar characteristic to the S1 and triplet states [58].

The generation of photocharges is an important fundamental issue in optoelectronic applications. In photovoltaic and electroluminescent devices, conversion between light and electricity is the core function. Carrier generation can be directly observed in experiments. There are experiments that suggest charge in PPV is generated because of excessive energy before excitons cool [59]. Arkhipov thought that excited states with high vibrational energy produce three kind of particles: excitons, charge pairs, and separated charges, via a two step process. Exciton dissociation is completed first by self-ionization, forming pairs of charges, which is associated with the polymer vibration cooling, occurring within a few picoseconds. Then charge pairs can dissociate to free carriers. Although the exciton generation process does not depend on the applied electric field, it will enhance the generation of carriers. An applied electric field can compete with internal Coulomb potential to prevent the formation of excitons or to affect excitons during its ionization [60, 61].

Figure 6.8 represents the time resolved differential curve of the exciton state (1.72 eV) and the charge induced absorption state (1.9 eV). Charge pairs continue to be produced within the lifetime of tens of picoseconds. From the dynamics of charge pairs we know the rate of charge generation decreases along with time.

In short, high binding energy in polymers, which induces a high potential barrier, can be overcome by applying an electric field, providing additional vibration energy, (> 1 eV, by high photon energy excitation), or by exciton annihilation. The external electric field threshold can be observed, indicating that a barrier exists between excitons and charge pairs. However, the external field assisted generation process is slow, and it can be observed in the entire process of the exciton lifetime. Due to the disorder of polymer systems, the binding energy and the time of charge transfer have a certain distribution, resulting in a charge generation rate changing over time. In this system, the ultrafast charge generation and vibrational relaxation process compete with each other. The charge generation process of hot excitons located in the S_1 state is therefore strongly time dependent. In the system with exciton binding energy close to zero, the

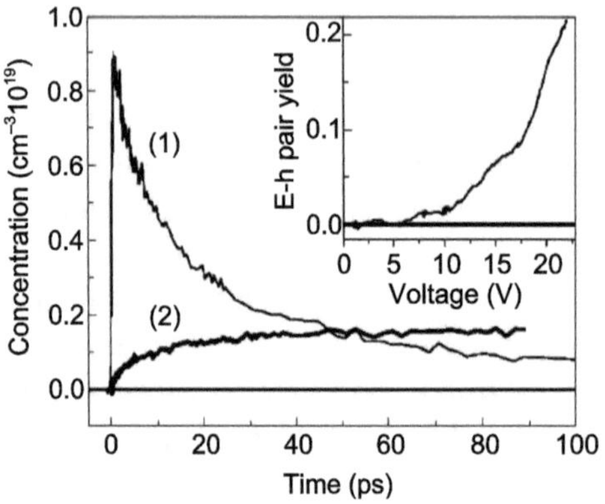

Fig. 6.8: The time resolved dynamics of exciton state (1.72 eV) and charge induced absorption state (1.9 eV) [62].

charge generation process depends on exciton motions, since excitons need to move to the charge separation sites to generate charge pairs. Therefore, the charge generated timescale is about 1 ps. This is significantly different from the slow generation process. Without an external electric field, the charge generation yield is low; conversely, the yield is increased by applying an external electric field.

When the exciton splits into charges, the gap of charges is several nanometers. Arkhipov observed exciton transitions between two quasistatic states. The two quasisteady states are produced in the conjugated chain segments under strong electric field, which are in neutral exciton state, and electron states located at both ends of the segments. In this study, charge separation space reached 6 nm; outside the region of Coulomb interaction and the barrier. Quantum mechanics calculations on excited states of these conjugated polymers show large charge separation distances of a few nanometers, even without an external field [63].

6.2.2.5 Charge transport in polymers

Whether and how the charge pairs separate into free charges are the key steps to understanding how photoelectric conversion of the polymer materials is achieved. However, controversy still exists around the intrachain and interchain charge separation mechanism. If we are using a one dimensional semiconductor approximation, photoexcited carriers delocalize on the polymer chain. Boundaries and defects of conjugated polymers can be considered as scattering and trapping centers of charges. According to this concept, the interchain charge transport is less important, but can reduce the probability of charge recombination. Others believe that charge separation between the chain and on the chain both happen. Meng and Hong suggested that the excitons are trapped in the boundary of the molecular chain, and the charges are generated in the adjacent molecular chains, so that charge recombination can be reduced by a barrier constituted by the boundary. Extending the lifetime provides charges with more chances to escape Coulomb interaction. This is similar to the mechanism in nonconjugated polymers and amorphous molecules in the crystal [64]. Plefer observed that the external electric field can facilitate charge generation when the orientation of the molecular chain of the PPV in a Langmuir–Blodgett (LB) film is strictly perpendicular to the field. Experiments observing the charge generation efficiency by changing the space between the molecular chains also confirmed the possibility of an interchain charge generation mechanism [65].

Exciton generation, transfer, and charge generation processes can be observed in real time via a variety of methods, but observation is much more difficult after charge generation, because between paired bound charge and free charge, there is very little difference in the spectra. Nonspectroscopic measurements, such as electronic ones, can be more effective in detecting space charge mobility characteristics. But the time resolution of these methods can only be nanoseconds. Time of flight (TOF) mass spectrometry methods can provide subnanosecond time resolution, and therefore can provide macroscopic carrier mobility information. Heeger, who used ultrashort electrical pulses to detect photoconductivity, made the time resolution reach tens of picoseconds. This research showed that light conductivity relaxation has a timescale of tens of picoseconds [66]. Noncontact time resolved microwave and terahertz measurements can be applied to the field with similar resolution [67].

6.2.2.6 Excitation quenching

In polymer film materials, the fluorescence quantum efficiency is usually from 5 to 30%. A large amount of the excitation energy is quenched by traps due to film morphology and impurities present in the film. It also strongly correlates with the excitation light intensity. Since there is flexibility in molecular chains and uncertainty of contact between chains, defects (nonradiation transition point, such as dimers, aggregates, etc.) will naturally be present. These defects and impurities together will produce the fluorescence quenching effect.

At low excitation light intensity, the interaction between the particles, such as singlet exciton annihilation, single-triplet exciton annihilation, and exciton-charge annihilation, can be ignored. Note that such a low excitation intensity relates to the pulse width and the particular polymer type. In general, continuous laser energy lower than 1 W/cm^2 can satisfy the above condition. Then the quenching of the excited state includes internal conversion, intersystem crossing, exciton dissociation and defect states quenching. For this polymer system, what is the main reason(s) for the quenching of the excited state? There is a dependency on the specific polymer and configuration. Even for the same polymer, since each molecular chain has different configurations and defects, there are differences even between the molecular chains.

The rate of intersystem crossing and internal conversion is associated with the chemical structure of the polymer molecules. Intersystem crossing efficiency can be more than 50% [68]. This makes intersystem crossing the main reason for singlet exciton deactivation. The absence of spin-orbit coupling makes intersystem crossing inefficient and dependent on the singlet-triplet energy level difference, which is similar to small organic molecule characteristics. For some special polymer without efficient spin-orbit coupling, intersystem crossing can still happen. For example, in MeLPPP, approximately 37% of the singlet excitons will transform into triplet excitons [69].

In the polymer film, energy can be efficiently transferred due to the excited state, and defects in the molecular chain structure can effectively quench excited states. Such defects can be caused by errors in the synthesis process. In the polymer PPV, the structural defects are mainly due to the main chain structure vinylidene double bond, which is not fully formed (single, triple bond instead of double bond), oxidation, the cis and trans isomers, and so on. These defects cut conjugated systems, causing charge transfer, and resulting in a decrease of fluorescence quantum yield. For instance, a MEH-PPV film stored in air will experience a few percent reduction of the quantum yield after oxidation. In addition, the impurity defects and aggregation defects can also lead to the quenching process.

6.2.3 Examples of polymer systems

6.2.3.1 Ultrafast photophysics of the branched polythiophene system

Polythiophene (PT) is a macromolecular compound produced by polymerization of thiophene monomers. By doping conjugated π orbits, it will have a certain conductivity. Therefore, polythiophene and similar organic conducting materials have been a hot research topic in recent years. In terms of applications, polythiophene based materials in organic photovoltaic devices have been widely studied. In 2000, the Nobel Prize in Chemistry was awarded to the three scientists, Alan Jay Heeger, Alan Graham MacDiarmid, and the Japanese scientist Hideki Shirakawa, for their contributions to organic conducting polymers.

Conductivity is the most important property of these conducting polymers. The conjugation crossing the molecular chain can let electrons delocalize along the conjugated polymer backbone, which is the path of charge extraction in photovoltaic devices. In addition to conductivity, the optical properties of polythiophene are also important. Conjugation and aggregation will generate broadened absorption in the visible region towards the solar spectrum, improving generation of excitons and separation of charges. Furthermore, since the conjugation is associated with the spectrum, changes in environmental factors, such as solvent, temperature, electric potential, and so on, mean the polythiophene main chain may have torsion and conjugation damage, which leads to changes of color that can be used as a sensor.

Since, in the polythiophene with fullerene doped systems, photoelectric conversion efficiency of about 5% can easily be obtained for the bulk heterojunction structure, a lot of attention has been paid to its photovoltaic applications. How to further enhance its photoelectric conversion efficiency is an important target of this study. The study focus is on whether the use of branched thiophene devices could create an internal charge path to enhance electron transportation, thereby improving conversion efficiency. Experiments show that attaching pendant groups can improve the photoelectric conversion efficiency. The one dimensional polythiophene derivative P3HT (Poly (3-hexylthiophene-2,5-diyl)), with a molecular weight of 68,000 was used; two dimensional polythiophene derivatives are obtained by chemical synthetic methods [70]. Figure 6.9 is the chemical structural formulas of P3HT and three branched polythiophene derivatives. The side chain of the three polythiophenes has one, two and three vinyl thiophenes, called P1, P2, and P3 respectively. The three materials can be dissolved in many organic solvents such as chlorobenzene, chloroform, and toluene. If the side chain density is defined as the ratio of thiophene units containing side chains to the total number of thiophene units in the main chain, for P1, P2 and P3, it is 50%.

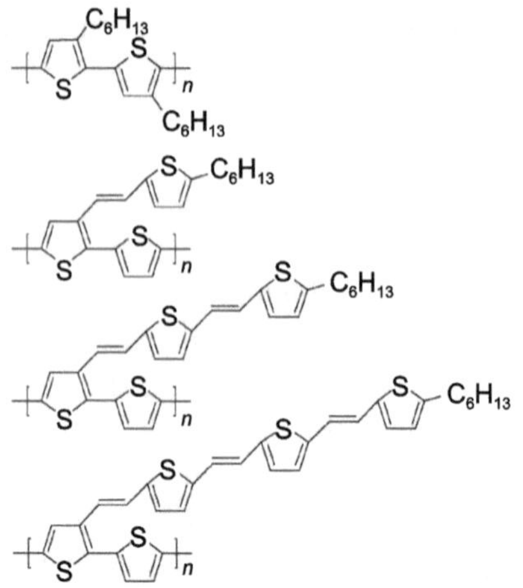

Fig. 6.9: P3HT and its branched derivatives.

(1) Steady state spectrum of branched molecules of polythiophenes

Research into polymer system dynamics using ultrafast spectroscopic methods requires a combination of transient state and steady state spectroscopy, and other methods. The material needs to be carefully characterized in order to study the effect of the side chain of two dimensional polythiophene and its length on the configuration, microstructure, conjugation, absorption, the excited state dynamics and other issues. The above four samples (P3HT, P1, P2, P3) in solution and in solid film have been detected by emission spectra of the steady state absorption, transient absorption, fluorescence spectroscopy, and X-ray diffraction spectra.

Figure 6.10 shows steady state absorption and emission spectra of P3HT, P1, P2, and P3 in chlorobenzene in the form of solution and film. The solid lines are the absorption spectra, dashed lines are the emission spectra.

For the steady state absorption spectra of a chlorobenzene solution, P3HT has a main absorption peak at about 456 nm due to polythiophene conjugated π-π * transition backbone; however, after adding side chains, the two dimensional samples showed two absorption peaks. The absorption is similar with π-π * transition of P3HT, but with relative redshift, which means that the energy level has been lowered after adding the side chain, as a result of the cooperation of the main chain and the side chain. The significant redshift of the absorption peaks from P3HT to P3 are shown, as at P3HT of 456 nm, P1 at 490 nm, P2 at 554 nm, and P3, it is not possible to read a clear location. However, according to the redshift of UV and visible absorption peaks it can be inferred that if the P3 visible absorption peak does occur, it will still appear as redshifted relative to P2. This redshift is due to the polythiophene conjugated side chain, which increases the overall degree of polymer conjugation. The coplanar of the thiophene ethylene side chain and the thiophene ring in the main chain reduce the energy level. The levels of P1 and P2 decrease to 2.0 and 1.8 eV respectively, which is advantageous in increasing the coverage of the solar spectrum to enhance the absorp-

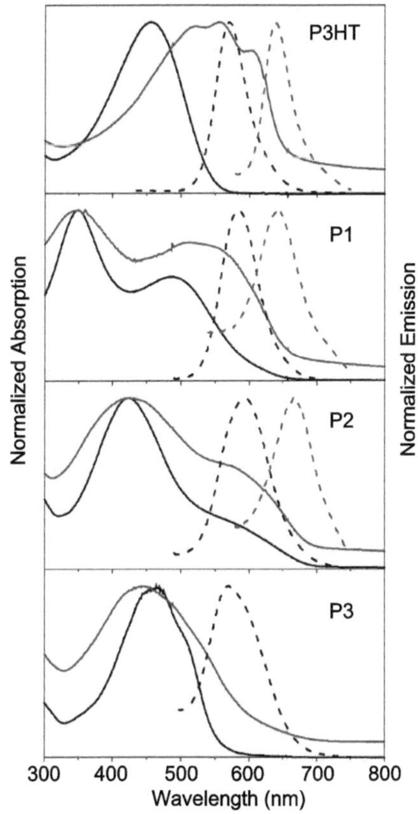

Fig. 6.10: The steady state absorption and emission spectra of P3HT, P1, P2 and P3 in chlorobenzene in the form of solution and film.

tion of the solar spectrum, thus helping to improve the power conversion efficiency. At the same time, along with redshift of the absorption spectrum, the high visible absorption peak gradually decreases, because the side chains are dense. When increasing the number of vinyls, the long side chains become entangled, and a larger steric hindrance affects the conjugation of main chain. This makes the backbone of the polymer chain difficult to extend in the original plane, resulting in distortion of the spatial configuration, thereby reducing their degree of conjugation and absorption intensity. For P3 containing three thiophene vinyl side chains, the side chain group greatly exceeds the scale of main chain periodicity connecting to the side chain, which can easily lead to a huge torsion of the main chain plane, resulting in a main chain conjugation drop. The corresponding absorption peak intensity then becomes small. Thus, large side chain structures would reduce the π-π conjugated degree of P3. According to reports, reducing the density of the side chain is an effective way to solve the steric hindrance, namely by inserting some thiophene units between the backbone units to increase the distance between the side chains [70, 71].

Another absorption peak is at the near-ultraviolet region owing to the absorption of conjugated thiophene vinyl side chains [72, 73]. For absorption peaks in the near-ultraviolet region, the absorption peak from P1 to P3 apparently redshifts (P1 located at 350 nm, P2 located at 424 nm, P3 located at 464 nm), which is due to the length of the side chain increasing, and the total conjugation degree gradually increasing. Notably, for the overall absorption, P3 is blueshifted compared with P3HT, but the efficiency of P3 based devices is higher than P3HT based devices.

For the measurement of steady state fluorescence spectra of solution samples, we can select two absorption peaks to excite. For example, excitation at 350 nm and 490 nm for P1 corresponds to the excitation of a side chain and a backbone chain. Experiments show that both excitations produce consistent fluorescence spectra. P2 and P3 have the same results, indicating that the excitation of side chains can lead to a two dimensional intramolecular energy migration towards the backbone chain [74]. In addition, the ratio of the relative fluorescence quantum efficiency of these samples is 36:12:3:1. That is, as the side chain length increases, the fluorescence quantum efficiency gradually decreases. This is because decreasing the molecular chain conjugation causes defects, leading to fluorescence quenching.

(2) Transient spectrum of branched polythiophene molecules in solution

By means of the transient pump probe technique, the excited state dynamics can be studied for materials of a wide spectral range. By single wavelength scanning, the complete transient spectrum from 450 nm to the infrared region can be obtained. Transient absorption spectra of P3HT, P1, P2 and P3 in chlorobenzene solution are shown in Figure 6.11. Delays for 0, 0.2, 0.5, 4, 10, and 100 ps of visible and near-infrared transient absorption spectra are shown. The pump wavelength is 415 nm. Before explaining the transient spectral information, we first illustrate several phenomena of transient spectra.

(1) Photobleaching (PB): After excitation, the population of electrons located in the ground state is reduced, so the transient absorption signal absorption decreases ($\Delta\text{OD} < 0$) or transmittance increases ($\Delta T/T > 0$); this is called ground state bleaching. This signal generally corresponds to the steady state absorption spectrum, usually located near the excitation wavelength. The observation of this signal represents the excitation conditions of the sample, and the dynamics of the excited state returning to the ground state. If there is a complex process of intermediate states, the relaxation of the excited state is inconsistent with the recovery ground state population.
(2) Simulated emission (SE): This process means a particle in an excited state can be induced to transit to the ground state by probe light, emitting a stimulated photon. The observed signal is at a similar position towards the steady state emission spectra, usually in the red side of ground state bleaching, appearing as absorbance

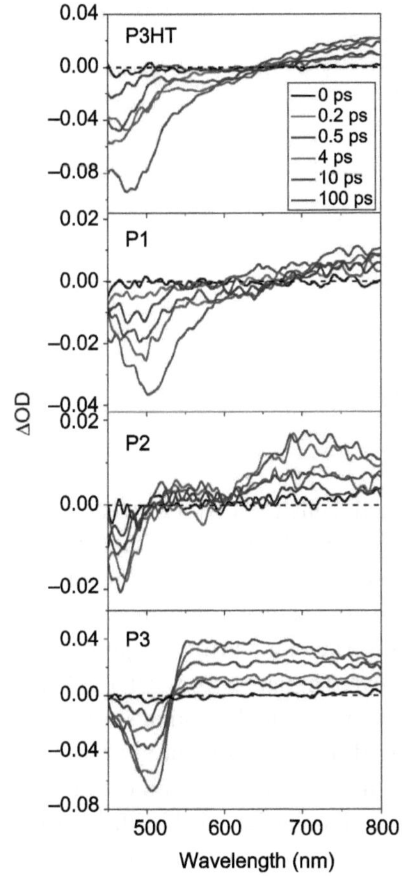

Fig. 6.11: Transient absorption of P3HT and its derivatives, pumped at 415 nm.

decreases ($\Delta OD < 0$) or transmittance increases ($\Delta T/T > 0$). These relaxation process and ultrafast fluorescence decay signals can confirm each other.

(3) Photoinduced absorption (PIA): PIA refers to the additional absorption feature due to the pump light. This absorption comes from the photoexcited electrons in an excited state, which further absorb photons in probe light and transit to a higher energy level, appearing as absorption increases ($\Delta OD > 0$) or transmittance decrease ($\Delta T/T < 0$). For the study of excited state dynamics applying transient absorption experiments, nonemissive states can be directly observed. At the same time, various photogenerated relaxation processes can be detected, besides the S_1 excited state.

Transient transition enhanced signals at 470, 500, 466, and 506 nm signal of P3HT, P1, P2, and P3 represent ground state bleaching, as shown in Figure 6.12. For P3HT, at 570 nm there is another relatively flat transition enhanced signal, which is stimulated emission, consistent with the steady state fluorescence emission peak of P3HT. For P1,

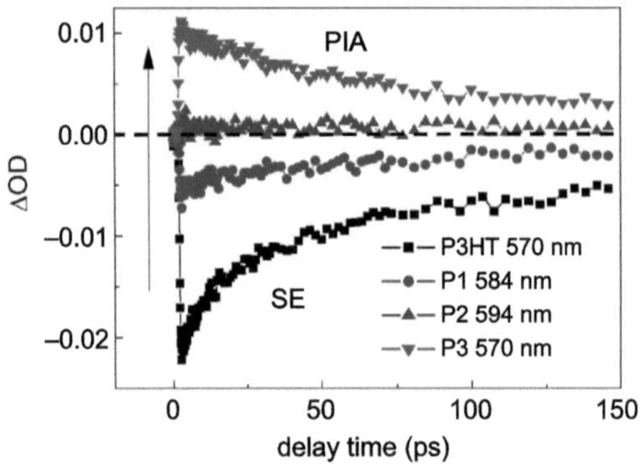

Fig. 6.12: Transient decay of P3HT and its branched derivatives, from stimulated emission to photoinduced absorption.

at 584 nm there is also a stimulated emission signal, consistent with its steady state fluorescence peak wavelength, whose intensity is decreased compared to that of P3HT. When the length of side chain is two, the intensity of the simulated emission (SE) signal of P2 is further lowered, even slightly reversed, and shown as enhanced absorption. For P3, the stimulated emission signal has been transformed into a light-induced absorption signal (570 nm); from negative to positive. Therefore, from P3HT to P3, as side chain length increased from zero to three, the transient absorption spectra of a stimulated emission signal of about 570 nm evolved into a light-induced absorption signal. From the evolution, it can be inferred that from P3HT to P3, with the increasing of side chains, the planar structure of the main chain is damaged by the effect of steric hindrance. Defects will gradually increase, so singlet excitons excited from the main chain or transferred from side chains through intramolecular energy transfer are quenched or trapped by such defects, resulting in the excited state quenching process. This result is also consistent with the decreasing of fluorescence quantum efficiency in the steady state spectrum.

By means of picosecond resolution streak camera technology, the singlet exciton lifetime of the four samples can be measured. As shown in Figure 6.13, fluorescence relaxation curves of four samples in solution are detected at each fluorescence peak by this measurement. The dotted lines represent the instrument response function (IRF). From the data we can see P3HT, P1, and P2 have similar fluorescence decay, while the decay of P3 is significantly accelerated compared with other three samples. By exponential fitting (solid line), the fluorescence decay curves of relaxation processes can be described by two exponential decays (τ_1 and τ_2 represent two lifetimes, a_1 and a_2 represent the amplitude of these two lifetimes respectively)

$$F(t) = a_1 \exp\left(\frac{-t}{\tau_1}\right) + a_2 \exp\left(\frac{-t}{\tau_2}\right) . \tag{6.15}$$

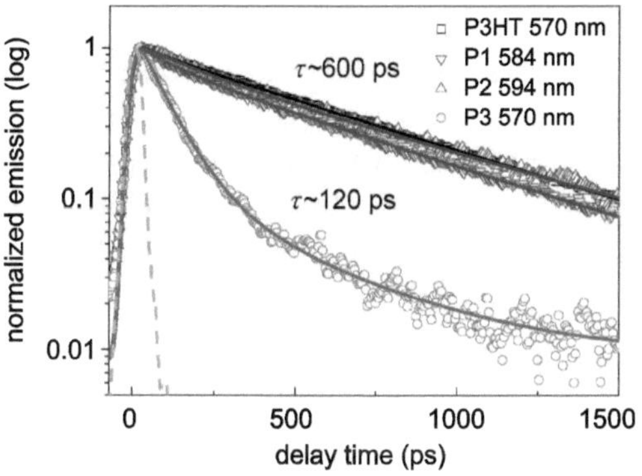

Fig. 6.13: The fluorescence relaxation dynamics of the four samples in chlorobenzene solution detected at the fluorescence emission peak. The excitation wavelength is 415 nm.

Tab. 6.1: The fitting results of fluorescence relaxation of the four samples in chlorobenzene solution.

	a_1	τ_1/ps	a_2	τ_2/ps
P3HT 570 nm	0.111 (10%)	76.8	0.997 (90%)	624.8
P1 584 nm	0.255 (23%)	119.5	0.836 (77%)	596.5
P2 594 nm	0.278 (25%)	190.0	0.816 (75%)	691.6
P3 570 nm	1.240 (90%)	86.0	0.136 (10%)	372.9

The fitting results are shown in Table 6.1. All the results contain a fast decay process of around 100 ps, which comes from the exciton quenching by the defects of spatial configuration of the main chain. Another slower relaxation process, ~600 ps, comes from the singlet exciton spontaneous emission process.

From P3HT to P2, the amplitude of the fast decay increases from about 10 to 25%. The introduction of a side chain causes weak steric hindrance to the planarization of the main chain and a weak increase in defect density. Therefore, the ratio of the first process increases slightly. According to equation (6.16) an average lifetime can be obtained:

$$\tau = \frac{a_1\tau_1 + a_2\tau_2}{a_1 + a_2}\,. \tag{6.16}$$

The average lifetime of the three samples from P3HT to P2 is about 600 ps, which is consistent with relative reports of polythiophene [22, 75, 76]. As for P3, a_1 increases to 90%, which takes the average lifetime to 120 ps. This is because the side chain length is more than the repetition period in the main chain, resulting in side chain entanglement and a significant weakening to the planar structure of the main chain. This damage almost destroys the conjugation structure, so that chromophores across several chains are relatively close to the defects, thereby enabling a sharp fluorescence quenching and the short fluorescence lifetime τ_2. Likewise, it can be inferred when the side chain length increases from zero to two, the side chain steric hindrance to planarization backbone, and to interconnection between the side chains, is small. Influence on the polymer backbone is gradually increased with side chain length. However, when the side chain contains three vinyl thiophenes, strong entanglement is produced between side chains, which directly leads to the occurrence of substantial distortion of the polymer backbone. Intense steric hindrance almost destroys the conjugation structure of the backbone, and the strong excited state quenching process occurs.

(3) Excited state evolution and charge recombination of polythiophene branched molecular thin films

By the above research, the difference in configuration of molecular chains of several polymers can be obtained. As photoelectric functional materials, research on the dynamic behavior of thin film devices and photoelectric conversion processes is more important than for other polymers. Reports on devices with two dimensional polythiophene photovoltaics have confirmed that they are more efficient than one dimensional P3HT based devices. As mentioned above, P3 has a greater defect density of molecular chains, which makes absorption bluer. Transient absorption experiments and X-ray diffraction spectra can reveal the effect of branches on the two dimensional polythiophenes on their charge recombination dynamics. This guides the design for new polymer photovoltaic materials.

Results of transient absorption measurements of P3HT, P1, P2, and P3 films are shown in Figure 6.14. For transient absorption spectra of thin film samples, besides ground state bleaching and stimulated emission, the most noticeable thing is the visible band from ~600 nm to the infrared region. As can be seen from the figure, four transient absorption spectra of the polymer film samples contain two photoinduced absorption bands. One is in the visible region (650–700 nm), and the other is in the vicinity of 900 nm in the near-infrared region. According to relevant reports [76], PIA in the visible region comes from absorption of interchain polarons, PA3, corresponding to the charge separation state; and the PIA in the near-infrared region is from absorption of intrachain exciton, PA2, as shown in Figure 6.15.

To verify the sources of photon induced absorption bands, the PIA signals at 890 nm and 665 nm were observed by pumping P3HT film with different excitation densities, as shown in Figure 6.16. This is because the two excited state dynamics have a different relationship with pump intensity. From the figure it can be easily seen that the PA2 relaxation process is associated with the pump density. For instance, at high energy conditions it performs a very fast relaxation process. The reason is that the exciton concentration is high at high pump energy density, and diffusion of the exciton in the material is applicable. During transfer, excitons collide with each other, resulting in singlet exciton-exciton annihilation. While under the same conditions, with the pump energy density gradually increasing from 100 $\mu J/cm^2$ up to 1000 $\mu J/cm^2$, the relaxation process of PA3 doesn't change significantly. This infers that the dynamics of the process associated with the PA3 is localized, independent

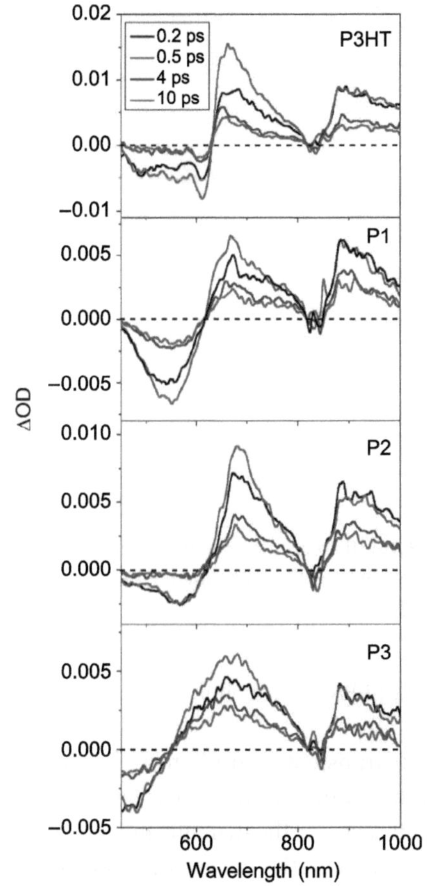

Fig. 6.14: PIA signals of P3HT, P1, P2, and P3 film. Pump wavelength is 415 nm with a fluence of 150 μJ/cm^2.

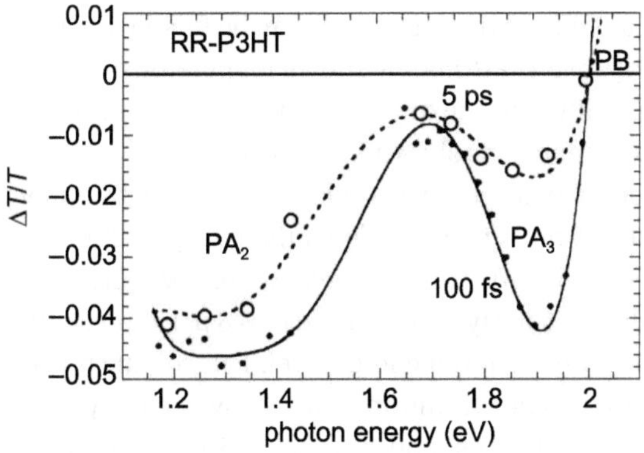

Fig. 6.15: Interchain polaron absorption of PA2 and intrachain exciton absorption of PA3 [76].

of excitation density. Its initial fast relaxation can be fitted with exponential decay ($t^{-1/2}$), which indicates the polaron twin composite [77]. The dependencies of PA2 and PA3 relaxation signals on pump energy density, and the exponential decay of PA3, are consistent with previous reports of Vardeny et al. [76, 78], confirming the two sources:

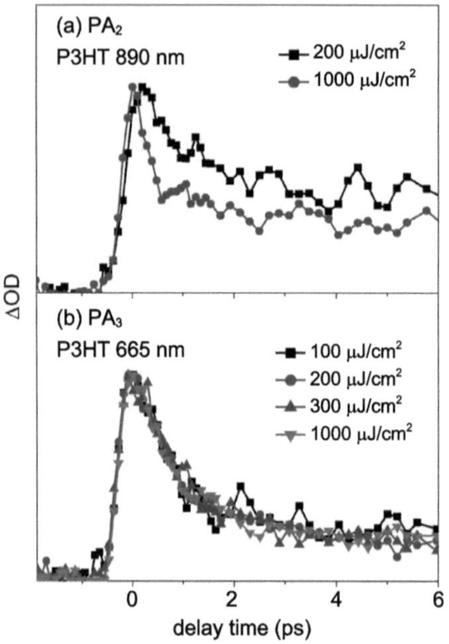

Fig. 6.16: The relaxation of photon induecd PA_2 and PA_3 of P3HT film with different excitation densities.

PA2 is from polythiophene intrachain excitons, and PA3 is from polythiophene chain interchain polaron pairs.

For different side chains, we need to compare transient absorption spectra of PA2 and PA3 of our four samples, which is shown in Figure 6.17. By comparison, PA2 relaxation kinetics are almost consistent for all samples, i.e. the dynamic process of intrachain excitons is similar. However, PA3 relaxation kinetics are associated with the side chain length, i.e. the longer the length, the slower the relaxation kinetics.

It is recognized that the PA3 relaxation in the first few ps can be fitted with an exponent function. The dynamics in Figure 6.17 are mainly shown for polaron dynamics. To compare, we fit the result curves with a multiexponential model. After fitting, we can get three exponential decay components, as shown in Table 6.2.

PA2 decay kinetics of the exciton relaxation process represent the polythiophene main chain. Because the pump light (415 nm) energy is higher than the polythiophene film steady state absorption band edge (650–700 nm) energy, excitons will first be excited to a higher energy level with ultrafast decay process τ_1 (0.6–0.8 ps), representing exciton transfer from a higher energy level to a lower energy level. This cooling process corresponds to excitons transiting from higher vibration states to lower energy states. Two slower relaxation processes, τ_2 (about 12 ps) and τ_3 (about 450 ps), are derived from isoenergetic transfer after ultrafast hot exciton relaxation (τ_1) [21]. From P3HT to P3, due to direct excitation of the main chain excitons and the similarity of the main chain structure, the intrachain exciton relaxation process is basically consistent.

However, for interchain relaxation of polarons, polythiophene samples of different side chain lengths exhibit different characteristics of relaxation. For PA3, ultrafast

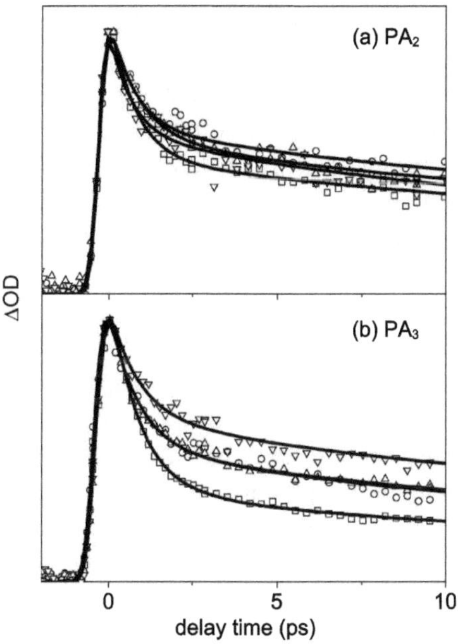

Fig. 6.17: Photon induced dynamics of P3HT (□, black), P1 (○), P2 (△) and P3 (▽); the solid lines are the fitting results. The pump wavelength is 415 nm with the excitation fluence of 150 μJ/cm^2. The relaxation of intrachain excitons PA2 is measured at 890 nm. The relaxation of interchain polarons of P3HT and P3 is measured at 665 nm, while that of P2 and P3 is measured at 690 nm.

Tab. 6.2: The fitting results of PA2 and PA3 of P3HT, P1, P2, and P3 films.

Samples and probe wavelengths	**Fitted parameters (ps)**		
	τ_1	τ_2	τ_3
PA_2 band			
P3HT 890 nm	0.7 (60%)	13 (20%)	464 (20%)
P1 890 nm	0.7 (49%)	13 (24%)	457 (27%)
P2 890 nm	0.8 (51%)	11 (22%)	421 (27%)
P3 890 nm	0.6 (50%)	12 (29%)	460 (21%)
PA_3 band			
P3HT 665 nm	0.9 (75%)	9 (13%)	254 (12%)
P1 665 nm	0.9 (60%)	11 (22%)	244 (18%)
P2 690 nm	0.9 (59%)	11 (21%)	242 (20%)
P3 690 m	0.9 (48%)	10 (24%)	247 (28%)

relaxation τ_1 (0.9 ps) characterizes the geminate recombination of polaron pairs, which has a high energy and high mobility. As the length of the side chain increases from zero (P3HT) to three (P3), the portion of geminate recombination decreases, from 75% for P3HT, to 48% for P3. For samples with longer thiophene vinyl side chains, the distance between the polymer backbones increases, which means longer spatial distance for geminate recombination, resulting in a reduced probability of recombination. Other polaron pairs are relatively localized and have lower energy. Their separation location is not far. Due to their lower energy, they cross the barrier for recombination at a smaller rate, resulting in lower kinetic relaxation rate with the corresponding fitting of τ_2 (9–11 ps) and τ_3 (about 250 ps) [78].

In reports, polymer photovoltaic solar cells based on two dimensional polythiophene, P2 and P3, have a PCE of 2.57 and 3.18% respectively, exceeding P3HT based devices (2.41%) under the same experimental conditions [70]. Yu et al. reported that [71] when the side chain contains only thiophene groups (excluding vinyl), the performance is also enhanced by improving the open circuit voltage compared with P3HT based devices.

This indicates that the chemical design of two dimensional polythiophene helps improve the performance of polymer photovoltaic devices and can be a direction for development of new polymer photovoltaic materials. Ultrafast spectroscopy studies show that the recombination probability of interchain delocalized polarons is reduced, which helps more polarons further separate into free charge by the built-in electric field effect, and this is the main reason for this photovoltaic cell device performance improvement. Meanwhile, large side chain steric hindrance may lead to serious weakening of conjugation. Therefore, we can consider altering the effects of steric hindrance on the conjugation of two dimensional polythiophene by adjusting the density of side chains.

Chemical modifications of two dimensional polythiophene change not only the probability of recombination but also its microstructure. Figure 6.18 shows X-ray diffraction (XRD) results for these four polythiophene films. P3HT has a characteristic diffraction peak for lattice at $2\theta = 5.4°$, which means an ordered layer structure of P3HT [79, 80]. According to the diffraction peak, we can calculate that the distance between P3HT layered structures is 1.63 nm. Nevertheless, the other three polythiophene materials P1, P2, and P3 have no diffraction peaks, indicating that the microstructure of the two dimensional polythiophene material is an amorphous structure. This amorphous structure has a consistent steady state and transient spectrum with solution samples in chlorobenzene: because the steric hindrance caused by the side chains in polythiophene leads to distortions in the space, the spatial arrangement cannot be in order.

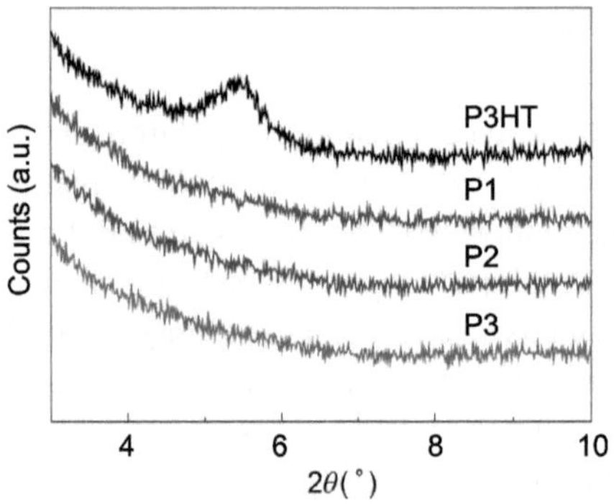

Fig. 6.18: The XRD results of P3HT, P1, P2, and P3 thin films.

In P3HT's ordered layer structure there exists accumulation and aggregation between layers, which will increase the possibility of recombination of polarons. Its microstructure also becomes amorphous with the addition of thiophene vinyl side chains, which increase the spatial distance between polymer chains and effectively reduces accumulation, aggregation and recombination. What's more, the side chains provide new charge transfer channels, thereby enhancing the charge transport ability, which is one of the factors of the performance improvement of two dimensional polythiophene photovoltaic devices.

6.2.3.2 Advances in thiazide polymers

Because the thiazine group has a six-membered ring containing electron-rich N atoms and an S atom, as shown in Figure 6.19, it has strong electron-donating properties and low oxidation potential [81]. In addition, because dihedral the angles formed by six-membered C-N-C bond and C-S-C bond are 140° and 141.9° respectively, it has a spatial configuration of the butterfly type in the ground state, which is significantly different from carbazole and fluorine groups [82, 83]. This effectively reduces the hydrodynamic volumes of corresponding polymers and π-π aggregation [84], and will greatly improve the optical properties of polymer solid films.

In recent years in the field of photovoltaic cells, due to this strong electron-donating property, thiazide group polymers have often been used in dye-sensitized solar cells [86–88] and polymer photovoltaic cells [89, 90], in which the thiazide acts as an electronic donor, with phenylquinoline as an electron acceptor. Electroluminance devices based on thiazide also achieved great progress [82, 91–93]. Daub and Shirdel et al. studied some intramolecular energy transfer and electron transfer kinetics from thiazide to pyrene and flavin groups in dimers and trimers [94–97]. Researches show that the electron donor-acceptor type in a polymer system can effectively reduce the bandgap, thus enhancing absorption of solar radiation. In these polymer structures, intramolecular charge transfer is a frequently occurring photophysical process in the excited state S_1 after excitation, which is often associated with torsion of certain groups around a single bond, forming the twisted intramolecular charge transfer (TICT) state [94, 95].

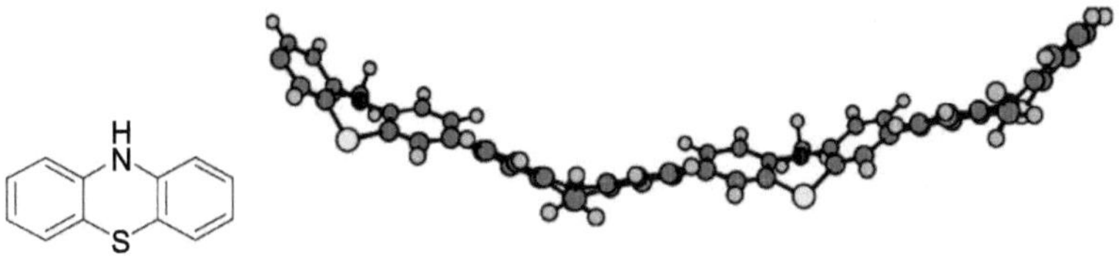

Fig. 6.19: The chemical structure formula and spatial configuration of the corresponding thiazine polymer [85].

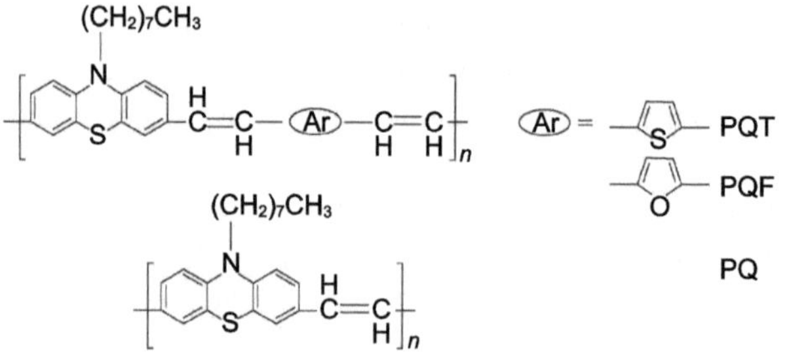

Fig. 6.20: Three polythiazide polymers.

For polymers containing thiazides, thiophene, furan, carbazole and other organic functional groups, many researchers have obtained the spatial configuration and electronic states of S_0 and S_1 by theoretical calculation [98–103]. The variation of spatial configuration of conjugated polymers affects their conjugation length, bandgap, and so on. The torsional relaxation of the spatial configuration also has a great impact on the performance of PV devices. By means of time resolved fluorescence spectroscopy, spatial torsional relaxation of a thiazinyl donor-acceptor (DA) polymer and the forming of TICT states are revealed.

We present several thiazine polymers synthesized by the Knoevenagel and Wittig method [104]. The chemical structures are shown in Figure 6.20. PQ is constituted by thiazine and ethylene groups, and PQT and PQF have introduced thiophene and furan groups on the base of the PQ, respectively. They form a DA structure, containing 6.8 repeating units.

Figure 6.21 is steady state absorption spectra of the three polymers in tetrahydrofuran (THF) solvent. They both show two absorption bands, one at 400–440 nm, and another at 300–330 nm. Absorption at 400–440 nm is derived from $\pi \rightarrow \pi^{*}$ transition of delocalized π electrons [105]. Qiu et al. studied polymers of different thiazine vinyl

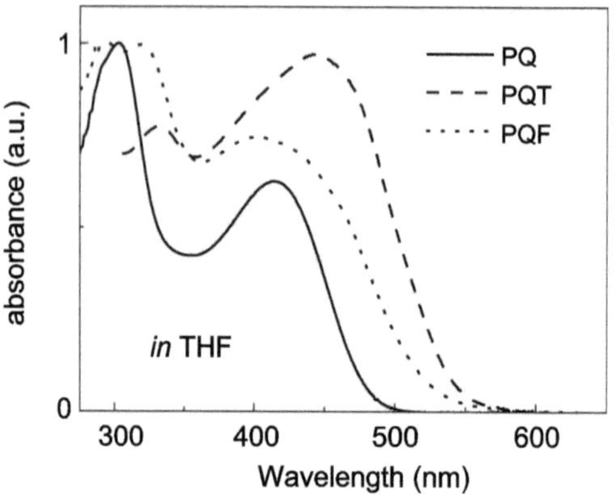

Fig. 6.21: Absorption of PQ, PQT, and PQF in THF.

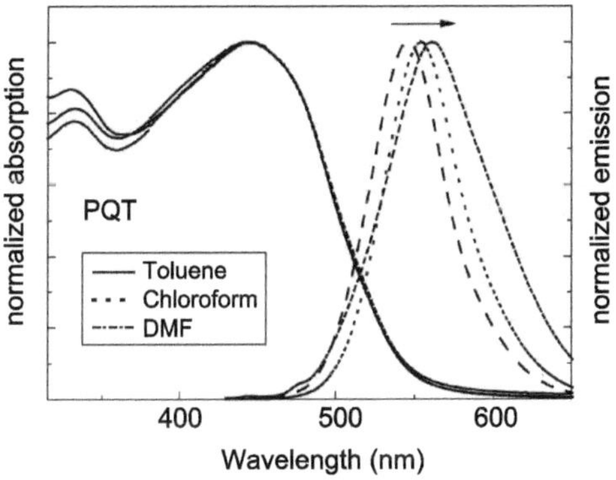

Fig. 6.22: The steady state absorption of PQT in toluene, chloroform and DMF.

chain lengths. With increasing chain length, the degree of conjugation increased, and the bandgap E_g also decreased [106]. PQ absorption is consistent with the reported polymer containing six repeating units, which confirms that the chain length of PQ is about six. The figure also shows that PQT and PQF have a redshifted and have wider absorption spectra compared with PQ due to increased conjugation length, $n \rightarrow \pi^*$ transition of nonbonding electrons in thiophene and furan S and O atoms [26], and intramolecular charge transfer in the DA structure. The addition of thienyl and furyl reduced the bandgap of phenothiazine.

The thiazine polymer can be dissolved in solvents with different polarities, e.g. toluene, chloroform, and dimethylformamide (DMF), of 1.23, 3.84 and 12.88×10^{-30} C · m respectively. In this series of solvents, steady state absorption spectra of PQT and PQF remain constant, while the emission spectra redshift with increasing solvent polarity, as shown in Figure 6.22. This shows that these types of polymers in the ground state do not have intramolecular charge transfer, and no intramolecular charge transfer (ICT) state forms. When the polymer is photoexcited to the excited state, intramolecular charge transfer occurs and the ICT state forms, resulting in redshifted emission, which is characteristics of DA molecular systems [91]. This is consistent with reports of Jenekhe et al. on a thiazide-phenylquinoline copolymer [91, 92].

Another group of solvents of similar polarity but different viscosity, tetrahydrofuran (THF), dichloroethane (DCE) and aniline, are chosen to compare the steady state absorption and emission spectra of PQT and PQF, taking coumarin-152 as a reference. Based on equation (6.17),

$$Q_s = \frac{S_s}{S_r} \frac{\text{Ab}_r}{\text{Ab}_s} \frac{n_s^2}{n_r^2} Q_r \tag{6.17}$$

PQT and PQF fluorescence quantum efficiency can be obtained (in which Q_s, S_s, Ab_s, n_s and Q_r, S_r, Ab_r, n_r are fluorescence quantum efficiency, the total fluorescence intensity, the absorption coefficient and refractive index of the sample and standard samples, respectively), as listed in Table 6.3. As the viscosity of the solvent increases, the

Tab. 6.3: Fluorescent quantum efficiency of PQT and PQF in THF, DCE, and aniline.

Solvent	Dipole (10^{-30}C · m)	Viscosity (mPa · S)	*QE* (%)	
			PQT	PQF
THF	5.70	0.550(20°C)	10.9	9.7
DCE	6.20	0.840(20°C)	9.6	5.9
Aniline	5.04	4.43(20°C)	1.5	1.0

fluorescence quantum efficiency of the sample decreases: fluorescence quantum efficiency of PQT and PQF decreased from 10.9% and 9.7% to 1.5% and 1.0% respectively.

The emission spectra in solvents of different polarities indicates that in the excited state intramolecular charge transfer will occur. Based on the understanding of polymer molecular structure, the viscosity of the polymer solvent will affect the configuration torsion, which is linked with the formation of the ICT state [107]. The decreasing fluorescence quantum efficiency with increasing viscosity of the solvent indicates that the presence of nonradiative relaxation channels related to spatial configuration after PQT and PQF are excited. At the same time, the intramolecular charge transfer is accompanied by torsion of the C-C single bond, forming a TICT state. The torsion consumes exciton energy, making the fluorescence quantum efficiency decrease.

To further illustrate the relaxation of the torsion process, we use streak camera fluorescence relaxation kinetics to compare the dynamics of the samples in different viscosity solvents. For PQ, its fluorescence relaxation can be fitted with a single exponential decay, obtaining the fluorescence lifetime of 1.7 ns, as shown in Figure 6.23. This is because the polymer of PQ is not a DA structure, without thiophene, furan, and other groups. PQ molecules in the excited state transit to the ground state via direct radiative relaxation without intramolecular charge transfer occurring.

But for PQT and PQF, because of the formation of a DA structure through the addition of thienyl and furyl groups, intramolecular charge transfer occurs after excitation. Therefore, multiexponential decay is needed to fit the fluorescence relaxation process. Figure 6.24 shows fluorescence relaxation of PQT and PQF in solvents of different viscosity: THF, DCE, and aniline. (a) and (b) represent the experimental results of 2.2 ns time window and fitting curves, (c) and (d) represent 160 ps time window and fitting results. The excitation wavelength is 415 nm, and the fitting wavelength is the fluorescence emission peaks of each sample. The solid lines represent the fitted curves obtained by the instrument response function IRF and deconvolution software.

From Table 6.4, it can be seen that PQT in three solvents contains two fluorescence relaxation components: a rapid relaxation of about 10 ps and a slow relaxation around 2 ns. The 2 ns relaxation process can generally be considered to be PQT exciton fluorescence lifetime. The 10 ps fast relaxation process gradually slows down with increasing solvent viscosity, from THF at 6.7 ps, to dichloroethane at 14.3 ps, and aniline at 17.7 ps. In PQT samples, the thiazide group has a strong electron-donating property,

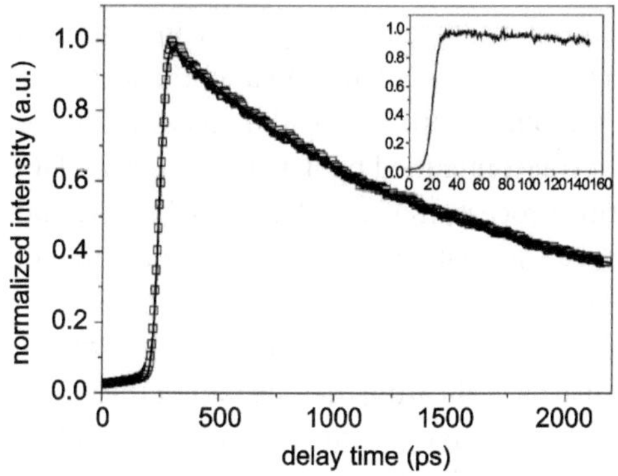

Fig. 6.23: Single exponential decay of PQ in THF.

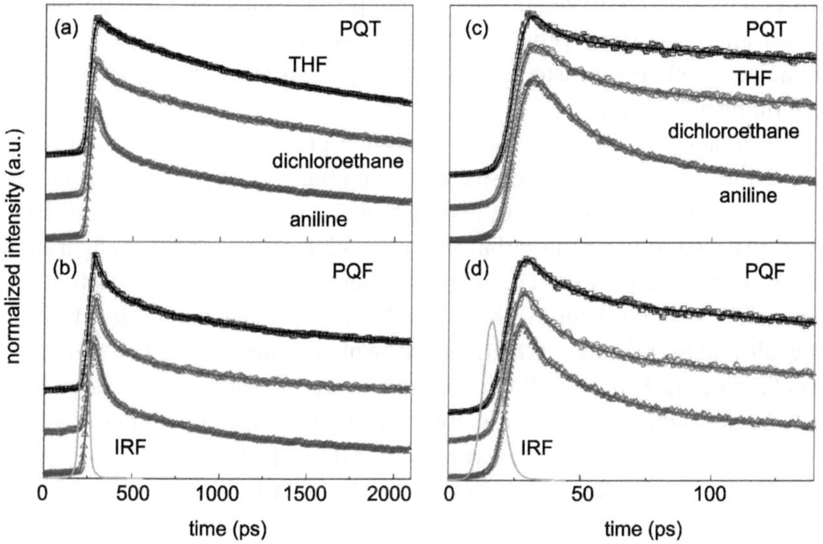

Fig. 6.24: Fluorescent peak decay of PQT and PQF in THF, DCE, and aniline.

Tab. 6.4: The fitting result of PQ, PQT, and PQF in solvents.

	THF	Dichloroethane	Aniline
PQ	1.7 ns	–	–
PQT	6.7 ps (38%)	14.3 ps (40%)	17.7 ps (55%)
	1.9 ns (62%)	2.0 ns (60%)	1.9 ns (45%)
PQF	2.0 ps (37%)	1.3 ps (72%)	1.2 ps (79%)
	12.4 ps (23%)	15.6 ps (13%)	22.8 ps (10%)
	4.2 ns (40%)	4.5 ns (15%)	2.2 ns (11%)

so electrons are prone to being transferred from the thiazide group to the intramolecular electron thiophene group, resulting in the formation of an ICT state. However, due to the single bond between thiazide and thiophene, intramolecular electron transfer processes in the molecular structure will be accompanied by twisting to form a TICT state. Due to the PQT fluorescence relaxation process, the fast relaxation time of about 10 ps slows down with increasing solvent viscosity, which infers configuration torsion along with the intramolecular electron transfer, meaning that the increasing solvent viscosity will slow down this process [107, 108]. In addition, the amplitude of fast relaxation increases as solvent viscosity increases, indicating that the higher the viscosity of the solvent, the more excitation energy consumed by the nonradiative channel of configuration torsion, which coincides with the decreasing fluorescence quantum efficiency. Studies have shown that [98, 100–102] thiophene polymers or molecules in the ground state configuration are generally nonplanar, while the excited state is planar with high torsional freedom. From the nonplanar configuration to the planer structure, the energy barrier is generally low (< 2 kcal/mol). This is also confirmed in PQT intramolecular charge transfer accompanied by the occurrence of configuration twisting, forming a TICT state.

In PQF samples this leads to a rapid relaxation process of configuration twisting, and with increasing solvent viscosity, relaxation time of this process is from THF at 12.4 ps, to dichloroethane at 15.6 ps, and aniline at 22.8 ps. However, studies have shown that, [101, 103, 109] for the carbazole-furan polymer, the spatial configuration of ground and excited states are flat or nearly flat. Here, the configuration twisting will not occur, which is contradictory with the results. This is likely due to the vinyl between thiazide and furan groups, which can cause photoisomerization between trans and cis configurations [110]. In addition, there exists a 1–2 ps ultrafast process of PQT, which may be derived from the nonradiative transitions induced by other solvents, and the ultrafast nonradiative processes and configuration twisting together consume exciton energy, making the fluorescence quantum efficiency PQF in the three different solvents lower than that of PQT.

6.3 Photophysics of polymer complexes

6.3.1 Background and development of organic photovoltaic devices

With advances in science and technology and the change in our way of life, we formed a severe reliance on energy. However, the oil depletion has always been a Sword of Damocles suspended over the human race. Traditional energy resources will dry up in the coming decades. Obtaining new energy resources is the world's premier issue for economic development. Solar energy is the key to solving this problem. Radiation energy provided by the sunlight in the atmosphere is $1.37\ \mathrm{kW/m^2}$. In 2008, humanity's total energy consumption was 143 PWH, less than one ten thousandth of the total solar radiation energy the earth received. Almost all the organic energetic materials on earth derive from the sun's energy, from coal, petroleum and natural gas, to the photosynthesis of green plants. Green plants' photosynthesis is one of nature's most efficient and most widely used energy harvesting and storage mechanisms. People have learned from this photosynthesis process, generating a new field of green energy: solar energy photoelectric conversion.

In 1839, French scientist Edmond Becquerel discovered the photovoltaic effect: he placed a platinum electrode in a solution of silver chloride, which displayed an electromotive force between the electrodes under light. In 1930, Schottky proposed the theory of the photovoltaic effect of Cu_2O. In same year, it was proposed for the first time that the photovoltaic effect could be used for solar cells, which directly convert solar energy into electricity. The world's first piece of inorganic solar cells was born in 1954. G. L. Pearson, C. S. Fuller, and D. M. Chapin in Bell Labs created the first solar battery with a silicon semiconductor. The energy conversion efficiency was 6% [111, 112]. Since then, inorganic solar cells were also developed, such as single crystal and polycrystalline devices like cadmium telluride and cadmium sulfide [113–115]. The energy conversion efficiency greatly progressed, being more than 25% under laboratory conditions [116].

The first piece of organic photovoltaic cells was born in 1958. Kearns and Calvin used magnesium phthalocyanine dyes (MgPc) sandwiched between two electrodes with different work functions to prepare the first organic photovoltaic cells. Electrons were extracted by electrode with low work function, and holes were filled by electrons from electrode high work function, which formed a photocurrent under illumination. These organic photovoltaic cells depend on the Schottky barrier from the contact of organic semiconductor films and two electrodes with different work functions, so they are called single Schottky type organic photovoltaic cells, whose energy conversion efficiency is very low. Since then, organic photovoltaic cells had made no breakthrough progress for a long period of time. In 1986 however, a landmark breakthrough was achieved by Dr Tang from Kodak. He proposed a double membrane heterojunction structure. This structure was similar to the p-n junction: the photoinduced excitons broke at the heterojunction interface and formed carriers. He used a derivative of four carboxyl perylene (PV, n-type) and copper phthalocyanine (CuPc, p-type) for a dual layer membrane, making an organic solar cell with energy conversion efficiency of 1% [117]. In 1992, Sariciftci and coworkers found that excitons in organic polymer and C_{60} interfaces could separate very efficiently. Fast electrons injected from the polymer into the C_{60} molecule with high efficiency, while the reverse electron transfer was of low speed. In 1993, they used MEH-PPV (polystyrene support) and C_{60} to make a dual layer heterojunction polymer photovoltaic cell, with MEH-PPV as electron donor and C_{60} as electron acceptor. The energy conversion efficiency was 0.04% [118]. Since then, C_{60} and its derivatives were widely used as electron acceptors in polymer photovoltaic cells [19, 20, 119–123]. However, in dual layer heterojunction devices, excitons can separate only at the interface of donor and acceptor. Since the exciton diffusion distance was only ~10 nm, the excitons away from the heterojunction interface recombined and were lost before migrating to this interface [124]. This limits the thickness of the organic layers to be less than 20 nm, which is insufficient for absorbing solar radiation. Therefore, it restricts the energy conversion efficiency of photovoltaic cell devices. To overcome this limitation, people developed the concept of bulk heterojunction: mixing films of donor and acceptor materials by evaporation or spin coating. Nanoscale donor-acceptor interfaces were formed within the microstructure of the organic layer. This greatly increased the heterojunction interface. This meant the photoinduced excitons could separate at nearby interfaces over a short distance to efficiently generate carriers. This design broke the device thickness limitation caused by the exciton diffusion distance. The device performance was greatly improved. Based on this concept of bulk heterojunction, polymer photovoltaic cells were greatly improved.

Now the structures of polymer photovoltaic cells are mostly bulk heterojunction structures, as shown in Figure 6.25. Above the anode ITO glass base there is a hole transporting layer (PEDOT-PSS film as commonly used) and an intermediate organic layer, which is an electron donor and electron acceptor hybrid film. On the mesoscopic scale these form a blending interpenetrating network structure. A metal electrode is evaporated above the film.

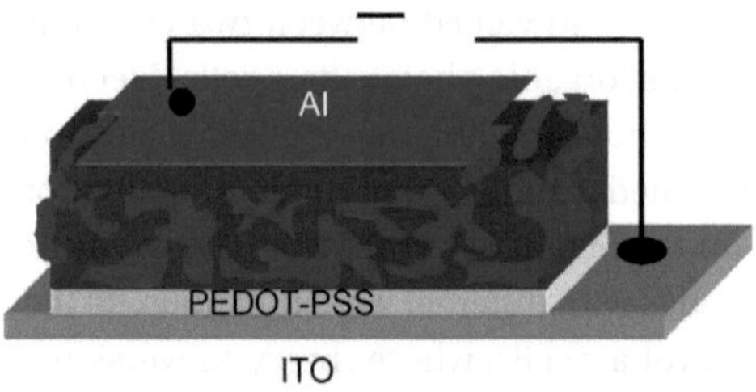

Fig. 6.25: Diagram of bulk heterojunction polymer photovoltaic cell device structure.

In the polymer photovoltaic cells, the photoelectronic conversion process can roughly be divided into four steps, as shown in Figure 6.26: [125, 126]

(1) After absorbing photons at certain energy, excitons are generated in the organic material;
(2) The excitons diffuse to the D-A interface;
(3) On the D-A interface, electrons transfer from the lowest unoccupied molecular orbit, (LUMO) of the donor to the lowest occupied molecular orbit (LOMO) of the acceptor, producing bound carrier pairs from excitons;
(4) Bound carriers further separate into free carriers, and electrons and holes are collected in the cathode and anode through the donor-acceptor interpenetrating network, generating photocurrent in an external circuit.

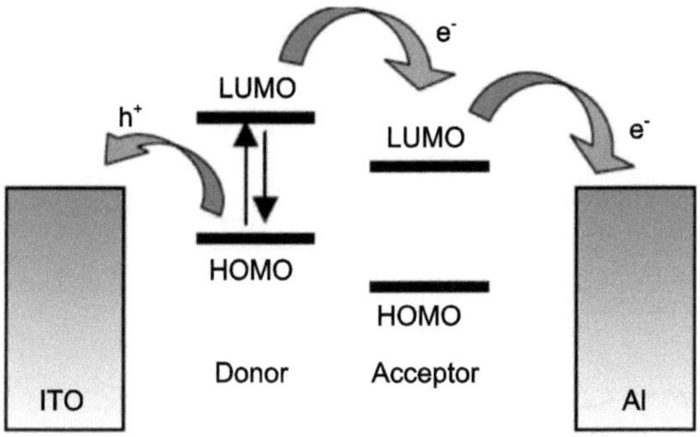

Fig. 6.26: The working principle of polymer photovoltaic cells.

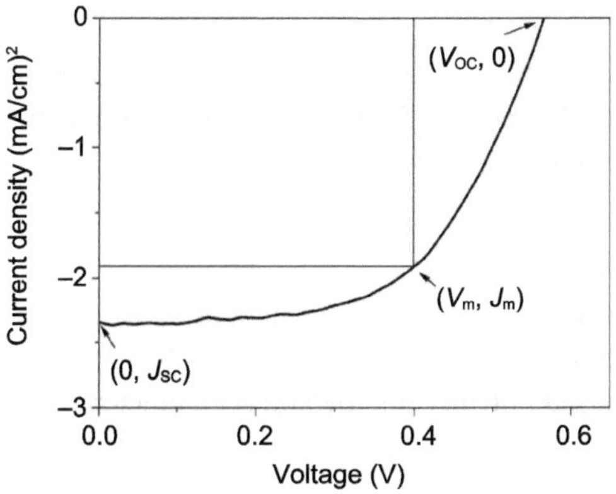

Fig. 6.27: Typical I-V curve for photovoltaic cells.

The performance of polymer photovoltaic cells devices is shown by open-circuit photovoltage (V_{OC}), short-circuit photocurrent (J_{SC}), fill factor (FF) and power conversion efficiency (PCE) as shown in Figure 6.27.

V_m and J_m are voltage and current corresponding to the maximum output power, and the filling factor FF refers to the ratio of the maximum output power photovoltaic battery and the product of the open circuit voltage and short circuit current:

$$\mathrm{FF} = \frac{V_m \cdot J_m}{V_{OC} \cdot J_{SC}} . \tag{6.18}$$

Photovoltaic energy conversion efficiency of η refers to the ratio of maximum output power and the input optical power:

$$\eta = \frac{P_{out}}{P_{in}} = \mathrm{FF}\frac{J_{SC} \cdot V_{OC}}{P_{in}} . \tag{6.19}$$

In the solar spectrum, most of the solar radiation energy is over 400 nm, as shown in Figure 6.28. For donor materials of polymer photovoltaic cells, in order to be able to

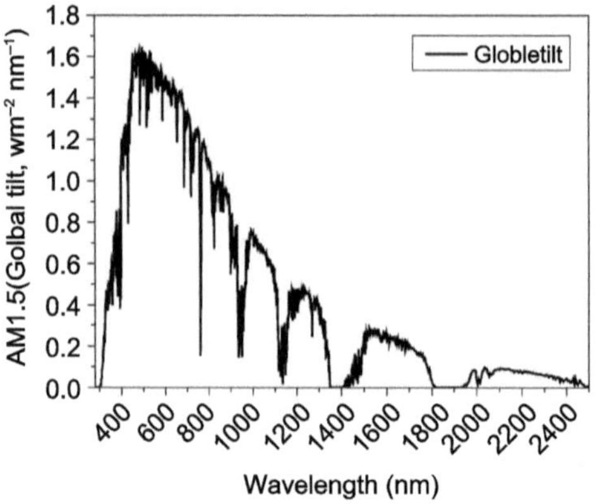

Fig. 6.28: Solar radiation spectrum [127].

absorb more solar radiation, finding and synthesizing low bandgap polymers is the basis for improving battery components. At present, there are two kinds of polymer used as donors in polymer photovoltaic cells.

(1) Poly-phenylenevinylene (PPV) and its derivatives

PPV and its derivatives, as shown in Figure 6.29, have good membranous and diverse synthetic methods [128]. They are widely used in the field of polymer photovoltaic cells [129, 130]. PPV-derivative based polymer photovoltaic cell efficiency can be ~2.5% [131].

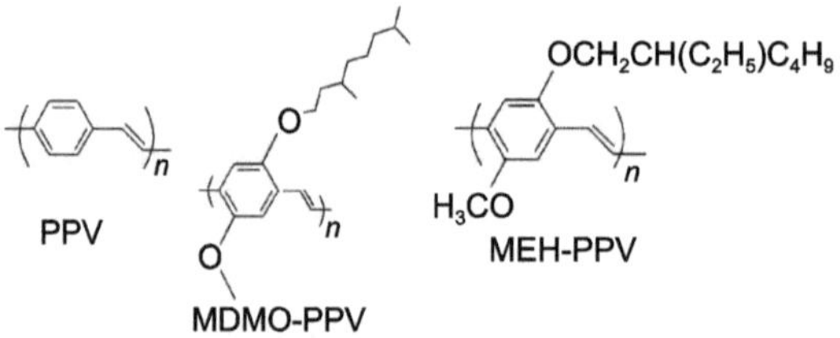

Fig. 6.29: PPV and its derivatives.

(2) Polythiophene and its derivatives

Polythiophene has higher carrier mobility than other polymers. For example, the carrier mobility of P3HT can be $3.1 \times 10^{-3}\,m^2V^{-1}s^{-1}$ [132]. Through chemical modification, polythiophene can reduce forbidden bands and its absorption spectrum can reach the near-infrared range. It has become a research hotspot in the field of polymer photovoltaic cells. Figure 6.30 shows some derivatives of polythiophene used in polymer photovoltaic cells [133]. Among them, the most widely used derivative, which has best performance, is poly P3HT, of which the energy conversion efficiency reaches 5% [134]. The research team led by Heeger reported a photovoltaic cell, made of thiophene copolymer and PC70BM, whose energy conversion efficiency was 6.1% [135]. An-

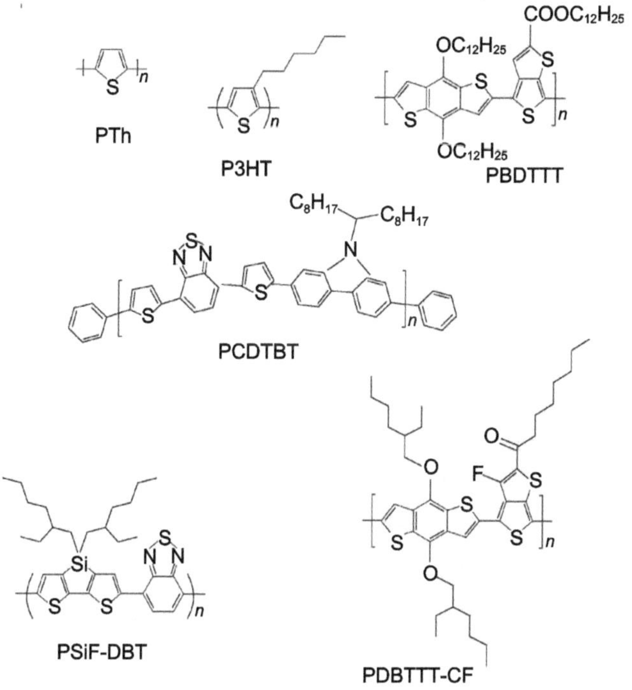

Fig. 6.30: Polythiophene and its derivatives [134–136, 138].

other polythiophene derivative with an improved electron acceptor broke the record with a conversion efficiency of 7.7% [136]. In addition, Li synthesized another kind of fullerene derivative, ICBA. Together with P3HT, the energy conversion efficiency reached 6.5% [137].

Polymer photovoltaic cells usually use fullerene (C_{60}) and its derivatives as electron acceptors. C_{60} is a kind of carbon whose atoms are in clusters, with 32 surfaces composed of 60 carbon atoms; it is also known as the football alkene. There are 60 π electrons at the surface of C_{60}, forming an electronic conjugate system with very strong reducibility and high electron affinity (2.6 eV to 2.8 eV). It can absorb up to six electrons. It has no strong absorption in the visible area, so it is an ideal electron transport material [128]. However, the solubility of C_{60} molecules is poor, easily aggregated, and not ready for films. Through chemical modification, we can get various fullerene derivatives. Among them, the most widely used is (6, 6)-Phenyl C_{61} butyric acid methyl ester (PCBM). In addition, through chemical modification we can also get other forms of derivatives, as shown in Figure 6.31 [19, 139–141].

Research on the dynamics and photochemistry of polymer photovoltaic materials plays an important role in the study of polymer photovoltaic cells, and it is the foundation of the development of polymer photovoltaic cells.

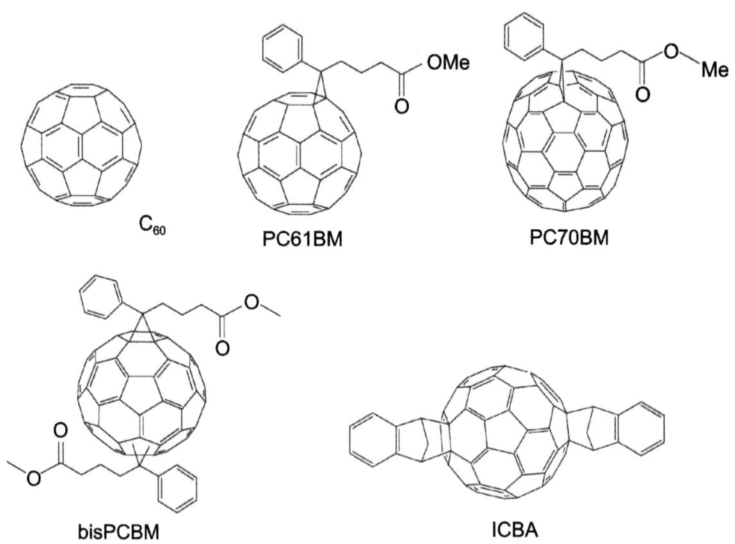

Fig. 6.31: Fullerene and its derivatives [19, 139–141].

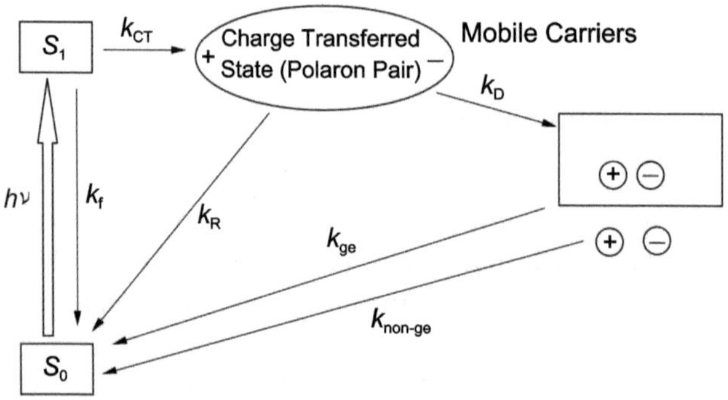

Fig. 6.32: Diagram of a polymer photovoltaic cell's physical process after absorbing photons.

In each step of the dynamics in polymer photovoltaic cells, the most important part is charge transfer between the acceptor-donor heterogeneous interface and bound charge generation, then the formation of free charges. Figure 6.32 illustrates the physical process of a polymer photovoltaic cell after absorbing a photon. k_f is the rate of excited states directly transiting back to the ground state, k_{CT} is the charge transfer rate, k_D is the rate of bound charge separating into free carrier, k_R is compound rate of polaron recombination, k_{ge} the free charge's geminate recombination, and k_{non-ge} is the nongeminate charge pair recombination [142, 143].

Electrons in the excited state of donors can transfer to the electron acceptor, or by other ways (including radiative and nonradiative tranfer), back to the ground state.

The charge transfer efficiency is [144]

$$\eta_{CT} = \frac{k_{CT}}{K_{CT} + k_f} . \tag{6.20}$$

For bulk heterojunction materials such as P3HT:PCBM, charge transfer is far greater than mechanisms, so the charge transfer efficiency is close to 1. According to the Marcus–Hush theory, the charge transfer rate is

$$k_{CT} = \frac{2\pi}{\hbar} E_B^2 \frac{1}{\sqrt{4\pi\lambda k_B T}} \exp\left(\frac{-(\lambda + \Delta G^0)^2}{4\lambda k_B T}\right) , \tag{6.21}$$

where E_B is Coulomb binding energy, λ is charge reorganization energy, ΔG_0 is the change of Gibbs free energy in the charge transfer process, k_B is the Boltzmann constant, and T is the absolute temperature. According to the Braun model, the probability of a polaron in the bound state separating into free charge depends on the electric field intensity E and absolute temperature T [145, 146]

$$P(T, E) = \frac{k_D(E)}{k_D(E) + k_R} , \tag{6.22}$$

where $k_D(E)$ is the rate of the bound charge separating into free carrier rate, and k_R is the recombination rate of bound charge rate having nothing to do with the electric field. On the basis of the Onsager theory [147], Braun obtained the relationship of charge separation rate $k_D(E)$ and electric field intensity E, carrier mobility $\langle\mu\rangle$, dielectric constant $\langle\varepsilon\rangle$, electron hole distance r and the absolute temperature T [146]

$$k_D(E) = v \exp\left(\frac{-E_B}{k_B T}\right)\left(1 + b + \frac{b^2}{3} + \frac{b^3}{18} + \frac{b^4}{180} + \ldots\right) , \tag{6.23}$$

in which,

$$v = \frac{3e\langle\mu\rangle}{4\pi\varepsilon_0\langle\varepsilon\rangle r^3} \tag{6.24}$$

$$b = \frac{e^3 E}{8\pi\varepsilon_0\langle\varepsilon\rangle k_B^2 T^2} \tag{6.25}$$

$$E_B = \frac{e^2}{4\pi\varepsilon_0\langle\varepsilon\rangle r} , \tag{6.26}$$

where ε_0 is permittivity of a vacuum. In bulk heterojunction polymer photovoltaic battery materials, the distance of the acceptor and the donor (the distance of the electron and hole: x) is not a constant, but has a certain distribution. Equation (6.22) should be written in the integral form of the electron hole distance x [148, 149]

$$P(T, E) = N_F \int_0^\infty P(x, T, E) F(x) dx , \tag{6.27}$$

in which

$$F(x) = x^2 \exp\left(-\frac{x^2}{a_0^2}\right) \tag{6.28}$$

$$N_F = \frac{4}{\pi^{\frac{1}{2}} a_0^3} \,. \tag{6.29}$$

So, in polymer photovoltaic cells, the free carrier generating rate is

$$\eta(\lambda) = \eta_A(\lambda)\eta_{\mathrm{CT}}P(T, E) \,, \tag{6.30}$$

in which $\eta_A(\lambda)$ is the polymer photovoltaic cell's absorptivity of different wavelengths.

In polymer photovoltaic cells, photoinduced current J_{ph} is the difference of the photocurrent J_{L} with light and the dark current J_{D} without light

$$J_{\mathrm{ph}} = J_{\mathrm{L}} - J_{\mathrm{D}} \,; \tag{6.31}$$

and without diffusion current [150],

$$J_{\mathrm{ph}} = eGL \,, \tag{6.32}$$

in which, e is the electron charge, G is the rate of carrier generation, and L is the thickness of the active layer. The built-in electron field in the device is

$$E = \frac{V_0 - V}{L} \,, \tag{6.33}$$

where V_0 is the voltage when the photoinduced current $J_{\mathrm{ph}} = 0$.

Considering the diffusion current, equation (6.32) gives [151]

$$J_{\mathrm{ph}} = eGL\left[\frac{\exp(eV/k_{\mathrm{B}}T) + 1}{\exp(eV/k_{\mathrm{B}}T) - 1} - \frac{2k_{\mathrm{B}}T}{eV}\right] . \tag{6.34}$$

The production rate of carrier G is also the function of the temperature and the internal electric field,

$$G(T, E) = G_{\max}P(T, E) \,, \tag{6.35}$$

where $G_{\max}$ is the production rate when all the polarons participate in the charge separation.

Combining equations (6.34), (6.35), and (6.27), we can get the equation of photoinduced current in the polymer photovoltaic cells

$$J_{\mathrm{ph}} = eG_{\max}\left[N_F \int_0^{\infty} P(x, T, E)dx\right] L\left[\frac{\exp(eV/k_{\mathrm{B}}T) + 1}{\exp(eV/k_{\mathrm{B}}T) - 1} - \frac{2k_{\mathrm{B}}T}{eV}\right] . \tag{6.36}$$

According to these theories of photocurrent, we can calculate the photoinduced current in polymer photovoltaic cells, which fits the experimental data well, as shown in Figure 6.33.

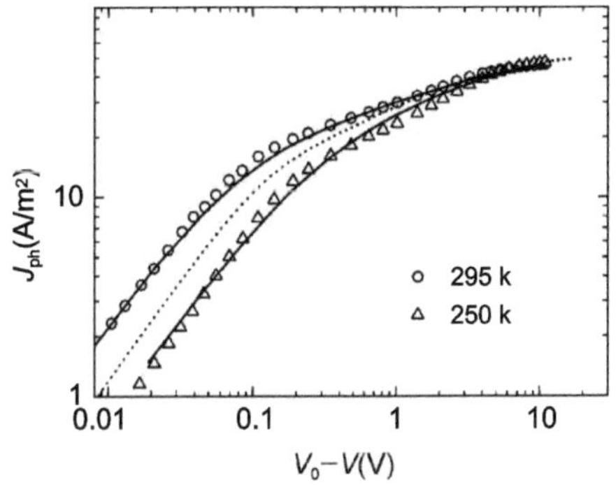

Fig. 6.33: The photoinduced current in the OC_1C_{10}-PPV:PCBM system [149].

6.3.2 Applications of ultrafast spectroscopy in polymer complexes

6.3.2.1 Photophysics of organic photovoltaic devices

Since the 1990s, the performance of polymer photovoltaic cells has been continuously surpassed. At the same time, research on the dynamics of polymer photovoltaic materials became deeper. In 1992, the discovery of electron transfer from conjugated polymers MEH-PPV to fullerene C_{60} opened a new era in the study of dynamics of polymer photovoltaic cells [152].

In 2001, Brabec et al. studied the MDMO-PPV to PCBM electron transfer rate by ultrafast spectroscopy, which was found to be about 45 fs, as shown in Figure 6.34. Subsequently, much research confirmed that the electron transfer rate of conjugated polymers to PCBM derivatives were within 100 fs. [24, 154–157] However, in the study of Hwang et al., it was found that the photoinduced electron transfer from oligo (thienylenevinylene) (OTV) to PCBM is 14 ps [158].

In addition, inside the polymer photovoltaic cells, carriers generated by electron transfer and the corresponding charged states have different relaxation lifetimes. They have relatively long electron diffusion and energy transfer times, which are generally

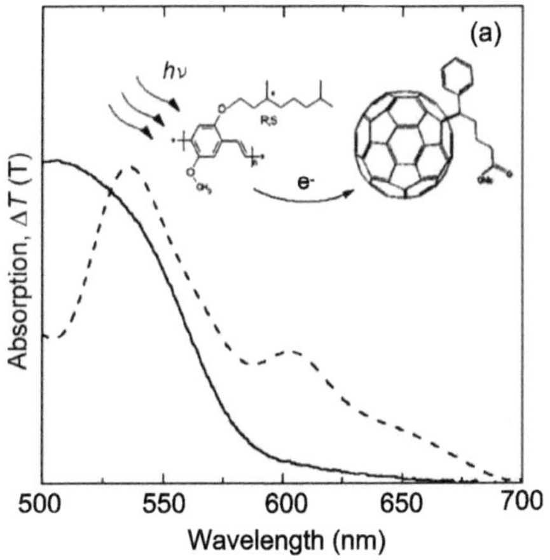

Fig. 6.34: Ultrafast electron transfer from MDMO-PPV to PCBM [153].

on ns-μs timescales, and even up to ms [74, 159]. For example, in polyfluorene and fullerene heterojunction materials, recombination of geminated carriers is in the 30 ns time range; free carrier transportation and recombination are on the ns and even μs timescale. Thus, in polymer photovoltaic cell materials, the photochemical and photophysical processes span a large timescale, from 10^{-15} to 10^{-3} s, across 12 orders of magnitude.

Spatially, in order to provide abundant donor acceptor interface and a richly connected blending interpenetrating network in bulk heterojunction polymer photovoltaic cells, two kinds of materials need to possess proper phase separation. If separation is too small, interpenetrating networks may short circuit; if the separation scale is too large, a lot of broken circuits will form, so that photoinduced exciton won't separate effectively. Figure 6.35 shows the typical topography and phase separating images of bulk heterojunction photovoltaic materials by atomic force microscopy (AFM), revealing the different phase separation scales of acceptor and donor. The choice of solvent, sample preparation environment, annealing temperature, annealing time, and other factors relate to this phase separation. From the photoexcitation

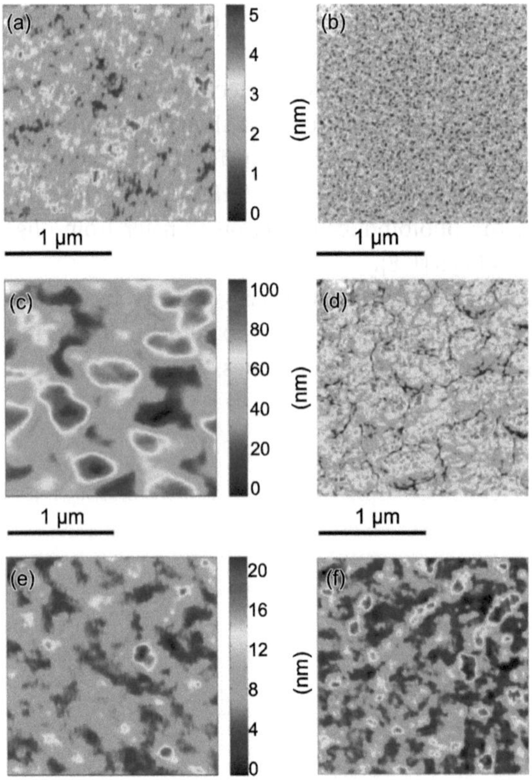

Fig. 6.35: The typical topography and phase separating images of bulk heterojunction materials [160].

and electron transfer to the free carrier collecting, the spatial scale for study is from the molecular to the nanometer scale; also across several orders of magnitude.

The cross-link of polymer photovoltaic cell dynamics in the time domain and space domain makes the research very complicated. Time resolved spectroscopy provides a powerful tool in this field. Using time resolved spectroscopy we can detect all kinds of photophysical processes. Ultrafast femtosecond lasers can access time resolutions down to ~10 fs, [153] which provides direct observation of the dynamics of polymer photovoltaic cells with reliable experimental basis.

6.3.2.2 Ultrafast photophysics of polymer-fullerene composites

The Sundstrom group studied the dynamics from ultrafast (30 fs) to microsecond (50 μs) timescales. They used femtosecond transient absorption technology and a nanoflash photolysis method (nanosecond flash photolysis) to study short and long dynamic process on the APFO3/PCBM system (APFO3, poly real [2,7-(9,9-dioctylfluorene)-Alt-5,5-(4',7'-di-2-thienyl-2',1',3-benzothiadiazole)]), with ratio of 1:1, 1:3 and 1:4 (weight ratio). When excited at 580 nm in the presence of PCBM, fluorescence quenching obviously occurred (in the ratio of 1:1, the fluorescence quantum efficiency was 0.4%) [161, 162], suggesting that PCBM directly affects the excited state of the polymer, that is, the quenching occurs before the energy transfer in the polymer. This quenching process can be confirmed by ultrafast fluorescence dynamics analysis: fluorescence quenching in the ultrafast form, which is an extremely short excited lifetime 200 fs. Without PCBM, this lifetime is ~50 ps.

The three transient absorption feature bands appear in ultrafast scales at 570 nm for ground state bleaching, 700 nm for stimulated radiation, and 900 nm for APFO3's excited state together with the carrier's contribution. In pure APFO3 film, the last one is the pure excited state absorption band, as shown in Figure 6.36 and 6.37 [163].

The complete dynamic measurement from the initial excitation to fully disappeared charge needs observation from femtoseconds to microseconds.

By observation of the complete timescale, the initial ~ 200 fs shows fast relaxation, followed by a 30 ps rise and 30 ns decay process. Under low pump light intensity ($< 6.6 \times 10^{13}$ ph/cm^2/pulse) the recombination process does not rely on the light intensity; with the increase of light intensity ($> 1 \times 10^{14}$ ph/cm^2/pulse), the recombina-

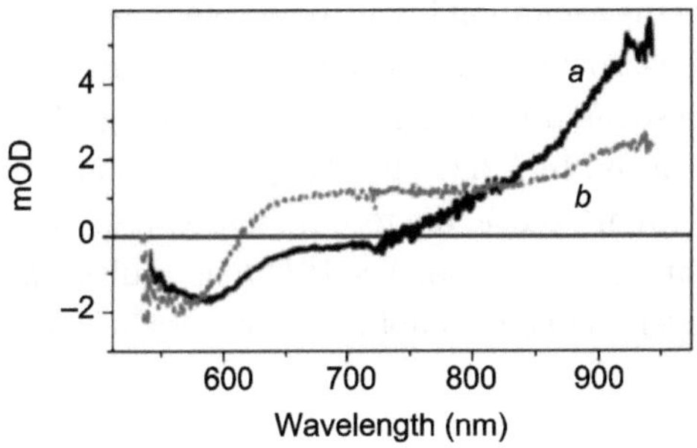

Fig. 6.36: (a) Pure APFO3's transient absorption spectra with 10 ps delay, pumped at 580 nm, light intensity 1.2×10^{14} photon/cm^2/pulse. (b) APFO3/PCBM mixture of 1:4, transient absorption under same conditions [163].

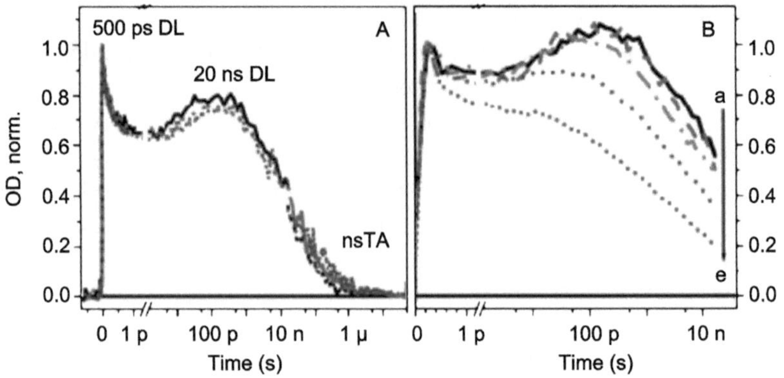

Fig. 6.37: APFO3/PCBM's transient absorption curve in the complete timescale; the irradiation intensity is (a) 2.3×10^{13}, (b) 6.6×10^{13}, (c) 1.1×10^{14}, (d) 2.3×10^{14}, and (e) 8.0×10^{14} (ph/cm^2 per pulse) [163].

tion process strongly depends on the pump intensity, the decay rate increasing quickly towards light intensity. Internal energy transfer processes of the polymer can be determined by the intensity dependent exciton-exciton annihilation process, with the average exciton hopping time $\tau_h \approx 6$ ps.

These timescales can be generally attributed to the ultrafast time process, with 200 fs as the ultrafast electron transfer process of the polymer from the excited state to PCBM, of which the efficiency is ~100%. In some systems, the charge transfer process from the polymer to fullerenes can reach 45 fs [153]. The result of electron transfer is that a hole is produced in the polymer, which is attracted by the electron in PCBM, forming a hole-electron pair, that is, the bound charge pair. Within ~1 ps this process is complete, indicating that the photon absorbed converts to carriers efficiently. Therefore, the charges are the only product of this process. This is shown as the absorption of 650–950 nm broadband in the transient spectrum. Meanwhile, because this timescale is obviously lower than the exciton's jump time in the polymer, 6 ps, this is a process that does not contain exciton diffusion. It also suggests full blending of PCBM and polymer.

Slow processes at the nanosecond scale represent charge recombination. These processes may be from geminated or nongeminated recombination. Geminated recombination is a local process, independent of the excitation intensity; nongeminated recombination is the process from free charges, so it is related to charge density and the excitation intensity. However, the two timescales are similar. Under high charge density, the nongeminated recombination is faster. In APFO3/PCBM systems, when the excitation intensity is less than 6×10^{13} ph/cm^2/pulse, the recombination process is time independent. When the light intensity is larger than 1×10^{14} /cm^2/pulse, the dependence starts to emerge, and the nongeminated recombination process becomes the main process.

Between ultrafast photocharge generation and charge recombination processes at nanosecond timescales, a hundred picosecond scale rising signal can be observed. This is a process from bound charge to free charges, through which free charge density increases, and competes with free charge recombination and Coulomb bound charge recombination. However, in this system, the recombination process of Coulomb bound charge is not significant.

When studying polymer/PCBM systems by ultrafast spectroscopy, we need to fit different dynamic parts, to distinguish the contribution of each component in the signal. For the initial charge transfer process from the polymer to PCBM, we can fit with a simple exponential; free charge generation also can be obtained by exponential fitting. The geminated charge recombination process cannot use single or double exponential or power law fitting. The Gaussian distribution fitting in the log k3 space (decay rate) can be applied. As a result, the rate constant k3 represents the average Gaussian distribution. Its distribution may be derived from the charge transfer active energy or transmission distance; in APFO3/PCBM systems it may include both contributions.

The pump intensity related recombination process includes two types: nongeminated charge recombination and collision of Coulomb charge pairs. The latter refers to a process that occurs when the Coulomb charge pair density is too large and charge pairs on the Onsager ball overlap in volume, interacting and leading to a charge recombination for one pair, while the neighboring pair still exists. These two processes are pump intensity dependent second order process. However, the decay rate is not time independent, but is a constant that can be defined as a time dependent rate $\gamma(t) = \gamma_0/t^{\alpha}$. Because the second order process usually depends on the motility, the decay rate represents the charge's motion in the system. The portion of geminate charge recombination and nongeminate recombination is shown in Figure 6.38.

The recombination of carriers can be studied in time domain [164–174] and frequency domain methods [175]. Time domain research methods include nanosecond-millisecond transient absorption studies. Scriciftei found PPV/fullerene's signal consists of two parts: an ultrafast process (< 20 ns) dependent on the pump intensity,

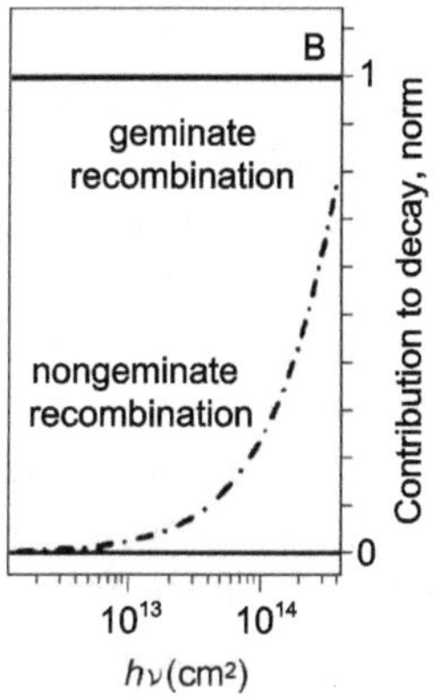

Fig. 6.38: The relative proportion of geminated charge recombination and nongeminated charge recombination in a 1:1 APFO3/PCBM system, as well as the relationship with the incident light intensity [163].

attributed to the motivated polaron recombination; and a slow process (100 ns–10 ms) as a complex process of confined polarons [166, 167]. The slow process indicates that the timescale of charge transport can compete and give certain quantum efficiency.

In the study above, confined polaron recombination is the main recombination process, while Janssen thought the ultrafast process was the main part of the charge recombination [170]. In addition, whether the ultrafast process is a pump strength dependent nongeminated charge recombination, or a pump independent geminate charge recombination process is debated. In a P3HT/PCBM system, Mesker revealed that 80% of the charge recombines within 30 ns after excitation is geminated, as studied by infrared time resolved spectrum, while the rest of the charges persist up to the millisecond timescale. The microwave conductivity test indicated that the charge recombination depends on the concentration of the PCBM, which can be geminate or nongeminate. Therefore, the charge recombination is a problem that is difficult clear, which may be due to the sample and difference in excitation conditions [172].

6.3.2.3 Absorption broadening of multicomponent polymers (PQ/MEH PPV)

Polymer photovoltaic cells with bulk heterojunction structures have a disadvantage: the conjugated polymers have limited spectral absorption towards solar radiation, while the absorption of fullerene derivatives, PCBM, is very weak. For example, for the most commonly used polymer material, P3HT, the bandwidth is 1.9 eV, which can absorb up to 22.4% of solar radiation photons [176], which means most of the radiation from the sun cannot be used. As a result, a variety of photovoltaic cell structures have been proposed to enhance absorption spectra of materials.

All-polymer solar cells are a kind of structure that is effective in improving the absorption. In these devices, some conjugated polymers take the position of electron acceptors. For example, the CN-ether-PPV [177], PCNEPV [178], DOCN-PPV [179], and F8TBT [180] are used as substituted materials for molecular acceptors like PCBM. These electron acceptor polymers play the role of PCBM. At the same time, they absorb solar radiation photons, which will provide significant improvement to the absorption of photovoltaic cells. PCNEPV and MDMO-PPV, for example, cause the absorption of the material to expand from 450–630 nm to 300–630 nm [181]. In all-polymer photovoltaic cells, the photon excites the polymer with wider forbidden band first, which then transfers the energy to another polymer and performs photoelectric conversion [130]. Recently, multicomponent polymer photovoltaic cells have been suggested to overcome the exciplex and triplet state in the polymer photovoltaic cells [182–184]. In this structure, the second electron donor material is mixed with the original one, increasing its ability to absorb photons. For example, a high and low degree of conjugated MEH-PPV are mixed, which shows extended absorption to 300–670 nm. The exciton transfer from low to high conjugated polymer is by resonance energy transfer [185]. In cells based on P3HT:PCBM bulk heterojunction cells, F4Th4 was mixed as another electron donor, which expanded the absorption to 300–700 nm. Energy

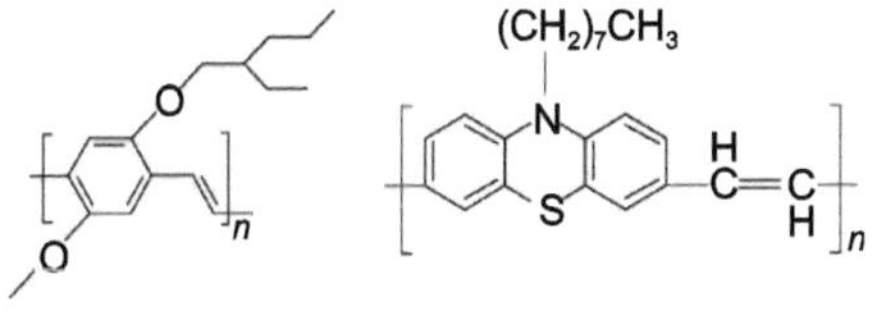

Fig. 6.39: The chemical structural formula of MEH-PPV and PQ.

transfer from F4Th4 to P3HT forms additional charge separation at the interface [186]. According to this, resonance energy transfer between the two donor polymers in multicomponent devices is an important dynamical mechanism. Multicomponent mixtures have also been used to improve the morphology of bulk heterojunction materials. Experiments confirmed that MDMO-PPV [187] and P3HT nanofibers [188] mixed in bulk heterojunction materials can improve the microstructure of the organic layer. In addition, the dye-polymer [189] and nanoparticle-polymer hybrid [190] methods can be used to improve the absorption efficiency and photoelectric conversion performance of the device.

Because polyphenylene vinylene (PPV) and its derivatives have good solubility, good film-forming, and other excellent photoelectric properties, they are widely applied in the field of polymer photovoltaic cells [129–131]. MEH-PPV is one PPV derivative. Since the electron acceptor PCBM has good energy level matching with MEH-PPV, this derivative is often used as an electron donor in the polymer photovoltaic cell. The chemical structure is shown in Figure 6.39 [191, 192]. Studies have shown that the mixed MEH-PPV and PCBM of 1:4 has the best photovoltaic properties, with energy conversion efficiency of 1.5–2% [131, 193]. However, because the MEH-PPV absorption spectrum is fairly narrow compared with the solar radiation spectrum, sunlight cannot be absorbed efficiently. This limits the possibility of improving MEH-PPV based photovoltaic cell device performance. A multicomponent polymer photovoltaic cell is suggested here: adding an electron donor polymer to the original bulk heterojunction polymer photovoltaic cell. This is a way to regulate the absorption of the organic layer, to expand its absorption bandwidth and match the solar spectrum. Solar radiation absorbed by the polymer needs to transfer to another heterojunction donor material (MEH-PPV) by means of resonance energy transfer. Finally, the excitons effectively separate into charge at the heterojunction interface, so as to improve the performance of polymer photovoltaic devices.

The two organic polymers chosen are MEH-PPV and PQ. Their absorption spectra neighbor to each other. The emission spectrum of the PQ overlaps well with the MEH-PPV absorption spectrum.

The dotted line in Figure 6.40 (a) is steady absorption spectrum of a MEH-PPV film and a PQ film. The MEH-PPV absorption peak is near 500 nm, while its absorption at 400 nm is very weak. The absorption peaks of PQ are located at ~400 nm. By mixing these two kinds of polymer materials, we expect to extend the absorption spectrum of the polymer materials. Mixing the PQ and MEH-PPV with 1:1 and 2:1 ratio, we get a hybrid film with steady state absorption as shown in the solid line in Figure 6.40 (a). The

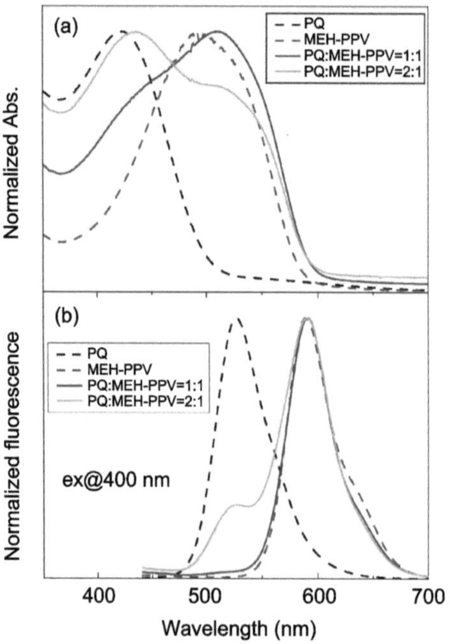

Fig. 6.40: (a) The steady state absorption spectrum; (b) the emission spectrum of MEH-PPV and PQ mixed films in different proportions.

absorption spectrum presents the sum of the two polymers, both in peak position and shape, indicating that there is no strong interaction in their ground state. The absorption spectrum has broadened effectively compared to the two polymers individually. The introduction of PQ compensates for the insufficient absorption of MEH-PPV at 400 nm. The shape of the absorption spectrum is tunable by adjusting the ratio of the two polymers, which allows it to better match the solar radiation spectrum.

Additionally, we show steady state fluorescence emission spectra of several thin film materials by 400 nm excitation, as shown in Figure 6.40 (b). It is shown that the PQ emission peak is 526 nm, with strong overlap to the MEH-PPV absorption. If the PQ emissive transition dipole moment and MEH-PPV absorption dipole moment are parallel, then there should be resonance energy transfer from PQ to MEH-PPV [182, 183, 194]. The overlapping of donor emission spectrum and acceptor absorption spectrum in FRET is shown as [195]

$$J(\lambda) = \frac{\int_0^\infty F_{\mathrm{D}}(\lambda)\varepsilon_A(\lambda)\lambda^4 d\lambda}{\int_0^\infty F_{\mathrm{D}}(\lambda) d\lambda}, \tag{6.37}$$

in which $F_{\mathrm{D}}(\lambda)$ is donor emission spectrum, $\varepsilon_A(\lambda)$ is absorption spectrum of acceptor, and transfer rate is determined by equations (6.12) and (6.13).

This is also confirmed by further measurements on steady state emission spectra. As the PQ-MEH PPV mixing ratio is 1:1, only the MEH-PPV fluorescence emission at 591 nm is detected. Because the excitation wavelength is 400 nm, the PQ will largely be excited, so when the PQ-MEH PPV mixing ratio is 1:1, almost all PQ absorption of the

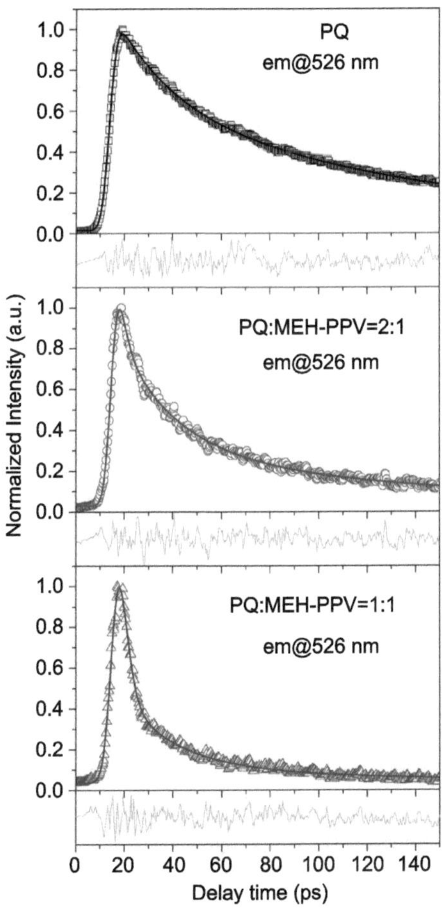

Fig. 6.41: Fluorescent relaxation dynamics of PQ blended with PQ/MEH-PPV at 526 nm, in different ratios.

photon energy transfer to MEH-PPV is by means of resonance energy transfer. When the ratio is 2:1, two peaks appear at 526 nm (PQ) and 591 nm (MEH-PPV). The amplitude of PQ is very low while in its original position. This shows that PQ does not form intermolecular aggregations with MEH-PPV, so resonance energy transfer can happen.

Thus, by mixing these two kinds of polymer, we extend the absorption bandwidth in the solar spectrum of MEH-PPV. By adjusting the ratio of these two kinds of polymer we can easily adjust the matching to the solar radiation spectrum. The photon energy absorbed by PQ may be transferred by resonance energy transfer to MEH-PPV. This is the foundation of multicomponent polymer photovoltaic cells.

Using a streak camera, the fluorescence relaxation kinetics of PQ-PQ/MEH-PPV mixed films in different proportions at 526 nm is shown in Figure 6.41. With increasing MEH-PPV composition proportion, 526 nm fluorescence relaxation speeds up compared with the PQ film fluorescence relaxation. The PQ fluorescence emission peak is 526 nm, so the addition of MEH-PPV makes PQ fluorescence quenching occur, which also confirms the assumption of resonance energy transfer.

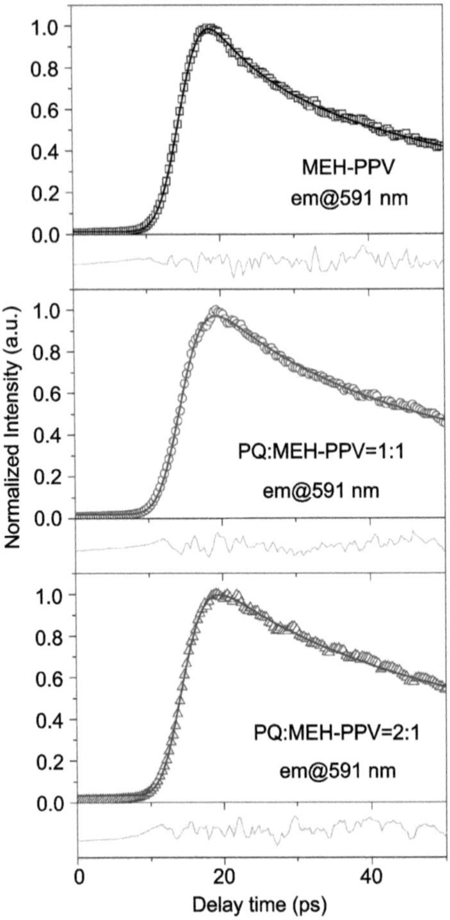

Fig. 6.42: The fluorescent relaxation dynamics of MEH-PPV blended with PQ/MEH-PPV films in different proportions at 591 nm.

Further deconvolution fitting of relaxation curves is shown in Figure 6.42, while the fitting results are summarized in Table 6.5. PQ film fluorescence relaxation processes can be fitted by a double exponential decay, whose time constants are 23.7 and 128.8 ps. These two decay rates can be regarded as the exciton relaxation process of PQ thin film samples. Fixing the longest time constant of 128.8 ps, then fitting the fluorescence relaxation of the two hybrid films, we get the third time constant: about 2 ps. In a PQ:MEH-PPV = 2:1 mixed film, the proportion of this ultrafast process is 57.8%; in a PQ:MEH-PPV = 1:1 mixed film, the proportion of this ultrafast process is 83.2%. That is, the addition of MEH-PPV, provides exciton relaxation in the PQ of the mixed film with one more ultrafast relaxation channel whose time constant is about 2 ps. The proportion of this ultrafast process is proportional to the content of MEH-PPV. This 2 ps relaxation channel is the resonance energy transfer process from PQ to MEH-PPV.

To finally verify this, it is necessary to analyze the MEH-PPV exciton relaxation process. Figure 6.42 shows the results of fluorescence relaxation of MEH-PPV mixed

Tab. 6.5: The convolution fitting results of fluorescent relaxation of MEH-PPV blended with PQ/MEH-PPV films in different proportions at 526 nm.

	τ_1 (ps)	a_1	τ_2 (ps)	a_2	τ_3 (ps)	a_3
PQ			23.7	41%	128.8	59%
PQ:MEH-PPV = 2:1	1.9	57.8%	23.8	29.3%	128.8	12.9%
PQ:MEH-PPV = 1:1	2.8	83.2%	29.8	15.4%	128.8	1.4%

Tab. 6.6: The convolution fitting results of fluorescent relaxation of MEH-PPV blended with PQ/MEH-PPV films in different proportions at 591 nm.

	τ_1 (ps)	a_1	τ_2 (ps)	a_2	τ_3 (ps)	a_3
MEH-PPV			14.0	51.6%	71.3	48.4%
PQ:MEH-PPV = 1:1	1.4	−49.7%	15.5	26.4%	71.3	23.9%
PQ:MEH-PPV = 2:1	1.4	−59.8%	16.1	13.6%	71.3	26.6%

with MEH-PPV films in different ratios at the MEH-PPV fluorescence peak (591 nm). The fitting results are summarized in Table 6.6.

Fluorescence relaxation of MEH-PPV film can be fitted by double exponential decay with time constants 14 ps and 71.3 ps. When PQ composition is mixed in different ratios, there will be a rising component of ~1.4 ps in the fluorescence decay in addition to the two exponential decays. The amplitude of the component is 49.7% when the PQ content is 50%, and 59.8% when the content of PQ is 67%, proportional to the content of PQ. This corresponds to the ultrafast decay process of PQ at 526 nm. Thus, from two aspects of donor and acceptor, it is confirmed that transfer rate is about 2 ps from the PQ to MEH-PPV resonance energy transfer process.

Bibliography

[1] T. H. Maiman. Stimulated optical radiation in ruby, Nature **187** (1960): 493–494.

[2] H. W. Mock and R. J. Collins. Mode Competition and self-locking effects in a Q-switched ruby laser. Appl. Phys. Lett. **7** (1965): 270–273.

[3] A. J. Demaria, D. A. Stetser and H. Heynau. Self mode-locking of lasers with saturable absorbers. Appl. Phys. Lett. **8** (1966): 174–176.

[4] R. L. Fork, B. I. Greene and C. V. Shank. Generation of optical pulses shorter than 0.1 psec by colliding pulse mode locking. Appl. Phys. Lett. **38** (1981): 671–672.

[5] P. Maine, D. Strickland, P. Bado, et al. Generation of ultrahigh peak power pulses by chirped pulse amplification. IEEE J Quantum Electron. **24** (1988): 389–403.

[6] D. Strickland and G. Mourou. Compression of amplified chirped optical pulses. Opt. Commun. **56** (1985): 219–221.

[7] D. E. Spence, P. N. Kean and W. Sibbett. 60-fsec pulse generation from a self-mode-locked Ti:sapphire laser. Opt. Lett. **16** (1991): 42–44.

[8] M. Nisoli, S. de Silvestri, O. Svelto, et al. A novel-high energy pulse compression system: generation of multigigawatt sub-5-fs pulses. Appl. Phys. B **65** (1997): 189–196.

[9] C. Rulliere, Femtosecongd laser pulses, Springer, 1998.

[10] F. Li, Ultrafast time-resolved optical parametric amplification system and its application in branched organic moleculars, PhD thesis, Peking University, 2008.

[11] X. F. Han, X. H. Chen, Y. X. Weng, et al. Ultrasensitive femtosecond time-resolved fluorescence spectroscopy for relaxation processes by using parametric amplification. J. Opt. Soc. Am. B **24** (2007): 1633–1638.

[12] J. S. Zhang, F. M. Li, S. F. Wang, et al. Femtosecond laser pumped conical emission and seeded ring amplification in BBO crystal. Chinese Physics Letter **22** (2005): 1652–1655.

[13] Q. Ding, K. Meng, H. Yang, et al. Femtosecond non-collinear parametric amplification for ultra-fast spectral dynamics. Opt. Commu. **284** (2011): 3110–3113.

[14] S. Günes, H. Neugebauer and N. S. Sariciftci. Conjugated polymer-based organic solar cells. Chem. Rev. **107** (2007): 1324–1338.

[15] H. Spangaard and F. C. Krebs. A brief history of the development of organic and polymeric photovoltaics. Sol. Energy Mater. Sol. Cells **83** (2004): 125–146.

[16] C. Zang, J. A. Stevens, J. J. Link, et al., Ultrafast proteinquake dynamics in cytochrome. J. Am. Chem. Soc. **131** (2009): 2846–2852.

[17] T. Kobayashi, T. Saito and H. Ohtani. Real-time spectroscopy of transition states in bacteriorhodopsin during retinal isomerization. Nature **414** (2001): 531–534.

[18] B. Valeur, Molecular fluorescence: principles and applications, Wiley-VCH Verlag GmbH, 2001.

[19] W. Knox and G. Mourou, A simple jitter-free picosecond streak camera. Opt. Commu. **37** (1981): 203–206.

[20] D. Y. Oberli, D. R. Wake, M. V. Klein, et al. Time-resolved Raman scattering in GaAs quantum wells. Phys. Rev. Lett. **59** (1987): 696.

[21] S. Cook, A. Furube and R. Katoh, Analysis of the excited states of regioregular polythiophene P3HT. Energy Environ. Sci. **1** (2008): 294–299.

[22] B. Valeur, Molecular fluorescence: principles and applications, Wiley-VCH Verlag GmbH, 2001.

[23] M. M. L. Grage, Y. Zaushitsyn, A. Yartsev, et al. Ultrafast excitation transfer and trapping in a thin polymer film. Phys. Rev. B **67** (2003): 205207(1–5).

[24] S. Westenhoff, C. Daniel, R. H. Friend, et al. Exciton migration in a polythiophene: Probing the spatial and energy domain by line-dipole Förster-type energy transfer. J. Chem. Phys. **122** (2005): 094903(1–8).

[25] M. M. L. Grage, T. Pullerits, A. Ruseckas, et al. Conformational disorder of a substituted polythiophene in solution revealed by excitation transfer. Chem. Phys. Lett. **339** (2001): 96–102.

[26] M. M. L. Grage, P. W. Wood, A. Ruseckas, et al. Conformational disorder and energy migration in MEH-PPV with partially broken conjugation. J. Chem. Phys. **118** (2003): 7644–7649.

[27] R. Kersting, B. Mollav, M. Rusch and M. M. L. Grage, Femtosecond site-selective probing of energy relaxing excitons in poly(phenylenevinylene): luminescence dynamics and lifetime spectra. J. Chem. Phys. **106** (1997): 2850–2864.

[28] K. Brunner, A. Tortschanoff, C. Warmuth, et al. Site torsional motion and dispersive excitation hopping transfer in π-conjugated polymers. J. Phys. Chem. B. **104** (2000): 3781–3790.

[29] V. Gulbinas, Y. Zaushitsyn, V. Sundström, et al. Dynamics of the electric field-assisted charge carrier photogeneration in ladder-type poly (para-phenylene) at a low excitation intensity. Phys. Rev. Lett. **89** (2002): 107401(1–4).

[30] J. G. Muller, U. Lemmer, J. Feldmann, et al. Precursor states for charge carrier generation in conjugated polymers probed by ultrafast spectroscopy. Phys. ReV. Lett. **88** (2002): 147401(1–4).
[31] J. Kalinowski, M. Cocchi, D. Virgili, et al. Charge photogeneration effect on the exciplex emission from thin organic films. Appl. Phys. Lett. **89** (2006): 011105(1–3).
[32] S. F. Alvarado, P. F. Seidler, D. G. Lldzey, et al. Direct determination of the exciton binding energy of conjugated polymers using a scanning tunneling microscope. Phys. Rev. Lett. **81** (1998): 1082–1085.
[33] D. Moses, J. Wang, A. J. Heeger, et al. Singlet exciton binding energy in poly (phenylene vinylene). Proc. Natl. Acad. Sci. U.S.A. **98** (2001): 13496–13500.
[34] I. G. Scheblykin, A. Yartsev, T. Pullerits, et al. Excited state and charge photogeneration dynamics in conjugated polymers. J. Phys. Chem. B **111** (2007): 6303–6321.
[35] W. J. D. Beenken and T. Pullerits, Spectroscopic units in conjugated polymers: A quantum chemically founded concept?. J. Phys. Chem. B. **108** (2004): 6164–6169.
[36] M. M-L. Grage, Y. Zaushitsyn, A. Yartsev, et al. Ultrafast excitation transfer and trapping in a thin polymer film. Phys. ReV. B **67** (2003): 205207(1–5).
[37] J. M. Guo, H. Ohkita, H. Benten, et al. Near-IR femtosecond transient absorption spectroscopy of ultrafast polaron and triplet exciton formation in polythiophene films with different regioregularities. J. Am. Chem. Soc. **131** (2009): 16869–16880
[38] R. Tautz, E. Como, C. Wiebeler, et al. Charge photogeneration in donor–acceptor conjugated materials: influence of excess excitation energy and chain length. J. Am. Chem. Soc. 2013, 135: 4282–4290
[39] A. Ruseckas, M. Theander, L. Valkunas, et al. Energy transfer in a conjugated polymer with reduced inter-chain coupling. J. Lumin. **1998**:76–**77**:474–477.
[40] T. Q. Nguyen, J. J. Wu, V. Doan, et al. Control of energy transfer in oriented conjugated polymer-mesoporous silica composites. Science **288** (2000): 652–656.
[41] D. Beljonne, G. Pourtois, C. Silva, et al. Interchain vs. intrachain energy transfer in acceptor-capped conjugated polymers. Proc. Natl. Acad. Sci. U.S.A. **99** (2002): 10982–10987.
[42] A. Ruseckas, M. Theander, M. R. Andersson, et al. Ultrafast photogeneration of inter-chain charge pairs in polythiophene films. Chem. Phys. Lett. **322** (2000): 136–142.
[43] J. E. Kroeze, T. J. Savenije, M. J. W. Vermeulen, et al. Contactless determination of the photoconductivity action spectrum, exciton diffusion length, and charge separation efficiency in polythiophene-sensitized TiO2 bilayers. J. Phys. Chem. B **107** (2003): 7696–7705.
[44] L. Lüer, H. J. Egelhaaf, D. Oelkrug, et al. Oxygen-induced quenching of photoexcited states in polythiophene films. Org. Electron. **5** (2004): 83–89.
[45] D. E. Markov, C. Tanase, P. W. M. Blom, et al. Simultaneous enhancement of charge transport and exciton diffusion in poly(p-phenylene vinylene) derivatives. Phys. Rev. B **72** (2005): 045217(1–6).
[46] P. E. Shaw, A. Ruseckas, I. D. W. Samuel. Exciton diffusion measurements in poly(3-hexylthiophene). Adv. Mater. **20** (2008): 3516–3520.
[47] A. J. Campillo, S. L. Shapiro, V. H. Kollman, et al. Picosecond exciton annihilation in photosynthetic systems. Biophys J. **16**(1) (1976): 93–97.
[48] A. J. Lewis, A. Ruseckas, O. P. M. Gaudin., et al. Singlet exciton diffusion in MEH-PPV films studied by exciton–exciton annihilation. Org. Electr. **7** (2006): 452–456.
[49] Y. R. Sun, N. C. Giebink, H. Kanno, et al. Management of singlet and triplet excitons for efficient white organic light-emitting devices. Nature **440**:908–912.
[50] B. J. Schwartz, Conjugated polymers as molecular materials: How chain conformation and film morphology influence energy transfer and interchain interactions. Annu. Rev. Phys. Chem. **54** (2003): 141–172.

[51] H. D. Burrows, J. S. de Melo, C. Serpa, et al. Triplet state dynamics on isolated conjugated polymer chains. Chem. Phys. **285** (2002): 3–11.

[52] J. M. Hodgkiss, S. Albert-Seifried, A. Rao, et al. Exciton-charge annihilation in organic semiconductor films. Adv. Funct. Mater. **22** (2012): 1567–1577.

[53] I. A. Howard, J. M. Hodgkiss, X. P. Zhang, et al. Charge recombination and exciton annihilation reactions in conjugated polymer blends. J. Am. Chem. Soc. **132**(1) (2010): 328–335.

[54] P. R. Hania and I. G. Scheblykin. Electric field induced quenching of the fluorescence of a conjugated polymer probed at the single molecule level. Chem. Phys. Lett. **414** (2005): 127–131.

[55] L. Zaikowski, P. Kaur, C. Gelfond, et al. Polarons, bipolarons, and side-by-side polarons in reduction of oligofluorenes. J. Am. Chem. Soc. **134** (2012): 10852–10863.

[56] T. D. Nguyen, G. Hukic-Markosian, F. J. Wang, et al. Isotope effect in spin response of π-conjugated polymer films and devices. Nat. Mater. **9** (2010): 345–352.

[57] D. R. McCamey, K. J. van Schooten, W. J. Baker, et al. Hyperfine-field-mediated spin beating in electrostatically bound charge carrier pairs. Phys. Rev. Lett. **104** (2010): 017601(1–4).

[58] M. Wohlgenannt, X. M. Jiang and Z. V. Vardeny. Confined and delocalized polarons in π-conjugated oligomers and polymers: A study of the effective conjugation length. Phys. ReV. B. **69** (2004): 241204(1–4).

[59] M. Yan, L. J. Rothberg, F. Papadimitrakopoulos, et al. Defect quenching of conjugated polymer luminescence. Phys. Rev. Lett. **72** (1994): 1104.

[60] C. Rothe, S. M. King and A. P. Monkman, Electric-field-induced singlet and triplet exciton quenching in films of the conjugated polymer polyspirobifluorene. Phys. Rev. B **72** (2005): 085220(1–7).

[61] F. Schindler, J. M. Lupton, J. Muller, et al. How single conjugated polymer molecules respond to electric fields. Nature Materials **2006**:141–146.

[62] V. Gulbinas, Y. Zaushitsyn, V. Sundström, et al. Dynamics of the electric field-assisted charge carrier photogeneration in ladder-type poly(para-phenylene) at a low excitation intensity. Phys. Rev. Lett. **89** (2002): 107401(1–4).

[63] A. Kohler, D. A. Dos Santos, D. Beljonne, et al. Charge separation in localized and delocalized electronic states in polymeric semiconductors. Nature **392** (1998): 903–906.

[64] H. F. Meng and T. M. Hong. Carrier generation mechanisms for photoconductivity in conjugated polymers. Phys. ReV. B. **61** (2000): 9913–9916.

[65] N. Pfeffer, D. Neher, M. Remmers, et al. Electric field-induced fluorescence quenching and transient fluorescence studies in poly (p-terphenylene vinylene) related polymers. Chem. Phys. **227** (1998): 167–178.

[66] D. Moses, H. Okumoto, C. H. Lee, et al. Mechanism of carrier generation in poly(phenylene vinylene): Transient photoconductivity and photoluminescence at high electric fields. Phys. Rev. B **54** (1996): 4748–4754.

[67] J. M. Warman, M. P. de Haas, G. Dicker, et al. Charge mobilities in organic semiconducting materials determined by pulse-radiolysis time-resolved microwave conductivity: π-bond-conjugated polymers versus π–π-stacked discotics. Chem. Mater. **16** (2004): 4600–4609.

[68] H. D. Burrows, J. S. Melo, C. Serpa, et al. S1~>T1 intersystem crossing in π-conjugated organic polymers. J. Chem. Phys. **115** (2001): 9601–9606.

[69] M. Wohlgenannt, W. Graupner, G. Leising, et al. Photogeneration and recombination processes of neutral and charged excitations in films of a ladder-type poly (para-phenylene). Phys. ReV. B. **60** (1999): 5321–5330.

[70] J. H. Hou, Z. A. Tan, Y. Yan, et al. Synthesis and photovoltaic properties of two-dimensional conjugated polythiophenes with bi(thienylenevinylene) side chains. J. Am. Chem. Soc. **128** (2006): 4911–4916.

[71] C. Y. Yu, B. T. Ko, C. Ting, et al. Two-dimensional regioregular polythiophenes with conjugated side chains for use in organic solar cells. Solar Energy Materials & Solar Cells **93** (2009): 613–620.

[72] J. H. Hou, Z. A. Tan, Y. J. He, et al. Branched poly(thienylene vinylene)s with absorption spectra covering the whole visible region. Macromolecules **39** (2006): 4657–4662.

[73] E. J. Zhou, C. He, Z. A. Tan, et al. Effect of side-chain end groups on the optical electrochemical, and photovoltaic properties of side-chain conjugated polythiophenes. J. Polymer Sci.: Part A: Polymer Chemistry **44** (2006): 4916–4922.

[74] H. Ohkita, S. Cook, Y. Astuti, et al. Charge carrier formation in polythiophene/fullerene blend films studied by transient absorption spectroscopy. J. Am. Chen. Soc. **130** (2008): 3030–3042.

[75] B. Kraabel, D. Mose and A. J. Heeger. Direct observation of the intersystem crossing in poly(3-octylthiophene). J. Chem. Phys. **103** (1995): 5102–5108.

[76] X. M. Jiang, R. Österbacka, O. J. Korovyanko, et al. Spectroscopic studies of photoexcitations in regioregular and regiorandom polythiophene films. Adv. Funct. Mater. **12** (2002): 587–597.

[77] Z. V. Vardeny, J. Strait, D. Mose, et al. Picosecond photoinduced dichroism in trans-(CH)x: direct measurement of soliton diffusion. Phys. Rev. Lett. **49** (1982): 1657–1660.

[78] O. J. Korovyanko, R. Österbacka, X. M. Jiang, et al. Photoexcitation dynamics in regioregular and regiorandom polythiophene films. Phys. Rev. B **64** (2001): 235122(1–6).

[79] Y. Kim, S. Cook, S. M. Tuladhar, et al. A strong regioregularity effect in self-organizing conjugated polymer films and high-efficiency polythiophene:fullerene solar cells. Nature Materials **5** (2006): 197–203.

[80] J. H. Hou, L. J. Huo, C. He, et al. Synthesis and absorption spectra of poly(3-(phenylenevinyl)thiophene)s with conjugated side chains. Macromolecules **39** (2006): 594–603.

[81] Y. H. Gao, J. F. Chen, X. J. Zhuang, et al. Proton transfer in phenothiazine photochemical oxidation: Laser flash photolysis and fluorescence studies. Chem. Phys. **334** (2007): 224–231.

[82] A. P. Kulkarni, X. X. Kong and S. Jenekhe. A high-performance organic light-emitting diodes based on intramolecular charge-transfer emission from donor-acceptor molecules: significance of electron-donor strength and molecular geometry. Adv. Func. Mater. **16** (2006): 1057–1066.

[83] X. X. Kong, A. P. Kulkarni and S. A. Jenekhe. Phenothiazine-based conjugated polymers: synthesis, electrochemistry, and light-emitting properties. Macromolecule **36** (2003): 8992–8999.

[84] M. Sailer, A. W. Franz and T. J. J. Müller. Synthesis and electronic properties of monodisperse oligophenothiazines. Chem. Eur. J. **14** (2008): 2602–2614.

[85] Y. Li, J. Feng and A. Ren. Theoretical study on electronic structure and optical properties of phenothiazine-containing conjugated oligomers and polymers. J. Org. Chem. **70** (2005): 5987–5996.

[86] W. J. Wu, J. B. Yang, J. L. Hua, et al. Efficient and stable dye-sensitized solar cells based on phenothiazine sensitizers with thiophene units. J. Mater. Chem., 2010:1772–1779.

[87] Z. B. Xie, A. Midya, K. P. Loh, et al. Highly efficient dye-sensitized solar cells using phenothiazine derivative organic dyes. Prog. Photovolt. **18** (2010): 573–581.

[88] H. N. Tian, X. C. Yang, R. K. Chen, et al. Phenothiazine derivatives for efficient organic dye-sensitized solar cells. Chem. Commu. **36** (2007): 3741–3743.

[89] W. Y. Wong, W. C. Chow, K. Y. Cheung, et al. Harvesting solar energy using conjugated metallopolyyne donors containing electron-rich phenothiazine-oligothiophene moieties. J. Organometallic Chem. **694** (2009): 2717–2726.

[90] H. Padhy, J. H. Huang, D. Sahu, et al. Synthesis and applications of low-bandgap conjugated polymers containing phenothiazine donor and various benzodiazole acceptors for polymer solar cells. J. Polymer Sci. Part A-Polymer Chemistry **48** (2010): 4823–4834.
[91] S. A. Jenekhe, L. D. Lu and M. M. Alam. New Conjugated Polymers with Donor-Acceptor Architectures: Synthesis and Photophysics of Carbazole-Quinoline and Phenothiazine-Quinoline Copolymers and Oligomers Exhibiting Large Intramolecular Charge Transfer. Macromolecules **34** (2001): 7315–7324.
[92] A. P. Kulkarni, P. T. Wu, T. W. Kwon, et al. Phenothiazine-phenylquinoline donor-acceptor molecules: effects of structural isomerism on charge transfer photophysics and electroluminescence. J. Phys. Chem. B **109** (2005): 19584–19594.
[93] G. W. Kim, M. J. Cho, Y. J. Yu, et al. Red emitting phenothiazine dendrimers encapsulated 2-{2-[2-(4-dimethylaminophenyl) vinyl]-6-methylpyran-4-ylidene} malononitrile derivatives. Chem. Matter. **9** (2007): 42–50.
[94] J. Daub, R. Engl, J. Kurzawa, et al. Competition between conformational relaxation and intramolecular electron transfer within phenothiazine-pyrene dyads. J. Phys. Chem. A **105** (2001): 5655–5665.
[95] N. Acar, J. Kurzawa, N. Fritz, et al. Phenothiazine-pyrene dyads: photoinduced charge separation and structural relaxation in the CT state. J. Phys. Chem. A **107** (2003): 9530–9541.
[96] J. Shirdel, A. Penzkofer, Z. Shen, et al. Photo-induced dynamics in a pyrene-isoalloxazine (flavin)-phenothiazine triad. Chem. Phys **337** (2007): 99–109.
[97] A. Stockmann, J. Kurzawa, N. Fritz, et al. Conformational control of photoinduced charge separation within phenothiazine-pyrene dyads. J. Phys. Chem. A **106** (2002): 7958–7970.
[98] N. D. Césare, M. Belletête, G. Durocher, et al. Towards a theoretical design of thermochromic polythiophenes. Chem. Phys. Lett. **275** (1997): 533–539.
[99] F. Liu, P. Zuo, L. P. Meng, et al. Configuration interaction study on the conformational and optical properties of excited furan oligomers. Theo. Chem. **726** (2005): 189–196.
[100] F. Liu, P. Zuo, L. P. Meng, et al. On the optical properties of thiophene oligomers: configuration interaction study on their ground (S0) and first singlet excited (S1) states. Theo. Chem. **726** (2005): 161–169.
[101] S. Suramitr, S. Hannongbua and P. Wolschann. Conformational analysis and electronic transition of carbazole-based oligomers as explained by density funtional theory. Theo. Chem. **807** (2007): 109–119.
[102] M. Leclerc, G. Dufresne, P. Blondin, et al. Molecular design of chromic functionalized conjugated polymers. Synth. Met. **119** (2001): 45–48.
[103] M. Belletête, M. Bédard, M. Lecletc, et al. Ground and excited state properties of carbazole-based dyads: correlation with their respective absorption and fluorescence spectra. Theo. Chem. **679** (2004): 9–15.
[104] Y. L. Liu, J. H. Li, H. Y. Cao, et al. Conjugated polymers containing phenothiazine moieties in the main chain. Polym. Adv. Technol. **17** (2006): 468–473.
[105] C. S. Krämer and T. J. J. Müller. Synthesis of functionalized ethynylphenothiazine fluorophores. Org. Lett. **2** (2000): 3723–3726.
[106] X. P. Qiu, R. Lu, H. P. Zhou, et al. Synthesis of linear monodisperse vinylene-linked phenothiazine oligomers. Tetrahedron Lett. **48** (2007): 7582–7585.
[107] W. Feng, Investigation on photo induced energy and charge transfer in functional organic molecules, PhD thesis, Peking University, 2009
[108] X. Zhao, Studies on ultrafast dynamics of carbazole based oligomers. Master degree thesis, Peking University, 2008
[109] J. Bouchard, M. Belletête, G. Durocher, et al. Solvatochromic properties of 2,7-carbazol-based conjugated polymers. Macromolecules **36** (2003): 4624–4630.

[110] J. E. Bushnell, P. R. Kemper, G. C. Bazan, et al. The determination of cis-trans conformations in tetrahedral p-phenylene vinylene oligomers. J. Phy. Chem A. **108** (2004): 7730–7735.
[111] D. M. Chapin, C. S. Fuller and G. L. Pearson. A new silicon p-n junction photocell for converting solar radiation into electrical power. J. Appl. Phys. **25** (1954): 676–677.
[112] M. B. Prince, Silicon solar energy converters. J. Appl. Phys. **26** (1955): 534–540.
[113] P. V. Meyers. Design of a thin-film cdte solar-cell. Solar Cells. **23** (1988): 59–67.
[114] K. W. Boer, The CdS/Cu2S Solar-cell. J. Crystal Growth. **59** (1982): 111–120.
[115] R. Fornari, Optimal growth conditions and main features of GaAs single crystals for solar cell technology: A review. Solar Energy Materials. **11** (1985): 361–379.
[116] B. C. Chung, G. F. Virshup, S. Hikido and N. R. Kaminar. 27.6% efficiency (1 sun, air mass 1.5) monolithic Al0.37Ga0.63As/GaAs two-junctioncascade solar cell with prismatic cover glass. Appl. Phys. Lett. **55** (1989): 1741–1743.
[117] C. W. Tang, Two-layer organic photovoltaic cell. Appl. Phys. Lett. **48** (1986): 183–185.
[118] N. S. Sariciftci, D. Braun, C. Zhang, et al. Wudl, Semiconducting polymer-buckminsterfullerene hetero-junctions: Diodes, photodiodes, and photovoltaic cells. Appl. Phys. Lett. **62** (1993): 585–587.
[119] T. Munters, T. Martens, L. Goris, et al. A comparison between state-of-the-art "gilch" and "sulphinyl" synthesised MDMO-PPV/PCBM bulk hetero-junction solar cells, Thin Solid Films **403** (2002): 247–251.
[120] F. L. Zhang, M. Johansson, M. R. Andersson, et al. Polymer solar cells based on MEH-PPV and PCBM, Syn. Met. **137** (2003): 1401–1402.
[121] W. L. Ma, C. Y. Yang, X. Gong, et al. Thermally stable, efficient polymer solar cells with nanoscale control of the interpenetrating network morphology, Adv. Fun. Mater. **15** (2005): 1617–1622.
[122] J. L. Segura, N. Martin and D. M. Guldi, Materials for organic solar cells: the C-60/pi-conjugated oligomer approach, Chem. Soc. Rew. **34** (2005): 31–47.
[123] S. Berson, R. Bettignies, S. Bailly, et al. Elaboration of P3HT/CNT/PCBM Composites for Organic Photovoltaic Cells. Adv. Fun. Mater. **17** (2007): 3363–3370.
[124] V. Gulbinas, I. Minevičiūtė, D. Hertel, et al. Exciton diffusion and relaxation in methyl-substituted polyparaphenylene polymer films. J. Chem. Phys. 2007, 127: 144907(1–8).
[125] J. M. J. Fréchet and B. C. Thompson, Polymer–fullerene composite solar cells. Angew. Chem. Int. Ed. **47** (2008): 58–77.
[126] A. C. Mayer, S. R. Scully, B. E. Hardin, et al. Polymer-based solar cell **10** (2007): 28–33.
[127] National Renewable Energy Laboratory (NREL), MS ExcelTM spreadsheet file downloaded from http://rredc.nrel.gov/solar/spectra/am1.5
[128] Fangwen Lu, Electron transfer properties of 8- hydroxy metal complexes and their applications in OLES, Master degree thesis, Peking University, 2009
[129] K. Lee and H. Kim, Polymer photovoltaic cells based on conjugated polymer–fullerene composites. Curr. Appl. Phys. **4** (2004): 323–326.
[130] P. A. C. Quist, J. Sweelssen, M. M. Koetse, et al. Formation and decay of charge carriers in bulk heterojunctions of MDMO-PPV or P3HT with new n-type conjugated polymers. J. Phys. Chem. C **111** (2007): 4452–4457.
[131] S. E. Shaheen, C. J. Brabec, N. S. Sariciftci, et al. 2.5% efficient organic plastic solar cells. Appl. Phys. Lett. **78** (2001): 841–843.
[132] J. E. Northrup, Atomic and electronic structure of polymer organic semiconductors: P3HT, PQT, and PBTTT. Phys. Rev. B. **76** (2007): 245202–245207.
[133] D. Mühlbacher, M. Scharber, M. Morana, et al. High photovoltaic performance of a low-bandgap polymer. Adv. Matt. **18** (2006): 2884–2889.

[134] G. Li, V. Shrotriya, J. Huang, et al. High-efficiency solution processable polymer photovoltaic cells by self-organization of polymer blends. Nat. Mater. **4** (2005): 864–868.
[135] S. H. Park, A. Roy, S. Beaupre, et al. Bulk heterojunction solar cells with internal quantum efficiency approaching 100%. Nature Photonics **3** (2009): 297–302.
[136] H. Y. Chen, J. H. Hou, S. Q. Zhang, et al. Polymer solar cells with enhanced open-circuit voltage and efficiency. Nature Photonics **3** (2009): 649–653.
[137] G. J. Zhao, Y. J. He and Y. F. Li, 6.5% efficiency of polymer solar cells based on poly(3-hexylthiophene) and indene-C_{60} bisadduct by device optimization. Adv. Mater. **22** (2010): 4355–4358.
[138] E. Wang, L. Wang, L. Lan, et al. High-performance polymer heterojunction solar cells of a polysilafluorene derivative. Appl. Phys. Lett. **92** (2008): 033307(1–3).
[139] Y. He, H. Y. Chen, J. Hou, et al. Indene-C_{60} bisadduct: a new acceptor for high-performance polymer solar cells. J. Am. Chem. Soc. **132** (2010): 1377–1382.
[140] N. Camaioni, L. Garlaschelli, A. Geri, et al. Solar cells based on poly(3-alkyl) thiophenes and [6], [0] fullerene:a comparative study. J. Mater. Chem. **12** (2002): 2065–2070.
[141] J. X. Li, N. Sun, Z. X. Guo, et al. Photovoltaic devices with methanofullerenes as electron acceptors. J. Phys. Chem. B **106** (2002): 11509–11514.
[142] M. Wojcik, P. Michalak and M. Tachiya, Geminate electron-hole recombination in organic solids in the presence of a donor-acceptor heterojunction. App. Phys. Lett. **96** (2010): 162102(1–3).
[143] J. G. Müller, J. M. Lupton and J. Feldmann, Ultrafast dynamics of charge carrier photogeneration and geminate recombination in conjugated polymer fullerene solar cells. Phys. Rev.B **72** (2005): 195208(1–10).
[144] S. Cook, R. Katoh and A. Furube, Ultrafast studies of charge generation in PCBM:P3HT blend films following excitation of the fullerene PCBM. J. Phys. Chem. C **113** (2009): 2547–2552.
[145] R. A. Marsh, C. R. McNeill, A. Abrusci, et al. A unified description of current–voltage characteristics in organic and hybrid photovoltaics under low light intensity. Nano Lett. **8** (2008): 1393–1398.
[146] C. L. Braun, Electric field assisted dissociation of charge transfer states as a mechanism of photocarrier production. J. Chem. Phys. **80** (1984): 4157–4161.
[147] L. Onsager, Deviations from Ohm's law in weak electrolytes. J.Chem. Phys. **2** (1934): 599–615.
[148] T. E. Goliber and J. H. Perlstein. Analysis of photogeneration in a doped polymer system in terms of a kinetic model for electric-field-assisted dissociation of charge-transfer states. J. Chem. Phys. **80** (1984): 4162–4167.
[149] V. D. Mihailetchi, L. J. A. Koster, J. C. Hummelen, et al. Photocurrent generation in polymer-fullerene bulk heterojunctions. Phy. Rev. Lett. **93** (2004): 216601(1–4).
[150] A. M. Goodman and A. Rose, Double extraction of uniformly generated electron-hole pairs from insulators with noninjecting contacts. J. Appl. Phys. **42** (1971): 2823–2830.
[151] R. Sokel and R. C. Hughes. Numerical analysis of transient photoconductivity in insulators. J. Appl. Phys. **53** (1982): 7414–7424.
[152] N. S. Sariciftci, L. Smilowitz, A. J. Heeger, et al. Photoinduced electron transfer from a conducting polymer to buckminsterfullerene. Science **258** (1992): 1474–1476.
[153] C. J. Brabec, G. Zerza and G. Cerullo. Tracing photoinduced electron transfer process in conjugated polymer and fullerene bulk heterojunctions in real time. Chem. Phys. Lett. **340** (2001): 232–236.
[154] E. Lioudakis, I. Alexandrou and A. Othonos. Ultrafast dynamics of localized and delocalized polaron transitions in P3HT/PCBM blend materials: the effects of PCBM concentration. Nanoscale Res. Lett. **4** (2009): 1475–1480.
[155] S. De and T. Kesti. Exciton dynamics in alternating polyfluorene/fullerene blends. Chemical Physics **350** (2008): 14–22.

[156] I. W. Hwang, D. Moses and A. J. Heeger, Photoinduced carrier generation in P3HT/PCBM bulk heterojunction materials. J. Phys. Chem. C **112** (2008): 4350–4354.
[157] Y. Zhao and G. Milnikov. Electron transfer rate in the Marcus inverted regime beyond the perturbation theory. Chemical Physics Letters **413** (2005): 362–366.
[158] I. W. Hwang, Q. H. Xu, C. Soci, et al. Ultrafast spectroscopic study of photoinduced electron transfer from semiconducting polymers to fullerene. Adv. Fun. Mater. **17** (2007): 563–598.
[159] S. Westenhoff, I. A. Howard, J. M. Hodgkiss, et al. Charge recombination in organic photovoltaic devices with high open-circuit voltages. J. Am. Chen. Soc. **130** (2008): 13653–13658.
[160] A. J. Moulé, K. Meerholz. Controlling morphology in polymer–fullerene mixtures. Adv. Mater. **20** (2008): 240–245.
[161] J. M. Beechem. Global analysis of biochemical and biophysical data. Methods Enzymol. **210** (1992): 37–54.
[162] K. G. Jespersen, W. J. D. Beenken, Y. Zaushitsyn, et al. The electronic states of polyfluorene copolymers with alternating donor-acceptor units. J. Chem. Phys. **121** (2004): 12613–12617.
[163] S. De, T. Pascher, M. Maiti, et al. Geminate charge recombination in alternating polyfluorene copolymer:fullerine blends. J. Am. Chem. Soc. **129** (2007): 8466–8472.
[164] J. Nelson, S. A. Choulis and J. R. Durrant. Charge recombination in polymer/fullerene photovoltaic devices. Thin Solid Films **451–452** (2004): 508–514.
[165] Nelson. Continuous-time random-walk model of electron transport in nanocrystalline TiO2 electrodes. J. Phys. ReV. B **59** (1999): 15374–15380.
[166] A. F. Nogueira, I. Montanari, J. Nelson, et al. Charge recombination in conjugated polymer/fullerene blended films studied by transient absorption spectroscopy. J. Phys. Chem. B **107** (2003): 1567–1573.
[167] I. Montanari, A. F. Nogueira, J. Nelson, et al. Transient optical studies of charge recombination dynamics in a polymer/fullerene composite at room temperature.Appl. Phys. Lett. **81** (2002): 3001–3003.
[168] S. C. J. Meskers, P. A. van Hal, A. J. H. Spiering, et al. Time-resolved infrared-absorption study of photoinduced charge transfer in a polythiophene-methanofullerene composite film. Phys. Rev. B **61** (2000): 9917–9920.
[169] T. Offermans, S. C. J. Meskers and R. A. Janssen. Monte-Carlo simulations of geminate electron–hole pair dissociation in a molecular heterojunction: a two-step dissociation mechanism. J. Chem. Phys. **308** (2005): 125–133.
[170] T. Offermans, S. C. J. Meskers and R. A. J. Janssen. Charge recombination in a poly(paraphenylene vinylene)-fullerene derivative composite film studied by transient, nonresonant, hole-burning spectroscopy. J. Chem. Phys. **119** (2003): 10924–10929.
[171] N. A. Schultz, M. C. Scharber, C. J. Brabec, et al. Low-temperature recombination kinetics of photoexcited persistent charge carriers in conjugated polymer/fullerene composite films. Phys. Rev.B **64** (2001): 245210.
[172] T. J. Savenije, J. E. Kroeze, M. M. Wienk, et al. Mobility and decay kinetics of charge carriers in photoexcited PCBM/PPV blends. Phys. Rev. B **69** (2004): 155205.
[173] A. J. Mozer, N. S. Sariciftci, L. Lutsen, et al. Charge transport and recombination in bulk heterojunction solar cells studied by the photoinduced charge extraction in linearly increasing voltage technique. Appl. Phys. Lett. **86** (2005): 112104.
[174] A. J. Mozer, G. Dennler, N. S. Sariciftci, et al. Time-dependent mobility and recombination of the photoinduced charge carriers in conjugated polymer/fullerene bulk heterojunction solar cells. Phys. Rev. B **72** (2005): 035217(1–10).
[175] H. Aarnio, M. Westerling, R. Osterbacka, et al. Recombination studies in a polyfluorene copolymer for photovoltaic applications. Synth. Met. **155** (2005): 299–302.

[176] E. Bundgaard and F. C. Krebs. Low band gap polymers for organic photovoltaics. Solar Energy Materials & Solar Cells **91** (2007): 954–985.
[177] T. Kietzke. H.-H. Hörhold and D. Neher, Efficient polymer solar cells based on M3EH-PPV. Chem. Mater. **17** (2005): 6532–6537.
[178] M. M. Mandoc, W. Veurman, L. J. A. Koster, et al. Charge transport in MDMO-PPV:PCNEPV all-polymer solar cells. J. Appl. Phys. **101** (2007): 104512.
[179] G. Y. Sang, Y. P. Zou, Y. Huang, et al. All-polymer solar cells based on a blend of poly[3-(10-n-octyl-3-phenothiazine-vinylene) thiophene-co-2,5-thiophene] and poly [1,4-dioctyloxyl-p-2,5-dicyanophenylenevinylene]. App. Phys. Letts. **94** (2009): 193302(1–3).
[180] C. R. McNeill, A. Abrusci, J. Zaumseil, et al. Dual electron donor/electron acceptor character of a conjugated polymer in efficient photovoltaic diodes. App. Phys. Letts. **90** (2007): 193506(1–3).
[181] S. C. Veenstra, J. Loos and J. M. Kroon. Nanoscale structure of solar cells based on pure conjugated polymer blends. Prog. Photovolt: Res. Appl. **15** (2007): 727–740.
[182] C. H. Yin, T. Kietzke, D. Neherb, et al. Photovoltaic properties and exciplex emission of polyphenylenevinylene-based blend solar cells. Appl. Phys. Lett. **90** (2007): 092117(1–3).
[183] C. H. Yin, T. Kietzke, M. Kumkec, et al. Relation between exciplex formation and photovoltaic properties of PPV polymer-based blends. Solar Energy Materials & Solar Cells **91** (2007): 411–415.(1–11)
[184] T. Offermans, P. A. Hal, S. C. J. Meskers, et al. Exciplex dynamics in a blend of π-conjugated polymers with electron donating and accepting properties MDMO-PPV and PCNEPV. Phy. Rew. B **72** (2005): 045213.
[185] D. Gupta, D. Kabra, N. Kolishetti, et al. An efficient bulk-heterojunction photovoltaic cell based on energy transfer in graded-bandgap polymers. Adv. Funct. Mater. **17** (2007): 226–232.
[186] Y. Zhao, X. Y. Wang and F. L. Shu. Investigation of multi-donor bulk-heterojunction photovoltaic cells based on P3HT:PCBM system. Solar Energy Materials & Solar Cells **95** (2011): 684–687.
[187] Y. Kim, M. Shin, H. Kim, et al. Influence of electron-donating polymer addition on the performance of polymer solar cells. J. Phys. D: Appl. Phys. **41** (2008): 225101(1–15).
[188] T. Salim, S. Y. Sun, L. H. Wong, et al. The role of poly(3-hexylthiophene) nanofibers in an all-polymer blend with a polyfluorene copolymer for solar cell applications. J. Phys. Chem. C **114** (2010): 9459–9468.
[189] M. Berggren, A. Dodabalapur, R. E. Slusher, et al. Light amplification in organic thin films using cascade energy transfer. Nature **389** (1997): 466–469.
[190] P. Suresh, P. Balaraju, S. K. Sharma, et al. Photovoltaic devices based on PPHT ZnO and dye-sensitized PPHT ZnO thin films. Solar Energy Materials & Solar Cells **92** (2008): 900–908.
[191] J. Wang, D. Wang, D. Moses, et al. Dynamic quenching of 5-(2-ethyl-hexyloxy)-p-phenylene vinylene (MEH-PPV) by charge transfer to a C_{60} derivative in solution. J. Appl. Polymer Sci. **82** (2001): 2553–2557.
[192] E. C. Chang, C. I. Chao and R. H. Lee, Enhancing the efficiency of MEH-PPV and PCBM based polymer solar cells via optimization of device configuration and processing conditions. J. Appl. Polymer Sci. **101** (2006): 1919–1924.
[193] H. Z. Yu, J. B. Peng and X. M. Zhou, The performance of solar cell based on blends of MEH-PPV and PCBM of various compositions. Acta Physica Sinica **57** (2008): 3898–3904.
[194] P. A. C. Quist, T. J. Savenije, M. M. Koetse, et al. The effect of annealing on the charge-carrier dynamics in a polymer polymer bulk heterojunction for photovoltaic applications. Adv. Fun. Mater. **15** (2005): 469–474.
[195] T. Förster. Intermolecular energy migration and fluorescence. Ann. Phys. **2** (1948): 55–75.

Index

https://doi.org/10.1515/9783110304558-007

Advances in Optical Physics

Qihuang Gong, Zhi Li, Limin Tong, Yipei Wang, Yufei Wang, Wanhua Zheng
Advances in Nanophotonics, 2017
ISBN 978-3-11-030431-2, e-ISBN (PDF) 978-3-11-030700-9,
e-ISBN (EPUB) 978-3-11-038288-4, Set-ISBN 978-3-11-030701-6

Liming Chen, Baifei Shen, Wei Yu, Zhinan Zeng
Advances in High Field Laser Physics, 2018
ISBN 978-3-11-030426-8, e-ISBN (PDF) 978-3-11-030441-1,
e-ISBN (EPUB) 978-3-11-038800-8, Set-ISBN 978-3-11-030442-8

Kelin Gao, Wuming Liu, Hailing Wang, Jin Wang, Jianping Yin
Advances in Precision Laser Spectroscopy, 2018
ISBN 978-3-11-030429-9, e-ISBN (PDF) 978-3-11-030447-3,
e-ISBN (EPUB) 978-3-11-038801-5, Set-ISBN 978-3-11-030448-0

Xianfeng Chen, Qi Guo, Weilong She, Heping Zeng, Guoquan Zhang
Advances in Nonlinear Optics, 2014
ISBN 978-3-11-030430-5, e-ISBN (PDF) 978-3-11-030449-7,
e-ISBN (EPUB) 978-3-11-038282-2, Set-ISBN 978-3-11-030450-3

Yong Deng, Zhenli Huang, Yu Li, Da Xing, Zhihong Zhang
Advances in Molecular Biophotonics, 2017
ISBN 978-3-11-030438-1, e-ISBN (PDF) 978-3-11-030459-6,
e-ISBN (EPUB) 978-3-11-038803-9, Set-ISBN 978-3-11-030460-2

Liangyao Chen, Ning Dai, Xunya Jiang, Kuijuan Jin, Hui Liu, Haibin Zhao
Advances in Condensed Matter Optics, 2014
ISBN 978-3-11-030693-4, e-ISBN (PDF) 978-3-11-030702-3,
e-ISBN (EPUB) 978-3-11-038818-3, Set-ISBN 978-3-11-030703-0

Xingcan Yao, Tiancai Zhang, Weiping Zhang, Kadi Zhu
Advances in Quantum Optics, 2018
ISBN 978-3-11-030435-0, e-ISBN (PDF) 978-3-11-030453-4,
e-ISBN (EPUB) 978-3-11-038802-2, Set-ISBN 978-3-11-030454-1